17th International Conference on Recent Progress in Many-Body Theories 2013

(MBT17)

Journal of Physics: Conference Series Volume 529

Rostock, Germany
8 – 13 September 2013

Editors:

H. Reinholz
J. Boronat

ISBN: 978-1-63439-310-2
ISSN: 1742-6588

Printed from e-media with permission by:

Curran Associates, Inc.
57 Morehouse Lane
Red Hook, NY 12571

Some format issues inherent in the e-media version may also appear in this print version.

Printed by Curran Associates, Inc. (2014)

For permission requests, please contact the Institute of Physics
at the address below.

Institute of Physics
Dirac House, Temple Back
Bristol BS1 6BE UK

Phone: 44 1 17 929 7481
Fax: 44 1 17 920 0979

techtracking@iop.org

Additional copies of this publication are available from:

Curran Associates, Inc.
57 Morehouse Lane
Red Hook, NY 12571 USA
Phone: 845-758-0400
Fax: 845-758-2634
Email: curran@proceedings.com
Web: www.proceedings.com

17th International Conference on Recent Progress in Many-Body Theories 2013

(MBT17)

Journal of Physics: Conference Series Volume 529

Rostock, Germany
8 – 13 September 2013

ISBN: 978-1-63439-310-2
ISSN: 1742-6588

IOP Institute of Physics

TABLE OF CONTENTS

Journal of Physics: Conference Series **529** (2014) 011001 doi:10.1088/1742-6596/529/1/011001

Preface

These are the proceedings of the XVII International Conference on Recent Progress in Many-Body Theories[1], which was held from 8–13 September 2013 in Rostock, Germany. The conference continued the triennial series initiated in Trieste in 1978 and was devoted to new developments in the field of many-body theories. The conference series encourages the exchange of ideas between physicists working in such diverse areas as nuclear physics, quantum chemistry, lattice Hamiltonians or quantum uids.

Many-body theories are an integral part in different fields of theoretical physics such as condensed matter, nuclear matter and field theory. Phase transitions and macroscopic quantum effects such as magnetism, Bose-Einstein condensation, super uidity or superconductivity have been investigated within ultra-cold gases, finite systems or various nanomaterials. The conference series on Recent Progress in Many-Body Theories is devoted to foster the interaction and to cross-fertilize between different fields and to discuss future lines of research. The topics of the 17th meeting were

- Cluster Physics
- Cold Gases
- High Energy Density Matter and Intense Lasers
- Magnetism
- New Developments in Many-Body Techniques
- Nuclear Many-Body and Relativistic Theories
- Quantum Fluids and Solids
- Quantum Phase Transitions
- Topological Insulators and Low Dimensional Systems.

109 participants from 20 countries participated. 44 talks and 61 posters werde presented. As a particular highlight of the conference, The Eugene Feenberg Memorial Medal[2] for outstanding results in the field of many-body theory and The Hermann Kümmel Early Achievement Award in Many-Body Physics[3] for young scientists in that field were awarded. The Feenberg Medal went jointly to Patrick Lee (MIT, USA) for his fundamental contributions to condensed-matter theory, especially in regard to the quantum Hall effect, to universal conductance uctuations, and to the Kondo effect in quantum dots, and Douglas Scalapino (UC Santa Barbara, USA) for his imaginative use and development of the Monte-Carlo approach and for his ground-breaking contributions to superconductivity. The Kümmel Award went to Max Metlitski (UC Santa Barbara) for remarkable advances in the theory of quantum criticality in metals. The nominations for the Kümmel Award were of such high standard that the Committee announced Honourable Mentions to Martin Eckstein (MPDS/U Hamburg, Germany) for his leading contributions in

[1] http://www.mbt17.de
[2] http://www.indiana.edu/~fpmbt/mediawiki/index.php?title=Main_Page#Feenberg_Memorial_Medal
[3] http://www.indiana.edu/~fpmbt/mediawiki/index.php?title=Main_Page#K.C3.BCmmel_Early_ Achievement_Award

the development of non-equilibrium dynamical mean field theory, Emanuel Gull (U Michigan, USA) for the development of the Continuous-Time Auxiliary-Field Quantum Monte Carlo Method and for its use in understanding the interplay of the pseudogap and superconductivity in the Hubbard model and Kai Sun (U Michigan, USA) for seminal contributions to the theory of topological effects in strongly correlated electron systems.

The Conference continues the series of conferences held before in Trieste, Italy (1979); Oaxtapec, Mexico (1981); Odenthal-Altenberg, Germany (1983); San Francisco, USA (1985); Oulu, Finland (1987); Arad, Israel (1989); Minneapolis, USA (1991); SchloßSegau, Austria (1994); Sydney, Australia (1997); Seattle, USA (1999); Manchester, UK (2001); Santa Fe, USA (2004); Buenos Aires, Argentina (2005); Barcelona, Spain (2007); Columbus, USA (2009) and Bariloche, Argentina (2011). It has been a great pleasure to prepare for the conference. We thank the IAC and in particular Susana Hernandez and David Neilson as well as the International Programme Committee for their great support and advice. Many more people have been involved locally in organizing this international meeting and thanks goes to them, in particular to the members of the LOC Sonja Lorenzen, Dieter Bauer, Niels-Uwe Bastian, Marina Hertzfeldt, Volker Mosert and Gerd Röpke. The next meeting[4] will take place in Buffalo, USA in 2015 and we look forward to yet another exciting exchange on Recent Progress in Many-Body Theories.

Heidi Reinholz and Jordi Boronat
Guest editors

[4] http://www.ubevents.org/event/manybody2015

Journal of Physics: Conference Series **529** (2014) 011001 doi:10.1088/1742-6596/529/1/011001

International Advisory Committee

Raymond F Bishop (Manchester, UK)
Jordi Boronat (U Politecnica, Barcelona, ES)
Charles E Campbell (Treasurer, Minneapolis, USA)
Joe Carlson (LANL, USA)
Siu A Chin (Texas A&M, USA)
John W Clark (St. Louis, USA)
Peter Fulde (Dresden, Germany)
Davide Galli (Milan, Italy)
Karen Hallberg (Bariloche, Argentina)
Morten Hjorth-Jensen (Oslo, Norway)
Susana Hernández (Chair, Buenos Aires, Argentina)
Eckhard Krotscheck (Buffalo, USA)
Efstratios Manousakis (Tallahassee, USA)
David Neilson (Secretary, Camerino, Italy)
Gerardo Ortiz (Indiana, USA)
Arturo Polls (Barcelona, Spain)
Heidi Reinholz (Rostock, Germany)
Mikko Saarela (Oulu, Finland)
Masahito Ueda (Tokyo, Japan)

Local Organizing Committee

Niels-Uwe Bastian
Dieter Bauer
Lena Boldt
Marina Hertzfeldt
Sonja Lorenzen
Volker Mosert
Heidi Reinholz (Chair)
Gerd Röpke

Cover Design

Wiebke Loseries

Sponsors

University of Rostock
Institute of Physics
Sonderforschungsbereich 652
Deutsche Forschungsgemeinschaft

Peer review statement

All papers published in this volume of *Journal of Physics: Conference Series* have been peer reviewed through processes administered by the proceedings Editors. Reviews were conducted by expert referees to the professional and scientific standards expected of a proceedings journal published by IOP Publishing.

Quantum spin liquid: a tale of emergence from frustration

Patrick A. Lee

Department of Physics, MIT, Cambridge, MA 02139, USA

E-mail: paleeQmit.edu

Abstract. I give a review of the current status of the quantum spin liquid, emphasizing the connection to lattice gauge theory, de-confinement and the emergence of novel particles and gauge fields in the low energy effective theory

The study of quantum spin liquid and frustrated magnetism is a field in many-body theory which has attracted a great deal of attention in recent years, due both to theoretical advances and the fact that promising candidate systems have been found experimentally. I would like to review some of the recent progress, but first begin with a bit of historical perspective. Louis Néel proposed the possibility of antiferromagnetic order back in 1936,[1] and by now Néel order is ubiquitous and seems second nature to condensed matter physicists. It is difficult for us to comprehend that Néel's ideas was initially met with skepticism and Landau was among the doubters. This was briefly mentioned in the obituary for Néel written by Jacques Friedel.[2]

... *Werner Heisenberg's work on exchange had just shown that short-range effects should be expected. Between 1931 and 1933, Néel observed those effects in the susceptibility of iron and alloys and in the specific heat of nickel. Then, assuming that short-range interactions could be antiparallel, Néel developed the concept of anti ferromagnetism, in which two interpenetrating atomic lattices are treated in a molecular field approximation. ... To describe anti ferromagnetism Lev Landau and Cornelius Gorter suggested quantum fluctuations to mix Néel's solution with that obtained by reversal of moments. ... Néel's difficulties with anti-ferromagnetism and inconclusive discussions in the Strassbourg international meeting of 1939 fostered his skepticism about the usefulness of quantum mechanics; this was one of the few limitations of this superior mind.*

To paraphrase Friedel, we can say that Landau's objection is that unlike a ferromagnet, Néel's up-down arrangement of spins is not an eigenstate of the Heisenberg exchange Hamiltonian, because the mutual spin flip term will reduce the sublattice order and may eventually disorder the Néel state. After all, Néel's solution is an uncontrolled mean-field theory, and the only exact solution known at the time is Bethe's solution of the Heisenberg spin chain,[3] which does not show long range order. Therefore in the absence of direct experimental evidence, one has reasons to be skeptical. Well, Nature has sided with Néel, and except for one dimensional spin systems, almost all experimental spin systems find a way to order at low temperatures. (Disorder and random exchange may lead to spin freezing, but let us leave spin glass out of this discussion.)

The possibility of quantum fluctuations destroying Néel order was raised again in 1973 by P.W. Anderson.[4] He focused on the singlet formation between two spins with energy gain of

$-S(S+1)J$. The origin of the factor unity in $(S+1)$ is quantum mechanics and $S=\frac{1}{2}$ enjoys the greatest energy gain. In a spin chain, a trial wave function of singlet dimers yields an energy of $-\frac{3}{8}J$ per bond, which is lower than the Néel value of $-\frac{1}{4}J$. Indeed, we now understand Bethe's solution as a linear superposition of singlets with varying range. Anderson suggested that a similar superposition may hold for higher dimensions, which he called the resonating valence bond (RVB) state. He further proposed that frustrated lattices such as the triangular lattice may help stabilize the RVB state vs Néel order. Unfortunately it was found that the Heisenberg model can partially relieve frustration by forming 120 degrees order.[5] The field laid dormant until 1987, when soon after the discovery of high temperature superconductors, Anderson argued that upon doping, the RVB state naturally forms a superconductor because the singlet pairs are now mobile and can be viewed as Cooper pairs.[6] This suggestion launched a great deal of theoretical activities. While there remains no consensus on the theory of high temperature superconductors, there has been a great deal of development in the "easier" problem of quantum disordered spin states, which we call spin liquids. The requirement for a spin liquid is that it is a charge insulator with an odd number of electrons per unit cell (i.e. a Mott insulator) which does not exhibit Néel order down to zero temperature, despite the presence of antiferromagnetic exchange. We now know that the spin liquid is more than just the absence of order. Theory predicts the emergence of new excitations such as particle carrying $S=\frac{1}{2}$ and no charge, called spinons. Spinons may be fermions or bosons, and the excitation may gapped or gapless. Furthermore, the spinons do not live by themselves, but are coupled to an emergent gauge field, which may be $U(1)$ or $Z2$. Current analytic theory is not powerful enough to tell us which of these possibilities is realized for a given Hamiltonian, but the different possibilities can be classified and their nature fully described. As we shall see, the spin liquid problem is closely related to lattice gauge theory coupled to matter field, and the spin liquid states correspond to the deconfined phases.

The field of spin liquids received a powerful boost several years ago with the discovery of promising experimental candidates. They include the organic molecular compounds and a mineral called Herbertsmithite, which is a $S=\frac{1}{2}$ Kagome lattice that can be synthesized in the laboratory.[7] The $S=\frac{1}{2}$ Kagome lattice is a highly frustrated two dimensional lattice which has long been suspected to harbor a spin liquid ground state based on numerical work.[8] What is interesting is that in addition to using frustration to stabilize the spin liquid as proposed by Anderson, the organic materials have taught us a second route, i.e., the proximity to the Mott transition. The two families of organic material consist of molecular dimers (called ET and dmit as abbreviation) which carry a single electron each, and form an approximate triangular lattice. It can be modeled as a Hubbard model on a triangular lattice with one electron per unit cell.[9, 10] Deep in the Mott insulator regime, the low energy physics is described by a nearest-neighbor Heisenberg model and the system is expected to order into the 120 degree phase. The spin liquid materials turn out to be insulators which turn into a metal (dmit) or superconductor (ET) under modest pressure. The proximity to the Mott transition means that while the low energy excitations may still be described by the spin degrees of freedom, the Hamiltonian is much more complicated than the nearest-neighbor Heisenberg. For example, ring exchange terms may be important.[9] We therefore focus on the nature of the ground state, and leave aside the question of what spin Hamiltonian can stabilize it.

These two quite distinct spin liquid candidates share the common feature that they both appear to be gapless. In the organic system, the spin susceptibility goes to a constant and the specific heat has a linear T term,[11] both characteristic of a metal with a Fermi surface, and most uncommon for an insulator. Thermo-conductivity κ was measured and it is found at least for the dmit salts that κ/T goes to a constant at low temperature.[12] All these strange properties are consistent with the emergence of fermonic spinons which form a Fermi surface. The situation with Herbertsmithite is less clear because there exists a significant amount of local

Journal of Physics: Conference Series **529** (2014) 012001 doi:10.1088/1742-6596/529/1/012001

moments between the Kagome planes which obscure the low temperature behavior. Recently a single crystal large enough for neutron scattering was grown.[13] The data show that the spin excitation is gapless and rather featureless as a function of energy and has been argued to support the picture of gapless spinons.

Now we describe how the Heisenberg model $H = J\sum_{<ij>} \boldsymbol{S}_i \cdot \boldsymbol{S}_j$ is mapped onto a lattice gauge theory. It is convenient to introduce fermions $f_{i\alpha}$ which carry spin index α on site and write the spin operator as

$$\boldsymbol{S}_i = \tfrac{1}{2} f^\dagger_{i\alpha} \boldsymbol{\sigma}_{\alpha\beta} f_{i\beta}. \tag{1}$$

The fermions must be subject to the constraint

$$f^\dagger_{i\uparrow} f_{i\uparrow} + f^\dagger_{i\downarrow} f_{i\downarrow} = 1, \tag{2}$$

which excludes empty or doubly occupied fermions. A similar representation can be made by introducing spin carrying bosons called Schwinger bosons instead of fermions. However, with bosonic representations, the mean field theory yields either Néel order or gapped spin liquid states and gapless spin liquid requires fine tuning to a quantum critical point. On the other hand, we will see that fermionic representation naturally accommodates gapless phases. Since the experimental systems observed so far appear to be gapless, we will use the fermionic representation. The Heisenberg term is quartic in the fermion operator and is given as[14]

$$\boldsymbol{S}_i \cdot \boldsymbol{S}_j = -\tfrac{1}{4} f^\dagger_{i\alpha} f_{j\alpha} f^\dagger_{j\beta} f_{i\beta} - \tfrac{1}{4}\left(f^\dagger_{i\uparrow} f^\dagger_{j\downarrow} - f^\dagger_{i\downarrow} f^\dagger_{j\uparrow}\right)\left(f_{j\downarrow} f_{i\uparrow} - f_{j\uparrow} f_{i\downarrow}\right) + \tfrac{1}{4} f^\dagger_{i\alpha} f_{i\alpha} \tag{3}$$

Eq.(2) invites mean field decoupling in the particle-hole and particle-particle channels

$$\chi_{ij} = < f^\dagger_{i\alpha} f_{j\beta} > \quad \text{and} \tag{4}$$

$$\Delta_{ij} = \langle (f_{j\downarrow} f_{i\uparrow} - f_{j\uparrow} f_{i\downarrow}) \rangle . \tag{5}$$

For simplicity, let us ignore the pairing channel and consider χ_{ij} only. Note that at the mean field level, the fermion acquires dynamics and is now free to hop on the lattice. Formally we introduce the Hubbard-Stratonovich field χ_{ij} and the Lagrange multiplier λ_i to enforce the constraint Eq.(2).

$$Z = \int d\chi d\lambda df df^\dagger e^{-S} \tag{6}$$

$$\begin{aligned} S \;=\; & \int d\tau \left[\sum_i \left[f^\dagger_{i\alpha} \partial_\tau f_{i\alpha} + i\lambda_i \left(f^\dagger_{i\alpha} f_{i\alpha} - 1 \right) \right] \right. \\ + \; & \left. \sum_{<ij>} J \left[\chi_{ij} \left(f^\dagger_{j\alpha} f_{i\alpha} + h.c. \right) + 2|\chi_{ij}|^2 \right] \right] \end{aligned} \tag{7}$$

The saddle point describes the mean field theory. Fluctuations around the saddle point are dominated by the phase fluctuation of χ_{ij} which becomes the compact $U(1)$ gauge field a_{ij} on link ij while $i\lambda$ becomes the time component. The low energy theory of Eq.(1) becomes a compact $U(1)$ lattice gauge theory coupled to fermions hopping on a lattice.[15]

It is not surprising that a gauge theory should result from a representation subject to constraints. Because the fermions introduced extra degrees of freedom, we see that there is a gauge redundancy in Eq.(1), i.e., $\boldsymbol{S}$ is invariant under a local gauge transformation $f_i \longrightarrow f_i e^{i\phi}$. Therefore a gauge theory must emerge.

Several mean field solutions are notable. If χ_{ij} is a constant, we will have a half-filled band with a Fermi surface. Another interesting class is called the flux phases where nonzero gauge fluxes penetrate the plaquettes. π-flux for the square lattice[16] and π flux through the hexagons in the Kagome lattice yield massless Dirac fermions.[17] Note the time reversal symmetry is preserved with π fluxes. If we include pairing we can gap the Fermi surface in a variety of ways. We do not know which mean field theory is closer to the truth for a given Hamiltonian, but we can discuss the consequences of each state and compare with experiments. This is why the availability of experiments is crucial for the advancement in the field. Recent advances in numerical methods such as DMRG[18] and projected wave functions plus Lanczos[19] are making important contributions to ascertain the domain of stability of spin liquids for a given Hamiltonian.

The enemy of the spin liquid is confinement. If the gauge theory is in a confined phase, the low lying degrees of freedom remain those of spin excitations, and the system is typically Néel ordered. On the other hand, if we are in the deconfined phase, the fermion and gauge fields emerge as new particles and fields at low energies. The fictitious fermions which we introduced as a formal device and had no dynamics at high energy and short distance take on a life of their own at low energies. This is a prime example of the notion of emergence. There are now several exactly solvable examples which show that this scenario plays out in certain spin Hamiltonians.[20, 21] The reader should not be concerned that the fermions are not gauge invariant. After all the electrons in our world are also not gauge invariant but we have no trouble thinking of them as real objects. The difference between electron and spinon is a qualitative one: the electrons are weakly coupled to the gauge field with a small fine structure constant whereas spinons couple with strength unity. Another difference is that in solids the electron velocity v_F is much less than the speed of light and the coupling to electric field is stronger than the coupling to magnetic field. In spin liquids both velocities are given by J and transverse gauge fluctuations which are unscreened become dominant. For example, for $U(1)$ spin liquids this gives rise to the prediction that the specific heat $\sim T^{2/3}$, i.e., non-Fermi liquid behavior,[22] a subject under intense studies today.[23, 24, 25, 26]

Pure gauge theory is always confining in two dimensions due to instanton fluctuations.[27] The coupling to matter fields can change that. It was shown that coupling to N components of Dirac spinons can suppress instantons if N exceeds a critical N_c.[28] In the π-flux state $N = 4$, and deconfinement is quite possible. In the case of the Fermi sea, it has been shown that there are so many gapless degrees of freedom that it is analogous to infinite N and deconfinement is always possible.[29]

In conclusion, the field of quantum spin liquid is entering an exciting new phase where dialogs between theory and experiment are possible. We can look forward to exciting discoveries ahead.

Acknowledgments
I acknowledge the support by the NSF under grant DMR-1104498.

References
[1] Néel L 1936 *Ann. Phys.* (Paris) **5** 232
[2] Friedel J 2001 *Physics Today* **54** 88
[3] Bethe H 1931 *Zeit. Physik* **71** 205
[4] Anderson P W 1973 *Mat. Res. Bull.* **8** 153
[5] Huse D A and Elser V 1988 *Phys. Rev. Lett.* **60** 2531
[6] Anderson P W 1987 *Science* **235** 1196
[7] For a review of some of the experimental systems, see Lee P A 2008 *Science* **321** 1306; Balents L 2010 *Nature* **464** 199
[8] Lecheminant P *et al.* 1997 *Phys. Rev.* B **56** 2521
[9] Motrunich O 2005 *Phys. Rev.* B **72** 045105

[10] Lee S S and Lee P A 2005 *Phys. Rev. Lett.* **95** 036403
[11] Yamashita S *et al.* 2008 *Nature Physics* **4** 459
[12] Yamashita S *et al.* 2010 *Science* **328** 1246
[13] Han T *et al.* 2012 *Nature* **492** 406
[14] Baskaran G, Zou Z and Anderson P W 1987 *Solid State Commun.* **63** 973
[15] Baskaran G and Anderson P W 1988 *Phys. Rev.* B **37** 580
[16] Affleck I and Marston J B 1988 *Phys. Rev.* B **37** 3774
[17] Ran Y, Hermele M, Lee P A and Wen X-G 2007 *Phys. Rev. Lett.* **98** 117205
[18] Yan S, Huse D and White S R 2011 *Science* **332** 1173
[19] Iqbal Y, Becca F, Sorella S and Poilblans D 2006, *Phys. Rev.* B **88** 060405
[20] Kitaev A 2006 *Annals of Physics* **321** 2
[21] Wen X-G 2003 *Phys. Rev. Lett.* **90** 016803
[22] Lee P A and Nagaosa N 1992 *Phys. Rev.* B **46** 5621
[23] Lee S S 2009 *Phys. Rev.* B **80** 165102
[24] Metlitski M and Sachdev S 2010 *Phys. Rev.* B **82** 075127
[25] Mross D, McGreevy J, Liu H and Senthil T 2010 *Phys. Rev.* B **82** 045121
[26] Davidovich D and Lee S S 2013 Cond-Mat arXiv: 1307.3170
[27] Polyabov A 1975 *Phys. Lett.* **59B** 82
[28] Hermele M *et al.* 2004 *Phys Rev.* B **70** 214437
[29] Lee S S 2008 *Phys. Rev.* B **78** 085129

17th International Conference on Recent Progress in Many-Body Theories (MBT17) IOP Publishing
Journal of Physics: Conference Series **529** (2014) 012002 doi:10.1088/1742-6596/529/1/012002

Quantum Monte Carlo and the Hubbard Model

Douglas Scalapino
Physics Department, University of California, Santa Barbara, CA 93106-9530, USA
E-mail: djs@physics.ucsb.edu

Abstract. Here we discuss quantum Monte Carlo and the Hubbard model. We note that while there is a fermion sign problem for the doped Hubbard model, Monte Carlo calculations have shown that the leading pairfield susceptibility has d-wave symmetry and that the pairing interaction is mediated by $S = 1$ particle-hole fluctuations.

Let me begin by thanking the Feenberg Memorial Metal Committee and the RPMBT for this award. It is of course an honor in light of Eugene Feenberg's seminal contributions to the strongly interacting quantum many-body problem. However, for me, it was particularly special because it brought back memories of attending Feenberg's course on "The Many-Body Problem" in 1961. At this time I was a graduate student at Stanford, so let me explain. In the late 50's the Stanford Mark IV electron accelerator was used to treat cancer patients during the day and I used it at night to study radiation produced free-radicals. Then in 1960 it was taken over for prototype design studies by Project M (which was to become SLAC). When this happened, I decided to switch to theory and Professor Edwin Jaynes agreed to become my advisor. Several months later, Ed informed me that he had accepted a Professorship at Washington University in St. Louis and encouraged me to come along. There in the physics building "Crow Hall" I shared an office with Feenberg's graduate students and attended his lectures. The warm reception by Feenberg and his students for an outsider from another graduate school is something that I will never forget.

Turning to physics, the award citation references Monte Carlo methods and superconductivity. After I left Washington University, I had the good fortune of learning about superconductivity working as a postdoc in 1962–64 with J.R. Schrieffer at the University of Pennsylvania. There along with John Wilkens, a PhD student of Bob's, we calculated the superconducting tunneling density of states for a model of Pb [1]. This was my first paper. In 1968 I moved to UCSB. There, in the first year of the ITP, 1979, Michael Creutz spoke about using Monte Carlo methods to study gluons on a lattice. Towards the end of his talk, he held up a stack of punchcards in one hand to illustrate how short the program was for these calculations. This made one wonder what might be possible for condensed matter lattice models. At that time the quarks were treated in the "quenched" approximation and there was interest in Monte Carlo methods that would include dynamic fermions. With Blankenbecler and Sugar, we developed a determinantal quantum Monte Carlo (DQMC) method [2] to treat fermions. Here the interacting problem was rewritten in terms of free fermions coupled to an auxiliary Hubbard-Stratonovich (HS) field. Using Wick's theorem, the fermions were integrated out leaving a determinant involving the HS field. The HS field was then summed over using a Monte Carlo method with the determinant giving the Boltzmann weight. We used this method to explore various electron-phonon models. Then in an insightful paper, J. Hirsch [3] introduced a discrete Hubbard-Stratonovich field and, combining a DQMC simulation with a finite scaling analysis, found that the groundstate of the half-filled Hubbard model had long range antiferromagnetic order. This work showed what could be done with the DQMC but at the same time set the bar very high for what was to come.

With the discovery of the high T_c cuprates [4] in 1986, there was intense interest in the doped 2D Hubbard model. There had been weak coupling predictions of d-wave superconductivity near the antiferromagnetic instability of the Hubbard model [5, 6]. However, little beyond RPA like calculations was known. It seemed like a perfect opportunity to apply the DQMC. However, for the doped Hubbard model the DQMC has what is called "a fermion sign problem". That is, the determinant of some HS field configurations can be negative. Furthermore, as the temperature is lowered relative to the bandwidth, the number of HS field configurations with positive determinants is found to be similar to the number with negative determinants. Then simple techniques, such as using the absolute value of the determinant as the Boltzmann weight and including the sign in the measurement, fail as the error grows exponentially at low temperatures.

So we do not have the same kind of DQMC results for the existence of long range d-wave superconductivity at low temperature for the doped 2D Hubbard model as we do for the zero temperature long range AF order of the 2D half-filled case. However, the results clearly show that the doped 2D Hubbard model contains d-wave pairing correlations [7] and that the pairing interaction responsible for these correlations is mediated by the exchange of $S = 1$ particle-hole fluctuations.

Fig. 1 shows the d-wave pairfield susceptibility

$$P_d(T) = \int_0^\beta dt \langle \Delta_d(t) \Delta_d^+(0) \rangle \tag{1}$$

with

$$\Delta_d^+ = \frac{1}{2\sqrt{N}} \sum_{i,\delta} (-1)^\delta c_{\ell\uparrow}^+ c_{\ell+\delta\downarrow}^+ \tag{2}$$

for a doped Hubbard model with a site filling $\langle n \rangle = 0.875$. Here $c_{\ell s}^+$ creates an electron with spin s on the ℓth lattice site. The sum i is over the N lattice sites, δ sums over the four near-neighbor sites of ℓ and $(-1)^\delta$ gives the characteristic d-wave $+-+-$ alternation. Although this calculation involved only a 4×4 lattice and the fermion sign problem prevented one from going to low temperatures, it is clear that short range d-wave pairfield correlations are present in the doped 2D Hubbard model. It is also clear from Fig. 1 that it it important to keep track of the sign of the determinant [8]. The results which show a decrease of $P_d(T)$ at low temperatures were obtained in a calculation in which the absolute value of the determinant was used as a Boltzmann weight but the sign was not included in the measurement.

The pairing interaction is given by the irreducible particle-particle vertex $\Gamma^{pp}(k, k')$ shown on the left-hand side of Fig. 2. It consists of all Feynman diagrams that can not be cut into two parts by cutting just two particle lines. Both DQMC [9] and a more recently developed dynamic cluster approximation (DCA) [10] have shown that the leading pairfield eigenfunction associated with $\Gamma^{pp}(k, k')$ has d-wave symmetry. Using the DCA, the pairing interaction $\Gamma^{pp}(k, k')$ was decomposed into a fully irreducible vertex $\Lambda_{\rm irr}$ and $S = 1$ (spin) and $S = 0$ (charge) particle-hole contribution. Plots of Γ^{pp} and these various contributions are shown in Fig. 3. The peak in the pairing interaction at large momentum transfer is what favors the d-wave pairing. From the decomposition, one clearly sees that the interaction is mediated by the $S = 1$ channel. Thus in spite of the fermion sign problem, one has been able to determine that the dominant pairing correlations in the doped 2D Hubbard model have d-wave symmetry and the pairing interaction that is responsible for these correlations arises from the $S = 1$ particle-hole channel.

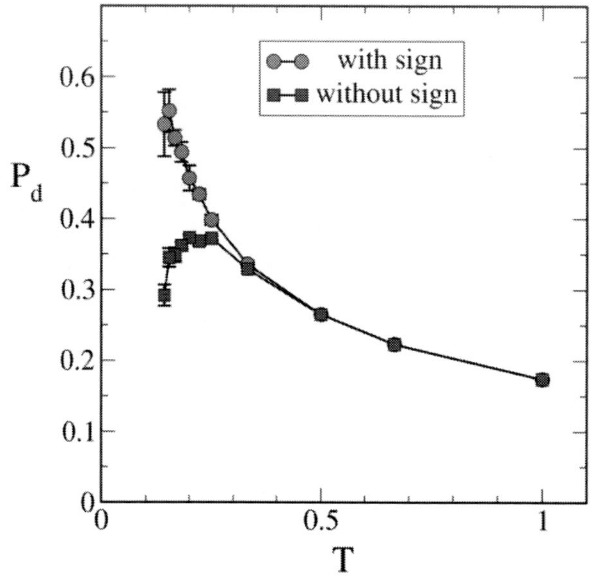

Figure 1. The d-wave pairfield susceptibility $P_d(T)$ (red circles) for a 4×4 lattice with $U = 4t$ and $\langle n \rangle = 0.875$ versus temperature T measured in units of the hopping t. The (blue squares) show the erroneous result that is found if the fermion sign is ignored (after Loh et al. [8]).

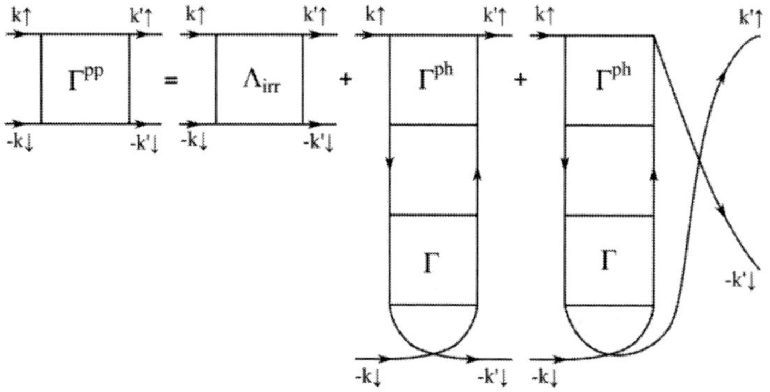

Figure 2. The pairing interaction is given by the irreducible particle-particle vertex Γ^{pp}. Here Γ^{pp} is decomposed into a fully irreducible two-fermion vertex $\Lambda_{\rm irr}$ plus contributions from the $S = 1$ and $S = 0$ particle-hole channels. Γ^{ph} are irreducible particle-hole vertices, Γ is the full vertex and the solid lines are fully dressed single particle propagators.

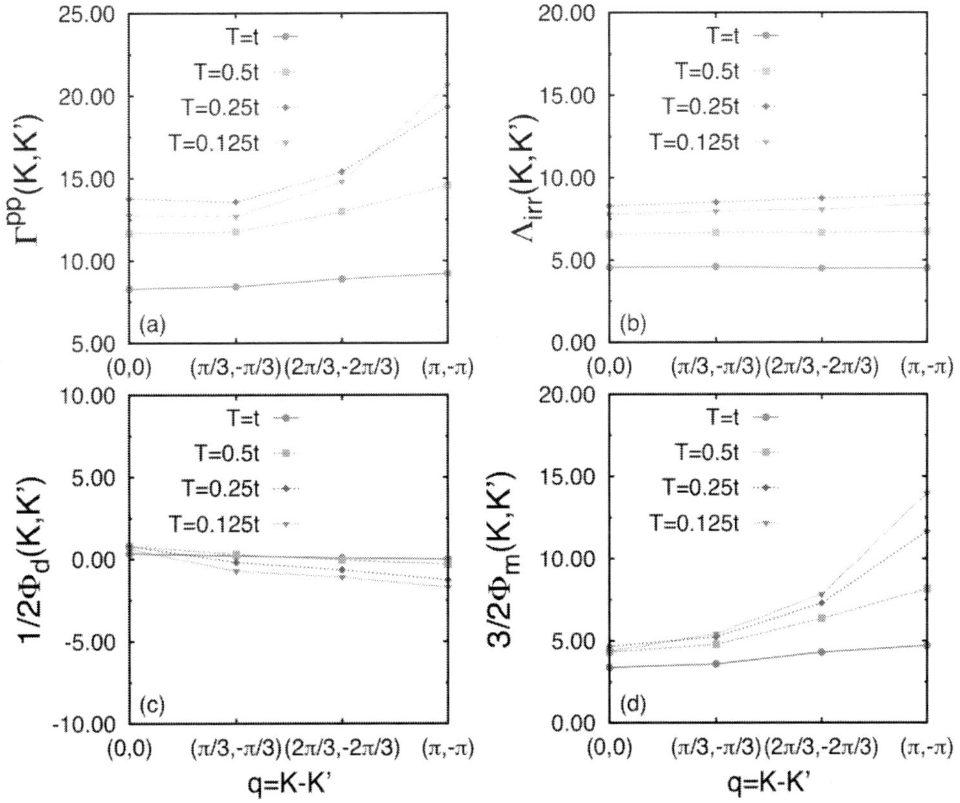

Figure 3. This figure illustrates the momentum dependence of the various contributions that make up the irreducible particle-particle pairing vertex Γ^{pp}. (a) The irreducible particle-particle vertex Γ^{pp} versus $q = K - K'$ for various temperatures with $\omega_n = \omega_{n'} = \pi T$. Note that the interaction increases with the momentum transfer as expected for a d-wave pairing interaction. (b) The q-dependence of the fully irreducible two-fermion vertex Λ_{irr}. (c) The q-dependence of the charge density ($S = 0$) channel for the same set of temperatures. (d) The q-dependence of the magnetic ($S = 1$) channel. Here, one sees that the increase in Γ^{pp} with momentum transfer arises from the $S = 1$ particle-hole channel (after Maier et al.[10]).

Although these results are for a single band Hubbard modeal, similar calculations have been carried out for the two-layer Hubbard model [11]. Here one finds that when the interlayer one-electron hopping becomes slightly larger than the intra layer hopping, the pairing shifts from d-wave to an $s^{\pm}$ state in which the gap switches sign between the bonding and anti-bonding Fermi surfaces, reminiscent of the Fe-pnictide/chalcogen superconductors. The Monte Carlo results for these models provide support for the idea that the unconventional, sign changing gap, superconducters share a common pairing mechanism mediated by the exchange of $S = 1$ spin-fluctuations.

Acknowledgments
I would like to begin by saying what a pleasure it is to win this award with Patrick Lee whose work and friendship have been important to me for many years. I want to acknowledge past graduate

students M. Jarrell, E. Loh, R. Scalettar and postdocs J. Hirsch, A. Moreo and S.R. White who played central roles in the early DQMC work and later allowed me to join them as a co-worker. I want to particularly acknowledge M. Jarrell and T.A. Maier for their development and application of the DCA to the Hubbard model. Finally, I want to thank my long-time colleagues R.L. Sugar and R. Blankenbecler. This work is supported by the Center for Nanophase Material Science at ORNL which is sponsored by the Division of Scientific User Facilities, U.S. DOE.

References

[1] Schrieffer J R, Scalapino D J and Wilkens J W 1963 *Phys. Rev. Lett.* **10** 336
[2] Blankenbecler R, Scalapino D J and Sugar R L 1981 *Phys. Rev.* D **24** 2278–86
[3] Hirsch J E 1985 *Phys. Rev.* B **31** 4403–19
[4] Bednorz J G and Muller K A 1986 *Z. Physik.* B **64** 189
[5] Miyake K, Schmitt-Rink S and Varma C M 1986 *Phys. Rev.* B **34** 6554
[6] Scalapino D J, Loh Jr. E and Hirsch J E 1986 *Phys. Rev.* B **34** 8190
[7] In addition, DCA results, carried out for a range of different cluster sizes and shapes find evidence for a transition to a d-wave superconducting phase at low temperature for the doped Hubbard model. Maier T, Jarrell M, Schutthess T C, Kent P R C and White J B 2005 *Phys. Rev. Lett.* **95** 237001
[8] Loh E Y, Gubernatis J E, Scalettar R T, White S R, Scalapino D J and Sugar R L 1990 *Phys. Rev.* B **41** 9301
[9] Bulut N, Scalapino D J and White S R 1993 *Phys. Rev.* B **47** R6157
[10] Maier T A, Jarrell M S and Scalapino D J 2006 *Phys. Rev. Lett.* **96** 047005
[11] Maier T A and Scalapino D J 2011 *Phys. Rev.* B **84** 180513.

Diagrammatical methods within the path integral representation for quantum systems

A Alastuey[1]

[1] Laboratoire de Physique, ENS Lyon/CNRS, 69364 Lyon Cedex 07, France

E-mail : angel.alastuey@ens-lyon.fr

Abstract The path integral representation has been successfully applied to the study of equilibrium properties of quantum systems for a long time. In particular, such a representation allowed Ginibre to prove the convergence of the low-fugacity expansions for systems with short-range interactions. First, I will show that the crucial trick underlying Ginibre's proof is the introduction of an equivalent classical system made with loops. Within the Feynman-Kac formula for the density matrix, such loops naturally emerge by collecting together the paths followed by particles exchanged in a given cyclic permutation. Two loops interact *via* an average of two-body genuine interactions between particles belonging to different loops, while the interactions between particles inside a given loop are accounted for in a loop fugacity. It turns out that the grand-partition function of the genuine quantum system exactly reduces to its classical counterpart for the gas of loops. The corresponding so-called magic formula can be combined with standard Mayer diagrammatics for the classical gas of loops. This provides low-density representations for the quantum correlations or thermodynamical functions, which are quite useful when collective effects must be taken into account properly. Indeed, resummations and or reorganizations of Mayer graphs can be performed by exploiting their remarkable topological and combinatorial properties, while statistical weights and bonds are purely c-numbers. The interest of that method will be illustrated through a brief description of its application to two long-standing problems, namely recombination in Coulomb systems and condensation in the interacting Bose gas.

1 Introduction

The standard formalism for studying the equilibrium properties of interacting quantum systems is the many-body perturbation theory [1]. Thermodynamic quantities are systematically expanded in powers of the interactions, and any contribution in the corresponding series can be represented by a Feynman graph built with loops describing free propagation, which are connected at vertices by interaction lines. The structure of Feynman graphs is determined by well-defined topological rules, while their contributions follow from conservation laws prescriptions at the vertices. The many-body perturbation theory is well-suited when interactions can be indeed treated as weak perturbations. This occurs for instance at high temperatures, or at high densities for fermions since the corresponding Fermi kinetic energy then becomes quite large.

In this paper, we will present another formalism, based on the Feynman path integral representation of quantum mechanics [2]. Within that representation, one can introduce an equivalent classical gas made of loops with interactions generated by imaginary-time averages of the genuine particle interactions along the paths defining the shapes of the loops. Interestingly, standard methods of classical statistical mechanics can be straightforwardly extended to the gas of loops, in particular the familiar Mayer series which are quite useful at low densities. Furthermore, the classical nature of the loops allows one to proceed to systematic reorganizations and resummations of the diagrammatical series, which are crucial for taking into account collective effects. Eventually, we stress that interactions are not treated perturbatively, an essential feature for dealing with phenomena like recombination for instance.

The usefulness of the loop formalism will be illustrated here through its applications to two long-standing problems in statistical mechanics of quantum systems. First, we will study recombination into atoms or molecules in a quantum plasma made with nuclei and electrons (see e.g. Ref. [3]). The second problem concerns the existence of Bose-Einstein (BE) condensation in an interacting Bose gas in the absence of external fields (see ch.2 in Ref. [4]).

The paper is organized as follows. In Section 2, we define the Hamiltonian of the general non-relativistic models with two-body interactions which can be studied within the path integral formalism. Then, we present the two examples considered further, namely the hydrogen plasma and an interacting Bose gas. In Section 3, after introducing the Feynman-Kac representation in the simple case of a single particle submitted to an external potential, we sketch the main steps of the transformations which lead to the introduction of a gas made with classical loops, which displays a grand-partition function identical to that of the genuine quantum system of interest. According to the corresponding so-called magic formula, we show that the equilibrium quantities of the quantum system can be represented by Mayer-like series for the equivalent gas of loops. The formalism is first applied to the hydrogen plasma in Section 4, where further resummations and reorganizations of Mayer series provide a series representations in terms of graphs made with particle clusters. The corresponding Screened Cluster Representation allows us to define and compute unambiguously the contributions of any recombined chemical species at finite temperatures. We turn to the second application in Section 5, namely the study of condensation in an interacting Bose gas. We show that, within the celebrated Kac scaling, the mean-field results can be readily retrieved from Mayer-like series. Contributions of fluctuations are also determined, but far from the critical region where a condensate might emerge. Eventually, we give some concluding comments in Section 6.

2 Non-relativistic quantum systems with two-body interactions

2.1 The model

In many situations encountered on Earth or in Astrophysics, a suitable description relies on the introduction of a quantum system made of point particles with Hamiltonian

$$H_N = -\sum_{i=1}^{N} \frac{\hbar^2}{2m_{\alpha_i}} \Delta_i + \frac{1}{2} \sum_{i \neq j} u_{\alpha_i \alpha_j}(|\mathbf{x}_i - \mathbf{x}_j|) \,. \tag{1}$$

Here, α_i denotes the species of the ith particle, while Δ_i is the Laplacian with respect to its position $\mathbf{x}_i$. In addition to the non-relativistic kinetic energy, the total potential energy is a sum of pairwise two-body instantaneous interactions $u_{\alpha_i \alpha_j}(|\mathbf{x}_i - \mathbf{x}_j|)$ which only depend on the relative distance $|\mathbf{x}_i - \mathbf{x}_j|$ between particles. Notice that the non-relativistic Hamiltonian (1) does not depend on the spins σ_{α_i} of the particles, which only determine their bosonic or fermionic nature.

2.2 Examples

A first example is the hydrogen plasma, which can be viewed, within the so-called physical picture, as a mixture of quantum point particles which are either protons or electrons, interacting via the instantaneous Coulomb potential. Protons and electrons have respective charges, masses, and spins, $e_\mathrm{p} = e$ and $e_\mathrm{e} = -e$, m_p and m_e, $\sigma_\mathrm{p} = \sigma_\mathrm{e} = 1/2$, while the two-body interaction reduces to

$$u_{\alpha_i \alpha_j}(|\mathbf{x}_i - \mathbf{x}_j|) = e_{\alpha_i} e_{\alpha_j} v(|\mathbf{x}_i - \mathbf{x}_j|) \tag{2}$$

with $v(r) = 1/r$. Here both species are fermions, and the two-body Coulomb interaction is long-ranged.

A second example is an atomic gas described as a system of neutral point particles with mass m and spin $\sigma = 0$. The corresponding two-body interaction potential $u(r)$ includes, in general, both a short-range repulsive part and a long-range attractive part, like the familiar Lennard-Jones potential. Now, the particles are bosons.

Notice that, at a more fundamental level, the atomic system should be also viewed as a quantum plasma made of point nuclei and point electrons with two-body Coulomb interactions, similarly to the case of the Hydrogen plasma. Nevertheless, the more phenomenological description in terms of an interacting Bose gas, is *a priori* sufficient at rather low temperatures and low densities. In fact, the difficult questions about recombination and effective interactions would be precisely investigated for the Hydrogen plasma.

3 Path integral representation and the equivalent gas of loops

3.1 The Feynman-Kac formula

For the sake of pedagogy, we introduce the path integral representation in the simple case of a single particle with mass m submitted to a potential $V(\mathbf{r})$. Its Hamiltonian reads

$$H = -\frac{\hbar^2}{2m} \Delta + V(\mathbf{r}) \,. \tag{3}$$

The corresponding density matrix at a given temperature T, namely the matrix element of Gibbs operator $\exp(-\beta H)$, is exactly given by Feynman-Kac formula [5, 6, 7, 8]

$$\langle \mathbf{r}_\mathrm{b}| \exp(-\beta H)|\mathbf{r}_\mathrm{a}\rangle = \frac{\exp[-(\mathbf{r}_\mathrm{b}-\mathbf{r}_\mathrm{a})^2/(2\lambda^2)]}{(2\pi\lambda^2)^{3/2}} \int \mathcal{D}(\boldsymbol{\xi}) \times \exp[-\beta \int_0^1 \mathrm{d}s\, V((1-s)\,\mathbf{r}_\mathrm{a} + s\,\mathbf{r}_\mathrm{b} + \lambda\,\boldsymbol{\xi}(s))] , \tag{4}$$

with thermal de Broglie wavelength $\lambda = (\beta\hbar^2/m)^{1/2}$. In the r.h.s. of (4), $\boldsymbol{\xi}(s)$ is a dimensionless Brownian bridge which starts from the origin at dimensionless time $s = 0$ and comes back at the origin at dimensionless time $s = 1$, *i.e.* $\boldsymbol{\xi}(0) = \boldsymbol{\xi}(1) = \mathbf{0}$. Functional measure $\mathcal{D}(\boldsymbol{\xi})$ is the normalized Gaussian Wiener measure which characterizes the Brownian process, and it is entirely defined by its covariance

$$\int \mathcal{D}(\boldsymbol{\xi})\; \xi_\mu(s)\; \xi_\nu(t) = \delta_{\mu\nu}\, \mathrm{inf}(s,t)\; (1-\mathrm{sup}(s,t)) . \tag{5}$$

The corresponding functional integration is performed over all Brownian bridges $\boldsymbol{\xi}(s)$. Representation (4) is the proper mathematical formulation of genuine Feynman's idea, which amounts to express the density matrix as a sum over all possible paths going from $\mathbf{r}_\mathrm{a}$ to $\mathbf{r}_\mathrm{b}$ in a time $\beta\hbar$, of weighting factors $\exp(-S/\hbar)$ where S is the classical action of a given path computed in potential $-V$. Here, such paths are parametrized according to

$$\boldsymbol{\omega}_\mathrm{ab}(s\beta\hbar) = (1-s)\,\mathbf{r}_\mathrm{a} + s\,\mathbf{r}_\mathrm{b} + \lambda\,\boldsymbol{\xi}(s) , \tag{6}$$

where $(1-s)\,\mathbf{r}_\mathrm{a} + s\,\mathbf{r}_\mathrm{b}$ describes the straight uniform path connecting $\mathbf{r}_\mathrm{a}$ to $\mathbf{r}_\mathrm{b}$. An example of path is shown in Fig. 1. Also, weighting factor $\exp(-S/\hbar)$ is splitted into the product of three terms. The first term, which arises from the kinetic energy of the straight uniform path, reduces to the Gaussian prefactor in front of the functional integral. The second term, associated with the kinetic contribution of the Brownian part of the path, is a Gaussian functional of $\boldsymbol{\xi}(s)$ embedded in Wiener measure $\mathcal{D}(\boldsymbol{\xi})$. The third and last term is rewritten as the Boltzmann-like factor associated with time average

$$\int_0^1 \mathrm{d}s\; V\,(\boldsymbol{\omega}_\mathrm{ab}(s\beta\hbar)) \tag{7}$$

of potential V along the considered path $\boldsymbol{\omega}_\mathrm{ab}$. We stress that, independently of the rather poetic introduction of path integrals by Feynman [2], representation (4) can be derived in a straightforward way by starting from the obvious identity $\exp(-\beta H) = [\exp(-\beta H/N)]^N$ combined with a suitable insertion of $(N-1)$ closure relations in position-space (see *e.g.* Ref. [9]). Feynman-Kac formula (4) then follows by taking the limit $N \to \infty$, as it has been proved for a wide class of potentials [5].

Feynman-Kac (FK) representation (4) perfectly illustrates the intrusion of dynamical features in equilibrium static quantities for quantum systems. Let us consider the diagonal density matrix $\langle \mathbf{r}_\mathrm{a}| \exp(-\beta H)|\mathbf{r}_\mathrm{a}\rangle$. Because of the non-commutativity of the kinetic and potential parts of H, that matrix element does not reduce to its classical counterpart

$$\frac{\exp(-\beta V(\mathbf{r}_\mathrm{a}))}{(2\pi\lambda^2)^{3/2}} , \tag{8}$$

Journal of Physics: Conference Series **529** (2014) 012003 doi:10.1088/1742-6596/529/1/012003

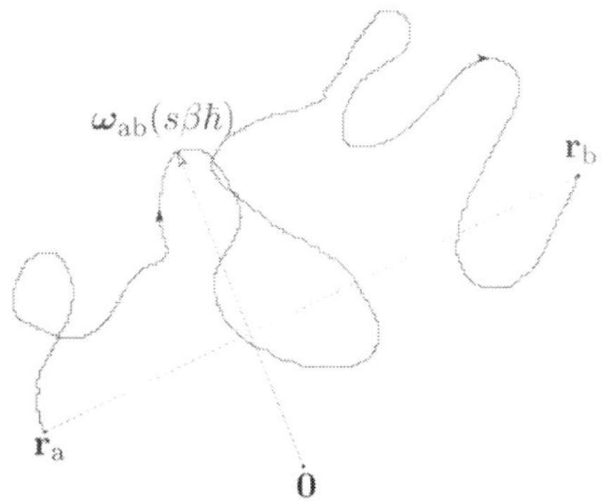

Figure 1: A Brownian path $\boldsymbol{\omega}_{ab}(s\beta\hbar)$. The dashed straight line is the uniform path connecting $\mathbf{r}_a$ to $\mathbf{r}_b$.

so it is not entirely determined by $V(\mathbf{r}_a)$. In fact, according to formula (4) specified to $\mathbf{r}_b = \mathbf{r}_a$, $\langle \mathbf{r}_a | \exp(-\beta H) | \mathbf{r}_a \rangle$ now depends on the potential landscape in some neigbourhood of $\mathbf{r}_a$ with size λ, which is explored in time-average (7) thanks to Brownian motion. Therefore, $\langle \mathbf{r}_a | \exp(-\beta H) | \mathbf{r}_a \rangle$ appears to be indeed generated by a dynamical process. Not surprisingly, particle mass m controls the importance of the corresponding dynamical effects. In particular, in the limit of an infinitely heavy particle $m \to \infty$, $\lambda = (\beta\hbar^2/m)^{1/2}$ vanishes and $\langle \mathbf{r}_a | \exp(-\beta H) | \mathbf{r}_a \rangle$ obviously tends to its classical counterpart (8) : dynamical effects do not intervene anymore in the potential contribution which takes its purely static form. Quantum corrections to classical formula (8) can be expanded in powers of $\hbar^2$ according to the well-known Wigner-Kirkwood expansion [10, 11, 12]. That expansion can be easily retrieved within FK representation (4), by expanding time-average (7) in power series of $\lambda\boldsymbol{\xi}$ and by applying Wick theorem to the calculation of the resulting moments of $\boldsymbol{\xi}$. Notice that the present dynamical considerations are the manifestation, in the framework of path integrals, of the Heisenberg uncertainty principle which prevents the particle to stay at $\mathbf{r}_a$. Hence, Brownian paths can be interpreted as describing intrinsic quantum fluctuations of position.

Remarkably, FK representation (4) involves only classical objects and c-numbers, so the operatorial structure of quantum mechanics is, in some sense, erased. That feature turns out to be particularly useful in the framework of the many-body problem, as described further. However, the intrinsic complexity of quantum mechanics is now hidden in the functional integration over all Brownian bridges, which remains a formidable task. In fact, explicit calculations can be performed in a few number of cases, as reviewed in Ref. [8]. Also, aymptotic Wigner-Kirkwood expansions of $\langle \mathbf{r}_a | \exp(-\beta H) | \mathbf{r}_a \rangle$ around the classical formula (8), can be derived for situations where λ becomes small compared to the characteristic length $[|\boldsymbol{\nabla} V(\mathbf{r}_a)|/|V(\mathbf{r}_a)|]^{-1}$ of variation of $V(\mathbf{r})$ in the neighbourhood of $\mathbf{r}_a$. Such situations occur for very heavy particles or for high temperatures, and also at large distances for potentials which decay as power laws at infinity. In the opposite limit where λ diverges, direct estimations of FK functional integral become rather cumbersome. Notice that, if H has a single isolated boundstate with energy E_0 and wavefunction ψ_0, in the zero-temperature limit, the asymptotic behaviour of $\langle \mathbf{r}_a | \exp(-\beta H) | \mathbf{r}_a \rangle$ is merely extracted from its spectral representation, *i.e.*

$$\langle \mathbf{r}_a | \exp(-\beta H) | \mathbf{r}_a \rangle \sim |\psi_0(\mathbf{r}_a)|^2 \exp(-\beta E_0) \quad \text{when} \quad T \to 0 \,. \tag{9}$$

As argued in Ref. [13], the relevant paths which provide the low-temperature behaviour (9) occupy a small piece of the whole functional phase space, and they are quite different from the typical paths with divergent size λ. Consequently, an exact direct estimation of their contribution remains an open problem in general.

3.2 The density matrix

Let us come back now to the many-body problem in the framework of the grand-canonical ensemble. The system is enclosed in a box with volume Λ, in contact with a thermostat at temperature T and a reservoir of particles that fixes the chemical potentials μ_α of the various species. The corresponding grand-partition function reads

$$\Xi_\Lambda = \mathrm{Tr}_\Lambda \exp\left(-\beta(H_N - \sum_\alpha \mu_\alpha N_\alpha)\right) . \tag{10}$$

The trace Tr_Λ is taken over a complete basis of N-body wavefunctions, which are symmetrized according to the statistics of each species, and satisfy Dirichlet boundary conditions at the surface of the box, while particle numbers N_α vary from 0 to ∞. The grand-partition function (10) can be expressed in the basis of positions and spins, where a given state is the suitably symmetrized Slater sum of products of one-body states $|\mathbf{x}\ \sigma_\alpha^z\rangle$. This provides a sum of diagonal and off-diagonal matrix elements of $\exp(-\beta H_N)$.

In the case of the hydrogen plasma, an example of above density matrix elements involved in the expression of the grand-partition function, is

$$\langle \mathbf{R}_1\mathbf{R}_3\mathbf{R}_2\mathbf{r}_2\mathbf{r}_3\mathbf{r}_1\mathbf{r}_4| \exp(-\beta H_{3,4}) |\mathbf{R}_1\mathbf{R}_2\mathbf{R}_3\mathbf{r}_1\mathbf{r}_2\mathbf{r}_3\mathbf{r}_4\rangle , \tag{11}$$

for the Hamiltonian $H_{3,4}$ of 3 protons and 4 electrons. Here, the positions of two protons are exchanged, as well as those of three electrons. Contributions of spins are factored out in simple degeneracy factors because the Coulomb Hamiltonian does not depend on the spins. For matrix element (11), that multipliyng degeneracy factor is 2^4, because the spin-states of the exchanged particles are necessarily identical.

The FK representation for each of the matrix elements of $\exp(-\beta H_N)$ takes a form similar to formula (4), with N_α paths $\boldsymbol{\omega}^\alpha$ for each species α, as well as a Boltzmann-like factor associated with the time average of the potential part of H_N. The paths associated with matrix element (11) can be collected into loops. In fact, that property holds for any matrix element of $\exp(-\beta H_N)$, because any permutation can always be decomposed as a product of cyclic permutations. A loop $\mathcal{L}$ is constructed by collecting q paths associated with q particles exchanged in a cyclic permutation. Accordingly, $\mathcal{L}$ is characterized by its position $\mathbf{X}$, which can be arbitrarily chosen among the extremities of paths $\boldsymbol{\omega}^\alpha$, and several internal degrees of freedom which are particle species (α), number q of exchanged particles, and shape $\lambda_\alpha\ \boldsymbol{\eta}$ obtained as the union of the q paths $\boldsymbol{\omega}^\alpha$. It turns out that $\boldsymbol{\eta}(s)$ is itself a Brownian bridge with flight time q, *i.e.* $\boldsymbol{\eta}(0) = \boldsymbol{\eta}(q) = \mathbf{0}$, distributed with the corresponding Wiener measure $\mathcal{D}(\boldsymbol{\eta})$.

In the FK representation, the time-average of the total interaction potential can be obviously rewritten as a sum of two-body interactions $\mathcal{V}$ between loops, plus a sum of loop self-energies $\mathcal{U}$, namely

$$\frac{1}{2}\sum_{i\neq j}\mathcal{V}(\mathcal{L}_i, \mathcal{L}_j) + \sum_i \mathcal{U}(\mathcal{L}_i) , \tag{12}$$

Figure 2: The four loops associated with matrix element (11). Loops $\mathcal{L}_1$ and $\mathcal{L}_2$ contain respectively 1 proton and 2 protons, while loops $\mathcal{L}_3$ and $\mathcal{L}_4$ contain respectively 3 electrons and 1 electron. The positions of the particles involved in the matrix element (11) are denoted by black disks with larger radius for the protons, and a smaller one for the electrons. The Brownian paths generated in the FK representation of (11), which go from one particle position to another one, are represented by thick and thin lines for the protons and the electrons respectively.

where loops associated with the considered matrix element are labelled as $\mathcal{L}_i$ with index i running from 1 to their total number N. For instance, four loops can be identified in the FK representation of matrix element (11), as shown in Fig. 2. Two-body potential $\mathcal{V}(\mathcal{L}_i, \mathcal{L}_j)$ between loops $\mathcal{L}_i$ and $\mathcal{L}_j$ reduces to a time-average along their respective shapes of the genuine two-body particle interaction $v(|\mathbf{X}_i + \lambda_{\alpha_i}\boldsymbol{\eta}_i(s) - \mathbf{X}_j - \lambda_{\alpha_j}\boldsymbol{\eta}_j(t)|)$ evaluated at times which differ by an integer value. Self-energy $\mathcal{U}(\mathcal{L}_i)$ for loop $\mathcal{L}_i$ is given by a similar average along its own shape of $v(|\lambda_{\alpha_i}\boldsymbol{\eta}_i(s) - \lambda_{\alpha_i}\boldsymbol{\eta}_i(t)|)$ evaluated at times which differ by a non-zero integer value, with a prefactor 1/2 which avoids double counting of genuine interactions between two exchanged particles.

3.3 The magic formula

At this stage, grand-partition function (10) is rewritten as a sum of Boltzmann-like factors associated with energies (12) multiplied by combinatorial factors and particle fugacities $z_\alpha/(2\pi\lambda_\alpha^2)^{3/2}$ with $z_\alpha = \exp(\beta\mu_\alpha)$, which have to be integrated over positions and shapes of the involved loops. It turns out that the whole sum can be rewritten as the grand-partition function of a classical gas of undistinguishable loops with suitable activities $z(\mathcal{L})$, namely

$$\Xi_\Lambda = \Xi_\Lambda^{(\mathrm{loop})} = \sum_{N=0}^{\infty} \frac{1}{N!} \int \prod_{i=1}^{N} \mathrm{d}\mathcal{L}_i \, z(\mathcal{L}_i) \prod_{i<j} \exp(-\beta\mathcal{V}(\mathcal{L}_i, \mathcal{L}_j)) \,. \tag{13}$$

Phase space measure $\mathrm{d}\mathcal{L}$ in the world of loops, involves discrete summations over species index α and exchanged-particle number q, spatial integration over position $\mathbf{X}$ inside Λ, and functional integration over shapes $\boldsymbol{\eta}$ with Wiener measure $\mathcal{D}(\boldsymbol{\eta})$ restricted to shapes such that $\mathbf{X} + \lambda_\alpha\boldsymbol{\eta}(s)$ remains inside Λ. Loop fugacity reads

$$z(\mathcal{L}) = (\pm 1)^{q-1} \frac{(2\sigma_\alpha + 1) z_\alpha^q}{q(2\pi q\lambda_\alpha^2)^{3/2}} \exp(-\beta\mathcal{U}(\mathcal{L})) \,, \tag{14}$$

the structure of which is easily intepreted as follows. For bosons, the sign is + and the fugacity is always positive. For fermions, the sign is − and factor $(-1)^{q-1}$ is the signature of a cyclic permutation of q objects, which is negative for q even. In front of the obvious particle-activity contribution z_α^q, factor $(2\sigma_\alpha+1)$ is the number of configurations of exchanged-particles spins which are all identical. Factor q in front of $(2\pi q\lambda_\alpha^2)^{3/2}$ is related to the q possible choices of position $\mathbf{X}$ among that of the q exchanged particles. The other factor q in front of λ_α^2, arises from the absorption of the remaining $(q-1)$ integrations over particle positions together with the q functional integrations over Brownian bridges into the single measure $\mathcal{D}(\boldsymbol{\eta})$. Notice that $q\lambda_\alpha^2$ is nothing but the square of de Broglie wavelength for inverse temperature $q\beta$.

We stress that identity (13) proceeds from remarkable combinatorial properties. For instance, in the case of the hydrogen plasma, there are various contributions in term $N=4$ of $\Xi_\Lambda^{(\text{loop})}$, which are identical to that of matrix element (11) in the Slater expansion of Ξ_Λ. Such contributions arise from the explicitation of the 4 phase-space measures $\mathrm{d}\mathcal{L}_i$ $(i=1,2,3,4)$ where 2 protonic loops and 2 electronic loops, carrying respectively 2 or 1 protons and 3 or 1 electrons, have to be chosen among 4 labelled loops. Also, there are several matrix elements in the Slater expansion of Ξ_Λ, involving 3 protons and 4 electrons, which provide identical contributions to that of matrix element (11). Thanks to a rather fortunate arrangment between both counting factors, the full respective contributions in both $\Xi_\Lambda^{(\text{loop})}$ and Ξ_Λ are indeed identical! According to those considerations, and also because of its synthetical form, identity (13) is oftenly called the magic formula.

Historically, Ginibre [14] was the first to introduce the notion of loops when studying the convergence of Mayer series for quantum gases with short-range forces. However, he did not write explicitely formula (13), which has been derived later by Cornu [15]. Recently, Martin [16] proposed an elegant and shorter derivation of that formula. Notice that identity (13) is valid for any kind of two-body interactions, and an arbitrary number of species with Fermi or Bose statistics.

Of course, and as quoted in the simple case of a single particle in an external potential, the intrinsic difficulty of quantum mechanics is now hidden in the functional integrations over loop shapes, so an exact calculation of loop grand-partition function remains far beyond human abilities...Nevertheless, magic formula (13) is quite useful because standard tools of classical statistical mechanics can be applied, as well as various transformations relying on simple properties of classical Boltzmann factors.

3.4 Mayer-like graphs

The structure of loop grand-partition (13) is identical to that of an ordinary classical system made of point particles with two-body interactions. Therefore, equilibrium quantities of loops can be represented by Mayer-like diagrammatical series, where points are replaced by loops. This provides Mayer series for particle quantities, like particle density ρ_α which reads

$$\rho_\alpha = \sum_{\mathcal{G}} \frac{1}{S(\mathcal{G})} \sum_{q_0=1}^{\infty} \int \mathcal{D}(\boldsymbol{\eta}_0)\, q_0\, z(\mathcal{L}_0) \int \prod_{i=1}^{n} \mathrm{d}\mathcal{L}_i\, z(\mathcal{L}_i) \left[\prod f\right]_{\mathcal{G}} . \tag{15}$$

Each graph $\mathcal{G}$ is constructed according to the standard Mayer rules [17, 18]. It is made with $(n+1)$ loops $\mathcal{L}_i$, $i=0,...,n$. Two loops i and j are connected at most by a single bond

$$f_{ij} = \exp(-\beta\mathcal{V}(\mathcal{L}_i,\mathcal{L}_j)) - 1\ , \tag{16}$$

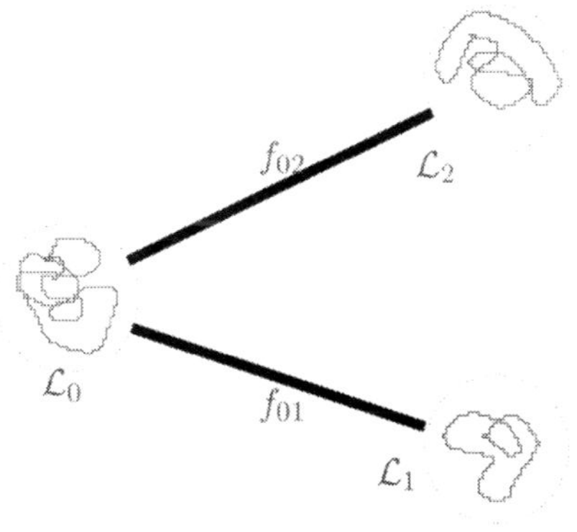

Figure 3: An example of graph $\mathcal{G}$ with the root loop $\mathcal{L}_0$ connected to two loops $\mathcal{L}_1$ and $\mathcal{L}_2$ *via* Mayer bonds f_{01} and f_{02}. The shapes of the loops are drawn inside bubbles, while the Mayer bonds are represented by thick straight lines connecting the bubbles.

and $[\prod f]_{\mathcal{G}}$ denotes the product of such bonds. Also, graph $\mathcal{G}$ is simply connected, namely it cannot be separated into two parts which are not connected by at least one bond f. Symmetry factor $S(\mathcal{G})$ is the number of permutations of black loops $\mathcal{L}_i$ with $i \geq 1$, which leave $\prod_{\mathcal{G}} f$ unchanged. Each loop is weighted by its fugacity $z(\mathcal{L}_i)$. The contribution of graph $\mathcal{G}$ is obtained by integrating over all degrees of freedom of black loops embedded in measure $\mathrm{d}\mathcal{L}_i$, while the position and species index of root loop $\mathcal{L}_0$ are fixed and only its shape and particle number are integrated over. The species index α_0 of root loop $\mathcal{L}_0$ is set equal to α. Eventually, $\sum_{\mathcal{G}}$ is performed over all topologically different unlabelled graphs $\mathcal{G}$. For instance, the contribution of graph $\mathcal{G}$ made with two field loops $\mathcal{L}_1$ and $\mathcal{L}_2$ only connected to root loop $\mathcal{L}$ (see Fig. 3), reads

$$\frac{1}{2}\sum_{q_0=1}^{\infty}\int \mathcal{D}(\boldsymbol{\eta}_0)\, q_0\, z(\mathcal{L}_0)\int \mathrm{d}\mathcal{L}_1 \mathrm{d}\mathcal{L}_2\, z(\mathcal{L}_1) z(\mathcal{L}_2) f_{01} f_{02}\,. \tag{17}$$

Let us first consider a short-range particle interaction $u_{\alpha_i\alpha_j}(|\mathbf{X}_i - \mathbf{X}_j|)$ which decays faster than some exponential when $|\mathbf{X}_i - \mathbf{X}_j| \to \infty$. Since the sizes of the loops are distributed with a Gaussian law, the loop-loop interaction potential $\mathcal{V}(\mathcal{L}_i, \mathcal{L}_j)$ is then bounded by an exponential decaying function of $|\mathbf{X}_i - \mathbf{X}_j|$. Since Mayer-like series (15) exhibit the same topological structure as their classical counterparts, the analysis of their convergence then becomes similar to that for classical systems with short-range interactions. This was exploited by Ginibre [14] who indeed proved the convergence at sufficiently low fugacities and high temperatures.

If the particle interaction is long ranged, the loop-loop interaction is also long ranged. In particular, for the hydrogen plasma or any Coulomb system, $\mathcal{V}(\mathcal{L}_i, \mathcal{L}_j)$ decays like the genuine Coulomb potential itself at large distances, for most of the loops which have a finite size controlled by the de Broglie wavelengths, namely

$$\mathcal{V}(\mathcal{L}_i, \mathcal{L}_j) \sim \frac{q_i e_{\alpha_i} q_j e_{\alpha_j}}{|\mathbf{X}_i - \mathbf{X}_j|} \quad \text{when} \quad |\mathbf{X}_i - \mathbf{X}_j| \to \infty\,. \tag{18}$$

Indeed, at large distances, loops can be shrinked to point charges, and further corrections to the monopolar term (18) can be expanded in multipolar power series of $\lambda_{\alpha_i}/|\mathbf{X}_i - \mathbf{X}_j|$ and $\lambda_{\alpha_j}/|\mathbf{X}_i - \mathbf{X}_j|$.

Like in the case of purely Coulombic interactions, the long-range nature of $\mathcal{V}(\mathcal{L}_i, \mathcal{L}_j)$ induces divergences in every Mayer-like graph because Mayer bonds behave as $-\beta\mathcal{V}(\mathcal{L}_i, \mathcal{L}_j)$ at large distances. Such divergences are removed *via* systematic chain resummations as exposed further.

4 Emergence of chemical species in the hydrogen plasma

In this Section, we consider the hydrogen plasma in contact with a thermostat at temperature T and a reservoir of particles that fixes the chemical potentials equal to μ_p and μ_e for protons and electrons respectively. Because the infinite system maintains local neutrality $\rho_\mathrm{p} = \rho_\mathrm{e}$ in any fluid phase, the bulk equilibrium quantities depend in fact solely on the mean

$$\mu = (\mu_\mathrm{p} + \mu_\mathrm{e})/2, \tag{19}$$

while the difference $\nu = (\mu_\mathrm{e} - \mu_\mathrm{p})/2$ is not relevant, as rigorously proved in Ref. [19]. Consequently, the common particle density $\rho = \rho_\mathrm{p} = \rho_\mathrm{e}$ depends only on T and μ.

We are interested in the emergence of usual chemical species, like atoms H or molecules H_2 within the grand-canonical description of the hydrogen plasma, namely the so-called physical picture. First, we provide rough estimations which allow us to identify their contributions at vanishing temperatures in the grand-partition function. Then, we introduce the well-known problem of estimating such contributions at finite temperature which has to face the divergent contributions of Rydberg states. The proper answer to that longstanding question requires a simultaneous and consistent treatment of both recombination and screening. We show that this can be achieved through a suitable reorganization of Mayer-like series. The corresponding screened cluster representation provides an unambiguous definition of finite-T internal partition functions for any chemical species without introducing arbitrary regularizations or adjustable parameters.

4.1 Rough estimations in the zero-temperature limit

In the zero-temperature limit at fixed negative chemical potential μ, the relative importance of the contributions of chemical species to Ξ_Λ given by the trace (10), follows from the entropy-energy balance between

- Exponentially vanishing fugacity factors $\exp(\beta\mu)$
- Exponentially growing Boltzmann factors $\exp(-\beta E^{(0)}_{N_\mathrm{p},N_\mathrm{e}})$

where $E^{(0)}_{N_\mathrm{p},N_\mathrm{e}}$ is the groundstate energy of $H_{N_\mathrm{p},N_\mathrm{e}}$. For instance, leading contributions of atoms H or molecules H_2 obviously arise from $H_{1,1}$ and $H_{2,2}$ with respective groundstate energies $E^{(0)}_{1,1} = E_\mathrm{H} = -e^2/(2a_\mathrm{B})$ where $a_\mathrm{B} = \hbar^2/(me^2)$ is the Bohr radius and $m = m_\mathrm{p}m_\mathrm{e}/(m_\mathrm{p}+m_\mathrm{e})$ is the reduced mass, and $E^{(0)}_{2,2} = E_{\mathrm{H}_2}$. The corresponding relative abundance of atoms and molecules depend not only on the relative values of the groundstate energies, but also on the chemical potential μ. In fact, thanks to the inequality

$$3E_\mathrm{H} < E_{\mathrm{H}_2} < 2E_\mathrm{H}\ , \tag{20}$$

the atomic weight $\exp(2\beta\mu)\exp(-\beta E_H)$ dominates the molecular weight $\exp(4\beta\mu)\exp(-\beta E_{H_2})$ if we choose μ sufficiently close to E_H. This suggests that an atomic phase should prevail under such conditions, in agreement with the chemical ionization equilibrium introduced by Saha [20].

Above rough estimations can be rigorously controlled within a suitable scaling low-temperature limit (SLT), $T \to 0$ with $\mu \to E_H$, introduced by Macris and Martin [21], who extended the pioneering work by Fefferman [22]. Those authors proved that, within the SLT which properly defines the so-called Saha regime, the hydrogen plasma becomes an ideal mixture of ionized protons, ionized electrons, and hydrogen atoms in their groundstate. The next important question is related to finite temperature effects: what about the contributions of excited atomic states and of other chemical species?

4.2 Divergence of internal partition functions at finite temperature

At finite temperatures, not only the atomic groundstate contributes to thermodynamic quantities, but also excited states. When T vanishes, an atom can be considered as frozen in its groundstate because of the finite gap between E_H and the first excited level $E_H/4$. This explains why the corresponding contribution to equilibrium quantities merely involves Boltzmann factor $\exp(-\beta E_H)$ as noticed just above. At finite temperatures, contributions of excited states must also be taken into account. For that purpose, it is natural to introduce the internal atomic partition function in the vacuum

$$Z_H^{(\mathrm{vac})} = 4\sum_{n=1}^{\infty} n^2 \exp(-\beta E_H/n^2) . \qquad (21)$$

In that definition, factor 4 arises from proton and electron spins degeneracy, factor n^2 accounts for orbital degeneracy of the level n with energy E_H/n^2, and summation extends over all boundstates up to $n = \infty$.

Obviously $Z_H^{(\mathrm{vac})}$ diverges because of the contributions of Rydberg states with $n \to \infty$. Similar divergences would occur in a naive description of recombination into molecules H_2, ions H^- and H_2^+, and also more complex entities like H_2^-, H_3^+, H_3, *etc*...In the literature, for atomic contributions, such divergences are removed within phenomenological prescriptions, in particular those leading to the so-called Brillouin-Planck-Larkin formula [23], [24], as described in the review [3] and commented in Ref. [25],

$$Z_{\mathrm{BPL}} = 4\sum_{n=2}^{\infty} n^2[\exp(-\beta E_H/n^2) - 1 + \beta E_H/n^2] . \qquad (22)$$

If the substracted terms in the expression (22) do ensure the convergence of the sum over n, their physical signification is not clear. In fact, the physical idea underlying that phenomenological approach, is that the divergent contributions of Rydberg states are screened by the free charges present in the system. Inside an Hydrogen atom, the proton-electron potential decays faster than $1/r$ at distances r larger than screening length λ_S. Thus, there is no accumulation of atomic boundstates near the continuum edge at $E = 0$, and all Rydberg states with n sufficiently large are suppressed. This gives some physical meaning to Brillouin-Planck-Larkin formula (22). At the same time, it suggests to introduce another phenomenological expression for the internal atomic

Figure 4: A convolution chain connecting two loops $\mathcal{L}_a$ and $\mathcal{L}_b$ *via* 3 intermediary loops. For convenience, the loops are schematically represented by disks, while the potential bonds $-\beta\mathcal{V}$ connecting two successive loops in the chain are represented by straight lines.

partition function, namely

$$Z_{\rm H}^{\rm (scr)} = 4\sum_{n=1}^{p_{\rm S}} n^2 \, \exp(-\beta E_{\rm H}/n^2) \, , \tag{23}$$

which explicitely accounts for screening effects by setting $p_{\rm S} \sim \sqrt{\lambda_{\rm S}/a_{\rm B}}$.

According to above considerations, the estimation of contributions from recombined entities cannot be disentangled from that of screened interactions between ionized charges. Thus, a fully consistent treatment of many-body effects is required, where the interplay between recombination and screening is properly taken into account. As described further, the Mayer-like series for the equivalent gas of loops turns out to be a suitable tool for that purpose. That formalism provides well-behaved prescriptions for computing the contributions of any recombined entity, not only those of atoms but also, for instance, those of molecules H_2 and ions H^-, H_2^+.

4.3 Recombination and screening within Mayer diagrammatical series

As mentioned previously, all Mayer-like graphs for the gas of loops diverge because of the long range nature of the loop interaction potential $\mathcal{V}(\mathcal{L}_i, \mathcal{L}_j)$. In order to account for the collective effects which remove those divergences, as well as for the recombination processes which leads to the formation of chemical species, we proceed to successive resummations and reorganizations which have been derived in detail in Ref. [27]. Here we only sketch the main steps of the method.

• Introduction of a screened effective potential

The first step consists in removing Coulomb divergences. All Mayer bonds f are first expanded in Taylor series (TS) with respect to $(-\beta\mathcal{V})$. This transforms the whole series (15) into a new set of analogous graphs $\mathcal{G}^{(TS)}$ where two loops can be connected by bonds which are either $-\beta\mathcal{V}$, or $(\beta\mathcal{V})^2/2$ or $-(\beta\mathcal{V})^3/6$, etc... Now, let us consider a graph $\mathcal{G}_{\rm ab}^{(TS)}$ where two given loops $\mathcal{L}_a$ and $\mathcal{L}_b$ are connected by the bond $-\beta\mathcal{V}(\mathcal{L}_a, \mathcal{L}_b)$. Such loops may be black loops or one of them may reduce to the root loop $\mathcal{L}_0$. Then, we consider all the other graphs $\mathcal{G}_{\rm ab,n}^{(TS)}$ identical to $\mathcal{G}_{\rm ab}^{(TS)}$ except that the direct simple bond $-\beta\mathcal{V}(\mathcal{L}_a, \mathcal{L}_b)$ between the two loops $\mathcal{L}_a$ and $\mathcal{L}_b$ is now replaced by a convolution chain of $(n+1)$ bonds $(-\beta\mathcal{V})$ connecting n intermediary loops $\mathcal{L}_i$, as shown in Fig. 4.

Thanks to remarkable combinatorial properties [27], the summation of the contributions of graphs $\mathcal{G}_{\rm ab,n}^{(TS)}$ to that of graph $\mathcal{G}_{\rm ab}^{(TS)}$, can be carried out at the local level, namely considering only loops $\mathcal{L}_a$ and $\mathcal{L}_b$ with the intermediary loops involved in the considered convolution chain. The result of that resummation does not depend on the global structures of the considered graphs, which accordingly are not shown in Fig. 4. That simplification holds for all further resummations and reexponentiations which are described below. Here, this leads to the introduction of the effective potential

$$e_{\alpha_a}e_{\alpha_b}\phi(\mathcal{L}_a,\mathcal{L}_b) = \mathcal{V}(\mathcal{L}_a,\mathcal{L}_b) + \sum_{n=1}^{\infty}(-1)^n\beta^n \int \prod_{i=1}^{n} \mathrm{d}\mathcal{L}_i\; z(\mathcal{L}_i)\; \mathcal{V}(\mathcal{L}_{i-1},\mathcal{L}_i)\; \mathcal{V}(\mathcal{L}_n,\mathcal{L}_b)\;, \tag{24}$$

with $\mathcal{L}_0 = \mathcal{L}_a$. That potential accounts for collective effects, because it involves contributions of an infinite number of loops. Interestingly, chain sum (24) can be reinterpreted as the complete perturbative expansion of the solution of integral equation

$$e_{\alpha_a}e_{\alpha_b}\phi(\mathcal{L}_a,\mathcal{L}_b) = \mathcal{V}(\mathcal{L}_a,\mathcal{L}_b) - \beta\int \mathrm{d}\mathcal{L}\; z(\mathcal{L})\; e_{\alpha_a}e_{\alpha}\phi(\mathcal{L}_a,\mathcal{L})\mathcal{V}(\mathcal{L},\mathcal{L}_b)\;. \tag{25}$$

Therefore, $e_{\alpha_a}e_{\alpha_b}\phi(\mathcal{L}_a,\mathcal{L}_b)$ is the total potential acting on loop $\mathcal{L}_b$, created by loop $\mathcal{L}_a$ plus a cloud of loops $\mathcal{L}$ distributed with density $-\beta z(\mathcal{L})\; e_{\alpha_a}e_{\alpha}\phi(\mathcal{L}_a,\mathcal{L})$. Its physical content is then strictly equivalent to that of the familiar classical Debye potential. Notice that in the classical limit $\hbar \to 0$, loops shrink to point charges so equation (25) becomes nothing but the integral version of classical linearized Poisson-Boltzmann equation. Thus, the effective potential $e_{\alpha_a}e_{\alpha_b}\phi(\mathcal{L}_a,\mathcal{L}_b)$ can be viewed as the quantum analogue of the Debye potential [26].

• Emergence of Boltzmann factors associated with the screened effective potential

The previous chain resummations can be repeated when two given loops in a graph $\mathcal{G}^{(TS)}$ are connected by the bond $[\beta\mathcal{V}(\mathcal{L}_a,\mathcal{L}_b)]^2$. Now, we have to consider all the other identical graphs where that bond is replaced by two convolution chains of bonds $(-\beta\mathcal{V})$ as shown in Fig. 5. The summation of all the correponding contributions gives raise to the bond $[\beta e_{\alpha_a}e_{\alpha_b}\phi(\mathcal{L}_a,\mathcal{L}_b)]^2/2$ where $\phi(\mathcal{L}_a,\mathcal{L}_b)$ is nothing but the effective potential (24).

The previous result is readily extended to any genuine bond $(-\beta\mathcal{V})^n/n!$, which gives raise to $[-\beta e_{\alpha_a}e_{\alpha_b}\phi(\mathcal{L}_a,\mathcal{L}_b)]^n/n!$ once all chain resummations have been performed. Then, the further summation of all bonds $[-\beta e_{\alpha_a}e_{\alpha_b}\phi(\mathcal{L}_a,\mathcal{L}_b)]^n/n!$ between two given loops $\mathcal{L}_a$ and $\mathcal{L}_b$ makes Boltzmann factors $\exp(-\beta e_{\alpha_a}e_{\alpha_b}\phi(\mathcal{L}_a,\mathcal{L}_b))$ emerge. As mentionned previously, all those resummations and reexponentiations can be performed independently of the global structures of the considered graphs [27].

• Introduction of renormalized fugacities

Chain resummations also apply to other substructures of graphs $\mathcal{G}^{(TS)}$, where to a given loop $\mathcal{L}_a$ are attached one or more rings made with convolution chains of bonds $(-\beta\mathcal{V})$ (see Fig. 6). The summation over all n intermediary loops inside a given convolution ring with n varying from 1 to ∞ provides $I_R(\mathcal{L}_a) = \beta(\mathcal{V}(\mathcal{L}_a,\mathcal{L}_a) - e_{\alpha_a}^2\phi(\mathcal{L}_a,\mathcal{L}_a))$. The sum over all numbers of ring attached to $\mathcal{L}_a$ transform its fugacity $z(\mathcal{L}_a)$ into the renormalized one $z_R(\mathcal{L}_a) = z(\mathcal{L}_a)\exp(I_R(\mathcal{L}_a)/2)$.

• Convergent prototype graphs

The topological structure of genuine Mayer graphs $\mathcal{G}$ is conserved through all the previous chain resummations and reexponentiations. Accordingly, we obtain a new set of prototype graphs $\mathcal{P}$, with

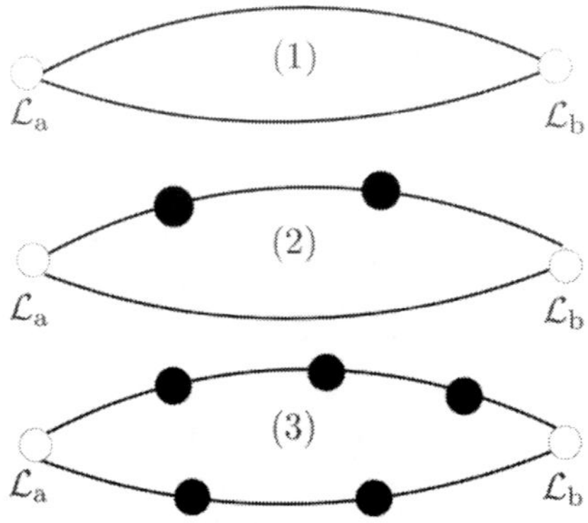

Figure 5: Three examples of double convolution chains connecting two loops $\mathcal{L}_a$ and $\mathcal{L}_b$: (1) chains reduce to the genuine direct bond $[\beta\mathcal{V}(\mathcal{L}_a, \mathcal{L}_b)]^2$ (2) one chain reduces to the direct bond $-\beta\mathcal{V}(\mathcal{L}_a, \mathcal{L}_b)$, while the other one involves 2 intermediary loops (3) both chains involve respectively 3 and 2 intermediary loops. The notations are the same as in Fig. 4.

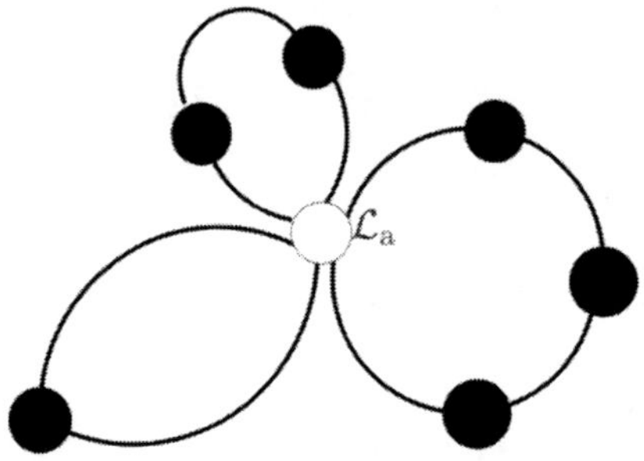

Figure 6: An example of three ring chains attached to a loop $\mathcal{L}_a$, involving respectively 1,2 and 3 intermediary loops. The notations are the same as in Fig. 4.

a topological structure identical to that of the genuine Mayer graphs. The bonds are now Mayer bonds associated with the effective potential ϕ, while the statistical weights are the renormalized fugacities $z_{\rm R}$, with specific restrictions to avoid double counting generated by the various chain and ring resummations [27].

Similarly to what happens in the classical case, collective effects embedded in the effective potential are expected to screen the bare loop-loop interaction. In fact, the static part $q_{\rm a}e_{\rm a}q_{\rm b}e_{\rm b}/|\mathbf{X}_{\rm a}-\mathbf{X}_{\rm b}|$ of $\mathcal{V}(\mathcal{L}_{\rm a},\mathcal{L}_{\rm b})$, which does not depend on loop shapes, is exponentially screened. On the contrary, the multipolar parts of $\mathcal{V}(\mathcal{L}_{\rm a},\mathcal{L}_{\rm b})$ which depend on loop shapes, are not exponentially screened. The resulting effective potential decays only as a dipolar interaction at large distances, namely [26]

$$\phi(\mathcal{L}_{\rm a},\mathcal{L}_{\rm b}) \sim \frac{\mathcal{A}_{\alpha_{\rm a}\alpha_{\rm b}}(\boldsymbol{\eta}_{\rm a},\boldsymbol{\eta}_{\rm b})}{|\mathbf{X}_{\rm a}-\mathbf{X}_{\rm b}|^3} \quad \text{when} \quad |\mathbf{X}_{\rm a}-\mathbf{X}_{\rm b}| \to \infty\,, \tag{26}$$

where $\mathcal{A}_{\alpha_{\rm a}\alpha_{\rm b}}(\boldsymbol{\eta}_{\rm a},\boldsymbol{\eta}_{\rm b})$ is an amplitude which depends only on loop species and shapes. The decay of $\phi(\mathcal{L}_{\rm a},\mathcal{L}_{\rm b})$ is at the border line for integrability at large distances. However, if loop shapes are first integrated over, all dangerous $1/R^3$-terms vanish. Thus, within that prescription for the order of integrations involved in measures $\mathrm{d}\mathcal{L}_i$, all long range Coulomb divergences are removed in the prototype graphs which provide convergent contributions.

- **Introduction of particle clusters**

The final step of the transformations consists in rewriting the protototype graphs $\mathcal{P}$ built with loops, as graphs G buit with particle clusters. For that purpose, we first identify sets of loops $\mathcal{E}_{\mathcal{L}} = \{\mathcal{L}_i\}$ such that all the effective interactions between loops inside $\mathcal{E}_{\mathcal{L}}$ are present in the product of Boltzmann factors $\exp(-\beta e_{\alpha_i}e_{\alpha_j}\phi(\mathcal{L}_i,\mathcal{L}_j))$. The explicitation of phase-space measures $\mathrm{d}\mathcal{L}_i$ as sums over species index and particle numbers allow us to identify particle clusters $C(N_{\rm p},N_{\rm e})$ in a given set $\mathcal{E}_{\mathcal{L}}$. The corresponding statistical weights $Z_\phi(C)$ account non-perturbatively for interactions between the particules through the presence of Boltzmann factors $\exp(-\beta e_{\alpha_i}e_{\alpha_j}\phi(\mathcal{L}_i,\mathcal{L}_j))$, so all recombination processes are exactly taken into account.

Once particle clusters have been identified, we proceed to substracting and adding counter terms to their statististical weights $Z_\phi(C)$. This defines truncated statistical weights $Z_\phi^T(C)$, which are well suited for studying the zero-density limit because they still provide finite contributions when ϕ is replaced by the bare Coulomb-like loop potential $\mathcal{V}$. The added counter terms define bonds F_ϕ built with powers of ϕ. They account for the resummation of *a priori* divergent long-ranged Coulomb contributions, which are ultimately finite thanks to the screening effects embedded in ϕ.

4.4 The screened cluster representation

The resummation machinery described above amounts to perform an exact transformation of the whole series (15), which can be rewritten as the Screened Cluster Representation (SCR),

$$\rho_\alpha = \sum_G \frac{1}{S(G)} \int \mathrm{d}C_0\, q_0\, Z_\phi^T(C_0) \int \prod_{i=1}^{n} \mathrm{d}C_i\, Z_\phi^T(C_i) \left[\prod F_\phi\right]_G\,, \tag{27}$$

where graphs G have the same topological structure as usual Mayer graphs. Now, ordinary points are replaced by particle clusters C_i, weighted by factors $Z_\phi^T(C_i)$. Two clusters are connected by

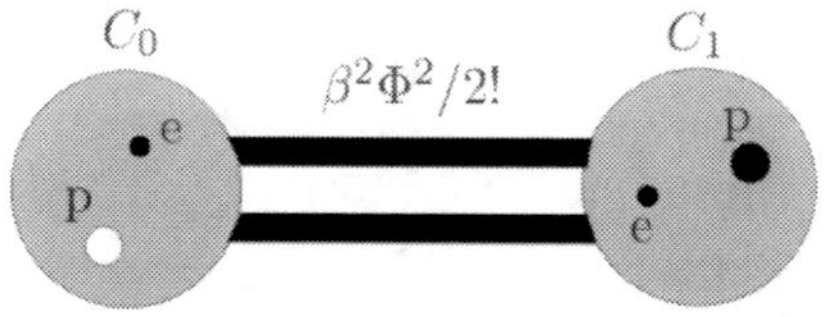

Figure 7: An example of graph G which contributes to the proton density ρ_p. The two clusters are represented by large grey disks. The bond $\beta^2\Phi^2/2!$ connecting the two clusters is represented by a double thick straight line. Inside the root cluster $C_0(1,1)$, the position of the proton, represented by a white disk, is fixed. The positions of the other proton inside $C_1(1,1)$ and of the electrons, represented by black disks, are integrated over.

at most one bond F_ϕ which can be either $-\beta\Phi$, $\beta^2\Phi^2/2!$ or $-\beta^3\Phi^3/3!$ with Φ the total effective potential between both clusters. Graphs G are simply connected, and symmetry factor $S(G)$ is computed as usual. Phase-space measure $\mathrm{d}C_i$ reduces to integrations over positions and shapes of loops belonging to set $\mathcal{E}_\mathcal{L}$ associated with cluster C_i, whereas in $\mathrm{d}C_0$ position of root particle α is not integrated over. Eventually, $\sum_G$ is performed over all topologically different graphs with some simple restrictions, in part avoiding double-counting of genuine particle interactions [27]. For instance, the contribution of graph G made with two clusters $C_0 = C(1,1)$ and $C_1 = C_1(1,1)$ connected by a bond $\beta^2\Phi^2/2!$ (see Fig. 7), reads

$$\int \mathrm{d}C_0\, Z_\phi^T(C_0) \int \mathrm{d}C_1\, Z_\phi^T(C_1)\, \frac{\beta^2\Phi^2(C,C_1)}{2}\,, \tag{28}$$

since $S(G) = 1$ and $q_0 = 1$. That graph arises from the introduction of the truncated weight $Z_\phi^T(C(2,2))$ for a cluster made with 2 protons and 2 electrons, according to the substraction and addition of counter terms to the full statistical weight $Z_\phi(C(2,2))$ as described above.

Remarkably, not only the genuine structure of Mayer graphs is conserved by the resummation machinery, but both statistical weights $Z_\phi^T(C)$ and bonds F_ϕ depend on the sole effective potential ϕ. Thanks to the sufficient fast decay of ϕ, all large-distances divergences are removed in every graph G, which does provide a finite contribution. The presence of complete interactions inside a given cluster $C(N_p, N_e)$ ensures that chemical species made with N_p protons and N_e electrons do emerge at low densities. Both features are detailed below. Notice that the physical ideas underlying the SCR, are close to those involved in the construction of the so-called ACTEX approach by Rogers [28].

Contrarily to Feynman graphs involved in standard many-body perturbation theory [1], graphs G in SCR account for non-perturbative effects in the Coulomb potential which are essential for a proper description of recombination. In fact, that representation provides the contributions, at finite temperature and finite density, of all chemical species and of their mutual interactions, without introducing arbitrary regularizations or adjustable parameters. Hence the SCR has been used for different purposes like

- The derivation of exact asymptotic expansions beyond Saha theory in the SLT regime [29]
- The analysis of the cross-over regime between molecular and atomic hydrogen [30]

In the next paragraph, we describe the structure of the contributions of the simplest chemical species like atoms H, molecules H_2, ions H^- and H_2^+, which are key ingredients in the previous works.

4.5 Contributions of chemical species at finite temperature

The ideal-like contribution of a given chemical species made with $N_{\rm p}$ protons and $N_{\rm e}$ electrons, emerges in the simple SCR diagram made with the single corresponding particle cluster. Indeed, at low densities, $\phi \to \mathcal{V}$, and $Z_\phi^T(C_0)$ gives raise to the cluster partition function

$$Z(N_{\rm p}, N_{\rm e}) = (2\pi\lambda^2_{N_{\rm p},N_{\rm e}})^{3/2} \lim_{\Lambda\to\infty} \frac{1}{\Lambda} {\rm Tr}[\exp(-\beta H_{N_{\rm p},N_{\rm e}})]^T_{\rm Mayer} \tag{29}$$

where $\lambda_{N_{\rm p},N_{\rm e}} = (\beta\hbar^2/(N_{\rm p}m_p + N_{\rm e}m_e))^{1/2}$ is the thermal de Broglie wavelength of the cluster. That partition function is defined in terms of bare Coulomb Hamiltonians, so it depends only on temperature and fundamental constants. It is finite thanks to the truncation procedure defining $[\exp(-\beta H_{N_{\rm p},N_{\rm e}})]^T_{\rm Mayer}$ and inherited from the procedure of subtracting and adding counter-terms involved in the resummation machinery. The own contributions of substracted counter terms arise in other SCR diagrams built with several particle clusters connected by bonds F_ϕ. Such contributions are indeed finite thanks to the screened nature of ϕ. We stress that, as far as thermodynamical properties are concerned, only the full contribution of $Z(N_{\rm p}, N_{\rm e})$ and of its related screened counter terms makes an unambiguous sense. This illustrates the interplay between recombination and screening deviced above on heuristic grounds.

Remarkably, the familiar chemical species naturally emerge from cluster partition function $Z(N_{\rm p}, N_{\rm e})$, which is intrinsic to the considered cluster in the vacuum. A given chemical species made with $N_{\rm p}$ protons and $N_{\rm e}$ electrons, is associated with bound states of bare Hamiltonian $H_{N_{\rm p},N_{\rm e}}$. In the zero-temperature limit, it provides the leading contribution to $Z(N_{\rm p}, N_{\rm e})$ which behaves as

$$\exp(-\beta E^{(0)}_{N_{\rm p},N_{\rm e}}) \tag{30}$$

apart from possible integer degeneracy factors. At finite temperatures, $Z(N_{\rm p}, N_{\rm e})$ involves not only contributions from thermally excited bound states, but also contributions from diffusive states describing the dissociation into various chemical species which cannot be disentangled. For instance $Z(2, 2)$ involves not only a molecular part, but it also includes the contributions of atom-atom interactions at short distances. Notice that the long range part of such atom-atom interactions, *i.e.* the familiar van der Waals interactions, are embedded in the contributions of graph G shown in Fig. 7.

Cluster partition functions $Z(N_{\rm p}, N_{\rm e})$ can be viewed as generalizations of Ebeling virial coefficients [31] introduced for dealing with contributions from two-particle clusters. Notice that $Z(1, 1)$ does not reduce exactly to the heuristic BPL partition function $Z_{\rm BPL}$, although both quantities are close at low temperatures [32]. Eventually, notice that the contributions of interactions between chemical species can be expressed also in terms of cluster functions similar to $Z(N_p, N_e)$, like that of graph G shown in Fig. 7. Thus, all contributions related to the formation of complex entities are properly taken into account.

5 Condensation in an interacting Bose gas

In this Section, we consider an interacting Bose gas made of a single species with zero spin, as described in Section 2. First, we recall the familiar criterion introduced for the ideal gas, namely the macroscopic condensation of particles in the same quantum state with zero momentum. We also present an equivalent criterion which is more suited for dealing with interactions, and provides the order parameter of the BE transition. Within the path integral representation in terms of the loop gas, that criterion is shown to be related to the behaviour of the density of open loops with large arbitrary sizes. Then, we turn to the Kac prescription, which amounts to consider the limit of an infinitely weak interaction potential with an infinite range. At leading order, such prescription provides a mean-field theory which predicts that the BE persists. Eventually, we show that Mayer-like series provide interesting preliminary results on the effects of fluctuations beyond that mean-field limit.

5.1 Order parameter of the Bose-Einstein condensation

The appearance of Bose condensate has been first derived for a perfect gas (for a thorough review see [33] and references given therein) showing that this phenomenon does not require any interaction. In the framework of the grand-canonical ensemble, where the given thermodynamical parameters are temperature T and chemical potential μ, the corresponding particle density ρ reads

$$\rho = \rho_0(\beta, \mu) = \frac{1}{(2\pi)^3} \int d\mathbf{k} \frac{1}{\exp(\beta(\epsilon(\mathbf{k}) - \mu)) - 1} . \tag{31}$$

Formula (31) describes the perfect gas thermodynamic function $\rho_0(\beta, \mu)$ for any $\rho < \rho_{0,c}(\beta)$ where

$$\rho_{0,c}(\beta) = \frac{1}{(2\pi)^3} \int d\mathbf{k} \frac{1}{\exp(\beta\epsilon(\mathbf{k})) - 1} \tag{32}$$

is the critical density. For $\rho > \rho_{0,c}(\beta)$, a macroscopic condensation of particles in state $\mathbf{k} = \mathbf{0}$ takes place.

It is well-known that the Bose-Einstein condensation is directly related to the appearance of off-diagonal long-range order (ODLRO) in the one-body density matrix, which can be viewed as the relevant order-parameter for that transition in the interacting system (see e.g. [33]). In the ideal case, it has been shown that

$$\lim_{|\mathbf{r}-\mathbf{r}'| \to \infty} \langle \mathbf{r} | \mathcal{D}_0^{(1)} | \mathbf{r}' \rangle = 0 \quad \text{for} \quad \rho < \rho_{0,c}(\beta) \tag{33}$$

and

$$\lim_{|\mathbf{r}-\mathbf{r}'| \to \infty} \langle \mathbf{r} | \mathcal{D}_0^{(1)} | \mathbf{r}' \rangle = \rho - \rho_{0,c}(\beta) \quad \text{for} \quad \rho > \rho_{0,c}(\beta) , \tag{34}$$

where $\mathcal{D}_0^{(1)}$ is the ideal one-body density matrix. For the interacting system, the eigenstates of the Hamiltonian no longer reduce to tensorial products of one-body states, so the possible persistence of the BE condensation has to be related to the existence of ODLRO for the one-body density matrix $\langle \mathbf{r}|\mathcal{D}_0^{(1)}|\mathbf{r}'\rangle$ above some critical density.

Within the equivalent gas of loops, the off-diagonal matrix element $\langle \mathbf{r}|\mathcal{D}_0^{(1)}|\mathbf{r}'\rangle$ reads [34]

$$\langle \mathbf{r}|\mathcal{D}_0^{(1)}|\mathbf{r}'\rangle = \sum_{q=1}^{\infty} \int \mathcal{D}(\boldsymbol{\eta})\; q\; \exp[-(\mathbf{r}-\mathbf{r}')^2/(2q\lambda^2)]\rho(\mathcal{L}_{\mathbf{r},\mathbf{r}'}) \; . \tag{35}$$

Here $\rho(\mathcal{L}_{\mathbf{r},\mathbf{r}'})$ is the density of open loops $\mathcal{L}_{\mathbf{r},\mathbf{r}'}$ starting from $\mathbf{r}$ and going to $\mathbf{r}'$ along a Brownian path

$$(1-(s/q))\,\mathbf{r} + (s/q)\,\mathbf{r}' + \lambda\,\boldsymbol{\eta}(s) \; , \tag{36}$$

with s varying from 0 to q. Accordingly, $\langle \mathbf{r}|\mathcal{D}_0^{(1)}|\mathbf{r}'\rangle$ can be represented by Mayer-like graphs analogous to those appearing in series (15) for the particle density where now the root loop $\mathcal{L}_0$ is replaced by the open loop $\mathcal{L}_{\mathbf{r},\mathbf{r}'}$. The emergence of ODLRO is related to the behaviour of $\rho(\mathcal{L}_{\mathbf{r},\mathbf{r}'})$ for large open loops with large values of q.

5.2 The mean-field Kac limit

Let us introduce a family of two-body potentials $u_\gamma(r)$ defined by the following parametrization

$$u_\gamma(r) = \gamma^3 v(\gamma r) \tag{37}$$

where γ is a real positive parameter and $v(r)$ is a given positive, integrable, and spherically symmetric pair potential. When dimensionless parameter γ approaches zero, the potential v being kept fixed, the potential $u_\gamma(r)$ becomes weak and long range, whereas the value of the integral

$$\int \mathrm{d}\mathbf{r}\, u_\gamma(r) = 4\pi \int_0^\infty \mathrm{d}r\, r^2 v(r) = a > 0 \tag{38}$$

remains constant.

In the strict limit $\gamma = 0$, one can expects that particles feel a mean-field constant potential $a\rho$. In other words, the system should then be driven by the so-called mean-field Hamiltonian [35]. That heuristic prediction has been rigorously proved by Pulé and Zagrebnov [36]. More precisely, ODLRO still persists, while the ideal critical density remains unchanged. Furthermore, below $\rho_{0,\mathrm{c}}(\beta)$, the density is the unique solution of the self-consistent equation

$$\rho = \rho_0(\beta, \mu - a\rho) = \frac{1}{(2\pi)^3} \int \mathrm{d}\mathbf{k} \frac{1}{\exp(\beta(\epsilon(\mathbf{k}) + a\rho - \mu)) - 1} \; . \tag{39}$$

which defines the mean-field equation of state $\rho = \rho_{\mathrm{mf}}(\beta, \mu)$. For $\rho > \rho_{0,\mathrm{c}}(\beta)$, that equation reduces to the simple linear law $\rho = \mu/a$. Notice the simple shift $a\rho$ in the one-body energy spectrum involved in the right hand side of formula (39), in agreement with the mean-field Hamiltonian picture.

Interestingly, the Mayer-like series (15) for ρ provide a self-consistent equation which takes the general form ([37])

$$\rho = F_\gamma(\beta, \mu - a\rho) \tag{40}$$

where function $F_\gamma(\beta,\nu)$ is defined by a series of multiconnected Mayer graphs, which do converge for ν sufficiently negative, and uniformly with respect to γ. For $\gamma = 0$, F_0 reduces to the sum of tree graphs, which is nothing but the ideal density function $\rho_0(\beta,\nu)$, so relation (40) does coincide with equation (39) as it should.

5.3 Contributions of fluctuations beyond the mean-field limit

Now the central question concerns the effects of fluctuations beyond the mean-field limit, and in particular their influence on the possible emergence of ODLRO. The answer requires to study the case of γ small but finite. Such a program has been partially accomplished by using the Mayer-like series for the equivalent gas of loops. For finite negative values of ν, $F_\gamma(\beta,\nu)$ can be expanded in powers of γ^3. The first correction to $F_0(\beta,\nu)$ of order γ^3, is non-trivial, because various resummations of infinite sets of graphs have to be performed in order to properly take into account collective effects arising from the long range nature of the two-body potential $u_\gamma(r)$. That correction reduces to the sum of contributions from tree and ring graphs [34],

$$\gamma^3\,[F^{(1)}_{\rm tree} + F^{(1)}_{\rm ring}] \tag{41}$$

with

$$F^{(1)}_{\rm tree} = \frac{\beta v(0)}{2}\,[f^{(1)}(\nu) - f^{(2)}(\nu)]\,, \tag{42}$$

$$F^{(1)}_{\rm ring} = \frac{\beta^2}{2(2\pi)^3}\int d\mathbf{q}\,\frac{[\hat{v}(|\mathbf{q}|)]^2}{1+\beta\,f^{(1)}(\nu)\,\hat{v}(|\mathbf{q}|)}\,f^{(1)}(\nu)\,f^{(2)}(\nu) \tag{43}$$

and

$$f^{(1)}(\nu) = \beta^{-1}\,\frac{\partial\rho_0}{\partial\nu}(\beta,\nu)\,,\; f^{(2)}(\nu) = \beta^{-2}\,\frac{\partial^2\rho_0}{\partial\nu^2}(\beta,\nu)\,. \tag{44}$$

Inserting the correction (41) into the self-consistent equation (40), one can compute the γ^3-corrections to the mean-field equation of state for fixed strictly negative values of $\nu = \mu - a\rho$. Unfortunately, the coefficients of the γ^3-terms diverges when $\nu \to 0$, *i.e.* when one approaches the critical region where a condensate might emerge. Thus, the above perturbative expansion in powers of γ^3 does not allow one to answer the question of existence of condensation for an arbitrarily small but fixed value of γ [38].

In order to analyze the behaviours of interest in the critical region for γ small but finite, various strategies using Mayer-like series for the equivalent loop gas have been proposed. A first insight should be gained by investigating an approximate version of the self-consistent equation (40), where $F_\gamma(\beta,\nu)$ is replaced by the contributions of tree graphs only,

$$F^{(\rm tree)}_\gamma(\beta,\nu) = \frac{1}{(2\pi\lambda^2)^{3/2}}\sum_{q=1}^{\infty}\frac{\exp(\beta q\nu)}{q}\int\mathcal{D}(\boldsymbol{\eta})\,\exp(-\beta\,\mathcal{U}(\mathcal{L}))\,. \tag{45}$$

The central problem would then be to determine the behaviour, for simultaneously large values of q and small values of γ, of the average over all loop shapes $\boldsymbol{\eta}$ of the Bolzmann factor associated with the self-energy $\mathcal{U}(\mathcal{L})$ of loop $\mathcal{L}$ made with q exchanged particles. Another strategy, more ambitious, would be to analyze exactly the possible emergence of ODLRO within formula (35). In that approach, one would have to determine the behaviour of the density of open loops $\rho(\mathcal{L}_{\mathbf{r},\mathbf{r}'})$, for both large values of $|\mathbf{r}-\mathbf{r}'|$ and q, and small values of γ.

6 Concluding remarks and perspectives

The classical nature of the loop gas allows one to perform exact and useful reorganizations of diagrammatical series with :

- Non-perturbative treatment of interactions
- Incorporation of collective effects
- Systematic account of fluctuations beyond mean-field approaches

Various applications are actually in progress. First, the ionic and molecular internal partition functions for hydrogen are determined within Quantum Monte Carlo numerical computations of their path integral reprsentations. Furthermore, the Screened Cluster Representation will be applied to the hydrogen-helium mixture in order to study the corresponding equation of state at low temperatures and low densities. For the interacting Bose gas, the preliminary results described in Section 5 are promising, and there is some reasonable hope to estimate contributions of large loops, by analogy with the exact results derived for self-repelling closed polymers. However, we notice that there exist alternative methods, like that based on the analysis of the hierarchy equations for the imaginary-time Green functions [39] which might be more efficient.

References

[1] Fetter A L and Walecka J D 1971 *Quantum theory of many particle systems* (McGraw-Hill, New York)

[2] Feynman R P and Hibbs A R 1965 *Quantum mechanics and path integrals* (McGraw-Hill, New York)

[3] Ebeling W, Kraeft D and Roepke G 2012 *Ann. Phys.* **524** 311

[4] Seiringer R 2010 *Hot topics in cold gases* XVIth International Congress on Mathematical Physics (Ed. Pavel Exner, World Scientific, Prague) pp 231-245.

[5] Simon B 1979 *Functional Integration and Quantum Physics* (Academic, New York)

[6] Schulman L S 1981 *Techniques and Applications of Path Integrals* (Wiley, New York)

[7] Roepstorff G 1994 *Path Integral Approach to Quantum Physics* (Springer, Berlin)

[8] Kleinert H 2004 *Path Integrals in Quantum Mechanics, Statistics, Polymer Physics and Financial Markets* (World Scientific)

[9] Alastuey A, Magro M and Pujol P 2008 *Physique et outils mathématiques : méthodes et exemples* (EDP Sciences, CNRS Editions)

[10] Wigner E P 1932 *Phys. Rev.* **40** 749

[11] Kirkwood J G 1934 *Phys. Rev.* **45** 116

[12] Landau L D and Lifchitz E M 1958 *Quantum mechanics* vol.5 of the course of Theoretical Physics (Pergamon Press, Oxford)

[13] Paulin S, Alastuey A and Dauxois T 2007 *J. Stat. Phys.* **128** 1391

[14] Ginibre J 1971 *Statistical Mechanics and Quantum Field Theory* edited by C. DeWitt and R. Stora (les Houches, Gordon and Breach)

[15] Cornu F 1996 *Phys. Rev. E* **53** 4562

[16] Martin P A 2003 *Acta Physica Polonica B* **34** 3629

[17] Mayer J E and Montroll E W 1941 *J. Chem. Phys.* **9** 2

[18] Hansen J P and Mc Donald I R 1986 *Theory of simple liquids* (Academic Press, London)

[19] Lieb E and Lebowitz J 1972 *Adv. Math.* **9** 316

[20] Saha M 1920 *Philos. Mag.* **40** 472

[21] Macris N and Martin P A 1990 *J. Stat. Phys.* **60** 619

[22] Fefferman C 1985 *Rev. Math. Iberoamericana* **1** 1

[23] Brillouin L 1930 *Les Statistiques Quantiques et leurs Applications* (Presse Univ. de France, Paris)

[24] Larkin A I 1960 *Zh. exp. teor. Fiz.* **38** 1896

[25] Balleneger V. 2012 *Ann. Phys.* **524** 103

[26] Ballenegger V, Martin P A and Alastuey A 2002 *J. Stat. Phys.* **108** 169

[27] Alastuey A, Ballenegger V, Cornu F and Martin P A 2003 *J. Stat. Phys.* **113** 455

[28] Rogers F J 1974 *Phys. Rev. A* **10** 2441

[29] Alastuey A, Ballenegger V, Cornu F and Martin P A 2008 *J. Stat. Phys.* **130** 1119

[30] Alastuey A and Ballenegger V 2012 *Cont. Plasma. Phys.* **52** 95

[31] Ebeling W 1967 *Ann. Phys. Leipzig* **19** 104

[32] Alastuey A and Ballenegger V 2012 *Phys. Rev. E* **86** 066402

[33] Ziff R M, Uhlenbeck G E and Kac M 1977 *Physics Reports* **32C** 169

[34] Martin P A and Piasecki J 2005 *Phys. Rev. E* **71** 016109

[35] Lewis J T 1986 *Lecture Notes on Physics* **257** pp 234 (Springer, New York)

[36] Pulé J V and Zagrebnov V A 2004 *J. Phys. A: Math. Gen.* **37** 8929

[37] Martin P A and Piasecki J 2003 *Phys. Rev. E* **68** 016113

[38] Martin P A 2005 *International Journal of Theoretical Physics* **44** 2021

[39] Alastuey A and Piasecki J 2011 *Phys. Rev. E* **84** 041122

Journal of Physics: Conference Series **529** (2014) 012004 doi:10.1088/1742-6596/529/1/012004

Autler-Townes Effects in Attosecond Circular Polarization Molecular Photoionization

Kai-Jun Yuan[1], Zhigang Sun[2,3], and André D Bandrauk[1]

[1] Laboratoire de Chimie Théorique, Faculté des Sciences, Université de Sherbrooke, Sherbrooke, Québec, Canada, J1K 2R1

[2] Center for Theoretical and Computational Chemistry and State Key Laboratory of Molecular Reaction Dynamics, Dalian Institute of Chemical Physics, Chinese Academy of Sciences, Dalian 116023, P.R. China

[3] Synergetic Innovation Center of Quantum Information and Quantum Physics, University of Science and Technology of China, Hefei, Anhui 230026, P.R. China

E-mail: kaijun.yuan@usherbrooke.ca; zsun@dicp.ac.cn; andre.bandrauk@usherbrooke.ca

Abstract. We present molecular photoionization simulations by intense ($I \sim 10^{16}$ W/cm^2) few cycle circularly polarized *attosecond* extreme ultraviolet laser pulses for aligned H_2^+ from numerical solutions of the corresponding time-dependent Schrodinger equation. With appropriate laser pulse parameters, circular attosecond coherent electron wave packets (CEWPs) are created in excited Rydberg states. Such CEWPs are spatially localized during ionization processes, thus resulting in sufficient population oscillations between the resonant excited Rydberg states and the initial ground state. Consequently Autler-Townes splitting in circular polarization energy spectra is predicted, which is shown to be critically sensitive to the pulse intensity, duration, and polarization. The resulting photoelectron angular distributions are rotated with respect to the molecular axis due to the nonspherical Coulomb potential of the molecule, resulting in different ionization rates at different laser polarization-molecular angles.

1. Introduction

Rapid developments in advanced laser pulse technology [1, 2] provide the tools necessary for investigating electron dynamics on the attosecond (1 asec=10^{-18} s) time scale and in the nonlinear nonperturbative regime. Such ultrashort laser pulses offer now the possibility of creating coherent electron wave packets (CEWPs) inside atoms and molecules on the attosecond time scale and sub-nanometer dimension [3-5]. Recently attosecond extreme ultraviolet (UV) light from high-order harmonics (HH) and free electron lasers (FEL), or combined with infrared (IR) laser pulses has been used to monitor electron dynamics in CEWPs by measuring photoelectron angular and energy distributions thus revealing effects of atomic and molecular orbital configurations [6], multi-pathway quantum interference [7-11] and electron correlations [13, 14]. UV-pump- UV-probe experiments have been adopted in molecular isomerization in acetylene cations [15]. It has also been shown that due to ac Stark effects in the continuum, dynamic interference of the photoelectrons emitted at different time can be observed with high frequency laser pulses [16]. These new experiments are done mainly with ultrashort intense *linearly* polarized pulses. Ultrashort circularly polarized laser pulses offer new avenues to investigate electron dynamics on attosecond time scales [17-21]. We have recently proposed

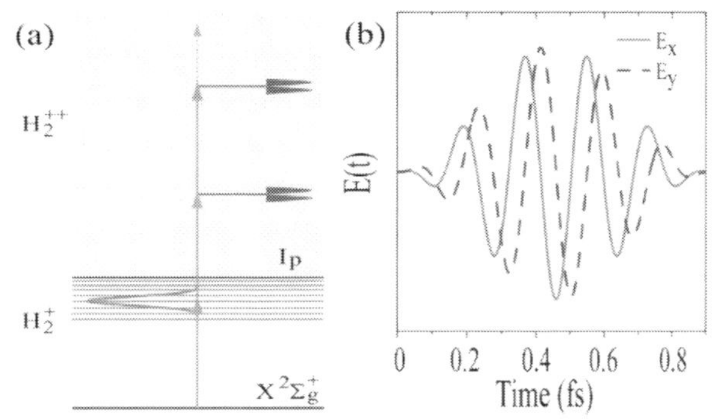

Figure 1. (Color online) (a) Schematic illustrations of multiphoton ionization pathways from the H_2^+ $^2\Sigma_g^+$ electron state at $R_e = 2$ a.u. in (b) intense five cycles circularly polarized FWHM=460 asec at $\lambda = 55$ nm ($\omega = 0.828$ a.u., $\tau = 184$ asec) UV laser fields $E(t)$.

methods of generating *circularly* polarized attosecond pulses from circularly polarized molecular HH [22, 23] using the nonspherical symmetry of molecules which allows for generating circularly polarized harmonics. Circularly polarized laser pulses have also been shown to be important to induce perturbatively circular electron ring currents in cyclic molecules, thus producing intense internal magnetic fields [24, 25] and to investigate recollision of electrons in double ionization [19, 20]. In [21], two color counter-propagating circularly polarized laser pulses are studied to study the scaling of the maximum HH spectrum and stabilization at high frequency in [26]. We have recently proposed that with intense few cycle circularly polarized UV pulses, strong attosecond magnetic field pulses with several tens of Teslas can be induced in molecules [27], thus suggesting a new experimental tool for nonlinear magneto-optics.

In this work we study photoionization of the single electron aligned H_2^+ with these new intense circularly polarized attosecond UV laser pulses, as schematically illustrated in figure 1. It is found that with proper conditions of the driving laser pulse, Autler-Townes (AT) splitting can be observed in molecular photoelectron energy distributions from localized circular CEWPs. AT splitting was first observed in radio-frequency transitions in atoms [28], which resulted from Rabi oscillations of the population between two specified atomic states. Molecular AT splitting resulting from vibrational coherent states with ultrashort femtosecond laser pulses has also been predicted in molecular ionization [29, 30]. Effects of molecular orientation and rotation on such splitting were also investigated [31]. To the best of our knowledge such observations of AT doublets resulting from coherent localized electronic wave packets have not been reported in photoionization due to the rapid electron motion on its natural time scale, the attosecond. A spatially localized electronic wave packet is difficult to confine and measure with linearly polarized laser pulses of moderate intensity and femtosecond (1fs=10^{-15} s) duration. Circularly polarized laser pulses differ from linear polarization by having a constant rotating electric field. We show that circularly polarized attosecond UV laser pulses produce "spinning" electrons, and the excited circular CEWPs can be localized on the sub-nanometer molecular dimension scale. Consequently, AT doublets are shown to occur in molecular ionization energy-resolved photoelectron angular distributions (MPADs) with pulse intensities near 10^{16} W/cm^2. Circularly polarized attosecond laser pulses are shown to be essential in observing new coherence effects in circular CEWPs, thus offering the possibility to observe AT splittings at high intensities due to circular polarization Rabi oscillations. The proposal of generating circularly polarized attosecond pulses [22, 23] make it thus possible to create spinning and localized circular electron wave packets in matter on attosecond time scale.

The paper is arranged as follows In section 2, we briefly describe the computational methods for time-dependent quantum electron wave packets obtained from the corresponding time-dependent Schrödinger equation (TDSE). The numerical results of photoelectron energy spectra by intense ultrashort few cycle circularly polarized UV pulses for H_2^+ are presented in section 3. Effects of the pulse intensity, duration, wavelength, and polarization on the AT splitting doublets are investigated. Finally, we summarize our findings in section 4. Throughout this paper, atomic units (a.u.) $e = \hbar = m_e = 1$ are used, for which the a.u. of intensity $I_0 = 3.51 \times 10^{16}$ W/cm^2 and corresponding electric field $E_0 = 5 \times 10^9$ V/cm.

2. Numerical methods

Simulations are performed on an aligned molecular ion H_2^+ by solving numerically the corresponding TDSEs [32, 33] within a static (Born-Oppenheimer approximation, BOA) nuclear frame,

$$i\frac{\partial}{\partial t}\psi(\mathbf{r},t) = [H_0 + V_L(\mathbf{r})]\psi(\mathbf{r},t), \tag{1}$$

where H_0 is the field free molecular Hamiltonian and $V_L(\mathbf{r})$ the interaction term. An intense few cycle circularly polarized attosecond UV laser pulses with wavelength $\lambda = 55$ nm (angular frequency $\omega = 2\pi c/\lambda = 0.828$ a.u.) is used to ionize the molecular ion H_2^+. We use a three dimensional (3D) model with static nuclei, enabling to go beyond perturbation theory and independent of gauge transformations [34, 35]. Such a fixed nuclei approach ignoring the vibrational and rotational degrees of freedom is appropriate due to the longer femtosecond/picosecond time scale of nuclear (vibrational/rotational) motions since the vibrational period of ground state H_2^+ is $\sim$ 15 fs. The 3D TDSE is solved using cylindrical coordinates (ρ, θ, z) with the molecular plane $x = \rho\cos\theta$ and $y = \rho\sin\theta$. The molecular Hamiltonian is expressed as

$$\begin{aligned} H_0(\rho,\theta,z) &= -\frac{1}{2\rho}\frac{\partial}{\partial\rho}\left(\rho\frac{\partial}{\partial\rho}\right) - \frac{1}{2\rho^2}\frac{\partial^2}{\partial\theta^2} \\ &\quad -\frac{1}{2}\frac{\partial^2}{\partial z^2} + V(\rho,\theta,z), \end{aligned} \tag{2}$$

where $V(\rho, \theta, z)$ is the two center electron-nuclear potential. The molecule ion is aligned along the x axis during ionization, which can be readily achieved with orientational laser technology [36]. The laser-electron radiative coupling is described by

$$V_L(\rho,\theta,t) = \mathbf{r}\cdot\mathbf{E}(t) = \rho\cos\theta E_x(t) + \rho\sin\theta E_y(t) \tag{3}$$

in the length gauge for circularly polarized laser pulses with forms,

$$\mathbf{E}(t) = Ef(t)[\mathbf{e}_x\cos(\omega t) + \mathbf{e}_y\sin(\omega t)], \tag{4}$$

where $e_{x/y}$ are the polarization directions. Such a field induces electric currents in the ring molecular (x, y) plane. For linear polarization, $E_y(t) = 0$ with the field parallel to the molecular x axis. A smooth $\sin^2(\pi t/n\tau)$ pulse envelope with maximum amplitude E for intensity $I = I_x = I_y = \frac{1}{2}c\varepsilon_0 E^2$ is adopted, where one optical cycle $\tau = 2\pi/\omega$. The laser electric field $\int E(t)dt = 0$ satisfies Maxwell's equation in the propagation region to mimic the electric field of the electromagnetic wave [1].

We numerically solve the 3D TDSE by a second order split operator method in the time step δt combined with a fifth order finite difference method and Fourier transform technique in the spatial steps $\delta\rho$, δz, and $\delta\theta$ [37, 38]. The time step is taken to be $\delta t = 0.01$ a.u.=0.24 asec. The spatial discretization is $\delta\rho = \delta z = 0.25$ a.u. for a radial grid range $0 \le \rho \le 128$ a.u. and $|z| \le 64$ a.u., and the angle grid size $\delta\theta = 0.025$ radian. The spatial grid step ensures the description of a maximum momentum $\Delta p_e = \pi/\Delta\rho \approx 13$ a.u. and energy $E_e = \Delta p_e^2/2{=}80$ a.u. for the photoelectron spectrum. To prevent unphysical effects due to the reflection of the wave packet from the boundary, we multiply $\psi(\rho,\theta,z,t)$ by a "mask function" or absorber in the radial coordinate with the form $\cos^{1/8}[\pi(\rho-\rho_a)/2\rho_{abs}]$. For all results reported here we set the absorber domain $\rho_a = \rho_{max} - \rho_{abs}{=}104$ a.u. with $\rho_{abs} = 24$ a.u., exceeding well the field induced electron oscillation $\alpha_d = E/\omega^2$ of the electron.

For an intense laser pulse, a quite large grid range must be used to obtain via Fourier transform the high energy of the ejected electron in the ionization spectra. In the present work we use an efficient method by calculating a radial flux (electron current density) $\mathcal{J}(t)$ to describe the ionization spectra, where the high kinetic energy of the free electron can be accurately calculated. At an asymptotic point $\rho_0 = 100$ a.u, the electron wave function $\psi(\mathbf{r},t)$ generates the flux $\mathcal{J}(t)$ in the molecular and laser polarization (x,y) plane before the grid absorption at boundary. For attosecond photoionization processes, the laser duration is very short. At such large asymptotic point ρ_0, the angular flux distributions can be ignored, i.e., $1/\rho_0\partial/\partial\theta|_{\rho_0}e_\theta \ll \partial/\partial\rho|_{\rho_0}e_\rho$. As a result we only need to consider the radial part of the electronic flux along the radial direction, $\partial/\partial\rho|_{\rho_f}e_\rho$. The time-independent energy-resolved angular differential yield (photoelectron spectra) is obtained by a Fourier transform

$$\begin{aligned}
\psi(\theta,E_e)|_{\rho_0} &= \int_{t_p}^{t_f}\psi(\theta,t)|_{\rho_0}e^{iE_et}dt,\\
\psi'(\theta,E_e)\,|_{\rho_0} &= \int_{t_p}^{t_f}\frac{\partial\psi(\theta,t)}{\partial\rho}|_{\rho_0}e^{iE_et}dt,\\
\mathcal{J}(\theta,E_e) &\sim \mathrm{Re}\left[\frac{1}{2i}\psi'^{*}(\theta,E_e)|_{\rho_0}\psi(\theta,E_e)|_{\rho_0}\right],
\end{aligned} \tag{5}$$

where t_p is time after pulses switch off and $t_f = 80\tau$ to insure that the photoelectrons pass through the asymptotic point ρ_0. $E_e = p_e^2/2$ (in a.u.) is the kinetic energy of a free electron with wave vector $k = p_e = 2\pi/\lambda_e$ (in a.u.), $p_e = (p_x^2+p_y^2)^{1/2}$ is the momentum of a photoelectron of wavelength λ_e. With the transformation $p_x = p_e\cos\theta$ and $p_y = p_e\sin\theta$, we then obtain the two dimensional momentum distributions of photoelectrons from equation (5).

3. Results and discussions

The excitation process and the laser pulse are illustrated in figure 1. The laser wavelength is chosen at $\lambda = 55$ nm (angular frequency $\omega = 0.828$ a.u., 1 optical cycle $\tau = 184$ asec), slightly below the equilibrium, $R_e = 2$ a.u., ionization potential $I_p = 1.1$ a.u., thus ensuring at least a two photon absorption to ionize H_2^+. Because the lifetimes of Rydberg states are significantly longer than the attosecond UV pulse duration, population then can be transferred from the ground $X^2\Sigma_g^+$ state to the Rydberg states after absorption of one photon, then returning to the ground state. Thus Rabi oscillations will occur between these states coupled strongly by a one photon transition, resulting in an AT splitting, figure 1, producing doublets in the MPAD. The full width at half maximum (FWHM) of the UV laser pulse is always fixed at 460 asec (5 cycles). Four different pulse intensities $I = 1\times10^{15}$ W/cm^2 ($E = 0.169$ a.u.), 5×10^{15} W/cm^2 (0.377 a.u.), 1×10^{16} W/cm^2 (0.534 a.u.), and 2×10^{16} W/cm^2 (0.755 a.u.) are used to ionize H_2^+, respectively. With such strong intensity pulses, the Keldysh parameter $\gamma = \sqrt{I_p/2U_p} \geq 2.3$ used to characterize strong field physics, where $U_p = I/4\omega$ is the ponderomotive energy, implies a multi-photon regime [1].

We show in figure 2 the angular distributions $\mathcal{J}(\theta,E_e)$ in the H_2^+ ground $X^2\Sigma_g^+$ state induced with five cycle circularly polarized attosecond UV laser pulses. Only the distributions with energies $E_e < 1.9$ a.u. are reported, i.e., two and three photon ionization photoelectron spectra are given and are termed molecular above threshold ionization (MATI) [11]. From figure 2 we

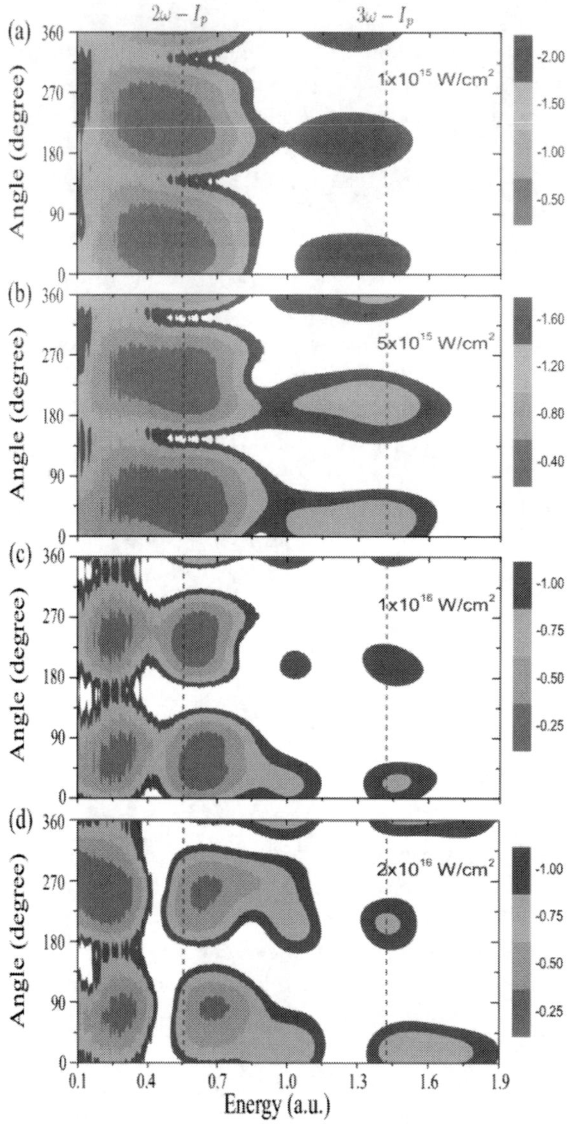

Figure 2. MPADs in H_2^+ with intense circularly polarized attosecond UV laser pulses at $\lambda = 55$ nm ($\omega = 0.828$ a.u.), FWHM=460 asec, and (a) $I = 1 \times 10^{15}$ W/cm^2 ($E = 0.169$ a.u.), (b) 5×10^{15} W/cm^2 (0.377 a.u.), (c) 1×10^{16} W/cm^2 (0.534 a.u.), (d) 2×10^{16} W/cm^2 (0.755 a.u.). Dotted lines present the classical kinetic energies $E_{en} = n\omega - I_p$ of photoelectrons.

see that the MPADs are critically sensitive to intensities of the attosecond UV laser pulses. In figure 2(a) at the lower laser intensity $I = 1 \times 10^{15}$ W/cm^2, MPADs with peaks at kinetic energies about $E_e = 0.5$ a.u. and 1.3 a.u. are observed, corresponding to direct two photon ($2\omega - I_p$) and three photon ($3\omega - I_p$) ionizations. Higher order MATI peaks can also be observed in UV laser fields and spectrum intensities decrease gradually as photoelectron kinetic energy increases [11, 12]. As the pulse intensity increases, the MPAD spectrum peaks become broad. At the pulse intensity $I = 1 \times 10^{16}$ W/cm^2 in figure 2(c), the two MATI peaks in figure 2(a) split into two separated sub-peaks with equal energy spacing $\Delta E_{AT} = 0.3$ a.u.. It is argued here that such doublets are in fact the AT splitting which comes from the Rabi oscillation between

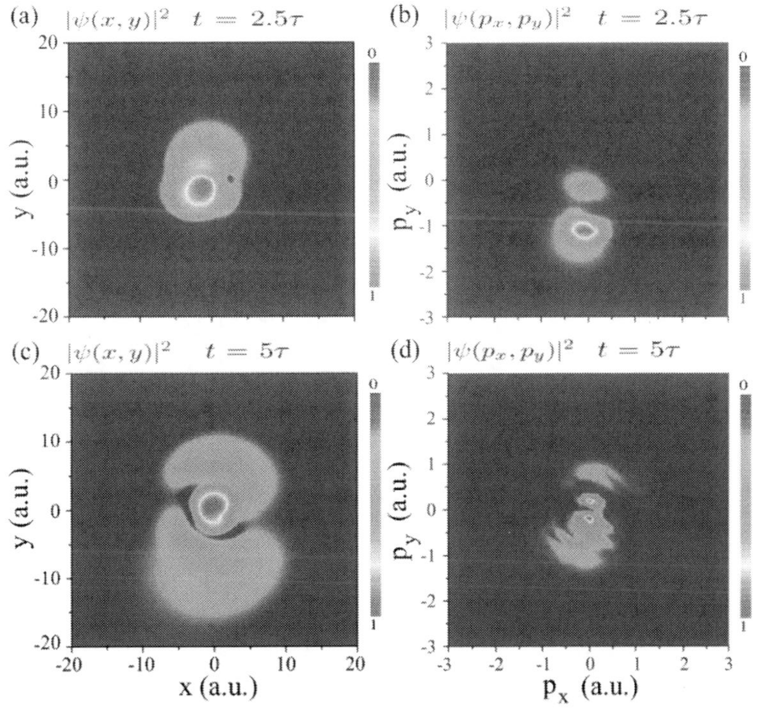

Figure 3. (Color online) Distributions of circular CEWPs, $|\psi(x,y)|^2$ and $|\psi(p_x,p_y)|^2$, in (a,c; left column) coordinate and (b,d; right column) momentum space at (a,b; upper row) $t = 2.5\tau = 460$ asec and (c,d; bottom row) $5\tau = 920$ asec with $I = 1 \times 10^{16}$ W/cm^2, $\lambda = 55$ nm, and FWHM=460 asec circularly polarized UV laser pulses, cf. figure 2(c).

the ground $X^2\Sigma_g^+$ and the coherent circular Rydberg states, i.e., the excited spatially localized CEWP. Increasing the pulse intensity further to $I = 2 \times 10^{16}$ W/cm^2 in figure 2(d) doublets with energy spacing 0.32 a.u. are also produced. Furthermore, we emphasize that MPADs occur nearly perpendicular to the molecular internuclear x axis, i.e., near 90° and 270° instead of 0° and 180°, a feature pointed out early by Corkum and coworkers for atoms [39].

The attosecond laser pulse has FWHM=460 asec, which has a spectral width at half maximum of about $\Delta\omega$ =0.3 a.u.. Assuming the excited states are dense enough with energy spacing much less than 0.3 a.u. [figure 2(c)], a well localized electronic wave packet can be expected to be produced from the excitation of multiple Rydberg states. Figure 3 shows distributions of the CEWPs, excited by the circularly polarized attosecond UV laser pulses at $I = 1 \times 10^{16}$ W/cm^2, corresponding to figure 2(c). At $t = 2.5\tau$=460 asec, the electric field has an extremum (figure 1(b)), i.e., in the x direction and is zero in the y direction. From the electromagnetic relation $E = -\partial A/\partial t$, then $A_x = 0$ whereas A_y is maximum. Furthermore since $p(t) = -\int E(t')dt' = A(t)$ and $A_y(t) < 0$ at $t = 2.5\tau$, one sees $p_y < 0$ and $p_x = 0$, from figure 1(b), thus confirming the dominant effect of the laser field. In figure 3(a,b) the 3D plot of the circularly polarized attosecond CEWP at time t=2.5τ is spatially well confined. The circular polarization of the laser pulse enables many more Rydberg states to be excited coherently as compared to linear polarization (see below). The spatial localization of the excited circular electronic wave packet enhances the Rabi oscillation between the ground $X^2\Sigma_g^+$ and the circular coherent Rydberg states, thus leading to AT splitting in the energy spectra. At the end of the pulse (t=5τ=920 asec), the circularly polarized attosecond CEWPs in the Rydberg states spread

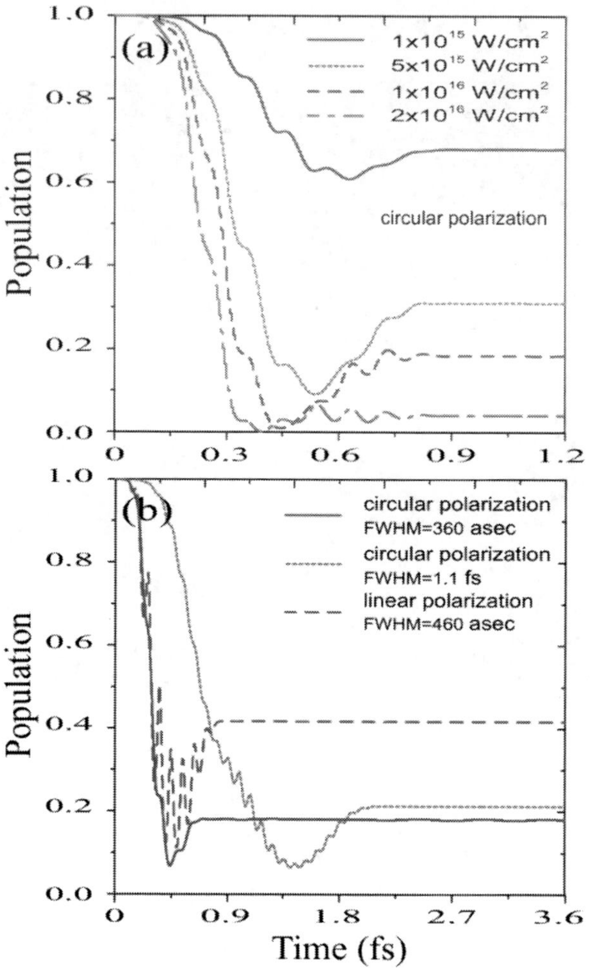

Figure 4. Population with time in the H_2^+ ground $X^2\Sigma_g^+$ state with $\lambda = 55$ nm attosecond UV laser pulses. (a) FWHM=460 asec and $I = 1 \times 10^{15}$ W/cm^2 (solid line), 5×10^{15} W/cm^2 (dotted line), 1×10^{16} W/cm^2 (dashed line), and 2×10^{16} W/cm^2 (dashed-dotted line) for circular polarization, cf. figure 2. (b) $I = 1 \times 10^{16}$ W/cm^2 and FWHM=360 asec (solid line) and 1.1 fs (dotted line) for circular polarization, and $I = 2 \times 10^{16}$ W/cm^2 and FWHM=460 asec (dashed line) for linear polarization, cf. figure 5.

and delocalize as illustrated in figure 3(c,d), as their radius increases with time [22].

In figure 4(a) we show evolutions of population in the ground $X^2\Sigma_g^+$ state at the four pulse intensities used in figure 2. Comparing to the weaker intensities $I = 1 \times 10^{15}$ W/cm^2 and 5×10^{15} W/cm^2 where there is weak population transfer out of the ground state, population first decreases completely to zero, and then increases to a fixed low value for the higher laser pulse intensity $I = 1 \times 10^{16}$ W/cm^2. At $I = 2 \times 10^{16}$ W/cm^2, complete ionization occurs. Thus partial Rabi oscillations occur for intensities $I \leq 1 \times 10^{16}$ W/cm^2 between the ground $X^2\Sigma_g^+$ and the coherent Rydberg states which results in AT splitting in MPADs, as shown in figure 2(c). For photoionization at low intensities, Rabi oscillations are too slow to induce the AT splitting in the photoelectron energy spectra, as shown in figures 2(a,b). After the laser pulse switch off at time t=920 asec, most of the excited electron wave packets emit, leading to a partial constant population in the ground state. As a result, a complete Rabi oscillation does not emerge, leading to a half oscillation only, as illustrated in figure 4(a). The energy spacings between the two sub-peaks displayed in figure 2(c) defines the Rabi frequency $\Omega = \mu E$, where μ is transition dipole [30]. In figure 2(c) at intensity $I = 1 \times 10^{16}$ W/cm^2, the energy spacing of the AT splitting doublets is 0.3 a.u.. Therefore the average transition dipole moment between the ground $X^2\Sigma_g^+$ state and the Rydberg states can be predicted as $\mu = \Omega/E_c = \Omega/\sqrt{2}E = 0.4$ a.u.. From figure 3 we see that that period of the Rabi oscillation is about $T_{rabi} = 600$ asec, slightly longer than the prediction from the Rabi frequency $2\pi/\Omega = 503$ asec. This difference mainly results from the increasing radius of the electron wave packet in the Rydberg states, as shown in figure 3. Moreover due to ionization only a partial Rabi oscillation is produced. At larger pulse intensities, ionization occurs at an early time, thus competing with Rabi oscillations. In figure 4(a) at $I = 2 \times 10^{16}$ W/cm^2 most of the electron is ionized before the pulse reaches its peak intensity. Rabi oscillations are partially suppressed and a larger energy distribution in the MPADs is obtained as illustrated in figure 2(d) corresponding to larger Rabi energy splitting.

We also consider the effect of duration of the attosecond UV laser pulses on the splitting in the energy spectra. Figure 5(a,b) shows MPADs with shorter and longer duration attosecond laser pulses at $I = 1 \times 10^{16}$ W/cm^2 and $\lambda = 55$ nm. The corresponding population evolutions in the H_2^+ ground $X^2\Sigma_g^+$ state are shown in figure 4(b). We see that in figure 5 the doublets in the MPAD peaks observed in figure 2(c) nearly vanish. For the shorter FWHM=360 asec photoionization in figure 5(a), the field-molecule interaction time is very short thus Rabi oscillations can not complete their cycle, as shown in figure 4(b) (black solid line). As a result, splitting in MPADs cannot be observed. Moreover, the shorter pulse spectral width is also very broad, resulting in broad photoelectron energy spectra and unresolved peaks. For longer FWHM=1.1 fs laser pulses, in figure 5(b) we see that the AT doublets are nearly absent as well. The MPAD peaks consist of two strongly asymmetric sub-peaks, one dominant and another weaker. Under such conditions, the coherent excited electron wave packets in the Rydberg states spread and delocalize during laser excitation processes, thus suppressing Rabi oscillations.

The laser polarization also plays an essential role in the CEWP generation and the AT splitting. Figure 5(c) shows the MPAD obtained with a parallel *linear* x polarization attosecond UV laser pulse at FWHM=460 asec, $I = 2 \times 10^{16}$ W/cm^2 ($E_L = E_c = \sqrt{2}E$), and $\lambda = 55$ nm. We see that strongly asymmetric split doublets are obtained in the energy spectra, similar to those with a circularly polarized laser pulse at FWHM=1.1 fs in figure 5(b). For linear polarization, the electron wave packets move away from the spatial resonance region quickly. A spatially localized attosecond electron wave packet cannot be efficiently produced, in contrast to the circular polarization results illustrated in figure 3(a,b). Thus Rabi oscillations are suppressed, as shown in figure 4(b) (dashed line). Therefore, MPADs exhibit strongly asymmetric splitting as shown in figure 5(c). Comparing to the results of circular polarization photoionization in figures 4(a) and 4(b), population of the ground electronic state in figure 4(b) (blue dashed line) exhibits a strong sub-oscillation from about 200 asec to 800 asec, corresponding to the driving pulse duration. This oscillation mainly results from the recollision of the free electron,following the classical three-step model [40]. Moreover one sees clearly that more angle nodes are observed in the two- and three-photon MATI peaks, which disappear with higher order ionization. This is mainly due to scattering effects of ejected electrons in photoionization processes [41].

We note furthermore that, in the MPADs with circularly polarized attosecond UV laser pulses in figures 2 and 5(a,b), the photoionization distributions are mainly localized along the y polarization ($\theta = 90°$ and $270°$) perpendicular to the molecular x-axis, as compared to MPADs parallel to the molecular x axis for linear polarization. This mainly results from the effects of the electronic charge clouds associated with each scattering center in the photoelectron ionization processes, leading to orientation-dependent ionization rates, which allows for tomographic imaging of molecular orbitals [34, 42]. Moreover, the MPAD is seen to be rotated with respect to

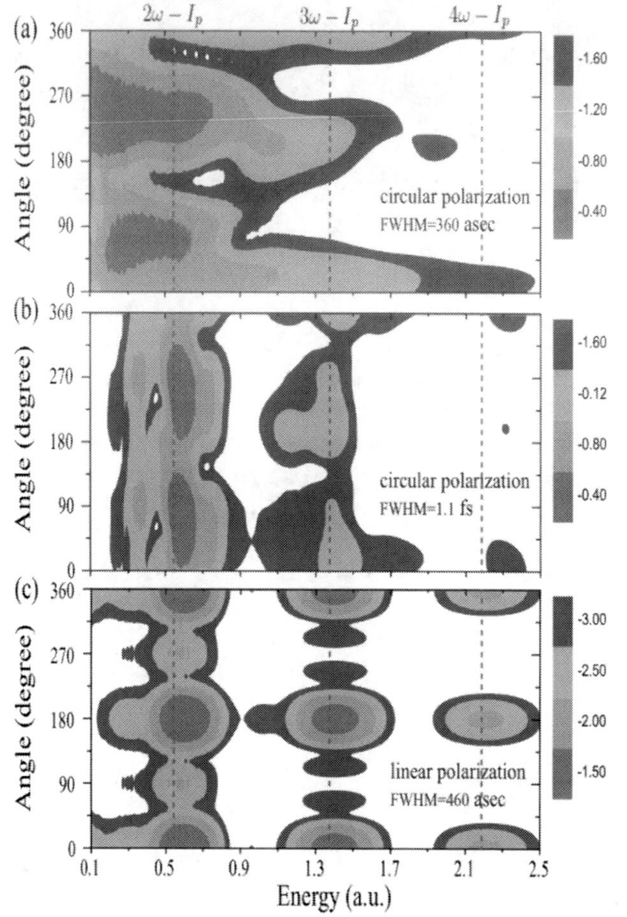

Figure 5. MPADs in H_2^+ with wavelength $\lambda = 55$ nm, (a,b) circularly and (c) linearly polarized attosecond UV laser pulses. The other laser parameters are (a) FWHM= 360 asec and intensity $I = 1 \times 10^{16}$ W/cm^2, (b) FWHM=1.1 fs and intensity $I = 1 \times 10^{16}$ W/cm^2, and (c) FWHM= 460 asec and intensity $I = 2 \times 10^{16}$ W/cm^2. Dotted lines present the classical kinetic energies $E_{en} = n\omega - I_p$ of photoelectrons.

the laser perpendicular y polarization axis due to the two center nonspherical Coulomb potential as well. Both orientation-dependent ionization rates and rotation angles of MPAD of aligned H_2^+ are shown to be sensitive to the photoelectron energies [32, 43, 44].

Of note is that in the photoelectron energy spectra illustrated in figures 2 and 5, MATI peaks exhibit a signature of energy shifts to low energy region relative to the classical kinetic energies $E_{en} = n\omega - I_p$ (dashed lines). This mainly results from the broad spectral width of the attosecond pulse since the photoelectron distributions are functions of the pulse spectral width and the Franck-Condon factor between the initial ground and the continuum electronic states. The energy shift has also been observed in the attosecond photoionization processes of H and He atoms with linearly polarized ultrashort few cycle UV laser pulses [45, 46].

4. Conclusions

In summary, AT doublets in circular polarization MPADs of H_2^+ have been reported from circular CEWPs. These are excited by a circularly polarized attosecond UV laser pulse at FWHM=460 asec and $\lambda = 55$ nm. An energy splitting of the AT-MPAD spectra is observed as a result of partial Rabi oscillations between spatially localized circular electron wave packets and the initial ground electronic state. Results from numerical solutions of appropriate TDSEs show that the circularly polarized attosecond CEWPs of Rydberg states created by a circularly polarized laser pulse with proper duration, intensity and wavelength is required to produce Rabi frequency and then to observe such splitting in energy spectra. The essence of the attosecond AT doublets is to induce spatially localized circular polarization CEWPs in the photoionization processes. Moreover, rotations of the angular distribution of the energy-resolved photoelectron with respect to the laser polarizations is observed as well, reflecting the influence of the nonspherical Coulomb potential in the molecule.

Acknowledgements
We thank RQCHP and Compute Canada for access to massively parallel computer clusters and NSERC for financial support in ultrafast science program. S is supported by the National Natural Science Foundation of China (rant No. 21222308, 2013CB922202, 21133006), the Chinese Academy of Sciences, and Ministry of Science and Technology of China.

References

[1] Krausz F and Ivanov M 2009 *Rev. Mod. Phys.* **81** 163
[2] Chang Z and Corkum P 2010 *J. Opt. Am. B* **27** B9
[3] Chelkowski S, Yudin G L and Bandrauk A D 2006 *J. Phys. B* **39** S409
[4] Niikura H, Villeneuve D M and Corkum P B 2005 *Phys. Rev. Lett.* **94** 083003
[5] Shao H C and Starace A F 2010 *Phys. Rev. Lett.* **105** 263201
[6] Kelkensberg F, Rouzée A, Siu W, Gademann G, Johnsson P, Lucchini M, Lucchese R R and Vrakking M J J 2011 *Phys. Rev. A* **84** 051404(R)
[7] Mauritsson J, Remetter T, Swoboda M, Klünder K, L'Huillier A, Schafer K J, Ghafur O, Kelkensberg F, Siu W, Johnsson P, Vrakking M J J, Znakovskaya I, Uphues T, Zherebtsov S, Kling M F, Lépine F, Benedetti E, Ferrari F, Sansone G and Nisoli M 2010 *Phys. Rev. Lett.* **105** 053001
[8] Ranitovic P, Tong X M, Hogle C W, Zhou X, Liu Y, Toshima N, Murnane M M and Kapteyn H C 2011 *Phys. Rev. Lett.* **106** 193008
[9] Choi N N, Jiang T F, Morishita T, Lee M H and Lin C D 2010 *Phys. Rev. A* **82** 013409
[10] Baggesen J C and Madsen L B 2011 *Phys. Rev. A.* **83** 021403(R)
[11] Yuan K J and Bandrauk A D 2012 *Phys. Rev. A.* **85** 013413
[12] Yuan K J, Lu H Z and Bandrauk A D 2013 *ChemPhysChem* **14** 1496
[13] Feist J, Nagele S, Pazourek R, Persson E, Schneider B I, Collins L A and Burgdörfer J 2009 *Phys. Rev. Lett.* **103** 063002
[14] Moore L R, Lysaght M A, Parker J S, van der Hart H W and Taylor K T 2011 *Phys. Rev. A* **84** 061404(R)
[15] Jiang Y H, Rudenko A, Herrwerth O, Foucar L, Kurka M, Kühnel K U, Lezius M, Kling M F, van Tilborg J, Belkacem A, Ueda K, Düsterer S, Treusch R, Schröter C D, Moshammer R and Ullrich J 2010 *Phys. Rev. Lett.* **105** 263002
[16] Demekhin P V and Cederbaum L S 2012 *Phys. Rev. Lett.* **108** 253001
[17] Wang X and Eberly J H 2009 *Phys. Rev. Lett.* **103** 103007
[18] Wang X and Eberly J H 2010 *Phys. Rev. Lett.* **105** 083001
[19] Mauger F, Chandre C and Uzer T 2010 *Phys. Rev. Lett.* **104** 043005
[20] Mauger F, Chandre C and Uzer T 2010 *Phys. Rev. Lett.* **105**, 083002
[21] Ceccherini F, Bauer D and Cornolti F 2003 *Phys. Rev. A* **68** 053402
[22] Yuan K J and Bandrauk A D 2012 *J. Phys. B* **45** 074001
[23] Yuan K J and Bandrauk A D 2013 *Phys. Rev. Lett.* **110** 023003
[24] Barth I, Manz J, Shigeta Y and Yagi K 2006 *J. Am. Chem. Soc.* **128** 7043
[25] Barth I and Manz J 2007 *Phys. Rev. A* **75** 012510
[26] Bauer D and Ceccherini F 2002 *Phys. Rev. A* **66** 053411
[27] Yuan K J and Bandrauk A D 2013 *Phys. Rev. A* **88** 013417
[28] Autler S H and Townes C H 1955 *Phys. Rev.* **100** 703
[29] Palacios A, Bachau H and Martín, F 2006 *Phys. Rev. A* **74** 031402(R)
[30] Sun Z and Lou N 2003 *Phys. Rev. Lett.* **91** 023002
[31] Yuan K J, Sun Z, Cong S L and Lou N 2006 *Phys. Rev. A* **74** 043421

[32] Yuan K J and Bandrauk A D 2012 *Phys. Rev. A* **85** 053419
[33] Yuan K J and Bandrauk A D 2012 *J. Phys. B* **45** 105601
[34] Kamta G L and Bandrauk A D 2005 *Phys. Rev. A* **71** 053407
[35] Kamta G L and Bandrauk A D 2006 *Phys. Rev. A* **74** 033415
[36] Lee K F, Villeneuve D M, Corkum P B, Stolow A and Underwood J G 2006 *Phys. Rev. Lett.* **97** 173001
[37] Bandrauk A D and Shen H 1993 *J. Chem. Phys.* **99** 1185
[38] Bandrauk A D and Lu H Z 200 *High-Dimensional Partial Differential Equations in Science and Engineering*, edited by Bandrauk A D, Delfour M and LeBris C, CRM Lecture Series Vol. 41 American Math. Soc. Philadelphia 1-15
[39] Corkum P B, Burnett N H and Brunel F 1989 *Phys. Rev. Lett.* **62** 1259
[40] Corkum P B 1993 *Phys. Rev. Lett.* **71** 1994
[41] Bauch S and Bonitz M 2008 *Phys. Rev. A* **78** 043403
[42] Meckel M, Comtois D, Zeidler D, Staudte A, Pavicic D, Bandulet H C, Pépin H, Kieffer J C, Dörner R, Villeneuve D M and Corkum P B 2008 *Science* **320** 1478
[43] Yuan K J, Chelkowski S and Bandrauk A D 2013 *J. Chem. Phys.* **138** 134316
[44] Yuan K J, Chelkowski S and Bandrauk A D 2014 *Chem. Phys. Lett.* **592** 334
[45] Peng L Y and Starace A F 2007 *Phys. Rev. A* **76** 043401
[46] Peng L Y, Pronin E A and Starace A F 2008 *New J. Phys.* **10** 025030

Journal of Physics: Conference Series **529** (2014) 012005 doi:10.1088/1742-6596/529/1/012005

Many-body Propagator Theory with Three-Body Interactions: a Path to Exotic Open Shell Isotopes

C Barbieri

Department of Physics, University of Surrey, Guildford GU2 7XH, UK

E-mail: C.Barbieri@surrey.ac.uk

Abstract. Ab-initio predictions of nuclei with masses up to A∼100 or more are becoming possible thanks to novel advances in computations and in the formalism of many-body physics. Some of the most fundamental issues include how to deal with many-nucleon interactions, how to calculate degenerate—open shell—systems, and pursuing ab-initio approaches to reaction theory. Self-consistent Green's function (SCGF) theory is a natural approach to address these challenges. Its formalism has recently been extended to three- and many-body interactions and reformulated within the Gorkov framework to reach semi-magic open shell isotopes. These exciting developments, together with the predictive power of chiral nuclear Hamiltonians, are opening the path to understanding large portions of the nuclear chart, especially within the *sd* and *pf* shells. The present talk reviews the most recent advances in ab-initio nuclear structure and many-body theory that have been possible through the SCGF approach.

1. Introduction

Microscopic first principle predictions of atomic nuclei are highly desirable since they can impact research of exotic isotopes. They could help in constraining extrapolations to higher mass regions [1] and to extreme proton-neutron asymmetries [2], including regions close to the drip lines where experimental data will be unavailable for the foreseeable future. Such an ambitious program requires advances in quantum many-body theory that include extending existing methods to deal with many-particle interactions, developing new approaches to calculate degenerate (open shell) fermionic systems, and calculating microscopic optical potentials.

Ab-initio methods such as coupled-cluster (CC) [3, 4], in-medium similarity renormalization group (IMSRG) [5, 6] or self-consistent Green's function (SCGF) theory based on the Dyson equation [7, 8, 9] (Dyson-GF) have accessed medium-mass nuclei up to A∼56 on the basis of realistic two-nucleon (NN) interactions. However, it has become clear that three-nucleon forces (3NFs) play a major role in determining crucial features of exotic isotopes, such as the evolution of magic numbers and the position of driplines [10, 11, 12, 13, 14]. Realistic NN and three-nucleon (3N) interactions based on chiral perturbation theory have been used in Lattice Effective Field Theory [15, 16] but also evolved to low momentum cutoffs, retaining both induced and pre-existing 3NFs [17, 18]. Systematic implementations of similar Hamiltonians within the above many-body methods are required to eventually achieve quantitative predictions of observables for medium-mass isotopes. This has been addressed only recently for CC [19, 20] and SCGF [21] methods. Based on these developments, closed shell nuclei up to Sn can now be approached [22].

A second and major challenge to ab-initio theory is that standard implementations of the above methods are limited to doubly closed (sub-)shell nuclei and to their immediate neighbors [23, 14]. As one increases the nuclear mass, longer chains of truly open shell nuclei emerge that cannot be accessed with these approaches. Many-body techniques that could tackle genuine open shell systems—or, at least, singly open shells—would immediately extend the reach of ab-initio studies from a few tens to several hundreds of mid-mass isotopes. Our collaboration has proposed to exploit ideas based on breaking particle-number symmetry in order to achieve this goal [24, 25, 26]. This has led to reformulating the SCGF in the Gorkov formalism (Gorkov-GF) [26, 27, 28], which will be discussed in this talk. Applications of Gorkov-GF have been successful for the Ca and neighbouring isotopes up to ^{54}Ti [29]. Recently, similar developments have also been introduced within the IMSRG framework [13].

The last challenge to theory is to provide consistent descriptions of the structure and reactions of nuclei in order to constrain and improve the analysis of experimental data. Microscopic calculations of elastic nucleon scattering were achieved, e.g., for a ^{16}O target in [30, 31]. Recently, ab-initio calculations, including cluster projectiles, have been possible for few-body targets by combining the ab-initio no-core shell model (NCSM) with the resonating-group method (RGM) [32, 33]. For larger masses, the SCGF becomes the method of choice due to the equivalence between the many-body self-energy and Fesbach theory of elastic scattering. This fact has guided recent advances of phenomenological dispersive optical models (DOM) [34, 35] and microscopic calculations of optical potentials [2, 30, 36]. Further developing these approaches to proper ab-initio methods will be crucial to advance our understanding of exotic radioactive beam experiments.

This talk reports about recent progress on the above topics within the SCGF approach to quantum many-body physics. The new details of the formalism are discussed first and results for finite nuclei are reported independently in sections 5 and 6.

2. Propagator theory

In Green's function (or propagator) theory one calculates the single particle propagator [37]:

$$g_{\alpha\beta}(\omega) = \sum_n \frac{\langle\Psi_0^A|c_\alpha|\Psi_n^{A+1}\rangle\langle\Psi_n^{A+1}|c_\beta^\dagger|\Psi_0^A\rangle}{\omega-(E_n^{A+1}-E_0^A)+i\eta} + \sum_k \frac{\langle\Psi_0^A|c_\beta^\dagger|\Psi_k^{A-1}\rangle\langle\Psi_k^{A-1}|c_\alpha|\Psi_0^A\rangle}{\omega-(E_0^A-E_k^{A-1})-i\eta}\,, \tag{1}$$

where α,β,..., label a complete orthonormal basis set. In Eq. (1), $|\Psi_n^{A+1}\rangle$, $|\Psi_k^{A-1}\rangle$ are the eigenstates, and E_n^{A+1}, E_k^{A-1} the eigenenergies of the $(A\pm1)$-nucleon system. Therefore, the poles of the propagator reflect experimental transfer energies for the addition or the separation of a nucleon to and from the ground state of the A-body nucleus. We obtain this information by solving the Dyson equation,

$$g_{\alpha\beta}(\omega) = g_{\alpha\beta}^{\mathrm{HF}}(\omega) + \sum_{\gamma\delta} g_{\alpha\gamma}^{\mathrm{HF}}(\omega)\,\Sigma^\star_{\gamma\delta}(\omega)\,g_{\delta\beta}(\omega)\,, \tag{2}$$

which recasts the many-body Schrödinger equation into an equation for the *dressed* (i.e. fully correlated) propagator $g(\omega)$. In Eq. (2), $g^{\mathrm{HF}}(\omega)$ represents the unperturbed propagator, which is taken to be an Hartree-Fock (HF) state for all applications discussed in the following. The information on the many-body dynamics is coded in the irreducible self-energy $\Sigma^\star(\omega)$. This has a Lehman representation similar to Eq. (1):

$$\Sigma^\star_{\alpha\beta}(\omega) = \Sigma^{\star,(\infty)}_{\alpha\beta} + \sum_{n\,n'} C_{\alpha n}\left[\frac{1}{\omega-M}\right]_{nn'} C^\dagger_{n'\beta} + \sum_{p\,p'} D_{\alpha p}\left[\frac{1}{\omega-N}\right]_{pp'} D^\dagger_{p'\beta} \tag{3}$$

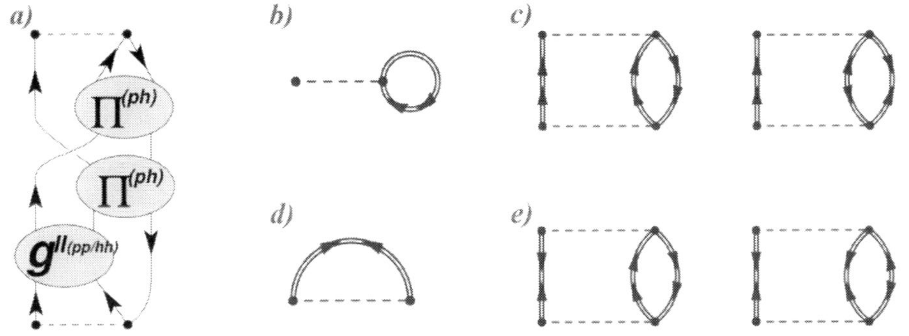

Figure 1. *Left*: Diagram a) gives one example of the particle-vibration coupling terms that enter the ADC(3) and FRPA expansions of the self-energy. *Center* and *right*: Contribution to the Gorkov irreducible self-energy up to second order, for the case of NN interactions only. Diagram b) is the first order contributions to the normal part of the static irreducible self-energy, $\Sigma^{11,(\infty)}$. This also appears in the standard Dyson formulation and extends the usual HF potential to the one corresponding to a fully correlated density matrix. Diagram d) is the anomalous first order contribution $\Sigma^{21,(\infty)}$, which corresponds to a correlated version of the pairing potential in the HFB equations. Diagrams c) are the normal second order terms and diagrams e) are the anomalous second order ones.

and provides the microscopic optical potential for a single nucleon interacting with the whole many-body system. In Eq. (3), M, N are interaction matrices in the $2h1p$ and $2p1h$ (or more complex) configuration spaces and C, D are the respective couplings to single-particle states. $\Sigma^{\star,(\infty)}$ represents the static (i.e., energy independent) contribution to the self-energy.

The self-energy is constructed starting from the Hamiltonian $H(A) = H - \hat{T}_{\rm c.o.m.}(A) = \hat{U}(A) + \hat{V}(A) + \hat{W}$, where we correct for the centre-of-mass kinetic energy. $\hat{U}$, $\hat{V}$ and $\hat{W}$ collect all one-, two- and three-body interactions, respectively. For finite closed-shell nuclei, the propagator (1) is calculated by first solving spherical HF equations. The $g^{\rm HF}(\omega)$ propagator is then used as a reference state to calculate the self-energy, using either the third-order algebraic diagrammatic construction [ADC(3)] [38, 39] or the Faddeev random phase approximation (FRPA) [40, 8] methods. Both the ADC(3) and the FRPA completely include all diagrams up to third-order and resum many others to all orders. Moreover, they completely accounts for particle-vibration diagrams as shown in Fig. 1a. Thus, in general, a full FRPA calculation involves the calculations of the particle-hole (ph) response and of the two-body spectral function. The latter has been employed, for example, to investigate nuclear correlations from two-nucleon emission experiments [41, 42, 43]. The difference between the two approaches is in the fact that vibrations in the ph and particle-particle/hole-hole (pp/hh) channels are calculated in Tamn-Dancoff approximation for ADC(3) and in random phase approximation for FRPA.

It is instructive to compare the perturbation theory content of the ADC(n) and FRPA methods to that of other many-body approaches. For the calculation of total energies, both ADC(3)/FRPA and the coupled cluster CCSD approaches retain in full all diagrams up to second order [44]. However, the Koltun energy sum rule in SCGF theory adds specific triple corrections. Our experience with both G-matrix and SRG-evolved interactions is that ADC(3) results are indeed systematically closer to CCSD(T) rather than to the simple CCSD [27]. When evaluating one nucleon addition and separation energies, the ADC(3) fully sums two-particle–one-hole ($2p1h$) and two-hole–one-particle ($2h1p$) configurations and should therefore be compared to the usual truncations of the excitation operator in particle-attached and particle-removed CC. This is usually sufficient to predict the dominant quasiparticle and quasihole fragments of the spectral distribution. Other small fragments—the so-called shakeup states—will generally require truncation schemes beyond $2p1h/2h1p$, such as the ADC(4) and ADC(5) [39].

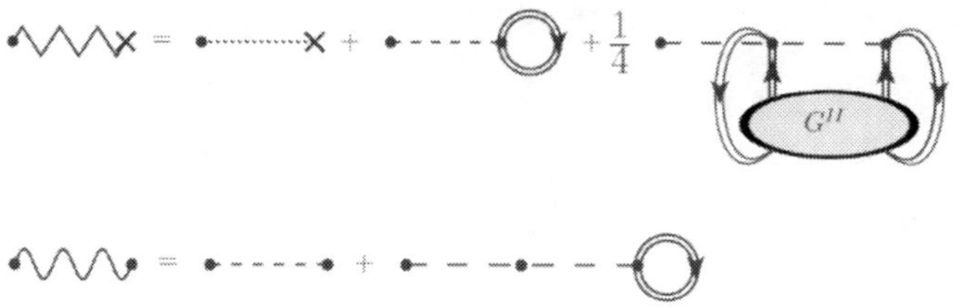

Figure 2. *Top:* Diagrammatic representation of the effective one-body interaction of Eq. (5). This is given by the sum of the original one-body potential $\hat{U}$ (dotted line), the two-body interaction $\hat{V}$ (dashed line) contracted with a dressed one-body propagator g (double line with arrow), and the three-body interaction $\hat{W}$ (long-dashed line) contracted with a dressed two-body propagator, g^{II}. The correct symmetry factor of 1/4 is also shown explicitly. *Bottom:* Diagrammatic representation of the effective two-body interaction of Eq. (6). This is given by the sum of the original two-body interaction and the three-body interaction contracted with a dressed propagator.

3. Extension of propagator theory to include three-body interactions

We have extended the Dyson SCGF formalism to the case of Hamiltonians which include three-body interactions. Full details are presented in Refs. [21, 45] and involve defining in-medium effective interactions that regroup specific sets of diagrams, such that only *interaction-irreducible* ones are retained[1]. Moreover, proper corrections to the Kotlun sum rule are required to calculate total binding energies. For a pure two-body Hamiltonian, the only possible interaction-reducible contribution to the self-energy is the generalised Hartree-Fock diagram of Fig. 1b[2]. However, many more appear when three- and many-body forces are present. This makes it useful to define effective interactions to group such contributions.

Hence, for a system with up to 3NFs, we define an effective Hamiltonian,

$$\widetilde{H} = \widetilde{U} + \widetilde{V} + \hat{W} \tag{4}$$

where $\widetilde{U}$ and $\widetilde{V}$ represent effective interaction operators. As long as only interaction-irreducible diagrams are calculated when using these effective interactions, no double counting of any diagram is possible [21]. The explicit expressions for the one- and two-body effective interaction operators are:

$$\widetilde{U} = \sum_{\alpha\beta} \left[U_{\alpha\beta} - i\hbar \sum_{\gamma\delta} V_{\alpha\gamma,\beta\delta}\, g_{\delta\gamma}(t-t^+) + \frac{i\hbar}{4} \sum_{\substack{\gamma\epsilon \\ \delta\eta}} W_{\alpha\gamma\epsilon,\beta\delta\eta}\, g^{II}_{\delta\eta,\gamma\epsilon}(t-t^+) \right] a^\dagger_\alpha a_\beta \,, \tag{5}$$

$$\widetilde{V} = \frac{1}{4} \sum_{\substack{\alpha\gamma \\ \beta\delta}} \left[V_{\alpha\gamma,\beta\delta} - i\hbar \sum_{\epsilon\eta} W_{\alpha\gamma\epsilon,\beta\delta\eta}\, g_{\eta\epsilon}(t-t^+) \right] a^\dagger_\alpha a^\dagger_\gamma a_\delta a_\beta \,, \tag{6}$$

where we have introduced specific orientations of the one-body, $g(\tau)$, and two-body, $g^{II}(\tau)$, green functions in time representation [46]. Eqs. (5) and (6) are shown diagrammatically in Fig. 2.

[1] A diagram is said to be *interaction-reducible* if it can be factorized in two lower order diagrams by cutting an interaction vertex or, equivalently, if it is connected and there exists a group of lines (whether interacting among themselves or not) that leave an interaction vertex and eventually all return to it.

[2] For the Gorkov formalism, one would also have the interaction-reducible Bogoliubov potential of Fig. 1d.

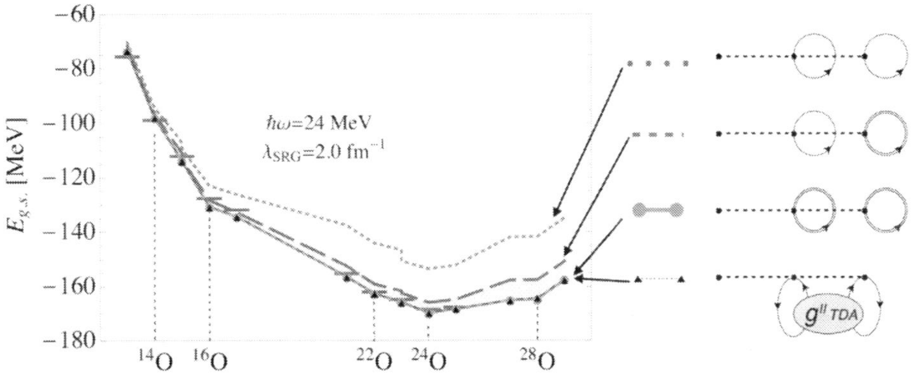

Figure 3. Calculated ground state energies of oxygen isotopes for different approximations of the 3NFs contributions to the static self-energy, see Eqs. (5) and (7). All results are obtained from a SRG evolved NN+3N interactions with cutoff λ_{SRG}=2 fm^{-1} and are based on the full ADC(3) self-energy [14]. However, the last term in Eq. (5) is calculated at different levels of self-consistency by contracting the 3NF with all unperturbed HF propagators (single line loops), by contracting with one dressed propagator (double line loops), or with two dressed propagators. The corresponding diagrams are shown on the right. Dressing self-consistently *all* the single-particle propagators is mandatory to obtain convergence with respect to the many-body truncation. Higher order contributions from ladder diagrams (TDA ladders in the figure) do not further contribute for the present interaction.

The use of effective interactions greatly reduces the number of diagrams to be accounted for in practical calculations. By construction, the only possible interaction-irreducible contribution at first-order is precisely given by Eq. (5), and it equals the static irreducible self-energy:

$$\Sigma^{\star,(\infty)}_{\alpha\beta} = \widetilde{U}_{\alpha\beta} \, . \tag{7}$$

Even though it only contributes to the self-energy at first order, the effective one-body potential $\widetilde{U}$ is very important. It represents the (energy-independent) mean field seen by each particle, including contributions from all many-body interactions. The simplest approximation to $\widetilde{U}$ would consists in performing the averages of Eq. (5) over Hartree-Fock or other Slater determinant reference states. This is equivalent to reducing $\widetilde{U}$ to the one-body term of a normal ordered Hamiltonian. Analogous treatments of the effective interaction have been employed in nuclear physics calculations up to date, including both finite nuclei [10, 18, 12, 47] and nuclear matter [48, 49, 50, 51] applications. For the particular case of *ab-initio* SCGF calculations, we find that it is mandatory to fully dress the reference propagators entering Eq. (5) [14]. This is shown in Fig. 3, where the last term of Eq. (5) is gradually improved. While dressing the propagators that enter the HF contribution coming from 3NFs is mandatory, additional NN correlations such as ladder series do not sensibly affect the final results[3].

At second order, the perturbative expansion of the irreducible self-energy generates five diagrams. By exploiting the two-body effective interaction $\widetilde{V}$, these are all grouped into the two interaction-irreducible terms show in Fig. 4. Likewise, there exist 53 diagrams at third order which reduce to the 17 interaction-irreducible ones shown in Fig. 5.

Although they have not all been investigated in actual calculations, the diagrams of Figs. 4 and 5 are expected to contribute very differently to the self-energy and other computed

[3] For methods like CC, such dressing of the mean field is generated implicitly when solving the CC equations iteratively. However, in the SCGF philosophy, this is included directly in the effective interaction $\widetilde{U}$ and the dressed propagator can be thought of as an improved reference state.

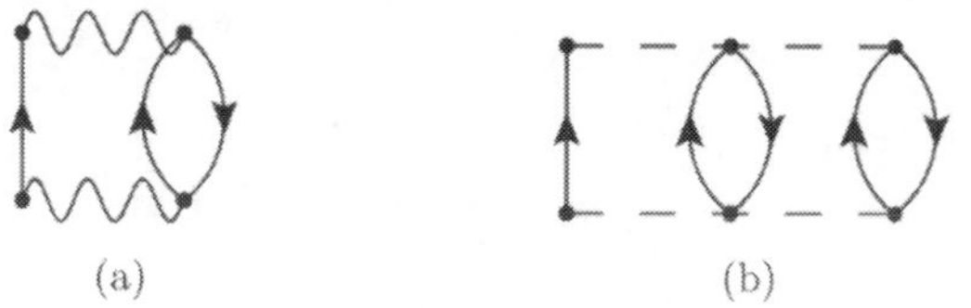

Figure 4. *Interaction-irreducible* diagrams contributing to the irreducible self-energy at second order. These are the only contributions that must be considered when using the effective Hamiltonian of Eq. (4).

quantities. In these figures, they are ordered in terms of their expected decreasing importance. Figs. 4a, 5a and 5b are the most relevant terms and are those already appearing in the case of only NN interactions. However, they account here for the effects of 3NFs at lowest order through the effective NN term of Eq. (6). These all involve $2p1h$ and $2h1p$ intermediate states and are the first contributions to the ADC(3) and FRPA expansions. Diagram 5c does not introduce any new intermediate state but describes the mixing among $2p1h$ and $2h1p$ configuration due to irreducible 3NFs. This is expected to be less important on the basis that 3NFs are generally weaker than the corresponding NN force (typically, $<\widehat{W}> \approx \frac{1}{10} <\widehat{V}>$ for nuclear interactions [52, 53]). Studies of normal-ordered Hamiltonians also clearly suggest that this term gives a small correction to the total energy compared to diagrams 5a and 5b [19, 18]. Note that diagrams 5a-5c contain the first-order terms of the all order summation that generates configuration mixing between $2p1h$ or $2h1p$ excitations. Nowadays, resummations of these configurations are performed routinely for the 5a the 5b terms in ADC(3) and FRPA calculations [8, 9, 54].

The diagrams in Figs. 4b and 5d to 5k introduce additional configurations of the $3p2h$ and $3h2p$ type. These are expected to affect sensibly the distribution of small fragments in the particle and hole spectral strength distribution but have smaller relevance for calculation of total binding energies and dominant quasipartcle and quasihole peaks. All these third order contributions provide the transition matrix elements connecting $2p1h$ to $3p2h$ (and $2h1p$ to $3h2p$). However, diagrams 5d to 5g may be more important than the remaining four, since they involve one more NN vertex. Note that the terms 4b and 5d to 5k together induce the same type of configurations that appear only from the fourth-order ADC(4) and beyond, when 3NFs are missing. Thus, they are expected to have similar impact on the many-body truncation. The remaining diagrams, 5l to 5q, describe the explicit mixing among $3p2h$ or $3h2p$ configurations and are expected to give contributions comparable to the NN-only ADC(5) approximation.

The energy sum rule originally proposed by Galitskii-Migdal [55] and by Koltun [56] needs to be extended for the presence of many-body interactions. When only up to three-body forces are present, this can be re-expressed in one of two equivalent forms:

$$E_0^A = \sum_{\alpha\beta} \frac{1}{4\pi i} \int_{C\uparrow} d\omega \;\; [U_{\alpha\beta} + \omega\delta_{\alpha\beta}] \; g_{\beta\alpha}(\omega) \; - \; \frac{1}{2}\langle\Psi_0^A|W|\Psi_0^A\rangle \tag{8}$$

and

$$E_0^A = \sum_{\alpha\beta} \frac{1}{6\pi i} \int_{C\uparrow} d\omega \;\; [2U_{\alpha\beta} + \omega\delta_{\alpha\beta}] \; g_{\beta\alpha}(\omega) \; + \; \frac{1}{3}\langle\Psi_0^A|V|\Psi_0^A\rangle \, . \tag{9}$$

Both Eqs. (8) and (9) are exact but they rely on the possibility of estimating the expectation value of either the two- or the three-body part of the Hamiltonian. Most recent SCGF calculations exploited Eq. (8) based on the consideration that absolute errors in evaluating $\langle\Psi_0^A|\hat{W}|\Psi_0^A\rangle$ should be smaller, since the contribution of the 3NF is about an order of magnitude less than that of the NN force [14, 51, 29]. For finite nuclei and SRG cutoffs of λ_{SRG}=2.0 fm^{-1},

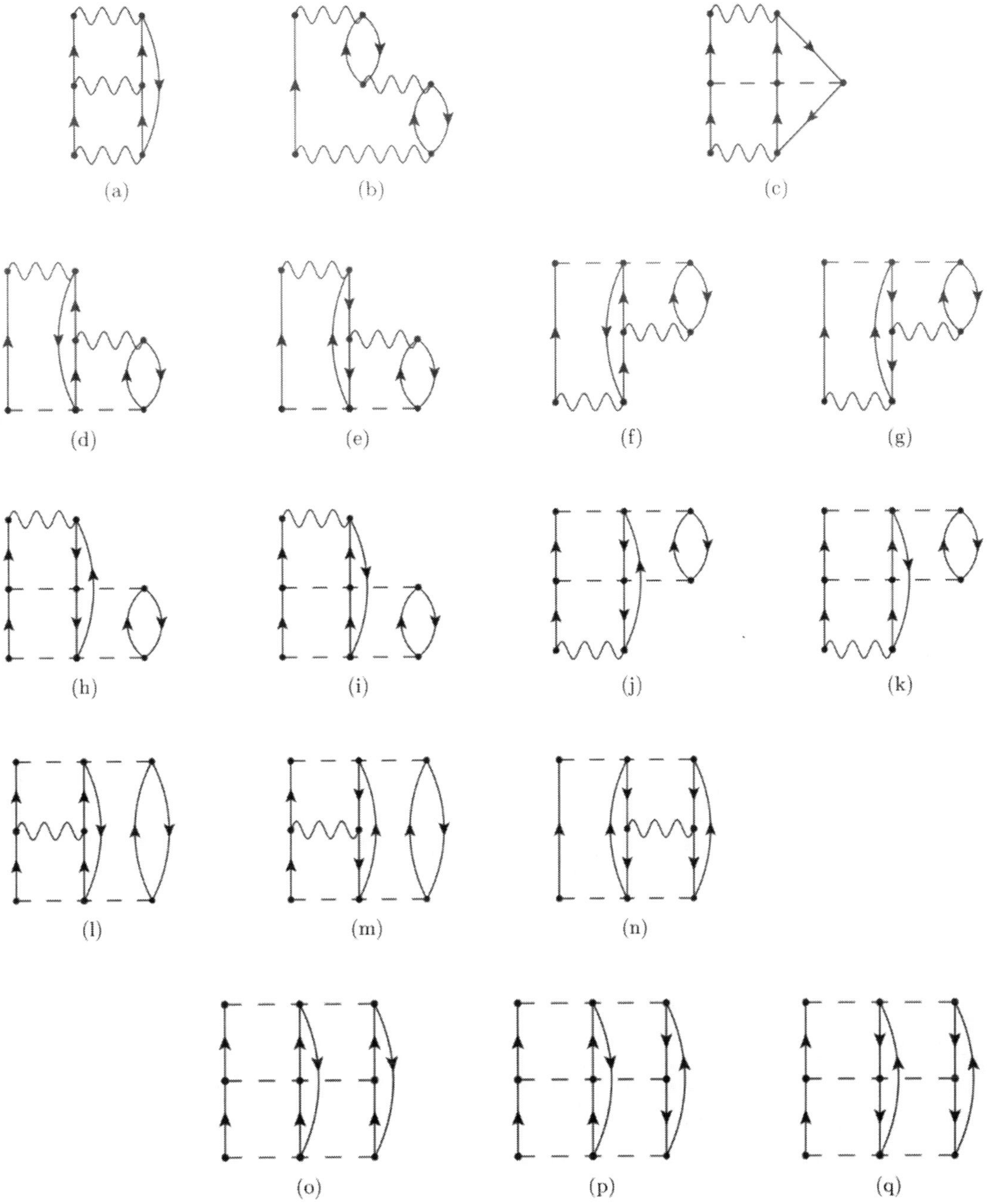

Figure 5. *Interaction-irreducible* diagrams that contribute to the irreducible self-energy at third order. These are the only third-order contributions that must be considered when using the effective Hamiltonian of Eq. (4).

we found that $\langle\Psi_0^A|\hat{W}|\Psi_0^A\rangle$ can be evaluated at first order in the *dressed* propagators and this is sufficient to converge the calculated binding energy with respect to many-body truncations [14]. This is the same level of approximation needed to converge the effective potential $\widetilde{U}$ with respect to many-body truncations (see Fig. 3 and its discussion above).

4. Gorkov formulation for semi-magic open shell systems

In Refs. [26, 27, 28], we have introduced an extension of the SCGF scheme that is based on the Gorkov's formalism [57] and that can address open shell nuclei. The Gorkov method is based on the idea of allowing the breaking of particle number symmetry in order to achieve an effective description of the pairing correlations, which play a crucial role in open shell systems and remove degeneracies in the reference state. In order to work with the correct number of protons and neutrons in average, chemical potentials are added to the Hamiltonian, that is, we consider the grand canonical Hamiltonian $\hat{\Omega}(A) = H - \hat{T}_{\rm c.o.m.}(A) - \mu_{\rm p}\hat{Z} - \mu_{\rm n}\hat{N}$. Additional (anomalous) propagators that account for the breaking and formation of Cooper pairs are introduced. The Lehman representation of the Gorkov propagators is

$$G^{11}_{\alpha\beta}(\omega) = \sum_k \left\{ \frac{\mathcal{U}^k_\alpha \mathcal{U}^{k*}_\beta}{\omega - \omega_k + i\eta} + \frac{\bar{\mathcal{V}}^{k*}_\alpha \bar{\mathcal{V}}^k_\beta}{\omega + \omega_k - i\eta} \right\}, \tag{10a}$$

$$G^{12}_{\alpha\beta}(\omega) = \sum_k \left\{ \frac{\mathcal{U}^k_\alpha \mathcal{V}^{k*}_\beta}{\omega - \omega_k + i\eta} + \frac{\bar{\mathcal{V}}^{k*}_\alpha \bar{\mathcal{U}}^k_\beta}{\omega + \omega_k - i\eta} \right\}, \tag{10b}$$

$$G^{21}_{\alpha\beta}(\omega) = \sum_k \left\{ \frac{\mathcal{V}^k_\alpha \mathcal{U}^{k*}_\beta}{\omega - \omega_k + i\eta} + \frac{\bar{\mathcal{U}}^{k*}_\alpha \bar{\mathcal{V}}^k_\beta}{\omega + \omega_k - i\eta} \right\}, \tag{10c}$$

$$G^{22}_{\alpha\beta}(\omega) = \sum_k \left\{ \frac{\mathcal{V}^k_\alpha \mathcal{V}^{k*}_\beta}{\omega - \omega_k + i\eta} + \frac{\bar{\mathcal{U}}^{k*}_\alpha \bar{\mathcal{U}}^k_\beta}{\omega + \omega_k - i\eta} \right\}, \tag{10d}$$

where the poles of the propagators are now given by $\omega_k \equiv \Omega_k - \Omega_0$, Ω_0 is the ground state of the even-A isotope and the index k labels odd-A eigenvalues of $\hat{\Omega}$: $\hat{\Omega}\,|\Psi_k\rangle = \Omega_k\,|\Psi_k\rangle$. The residues of pole ω_k relate to the probability amplitudes $\mathcal{U}^k$ ($\mathcal{V}^k$) to reach state $|\Psi_k\rangle$ by adding (removing) a nucleon to (from) the even-A ground state $|\Psi_0\rangle$.

Formally, Eqs. (2) and (3) still hold with all quantities (propagators and self-energies) now being matrices in a 2×2 Gorkov space. The self-energy still splits into static and dynamic contributions, $\Sigma^{g_1 g_2}(\omega) = \Sigma^{g_1 g_2,(\infty)} + \Sigma^{g_1 g_2,(dyn)}(\omega)$ (where g_i=1,2 are the Gorkov indices). The working equations for the Gorkov formalism have been derived in full up to second order and NN interactions [26]. The corresponding self-energy diagrams are shown Figs. 1b to 1e. In the present implementation of the Gorkov formalism, we add 3NFs as described in Sec. 3 by exploiting the effective interactions of Eqs. (5) and (6) and adding the anomalous term corresponding to $\tilde{U}$ [29]. The extension of the Gorkov framework to FRPA and ADC(3) type of self-energies is currently underway.

5. Studies of correlations and optical potentials

In Eq. (2), the irreducible self-energy $\Sigma^\star(\omega)$—or, equivalently, the Gorkov's normal self-energy $\Sigma^{11}(\omega)$—acts as an energy dependent non-local single particle potential. In fact, it can be proven that $\Sigma^\star(\omega)$ is exactly the Fesbach microscopic optical potential, once this is extended to treat both nucleon-nucleus states (above the Fermi surface, E_F) and hole-nucleus states (below E_F) [58, 59]. This provides a way to calculate optical potentials from *ab-initio* theory.

In order to investigate the microscopic properties of optical models, we transform Eq. (3) to coordinate space and separate its central and spin-orbit angular momentum components as follows [2]:

$$\begin{aligned}\Sigma^\star(\boldsymbol{x},\boldsymbol{x}';E) &= \sum_{\ell j m_j \tau} \mathcal{I}_{\ell j m_j}(\Omega,\sigma)\left[\sum_{n_a,n_b} R_{n_a\ell}(r)\Sigma^\star_{ab}(E)R_{n_b\ell}(r')\right](\mathcal{I}_{\ell j m_j}(\Omega',\sigma'))^* \\ &= \sum_{\ell j m_j \tau} \mathcal{I}_{\ell j m_j}(\Omega,\sigma)\left[\Sigma^\ell_0(r,r';E) + (\vec{\boldsymbol{\ell}}\cdot\vec{\boldsymbol{\sigma}})\,\Sigma^\ell_{\ell s}(r,r';E)\right](\mathcal{I}_{\ell j m_j}(\Omega',\sigma'))^*, \quad (11)\end{aligned}$$

where $\boldsymbol{x} \equiv \boldsymbol{r},\sigma,\tau$. In (11), σ represents the spin variable, τ is the isospin, n is the principal quantum number of the harmonic oscillator basis used in the calculations, and $a \equiv (n_a,\ell,j,\tau)$ (note that for a $J = 0$ target the self-energy is independent of m_j). The standard radial harmonic-oscillator function is denoted by $R_{n\ell}(r)$, while $\mathcal{I}_{\ell j m_j}(\Omega,\sigma)$ represents the j-coupled angular-spin function. We calculated the potential (11) in Ref. [2] by using FRPA and the Argonne V18 NN interaction [60].

The most recent implementations of the dispersive optical model (DOM) are substantially phenomenological optical potentials whose analytical structure is, however, constrained by our best knowledge of ab-initio calculated self-energies [34, 61, 62, 35]. In fact, the DOM can be seen as a phenomenological parameterisation of $\Sigma^\star(\omega)$ that allows for data driven extrapolations to exotic isotopes. Note that the analytic structure of Eq. (3) implies a dispersive relation that links the real and imaginary parts of $\Sigma^\star(\omega)$ and that it encodes the causality principle. This relation is at the core of the DOM approach. In fitting optical potentials, it is usually found that volume integrals are better constrained by the experimental data [63, 64]. Thus, we focus our attention on the imaginary part of the central self-energy, which corresponds to the total absorption. Its volume integral is given by

$$J^\ell_W(E) = 4\pi \int dr r^2 \int dr' r'^2 \,\mathrm{Im}\, \Sigma^\ell_0(r,r';E)\,. \quad (12)$$

The left panel of Fig. 6 compares the ℓ-averaged FRPA volume integrals (J^{avg}_w) with the corresponding global DOM potentials of Ref. [61]. This local form of the DOM has been corrected by the effective mass [63, 65] to account for non-locality. The overall effect of this correction is to enhance the absorption. The FRPA exhibits different behavior above and below E_F than the DOM. Microscopic calculations predict significantly less absorption below E_F, whereas in the DOM fit, the absorption was assumed to be roughly symmetric up to about 50 MeV away from E_F [63, 61, 62]. Since only the absorption above the Fermi energy is strongly constrained by elastic scattering data, it is encouraging that the ℓ-averaged FRPA result is reasonably close to the DOM fit for both nuclei, up to $E - E_F < 80$ MeV where the FRPA results are not sensibly affected by the truncation of the model space.

The simplifying assumptions of a symmetric absorption around E_F and locality in the DOM imply an unrealistic occupation of higher ℓ-values below the Fermi energy which is not obtained in the FRPA. This is shown in the right panel of Fig. 6 for scattering of neutrons against ^{40}Ca. Below the Fermi energy the contribution from s, p and sd orbits dominates while higher ℓ-values are suppressed. This is a clear indication of the nucleon occupations in the mean field shell structure of the target nucleus. The dash-double-dotted curve illustrates the same DOM fit also shown in the left panel. This compares favourably to FRPA for the low-ℓ partial waves below the Fermi surface. Recent fits of the DOM include non-locality effects explicitly and find that this is required in order to reproduce the features of spectral distribution of single particle strength below E_F, including the experimentally observed position of quasiparticle peaks [35]. This also allow for the description of high-momentum components that is suggested by $(e,e'p)$ observations at Jefferson Laboratory [66, 67, 68].

Journal of Physics: Conference Series **529** (2014) 012005 doi:10.1088/1742-6596/529/1/012005

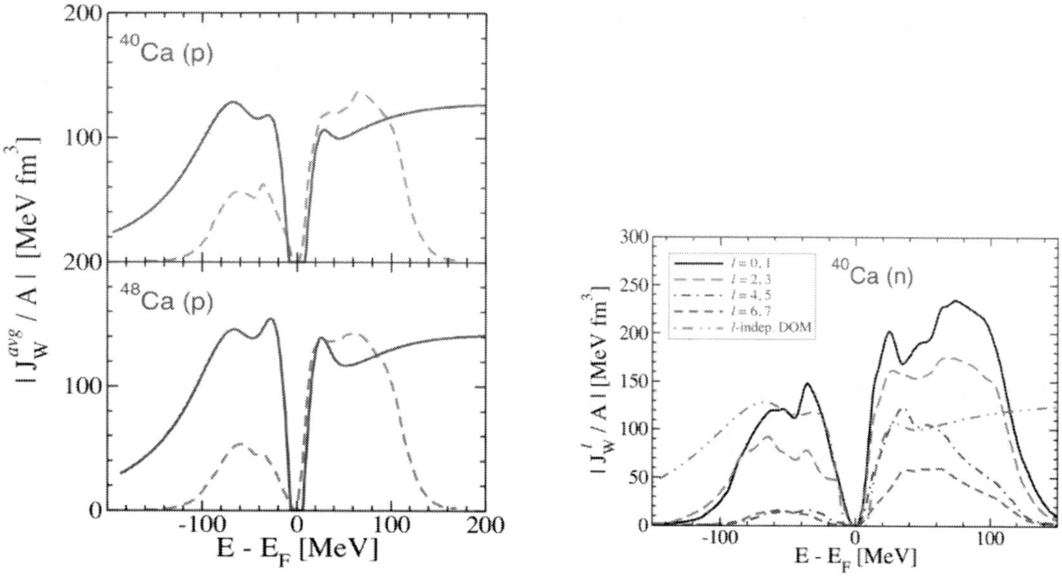

Figure 6. *Left*: The FRPA results for the volume integral averaged over all ℓ-channels J^{avg}_w(dashed line) are compared with the DOM result (solid line), corrected for non-locality, for proton scattering on ^{40}Ca and ^{48}Ca. *Right*: Separate partial wave contributions to J_W averaged over adjacent ℓ-channels in the model space. This plot is for neutrons in ^{40}Ca. The dash-double-dotted curve represents the DOM result. All calculation are based on the NN-only AV18 realistic interaction.

The behavior of the nuclear self-energy with changing proton-neutron asymmetry ($\alpha = (N - Z)/A$) is very important in seeking for *global* parametrizations of the DOM. Hence its understanding is pivotal in predicting unstable isotopes. The changes in the calculated J^{avg}_W for isotopes going from ^{40}Ca to ^{60}Ca are shown in the top-left panels of Fig. 7. These cover a range of asymmetries form α=0 to 1/3. Microscopic FRPA predicts an opposite behavior of protons and neutrons above E_F, with the proton (neutron) potential increasing (decreasing) when more neutrons are added, qualitatively in agreement with expectations from the Lane potential model [69]. The DOM fit of Ref. [62] includes several isotopes in the Ca and Sn chains and employs a similar trend for the volume integrals.

The characteristics of the above asymmetry dependence indicate that the nuclear interaction between protons and neutrons plays a major role. The tensor force of the nuclear interaction can provide one such mechanism since it is particularly strong in the isospin $T = 0$ sector. Moreover, it has already been shown to influence the evolution of single-particle energies at the Fermi surface [70]. To investigate its implication for single-particle properties at energies farther removed from E_F, we recalculated J^{avg}_W by suppressing the tensor component of the AV18 interaction. It is apparent that this has a very significant effect on the correlations far from the Fermi surface for nucleon-nucleus scattering. On the other hand, absorption near the Fermi surface are dominated by correlations other than tensor.

In the right panel of Fig. 7 (dot-dashed line), we have also calculated the FRPA self-energy by suppressing charge-exchange excitations in the polarization propagator $\Pi^{(ph)}$ (see Fig. 1a). These contributions correspond to a mechanism in which the proton (neutron) projectile is exchanged with a neutron (proton) in the target. This includes Gamow-Teller resonances whose strength increase with asymmetry as $\approx 3(N - Z)$ [61]. Ab-initio FRPA calculations indicate that such correlations do not generate significant absorption in scattering processes.

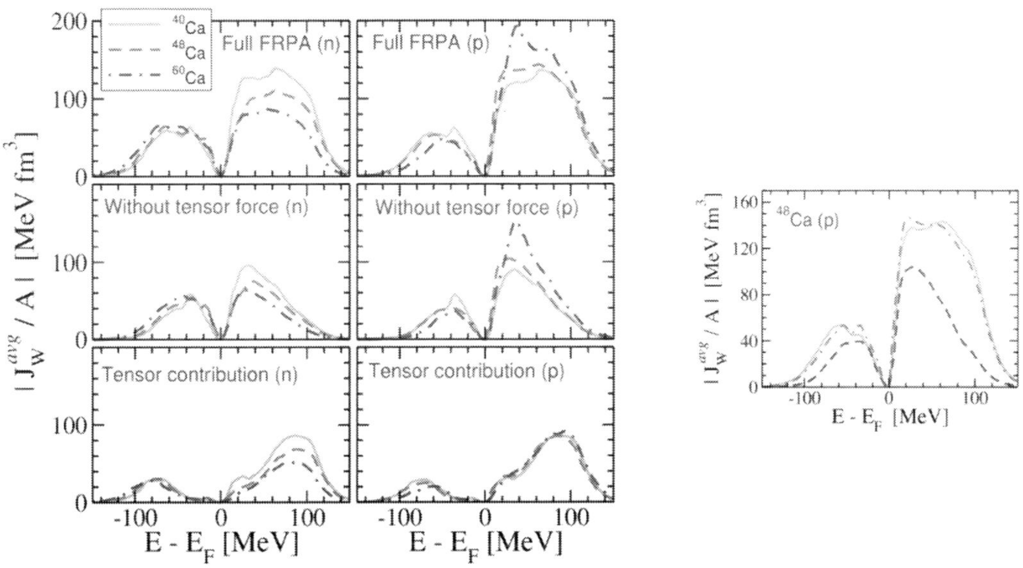

Figure 7. *Left*: Asymmetry dependence of the absorption for neutrons and protons and dependence on tensor correlations. J_W^{avg} for ^{40}Ca (solid) is compared with the results for ^{48}Ca (dashed), and ^{60}Ca (dot-dashed). The figures in the lowest row are the difference between the full calculation and the middle row figures, to show the effect of correlations arising from the tensor force. *Right*: Effect of the tensor force and charge exchange correlations on the proton-^{48}Ca self-energy. The solid curve is the imaginary volume integral J_W^{avg} from the full FRPA calculation, while the dashed curve results from removing the tensor term in the AV18 interaction. The dash-dotted curve is obtained by excluding charge exchange from the full calculation. Similar results are found for neutrons and the other Ca isotopes. All calculation are based on the NN-only AV18 realistic interaction.

6. Role of chiral three-nucleon forces in medium mass isotopes

Calculations were performed using chiral NN and 3N forces evolved to low momentum scales through free-space similarity renormalization group (SRG) techniques [17]. The original NN interaction is generated at next-to-next-to-next-to-leading order (N^{3}LO) with cutoff Λ_{NN}=500 MeV [71, 72] and is supplemented by a local NNLO 3NF [73] with a reduced cutoff of Λ_{3N}=400 MeV, as in Ref. [18]. The chiral NNLO 3NF contain the two-pion exchange Fujita-Miyazawa contribution. The SRG evolution of the sole chiral NN interaction already generates 3N operators in the Hamiltonian, which we refer to hereafter as the "induced" 3NF. When the pre-existing chiral 3N interaction is also included, we refer to it as the "full" 3NF. A SRG cutoff λ_{SRG}=2.0 fm^{-1} was used for most calculations presented in this section.

Fully ab-initio calculations based on evolved NN plus full 3NFs can accurately reproduce ground state energies of the entire oxygen isotopic chain, as shown in Fig. 3 and reported in Ref. [77]. This was subsequently confirmed by extended calculations across different many-body methods, which estimated in details the errors associated with the SRG evolution and the many-body truncations schemes [13, 14]. These results give a first principle confirmation of the repulsive effects of the two-pion exchange Fujita-Miyazawa interaction, which was found to be at the origin of the anomalous dripline at ^{24}O [10]. This is seen in the left panel of Fig. 8, which shows the predicted evolution of neutron single particle spectrum (addition and separation energies) for oxygen isotopes. Induced 3NFs reproduce the overall trend but predict a bound $d_{3/2}$ when the *sd* shell is filled. Adding pre-existing 3NFs—the full Hamiltonian—raises this

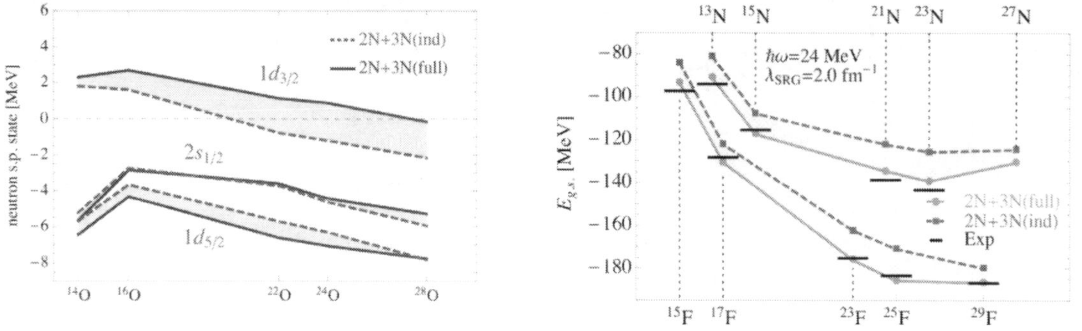

Figure 8. *Left*: Energies for the addition and separation of a neutron to and from various oxygen isotopes. These correspond to the dominant quasiparticle peaks that can be observed in the final Z=8 and N±1 isotopes. *Right*: Ground-state energies of nitrogen and fluorine isotopes calculated for induced and full interactions. Experimental data are from [74, 75, 76]. Calculations have been performed with chiral two- and three-body interactions evolved to $\lambda_{SRG} = 2.0\,\text{fm}^{-1}$ by means of SRG techniques [14]. In both plots, the dashed and full lines join results for the induced and full interactions, respectively.

orbit above the continuum also for the highest masses.

The same mechanism affects neighboring isotopic chains, as demonstrated in the right panel of Fig. 8 for the odd-even semi-magic nitrogen and fluorine. Induced 3NF forces consistently under bind these isotopes and even predict a ^{27}N close in energy to ^{23}N. This is fully corrected by full 3NFs that strongly binds ^{23}N with respect to ^{27}N, in accordance with the experimentally observed dripline. The repulsive effects of filling the $d_{3/2}$ is also observed in ^{29}F but it is counter balanced by attractions due to the inclusion of an extra proton. This leads to a slightly bound ^{29}F in accordance with observation [76]. Eventually, leading-order (NNLO) original three-body terms are crucial both to bring calculated energies close to the experiment and to yield a correct description of the drip lines.

Both Dyson and Gorkov formalisms yield the single particle spectral function. Its diagonal part,

$$S_\alpha(\omega) = \frac{1}{\pi}|\text{Im } g_{\alpha\alpha}(\omega)| = \frac{1}{\pi}|\text{Im } G^{11}_{\alpha\alpha}(\omega + \mu_\alpha)| \,, \tag{13}$$

describes the energy distribution of the spectral strength (i.e. spectroscopic factors) for the addition or removal of one particle. This is demonstrated in Fig. 9 for second-order Gorkov calculations of ^{44}Ca. The left panel shows the spectral distribution for different partial waves. States below (above) the Fermi surface refer to the separation (addition) of a neutron, leading to final states of ^{43}Ca (^{45}Ca). These results refer to calculations based on evolved NN interactions only, with all (induced and full) 3NFs discarded. The effects of 3NFs on the spectral distribution in the *sd* and *pf* shells is shown in the right panel, and it is compared to the NN-only results from the left. We find that the NN plus full 3NFs Hamiltonian is fundamental to predict the correct location of the Fermi surface with respect to the single-particle continuum. 3NFs also reduce the average energy separation of the two major valence shell, *sd* and *pf*, by a factor of two and bring these in closer agreement with the experiment. In our calculations we find that the neutron separation energy to the $J^\pi = 3/2^+$ state of ^{43}Ca is about -16 MeV. This is close to the phenomenological predictions based on the DOM of Ref. [61].

The newly introduced Gorkov-GF method allows for the first time ab-initio calculations of adjacent isotopes within large portions of the nuclear chart. The importance of leading order chiral 3NFs in reproducing the correct trend of binding energies for the oxygen nuclides is then

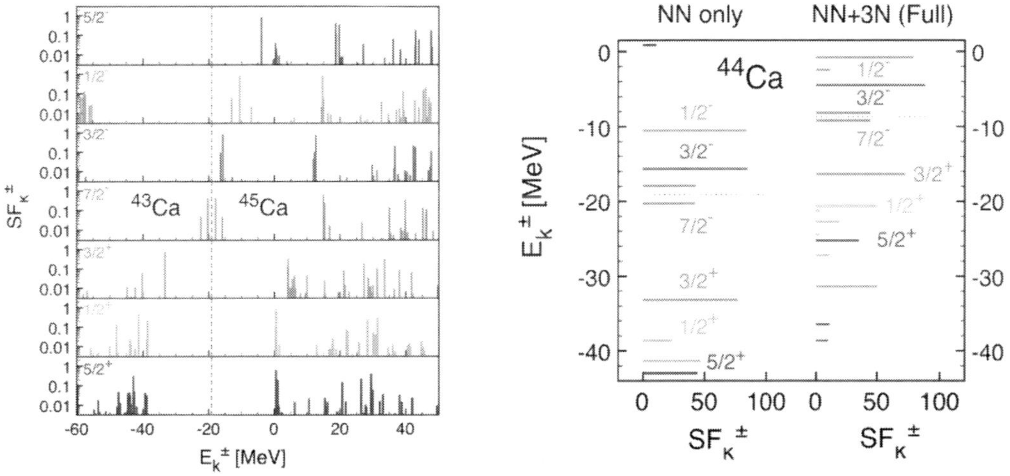

Figure 9. Spectral strength distribution for the removal and addition of a neutron from and to ^{44}Ca, calculated using Gorkov-GF at second order [27]. *Left*: Distribution of different partial waves calculated with an evolved NN interaction only. *Right*: Distribution in the energy region of the *sd* and *pf* nuclear shells, obtained from NN-only and NN plus full 3NFs. The experimental neutron chemical potential (corresponding to the Fremi surface) is $\mu_n(^{44}\text{Ca})$=-9.27 MeV and it is well reproduced by full 3NFs.

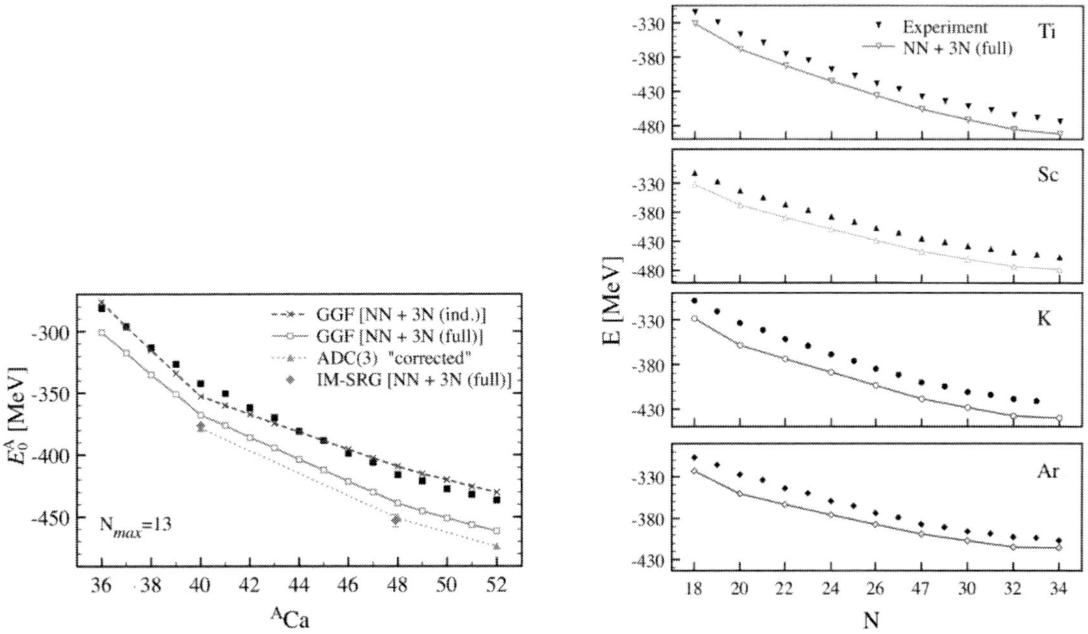

Figure 10. [27]. *Left*: Binding energies of Ca isotopes obtained from induced and full 3NF Hamiltonians, compared to the experiment. The triangle (with dotted light-blue line) and diamond symbols refer to Dyson-ADC(3) and IM-SRG [6] results for the closed shell cases. These confirm that the effects of the second order truncation in Gorkov calculations amount to a small constant shift of binding energies, which does not invalidate our general conlcusions. *Right*: Predicted binding energies for Ar, K, Sc and Ti isotopic chains for the full 3NF Hamiltonian (full lines). In all plots, experimental values are indicated by full black symbols and are taken from [78, 79, 80]. Calculations are from second order Gorkov-GF, except where indicated, and with chiral interactions evolved to $\lambda_{SRG} = 2.0\,\text{fm}^{-1}$ [29].

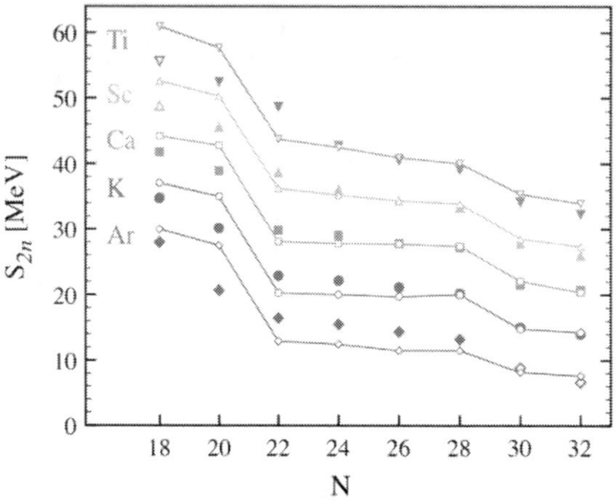

Figure 11. Two-neutron separation energies, S_{2n}, along Ar, K, Ca, Sc and Ti isotopic chains. The experimental values (isolated symbols) [78, 79, 80] are compared to predictions for the full 3NF Hamiltonian with a SRG cutoff of $\lambda_{SRG} = 2.0\,\text{fm}^{-1}$ (full lines). Results are for second order Gorkov calculations and the values for K, Ca, Sc and Ti are respectively shifted by +5 MeV, 10 MeV, 15 MeV and 20 MeV for display purposes. From Ref. [29].

confirmed for heavier isotopes. This is shown in the left panel of Fig. 10, which compares induced and full Hamiltonians for Ca isotopes. The trend of the binding energies is predicted incorrectly by the induced 3NFs alone (blue dashed line), which generate a wrong slope and exaggerate the kink at ^{40}Ca. This problem is fully amended by the inclusion of NNLO chiral 3NFs (red full line). However, the latter introduce additional attraction that results in a systematic over binding of ground-states throughout the whole chain. Analogous results for Ar, K, Sc and Ti isotopic chains are shown in the left panel for full 3NFs and lead to the same conclusion regarding the role of the initial chiral 3NF: it provides the correct trend of binding energies but it generates a rather constant over binding throughout this mass region.

The systematic of S_{2n} obtained with the NN plus full 3NF is displayed in Fig. 11 along Ar, K, Ca, Sc and Ti isotopic chains, up to N=32. When the neutron chemical potential lies within the *pf* shell, predicted S_{2n} reproduce the experiment to good accuracy. The results are quite remarkable considering that NN+3N chiral interactions have been fitted solely to few-body data up to $A = 4$. Still, the quality slightly deteriorates as the proton chemical potential moves down into the *sd* shell, that is, going from Ca to K and Ar elements. The increasing underestimation of the S_{2n} is consistent with a too large gap between proton *sd* and *pf* major shells, which prevents quadrupole neutron-proton correlations to switch on when the two chemical potentials sit on both sides of the gap. This is supported by the results found for NN-N^{3}LO potential in Refs. [9, 81] and in Fig. 9. The too large jump of the S_{2n} between N=20 and N=22 is visible for all elements and becomes particularly pronounced as one moves away from the proton magic ^{40}Ca nucleus where the experimental jump is progressively washed out. At N=18, the situation deteriorates going from ^{38}Ca to ^{39}Sc and ^{40}Ti, when the proton chemical potential moves up into the *pf* shell (but not going to ^{37}K and ^{36}Ar). This is again consistent with an exaggerated shell gap between *sd* and *pf* shells. These finding suggest that repulsive effects form higher chiral cutoffs or improved 3NFs may become important in heavy nuclei.

We note that the present calculations may be sensible to the truncations in three-body space performed during the SRG evolution. Nevertheless, the error introduced is expected to be small and only at large values of N for isotopes around Ca [22].

7. Conclusions

This talk reviewed the recent developments in ab-initio calculations of finite nuclei based on the self-consistent Green's function method. The SCGF formalism offers important opportunities in advancing the theory of exotic isotopes by providing direct theoretical links between nuclear structure and reactions. Such calculations yield both binding energies and the single particle spectral function, which is probed in most nuclear structure experiments. At the same time they provide a path to generate optical potentials from first principles and at intermediate energies (up to $\approx$100 MeV), where ab-initio methods are currently lacking.

Successful ab-initio applications based on SCGF have become possible very recently following a series of technical advances [8, 9, 27, 14] and the availability of chiral interactions evolved to low-momentum scales [17, 18]. Of particular relevance are the recent extension to three- and many-body forces [21, 51, 14] and the introduction of the Gorkov SCGF formalism that allows for the first time calculations of semi-magic nuclei through large portions of the nuclear chart [26, 27, 28]. Applications to scattering have so far been limited to pilot studies of proton-nucleus scattering [30] and microscopic optical potentials [2].

The SCGF self-energies were calculated for ^{40}Ca, ^{48}Ca and ^{60}Ca and compared with potentials from the DOM that were obtained from fitting elastic-scattering and bound-state data. The microscopic FRPA results explain many features of the empirical DOM potentials and provides several suggestions to improve their functional, including the exploitation of parity and angular momentum dependence, which stem from the different fillings of core orbits. The non-locality of the calculated self-energies has also suggested further developments of the DOM potential [35]. The NN tensor force has also been seen to strongly enhance the absorption in elastic scattering at energies above 20-30 MeV.

The recent ab-initio calculations of Ref. [14, 13, 29] nicely validate chiral Hamiltonians within a wide range of masses. In particular, these prove the capability of the leading order initial 3NFs (up to NNLO) to predict the trend of binding energies along full chains of open shell isotopes and at large neutron-proton asymmetry. To our knowledge, Ref. [29] provides for the first time a study of several adjacent isotopic chains based on first principles. These results make evident that presently available interactions generate a systematic over binding of $\approx$1 MeV/A throughout the Ca region and this is further confirmed for heavier closed shell nuclides in Ref. [22]. Overall, this indicates that repulsive effects form higher chiral cutoffs or from four-body interactions may become important in heavy nuclei. Improved 3NFs, at N^{3}LO and beyond, may also have an impact on improving this situation.

To date, the ab-initio SCGF is a very promising approach which is being extended in scope. A most compelling challenge for the coming future will be to provide reliable (and converging) microscopic calculations of optical potentials, so that these can be exploited in the analysis of nuclear structure experiments. Other possible directions can involve linking the SCGF formalism to the phenomenological shell model. In this respect, the ab-initio calculations of effective interactions is an application of current interest [82, 83] that has already been employed in approximate form within the SCGF formalism [81]. A consistent approach can also be used to extract corresponding effective charges [84].

Acknowledgments

It is a pleasure to thank A. Carbone, A. Cipollone, M. Degroote, W. H. Dickhoff, T. Duguet, M. Hjorth-Jensen, P. Navrátil, A. Polls, A. Rios-Huguet, V. Somà, D. Van Neck and S. Waldecker for several discussions and for all the fruitful collaborations that led to the above results. This work was supported by the United Kingdom Science and Technology Facilities Council (STFC) under Grants No. ST/I003363/1, ST/J000051/1 and ST/L005816/1. Calculations were performed in part using HPC resources from the DiRAC Data Analytic system at the University of Cambridge (BIS National E-infrastructure capital grant No. ST/K001590/1 and STFC grants No. ST/H008861/1, ST/H00887X/1, and ST/K00333X/1).

References

[1] Erler J, Birge N, Kortelainen M, Nazarewicz W, Olsen E, Perhac A M and Stoitsov M 2012 *Nature* **486** 509
[2] Waldecker S, Barbieri C and Dickhoff W H 2011 *Phys. Rev.* C **84** 034616
[3] Hagen G, Papenbrock T, Dean D J and Hjorth-Jensen M 2010 *Phys. Rev.* C **82** 034330
[4] Binder S, Langhammer J, Calci A, Navrátil P and Roth R 2013 *Phys. Rev.* C **87** 021303
[5] Tsukiyama K, Bogner S K and Schwenk A 2011 *Phys. Rev. Lett.* **106** 222502
[6] Hergert H, Bogner S K, Binder S, Calci A, Langhammer J, Roth R and Schwenk A 2013 *Phys. Rev.* C **87** 034307
[7] Dickhoff W H and Barbieri C 2004 *Prog. Part. Nucl. Phys.* **52** 377
[8] Barbieri C, Van Neck D and Dickhoff W H 2007 *Phys. Rev.* A **76** 052503
[9] Barbieri C and Hjorth-Jensen M 2009 *Phys. Rev.* C **79** 064313
[10] Otsuka T, Suzuki T, Holt J D, Schwenk A and Akaishi Y 2010 *Phys. Rev. Lett.* **105** 032501
[11] Holt J D, Otsuka T, Schwenk A and Suzuki T 2012 *J. Phys.* G **39** 085111
[12] Hagen G, Hjorth-Jensen M, Jansen G, Machleidt R and Papenbrock T 2012 *Phys. Rev. Lett.* **108** 242501
[13] Hergert H, Binder S, Calci A, Langhammer J and Roth R 2013 *Phys. Rev. Lett.* **110** 242501
[14] Cipollone A, Barbieri C and Navrátil P 2013 *Phys. Rev. Lett.* **111** 062501
[15] Lähde T A, Epelbaum E, Krebs H, Lee D, Meißner U G and Rupak G 2014 *Phys. Lett.* B **732** 110
[16] Epelbaum E, Krebs H, Lähde T A, Lee D, Meißner U G and Rupak G 2014 *Phys. Rev. Lett.* **112** 102501
[17] Jurgenson E D, Navrátil P and Furnstahl R J 2009 *Phys. Rev. Lett.* **103** 082501
[18] Roth R, Binder S, Vobig K, Calci A, Langhammer J and Navrátil P 2012 *Phys. Rev. Lett.* **109** 052501
[19] Hagen G, Papenbrock T, Dean D J, Schwenk A, Nogga A, Włoch M and Piecuch P 2007 *Phys. Rev.* C **76** 034302
[20] Binder S, Piecuch P, Calci A, Langhammer J, Navrátil P and Roth R 2013 *Phys. Rev.* C **88** 054319
[21] Carbone A, Cipollone A, Barbieri C, Rios A and Polls A 2013 *Phys. Rev.* C **88** 054326
[22] Binder S, Langhammer J, Calci A and Roth R 2013 (*Preprint* arXiv:1312.5685)
[23] Jansen G R, Hjorth-Jensen M, Hagen G and Papenbrock T 2011 *Phys. Rev.* C **83** 054306
[24] Somà V, Duguet T and Barbieri C 2011 *J. Phys.: Conf. Ser.* **321** 012039
[25] Somà V, Duguet T and Barbieri C 2012 *J. Phys.: Conf. Ser.* **337** 012001
[26] Somà V, Duguet T and Barbieri C 2011 *Phys. Rev.* C **84** 064317
[27] Somà V, Barbieri C and Duguet T 2013 *Phys. Rev.* C **87** 011303
[28] Somà V, Barbieri C and Duguet T 2014 *Phys. Rev.* C **89** 024323
[29] Somà V, Cipollone A, Barbieri C, Navrátil P and Duguet T 2014 *Phys. Rev.* C **89** 061301
[30] Barbieri C and Jennings B K 2005 *Phys. Rev.* C **72** 014613
[31] Hagen G, Papenbrock T and Hjorth-Jensen M 2010 *Phys. Rev. Lett.* **104** 182501
[32] Navrátil P and Quaglioni S 2012 *Phys. Rev. Lett.* **108** 042503
[33] Baroni S, Navrátil P and Quaglioni S 2013 *Phys. Rev. Lett.* **110** 022505
[34] Charity R J, Sobotka L G and Dickhoff W H 2006 *Phys. Rev. Lett.* **97** 162503
[35] Mahzoon M H, Charity R J, Dickhoff W H, Dussan H and Waldecker S J 2014 *Phys. Rev. Lett.* **112** 162503
[36] Dussan H, Waldecker S J, Dickhoff W H, Müther H and Polls A 2011 *Phys. Rev.* C **84** 044319
[37] Fetter A L and Walecka J D 1971 *Quantum Theory of Many-Particle System* 1st ed (McGraw-Hill, New York)
[38] Schirmer J 1982 *Phys. Rev.* A **26** 2395
[39] Schirmer J, Cederbaum L S and Walter O 1983 *Phys. Rev.* A **28** 1237
[40] Barbieri C and Dickhoff W H 2001 *Phys. Rev.* C **63** 034313
[41] Barbieri C, Giusti C, Pacati F D and Dickhoff W H 2004 *Phys. Rev.* C **70** 014606
[42] Middleton D, Annand J, Barbieri C, Barneo P, Blok H, Bhm R, Distler M, Friedrich J, Giusti C, Glazier D, Grabmayr P, Hehl T, Heim J, Hesselink W, Jans E, Kohl M, Lapikás L, MacGregor I, Martin I, McGeorge J, Merkel H, Moschini F, Rosner G, Seimetz M, de Vries H, Walcher T, Watts D and Zihlmann B 2006 *Eur. Phys. J.* A **29** 261
[43] Middleton D, Annand J, Barbieri C, Giusti C, Grabmayr P, Hehl T, MacGregor I, Martin I, McGeorge J, Moschini F, Pacati F, Schwamb M and Watts D 2010 *Eur. Phys. J.* A **43** 137
[44] Trofimov A B and Schirmer J 2005 *The Journal of Chemical Physics* **123** 144115
[45] Carbone A 2014 Ph.D thesis, Universitat de Barcelona, Spain

[46] Dickhoff W H and Van Neck D 2008 *Many-body theory exposed!* 2nd ed (World Scientific Publishing, London)
[47] Hagen G, Hjorth-Jensen M, Jansen G R, Machleidt R and Papenbrock T 2012 *Phys. Rev. Lett.* **109** 032502
[48] Somà V and Bożek P 2008 *Phys. Rev.* C **78** 054003
[49] Somà V and Bożek P 2009 *Phys. Rev.* C **80** 025803
[50] Hebeler K, Bogner S K, Furnstahl R J, Nogga A and Schwenk A 2011 *Phys. Rev.* C **83** 031301
[51] Carbone A, Polls A and Rios A 2013 *Phys. Rev.* C **88** 044302
[52] Grangé P, Lejeune A, Martzolff M and Mathiot J F 1989 *Phys. Rev.* C **40** 1040
[53] Epelbaum E, Hammer H W and Meissner U G 2009 *Rev. Mod. Phys.* **81** 1773
[54] Ortiz J V 2013 *Wiley Interdisciplinary Reviews: Computational Molecular Science* **3** 123
[55] Galitskii V M and Migdal A B 1958 *Sov. Phys.-JETP* **7** 96
[56] Koltun D S 1972 *Phys. Rev. Lett.* **28** 182
[57] Gorkov L P 1958 *Sov. Phys. JETP* **7** 505
[58] Capuzzi F and Mahaux C 1996 *Annals of Physics* **245** 147
[59] Escher J and Jennings B K 2002 *Phys. Rev.* C **66** 034313
[60] Wiringa R B, Stoks V G J and Schiavilla R 1995 *Phys. Rev.* C **51** 38
[61] Charity R J, Mueller J M, Sobotka L G and Dickhoff W H 2007 *Phys. Rev.* C **76** 044314
[62] Mueller J M, Charity R J, Shane R, Sobotka L G, Waldecker S J, Dickhoff W H, Crowell A S, Esterline J H, Fallin B, Howell C R, Westerfeldt C, Youngs M, Crowe B J and Pedroni R S 2011 *Phys. Rev.* C **83** 064605
[63] Mahaux C and Sartor R 1991 *Adv. Nucl. Phys.* **20** 1
[64] Greenlees G W, Pyle G J, and Tang Y C 1968 *Phys. Rev.* **171** 1115
[65] Dickhoff W H, Van Neck D, Waldecker S J, Charity R J and Sobotka L G 2010 *Phys. Rev.* C **82** 054306
[66] Rohe D, Armstrong C S, Asaturyan R, Baker O K, Bueltmann S, Carasco C, Day D, Ent R, Fenker H C, Garrow K, Gasparian A, Gueye P, Hauger M, Honegger A, Jourdan J, Keppel C E, Kubon G, Lindgren R, Lung A, Mack D J, Mitchell J H, Mkrtchyan H, Mocelj D, Normand K, Petitjean T, Rondon O, Segbefia E, Sick I, Stepanyan S, Tang L, Tiefenbacher F, Vulcan W F, Warren G, Wood S A, Yuan L, Zeier M, Zhu H and Zihlmann B (E97-006 Collaboration) 2004 *Phys. Rev. Lett.* **93** 182501
[67] Barbieri C and Lapikás L 2004 *Phys. Rev.* C **70** 054612
[68] Barbieri C, Rohe D, Sick I and Lapikás L 2005 *Phys. Lett.* B **608** 47
[69] Lane A M 1962 *Nucl. Phys.* **35** 676
[70] Otsuka T, Suzuki T, Fujimoto R, Grawe H and Akaishi Y 2005 *Phys. Rev. Lett.* **95** 232502
[71] Entem D R and Machleidt R 2003 *Phys. Rev.* C **68** 041001
[72] Machleidt R and Entem D R 2011 *Phys. Rep.* **503** 1
[73] Navrátil P 2007 *Few-Body Systems* **41**(3-4) 117
[74] Audi G, Wapstra A and Thibault C 2003 *Nuc. Phys. A* **729** 337
[75] Jurado B, Savajols H, Mittig W, Orr N, Roussel-Chomaz P, Baiborodin D, Catford W, Chartier M, Demonchy C, Dlouhý Z, Gillibert A, Giot L, Khouaja A, Lépine-Szily A, Lukyanov S, Mrazek J, Penionzhkevich Y, Pita S, Rousseau M and Villari A 2007 *Phys. Lett.* B **649** 43
[76] Gaudefroy L, Mittig W, Orr N A, Varet S, Chartier M, Roussel-Chomaz P, Ebran J P, Fernández-Domínguez B, Frémont G, Gangnant P, Gillibert A, Grévy S, Libin J F, Maslov V A, Paschalis S, Pietras B, Penionzhkevich Y E, Spitaels C and Villari A C C 2012 *Phys. Rev. Lett.* **109** 202503
[77] Barbieri C, Cipollone A, Somà V, Duguet T and Navrátil P 2012 *Summary Report of EURISOL Topical and Town Meetings, Lisbon, October 15-19, 2012.* (*Preprint* `arXiv:1211.3315`)
[78] Wang M, Audi G, Wapstra A, Kondev F, MacCormick M, Xu X and Pfeiffer B 2012 *Chinese Physics C* **36** 1603
[79] Gallant A T, Bale J C, Brunner T, Chowdhury U, Ettenauer S, Lennarz A, Robertson D, Simon V V, Chaudhuri A, Holt J D, Kwiatkowski A A, Mané E, Menéndez J, Schultz B E, Simon M C, Andreoiu C, Delheij P, Pearson M R, Savajols H, Schwenk A and Dilling J 2012 *Phys. Rev. Lett.* **109** 032506
[80] Wienholtz F, Beck D, Blaum K, Borgmann C, Breitenfeldt M *et al.* 2013 *Nature* **498** 346
[81] Barbieri C 2009 *Phys. Rev. Lett.* **103** 202502
[82] Bogner S, Hergert H, Holt J, Schwenk A, Binder S *et al.* 2014 (*Preprint* `arXiv:1402.1407`)
[83] Jansen G, Engel J, Hagen G, Navratil P and Signoracci A 2014 (*Preprint* `arXiv:1402.2563`)
[84] Barbieri C and Otsuka T In preparation

Condensation of bosons with Rashba-Dresselhaus spin-orbit coupling

Gordon Baym[1] **and Tomoki Ozawa**[1,2]

[1]Department of Physics, University of Illinois, 1110 W. Green Street, Urbana, IL 61801 U.S.A.
[2]INO-CNR BEC Center and Dipartimento di Fisica, Università di Trento, I-38123 Povo, Italy

E-mail: gbaym@illinois.edu, ozawa@science.unitn.it

Abstract. Cold atomic Bose-Einstein systems in the presence of simulated Rashba-Dresselhaus spin-orbit coupling exhibit novel physical features. With pure in-plane Rashba coupling the system is predicted in Bogoliubov-Hartree-Fock to have a stable Bose condensate below a critical temperature, even though the effective density of states is two-dimensional. In addition the system has a normal state at all temperatures. We review here the new physics when the system has such spin-orbit coupling, and discuss the nature of the finite temperature condensation phase transition from the normal to condensed phases.

1. Introduction

The ability to simulate gauge fields in neutral ultracold atom laboratory systems, reviewed in [1], has opened a wide variety of opportunities to create and study novel physical systems. Not only can one create artificial Abelian gauge fields, which provide analogs of conventional magnetic fields, one can in principle produce non-Abelian gauge fields; the prospect of simulating lattice gauge theories, reviewed in [2], with potential applications to quantum chromodynamics, is a remarkable new direction.

Among the simplest such non-Abelian fields are artificially induced spin-orbit couplings of the linear momentum, $\mathbf{p}$, of a particle to its "spin" $\boldsymbol{\sigma}$ – which in cold atom systems is more generally a hyperfine multiplet (reviewed in [3]). Of particular interest is the in-plane Rashba-Dresselhaus interaction [4, 5] with two hyperfine states, (a, b), with the single particle Hamiltonian in the (a, b) spinor basis,

$$H_{RD} = \frac{p^2 + \kappa^2}{2m} + \frac{\kappa}{m}(\sigma_x p_x + \eta\sigma_y p_y); \tag{1}$$

here m is the atomic mass, κ is the coupling constant, σ_x and σ_y are Pauli matrices, and η between 0 and 1, measures the anisotropy of the interaction; $\eta = 1$ corresponds to a pure Rashba interaction, and $\eta = 0$ to an equal mixture of Rashba and Dresselhaus couplings. This latter situation, together with a Zeeman coupling term $\propto \sigma_z$, was first realized experimentally by Lin et al. [6] and then by Zhang et al. [7] in bosonic systems, and P. Wang et al. in fermions [8].

Here we focus on two component bosonic systems with a pure Rashba interaction, $\eta = 1$, in which the two body interactions between particles in the same hyperfine states are described by short range pseudopotentials $v_{aa}(\mathbf{r}) = g_{aa}\delta(\mathbf{r})$, and $v_{bb}(\mathbf{r}) = g_{bb}\delta(\mathbf{r})$, and the interaction between

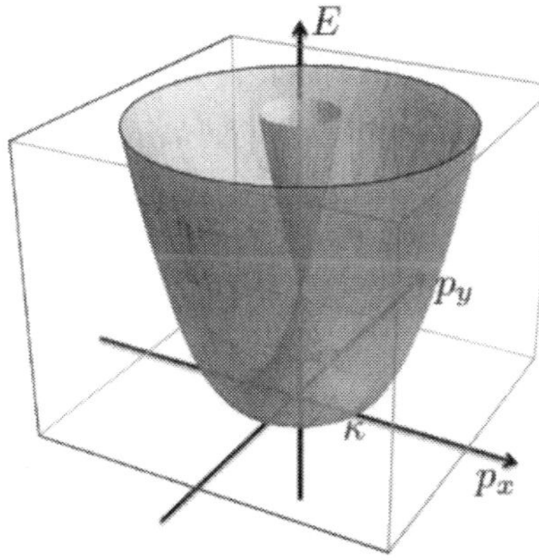

Figure 1. Spectrum of the single particle Hamiltonian with Rashba in-plane spin-orbit coupling.

different hyperfine states by $v_{ab}(\mathbf{r}) = g_{ab}\delta(\mathbf{r})$. As we shall see, this system presents intriguing questions not encountered in simple Bose-Einstein condensation. At $\eta = 1$, the single particle Hamiltonian can be written as

$$H_R = \frac{(\boldsymbol{p}_\perp + \kappa\boldsymbol{\sigma})^2 + p_z^2}{2m}. \tag{2}$$

where $\mathbf{p}_\perp \equiv (p_x, p_y)$, $\boldsymbol{\sigma} \equiv (\sigma_x, \sigma_y)$. The single particle eigenstates are naturally constructed by diagonalizing the Hamiltonian, in which basis the single particle eigenenergies become

$$\epsilon_{\mathbf{p}} = \frac{(p_\perp \pm \kappa)^2 + p_z^2}{2m}; \tag{3}$$

the ground state branch $(-)$ has a circle of minima at $p_\perp = \kappa$, while the excited branch $(+)$ is connected to the ground state branch by a Dirac point at $p_\perp = 0$. See Fig. 1.

At zero temperature the system can be condensed in any linear combination of states on the circle. The preferred condensates depend, however, on the interactions [9]. For $g_{aa} = g_{bb} > g_{ab}$, i.e., greater repulsion between like than different hyperfine states, the preferred condensate within mean-field theory is a plane wave state, e.g.,

$$\Psi_p(\mathbf{r}) = \sqrt{\frac{n}{2}} e^{i\kappa x} \begin{pmatrix} 1 \\ -1 \end{pmatrix}. \tag{4}$$

Here the wave vector of the condensate, $\boldsymbol{\kappa} = \kappa\hat{x}$, points along the x axis, but it could point anywhere on the circle of minima. In the opposite regime, $g_{aa} < g_{ab}$ the preferred condensate is a striped state, e.g.,

$$\Psi_s(\mathbf{r}) = \sqrt{n} \begin{pmatrix} \cos\kappa x \\ -i\sin\kappa x \end{pmatrix}, \tag{5}$$

where again the wavevectors in the condensate, $\pm\kappa\hat{x}$, are chosen arbitrarily to lie along the x axis. Condensate with a larger number of wavevectors lying on the degenerate circle have higher energy.

In the remainder of this paper we limit the discussion, for simplicity, to the plane wave condensate with isotropic interactions, $g_{aa} = g_{bb} = g_{ab}$. We first ask whether the condensate is stable against quantum fluctuations at zero temperature, and thermal fluctuations. It is not immediately obvious that the system can have a stable condensate, since the density of single particle states in the low energy limit is two dimensional in nature, $\sim m\kappa/2\pi$. Thus one could imagine that, as in two-dimensional Berezhinskii-Kostelitz-Thouless (BKT) systems [10], the condensate is unstable owing to the number of excited particles diverging. As we shall see, the condensate is in fact stabilized by interactions. A second important feature of this system is that the normal state is not kinematically forbidden, as in usual Bose-Einstein condensation below the transition temperature. We also ask for the nature of the finite temperature phase transition to the normal state.

2. Stability of the Bose-Einstein condensate

Unlike in BKT systems, condensates at zero temperature are stabilized by the interparticle interactions [11]. As we now discuss, the number fluctuations at $T = 0$ are finite and of order $\sqrt{(mg)^3 n} \sim \sqrt{na^3}$ in weak coupling, similar to usual Bose gases in the Bogoliubov approximation, where we write $g = 4\pi\hbar^2 a/m$ [owing to renormalization effects in the presence of spin-orbit interactions, the length a is not simply related to the actual scattering length [12]]. The modes of the condensate, and the corresponding condensate depletion, are readily derived within Bogoliubov mean-field from the single particle matrix Green's function with anomalous components,

$$\mathbf{G}(\mathbf{q}, t_1 - t_2) \equiv -i\langle T\left(\Psi_{\mathbf{q}}(t_1)\Psi^\dagger_{\mathbf{q}}(t_2)\right)\rangle, \tag{6}$$

where

$$\Psi_{\mathbf{q}}(t) \equiv \left(\psi_{-,\boldsymbol{\kappa}+\mathbf{q}}(t), \psi^\dagger_{-,\boldsymbol{\kappa}-\mathbf{q}}(t), \psi_{+,\boldsymbol{\kappa}+\mathbf{q}}(t), \psi^\dagger_{+,\boldsymbol{\kappa}-\mathbf{q}}(t)\right); \tag{7}$$

here $\psi_{\pm,\mathbf{p}} \equiv (\mathbf{a_p} \pm \mathbf{b_p})/\sqrt{2}$, with $\mathbf{a_p}$ and $\mathbf{b_p}$ removing particles of momentum $\mathbf{p}$ from the a and b hyperfine states, respectively. In the Bogliubov approximation, where $\psi_{-,\boldsymbol{\kappa}} \to \sqrt{N_0}$ with $N_0 = n_0 V$ the number of condensate particles, and with Hartree-Fock energies included, one finds the Fourier transform in time of the inverse Green's function,

$$\mathbf{G}^{-1}(\mathbf{q}, z) = \begin{pmatrix} z - A & -gn_0 & i(\kappa/m)q_y & 0 \\ -gn_0 & -z - A & 0 & i(\kappa/m)q_y \\ -i(\kappa/m)q_y & 0 & z - B & 0 \\ 0 & -i(\kappa/m)q_y & 0 & -z - D \end{pmatrix}, \tag{8}$$

where

$$\begin{aligned} A(\mathbf{q}) &\equiv q^2/2m - \mu + g(2n_0 + 2n_- + n_+) \\ B(\mathbf{q}) &\equiv (2\boldsymbol{\kappa} + \mathbf{q})^2/2m - \mu + g(n_0 + n_- + 2n_+) \\ D(\mathbf{q}) &\equiv B(-\mathbf{q}).. \end{aligned} \tag{9}$$

In leading order, the chemical potential is $\mu_0 = \partial\langle\mathcal{H}\rangle/\partial N_0 = gn_0 + 2gn_- + gn_+$, where

$$n_\mp = \frac{1}{V}\sum_{\mathbf{p}\neq\boldsymbol{\kappa}}\langle\psi^\dagger_{\mp,\mathbf{p}}\psi_{\mp,\mathbf{p}}\rangle \tag{10}$$

are the number of non-condensate particles in the $(-)$ and $(+)$ states.

The modes of the condensate, found as zeroes of $\det\mathbf{G}^{-1}(\mathbf{q}, z)[= \det\mathbf{G}^{-1}(-\mathbf{q}, -z)]$ come in pairs: two of positive and two of negative frequency, for each $\mathbf{q}$. One excitation is

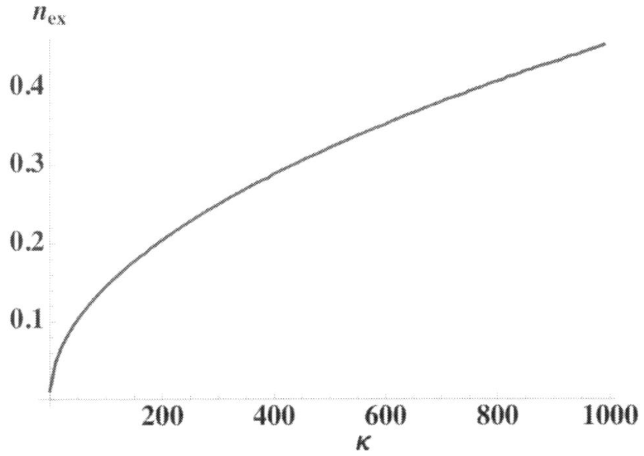

Figure 2. The number of excited particles, in units of $(2mgn_0)^{3/2}$ as a function of the spin-orbit coupling strength κ in units of $\sqrt{2mgn_0}$.

gapless as $\mathbf{q} \to 0$, and the other is gapless as $\mathbf{q} \to -2\boldsymbol{\kappa}$. With strong spin-orbit coupling, $\kappa^2/m \gg g|n_+ - n_-| \equiv g|\Delta n|$, the spectrum to leading order for $|\mathbf{q}| \ll \kappa$ is

$$\epsilon_1(\mathbf{q}) \approx \sqrt{2gn_0}\left[\frac{q_x^2+q_z^2}{2m} + \frac{q_y^2}{4\kappa^2}\left(g\Delta n + \frac{q_y^2}{2m}\right)\right]^{1/2}; \tag{11}$$

since $q_y^2/2m$ is generally larger than $g|\Delta n|$ in typical experimental setups [6], the dispersion is essentially quadratic for $q_x = q_z = 0$. Similarly, the gapless spectrum for $\mathbf{q}' \equiv \mathbf{q} + 2\boldsymbol{\kappa} \ll \kappa$ is quadratic and free-particle like,

$$\epsilon_2(\mathbf{q}') = \frac{q_x'^2 + q_z'^2}{2m} + \frac{gn_0}{\kappa^2/m + gn_0}\frac{q_y'^2}{4m}. \tag{12}$$

The condensate depletion at $T = 0$ is given by

$$n_{ex} = n_- + n_+ = i\int \frac{dz d^3q}{(2\pi)^4}\left(G_{11}(\mathbf{q}, z) + G_{33}(\mathbf{q}, z)\right), \tag{13}$$

with the z integration contour surrounding the negative poles in the positive sense; the integral converges in the ultraviolet. Figure 2 plots the depletion evaluated numerically. Generally $n_- \gg n_+$ and $n_{ex} \sim n_0\sqrt{(mg)^3 n_0} \sim n_0\sqrt{n_0 a^3}$. The leading behavior in κ is $n_{ex} \sim \sqrt{\kappa}$, as shown analytically in Ref. [13].

The shift in the ground state energy arising from fluctuations, $\Delta E \equiv E - gn^2/2$, is of order $gn(2mgn)^{3/2}$ for weak coupling [11]. The energy decreases with increasing κ, and ΔE changes from positive to negative at $\kappa \sim 0.6\sqrt{2mgn_0}$.

We next show that the condensate remains stable under thermal fluctuations. In terms of the mode energies (11) and (12), the infrared contribution to the condensate depletion is

$$n_{ex} \sim T\int \frac{d^3q}{(2\pi)^3}\left(\frac{gn_0}{\epsilon_1(\mathbf{q})^2} + \frac{1}{\epsilon_2(\mathbf{q})} + C\right), \tag{14}$$

where C is a constant as $\mathbf{q} \to 0$. The integral converges in the infrared, and condensate depletion does not destroy the Bose-Einstein condensate at finite temperature.

Unlike in usual weakly interacting Bose systems in three dimensions in the absence of spin-orbit coupling, the normal state ($n_0 = 0$) of the system is kinematically allowed at low temperatures. Is the normal state preferred over the condensate at low temperature? To answer this we evaluate the Helmholtz free energy density of the normal state,

$$\mathcal{F} = \mu n - \frac{3}{4} g n^2 + \frac{1}{\beta V} \sum_{\mathbf{p}} \left\{ \ln\left(1 - e^{-\beta\xi_-(\mathbf{p})}\right) + \ln\left(1 - e^{-\beta\xi_+(\mathbf{p})}\right) \right\}, \tag{15}$$

where $\xi_\pm(\mathbf{p}) \equiv \{(p_\perp \pm \kappa)^2 + p_z^2\}/(2m) - \mu + 3gn/2$; the chemical potential is determined by the number equation $n = (1/V)\sum_{\mathbf{p}} \{f(\xi_-(\mathbf{p})) + f(\xi_+(\mathbf{p}))\}$, where $f(x) \equiv 1/(e^{\beta x} - 1)$. To a good approximation, with $\Delta\mu = \mu - 3gn/2$,

$$n(\mu) \approx -\frac{m\kappa T}{2\pi} \ln\left(-\Delta\mu/T\right), \tag{16}$$

which is essentially the mean field result for a two-dimensional BKT system [10]. Unlike for free bosons in three dimensions, there always exists a μ which satisfies the number equation, thus the non-condensed state is not kinematically forbidden at any non-zero temperature.

In the normal state, as $T \to 0$, $\mu \to 3gn/2$, and $\mathcal{F} \to 3gn^2/4$. (Note that the point $T = 0$, $\mu = 3gn/2$ is highly singular.) This energy is larger than the ground state energy in the condensed phase, $(gn^2/2)(1+\mathcal{O}(\sqrt{(2mg)^3 n}))$, indicating that at sufficiently low temperature, the condensate is energetically preferred. However, this calculation does not exclude the possibility that the system has a correlated normal state with energy lower than that of the condensate, and thus the system would not condense.

3. Phase transition between the normal and condensed states

As we see, both the normal state and condensed state at the present level of mean-field calculation are stable at zero temperature, where the condensate, of lower energy, is favored. However, at sufficiently high temperature the condensate goes away. Thus we now ask, what is the nature of the phase transition from condensate to normal; first or second order? See Fig. 3. As we find in mean-field Bogoliubov-Hartree-Fock, the transition is first order [14].

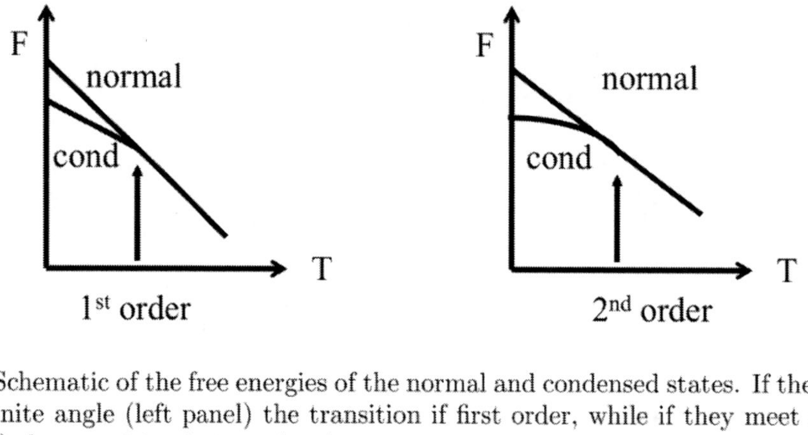

Figure 3. Schematic of the free energies of the normal and condensed states. If the free energies meet at a finite angle (left panel) the transition if first order, while if they meet at zero angle (right panel) the transition is second order.

This first order transition is distinct from the spurious first order phase transition one finds within the Bogoliubov-Hartree-Fock in usual Bose gases [15] (see also [16]). There the spurious transition is driven by order parameter fluctuations which lead to a density of particles excited out of the condensate near the transition temperature,

$$n_{ex}(n_0) = \int \frac{d^3q}{(2\pi)^3}\left[\left(\frac{1}{e^{\beta E_q}-1}+\frac{1}{2}\right)\frac{q^2/2m+gn_0}{E_q}-\frac{1}{2}\right] = n_{ex}(0) - \frac{1}{\lambda^2}\sqrt{mgn_0} + \mathcal{O}(n_0^{3/2}); \quad (17)$$

here $\lambda = \sqrt{2\pi/mT}$ is the thermal wavelength and $E_q \equiv \sqrt{(q^2/2m)^2 + gn_0q^2/m}$. The $\sqrt{n_0}$ term, which is non-analytic in the condensate fraction, leads to a $-n_0^{3/2}$ term in the free energy, and as a consequence the system, as cooled from above T_c^0 – the transition temperature of the ideal Bose gas – undergoes a first order transition, with a jump in n_0, at a temperature $T > T_c^0$ (cf. [16]).

This first order transition is removed when correctly determining the critical behavior at the phase transition. The issue is that the point of vanishing scattering length $a = mg/4\pi \to 0$ and deviation from the transition temperature $t = (T_c - T)/T_c \to 0$ is highly singular. The Bogoliubov-Hartree-Fock approximation is valid in the limit $a \to 0$ at non-zero t, while to study the order of the transition, one needs $t \to 0$ at non-zero a. These two limits are not equivalent. Near the transition in a weakly interacting gas, one has the scaling structure for the condensate fraction in terms of a dimensionless scaling function $h(x)$:

$$\frac{n_0}{n} = \frac{a}{\lambda}h\left(\frac{t\lambda}{a}\right), \quad (18)$$

where $h(x\to\infty) \sim x$, the non-interacting gas limit, and for $x \to 0$, as one approaches the transition at finite a, one has $h \sim x^{2\beta}$, with $\beta \approx 1/3$, the critical exponent for the order parameter.

The spurious first order transition does not occur in the presence of spin-orbit coupling. In the simple Bose gas the relevant momentum scale is $\sqrt{an_0}$. However, Rashba spin-orbit coupling introduces a second scale, κ, and as a consequece the density of excited particles is analytic in n_0 for $\kappa^2 \gg an_0$.

To determine the nature of the phase transition, we calculate the free energy $\mathcal{F}$ in terms of the Green's function using,

$$\frac{\partial \mathcal{F}}{\partial n_0} = -\mu + gn_0 - gT\sum_\nu \int \frac{d^3q}{(2\pi)^3}\left(2G_{11}(\mathbf{p},z_\nu) + G_{33}(\mathbf{p},z_\nu) + \frac{1}{2}\left(G_{21}(\mathbf{p},z_\nu) + G_{12}(\mathbf{p},z_\nu)\right)\right). \quad (19)$$

Expanding in small $gn_0 \ll \kappa^2/m$, as detailed in [14], one obtains

$$\frac{\partial \mathcal{F}}{\partial n_0} = -\mu + \frac{3}{2}gn(\mu) + Xgn_0 + Y(gn_0)^2 + \dots, \quad (20)$$

where

$$X(\mu,T) \equiv 1 - \frac{4m^2gT}{\kappa}\alpha\left(\frac{\Delta\mu}{\epsilon_\kappa}\right), \quad Y(\mu,T) \equiv \frac{4m^2gT}{\kappa\epsilon_\kappa}\beta\left(\frac{\Delta\mu}{\epsilon_\kappa}\right), \quad (21)$$

$\epsilon_\kappa \equiv \kappa^2/2m$, and $\alpha(x)$ and $\beta(x)$ are dimensionless functions of x, with the asymptotic forms as $x \to 0^-$, $\alpha(x) \simeq -19/32\pi x$ and $\beta(x) \sim 0.16/x^2$, and approaching 0 as $x \to -\infty$.

Integrating with respect to n_0 we have

$$\mathcal{F}(n_0) = \mathcal{F}_\mathrm{n} - \Delta\mu\, n_0 + X\frac{g}{2}n_0^2 + Y\frac{g^2}{3}n_0^3, \quad (22)$$

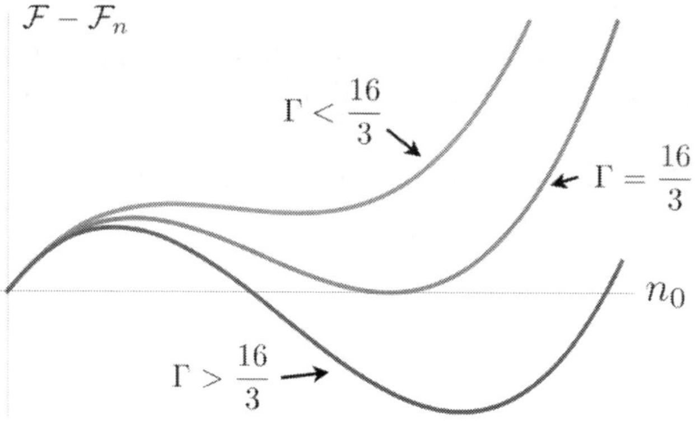

Figure 4. The free energy difference of the normal and condensed phases, showing the first order phase transition from the normal to condensed phase at $\Gamma = 16/3$. The top line is for $T > T_c$; the middle line is at T_c; and the bottom line is for $T < T_c$.

where $\mathcal{F}_{\rm n} = \mathcal{F}(n_0 = 0)$ is the free energy in the normal phase. The coefficients of n_0 and n_0^3 are both positive; however,for given $\mu > 0$, X goes from positive at high T to negative at low T, and decreases continuously with decreasing temperature.

This change in the sign of X drives a first order phase transition, since at sufficiently small T, the two conditions for the transition $\mathcal{F}(n_0) = \mathcal{F}_{\rm n}$ and $\partial\mathcal{F}(n_0)/\partial n_0 = 0$ become simultaneously satisfied at T_c where $X^2 = -16Y\Delta\mu/3$. At this temperature the system undergoes a transition to the condensed phase, as illustrated schematically in Fig. 4. The combination $X^2/(-Y\Delta\mu) \equiv \Gamma$ monotonically decreases with T, as long as $X < 0$. At the transition, n_0 jumps from zero to $|3X/4Yg| > 0$ on the condensate side.

As derived in [14], the transition temperature is given by

$$T_c \approx \frac{2\pi n(\mu)}{m\kappa} \frac{1}{|\ln(2m\kappa g)| + C}, \tag{23}$$

where $C \sim 3.4$. The transition temperature depends linearly on the density, and as expected approaches zero as $g \to 0$. In addition, the jump in the condensate density at the transition is $n_0/n(\mu) \sim 0.32/(|\ln(2m\kappa g)| + C)$, approaching zero as $g \to 0$, and increasing with increasing g.

A deep issue is whether the present transition remains first order at a higher level of approximation. On the one hand, this system is similar to other bosonic systems with continuously degenerate single-particle minima, such as a weak-crystallization model [17] and magnon systems [18], in which condensation transitions are predicted to be first order.

On the other hand, the single particle density of states closely resembles that in a two-dimensional BKT system hinting at a second order transition. In finite geometry the condensate fraction is discontinuous at the BKT transition, but in a macroscopic system correlation corrections change the transition from first order to continuous, as one sees from scaling arguments [10]. To fully address this issue requires scaling or renormalization group analyses; Monte-Carlo calculations of the transition would also be useful.

Several additional factors need to be taken into account in the analysis. The first is the dependence of the order of the transition on the geometry of the order parameter, plane wave, stripes, or with higher symmetry, e.g., triangular. A second is the effect of the renormalization of the interparticle interactions in the presence of spin-orbit coupling [12]

Finally we ask whether one can construct normal (non-condensed) states of lower free energy than the optimal condensed phase. A key feature of the Rashba spin-orbit coupling considered here is the circle of degenerate single particle states, which the normal state does not take particular advantage of. For example. rapidly rotating Bose condensates, develop a set of nearly degenerate (lowest Landau level) single particle excitations; this degeneracy allows the formation of highly correlated normal, or totally fragmented condensates, with low occupation per level, at sufficiently rapid rotation. Whether a similar highly correlated state is better than a Bose-Einstein condensate remains an open question (see, e.g, [19]).

Acknowledgments
The research reported here was supported in part by U.S. National Science Foundation Grants PHY07-070161, PHY09-69790, and PHY13-05891. Author GB is grateful to the Aspen Center for Physics, supported in part by NSF Grant PHY10-66293, where part of this work was carried out. Author TO is supported by the ERC through the QGBE grant and by Provincia Autonoma di Trento. We thank Jason Ho and Markus Holzmann for many helpful insights and discussions.

References

[1] Dalibard J, Gerbier J, Juzeliūnas G, and Öhberg P 2011 *Rev. Mod. Phys.* **83** 1523-43
[2] Wiese U-J 2013 *Ann. d. Physik* **525** 777-96
[3] Galitski V and Spielman I B 2013 *Nature* **494** 49-54
[4] Rashba E I 1960 *Fiz. Tverd. Tela* **2** 1224-38 [1960 *Sov. Phys. Solid State* **2** 1109-22]
[5] Dresselhaus G 1955 *Phys. Rev.* **100**, 580-6
[6] Lin Y-J, Jiménez-García K, and Spielman I B 2011 *Nature* **471** 83-6
[7] Zhang J-Y *et al.* 2012 *Phys. Rev. Lett.* **109** 115301
[8] Wang P, Yu Z-Q, Fu Z, Miao J, Huang L, Chai S, Zhai H, and Zhang J, 2012 *Phys. Rev. Lett.* **109** 095301
[9] Wang C, Gao C, Jian C-M, and Zhai H 2010 *Phys. Rev. Lett.* **105** 160403
[10] Holzmann M, Baym G, Blaizot J-P, and Laloë F 2007 *Proc. Natl. Acad. Sci. USA* **104** 1476-81
[11] Ozawa T and Baym G 2012 *Phys. Rev. Lett.* **109** 025301
[12] Ozawa T and Baym G 2011 *Phys. Rev.* A **84** 043622
[13] Cui X and Zhou Q 2013 *Phys. Rev.* A **87** 031604(R)
[14] Ozawa T and Baym G 2013 *Phys. Rev. Lett.* **110** 085304
[15] Holzmann M and Baym G 2003 *Phys. Rev. Lett.* **90** 040402
[16] Baym G and Grinstein G 1977 *Phys. Rev.* D **15** 2897-912
[17] Brazovskii S A 1975 *Zh. Eksp. Teor. Fiz.* **68** 175-85 [1975 *Sov. Phys. JETP* **41** 85-9]
[18] Jackeli G and Zhitomirsky M E 2004 *Phys. Rev. Lett.* **93** 017201
[19] Sedrakyan T A, Kamenev A, and Glazman L I. 2012 *Phys. Rev.* A **86**, 063639

Shape Resonances in superconducting gaps in a 2DEG at oxide-oxide interface

A Bianconi [1], D Innocenti [1], A Valletta [2], A Perali [3]
[1] RICMASS, Rome International Center for Materials Science Superstripes,
Via dei Sabelli 119A, 00185 Rome, Italy
[2] Department of Physics, Institute for Microelectronics and Microsystems, IMM CNR,
Via del Fosso del Cavaliere 100, 00133 Roma, Italy
[3] Physics Unit, School of Pharmacy, University of Camerino, 62032 Camerino, Italy

E-mail: antonio.bianconi@ricmass.eu;

Abstract. In multiband superconductivity, the case where the single electron hopping between different Fermi surface spots of different symmetry is forbidden by selection rules is recently attracting a large interest. The focus is addressed to superconductivity made of multiple condensates with different symmetry where the chemical potential crosses a 2.5 Lifshitz transition. This can now be investigated experimentally by fine-tuning of the chemical potential in the range of tens meV around a band edge using gate voltage control. We discuss here the case of a superconducting two-dimensional electron gas (2DEG), at the interface between two insulating oxides confined within a slab of 5 nanometers thickness, where the electronic structure is made of subbands generated by quantum size effects. We obtain shape resonances in the superconducting gaps, characterisc gaps to T_c ratios and the BCS-BEC crossover in the upper subband for different pairing strength in the shallow Fermi surface, pointing toward the best configurations for enhanced superconductivity in 2DEG.

1. Introduction

There is growing agreement that advances in quantitative protocols for material design of high temperature superconductors should be based on the control of a) lattice parameters, b) defects distribution and c) tuning of the chemical potential, in heterostructures at atomic limit [1-13]. Since 1993 [1] it has been proposed that it is possible to increase the superconducting critical temperature by selecting an optimal geometrical architecture of composite materials made of atomic layers (or wires or stripes) intercalated by nanoscale spacers. The interest is focusing on tuning the chemical potential near a band edge where "shape resonances" in superconducting gaps appear in multi-condensates superconductivity [1-13].

The "shape resonance" is a type of Fano resonance in the configuration interaction between open and closed scattering channels, well known in atomic and nuclear physics [14-17] which occurs in the pairing processes in multi-condensates superconductors by the tuning the chemical potential near a band edge, as it was first recognized by Blatt [18,19]. It was first shown in 1996 [3-5] that the "shape resonance" gives the maximum of the critical temperature where the chemical potential is tuned above the band edge so that the Fermi energy in the small Fermi surface is of the order of one-two times the the energy scale of the pairing interaction. Therefore the shape resonance in the superconducting gaps gives a maximum Tc where the interacting Fermi liquid in the upper band is in the crossover regime from a Bose-like to a Fermi-like behavior [20-25,10].

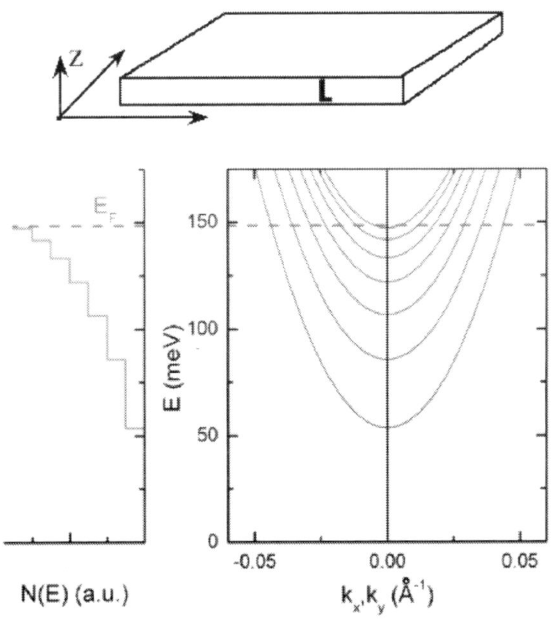

Figure 1. Upper panel: the 2D metallic slab hosting the 2DEG of thickness L. The lower panels from left to right: the left panel shows the total density of states N(E) for the sub-bands where the Fermi level (E_F) of the upper n-th subband is close to band edge. The chemical potential is at 100 meV above the bottom of the first subband. The right panel shows the energy dispersion in each subband in the $k_{//}(k_x,k_y)$ plane. The Fermi wave-vector for electrons in the first subband is near 0.04 Å^{-1}.

Superconductivity in thin metallic films with a thickness at atomic limit has been shown to persist to thicknesses much below the superconducting coherence length and even down to a single layer films [26-32]. Interestingly, in low dimensional superconductors, the superconducting coherence length (ξ) or the magnetic penetration depth (λ) do not define the critical length scale for the destruction of the superconducting state.

Layered complex metal oxides made of different modules provide systems with unique electronic, magnetic, ferroelectric and superconducting properties due to lattice heterogeneneity, valence fluctuations, lattice misfit strain, photoinduced effects, complex phase separation high temperature superconductivity [33-43].

At interfaces between complex oxides, electronic systems with unusual electronic properties can be generated [44,45]. Evidence of a two-dimensional electron gas (2DEG) at interfaces between two oxides $LaAlO_3$ and $SrTiO_3$ has been reported by several groups [46-49]. This 2DEG provides a superconducting layer that is a realization of a two-dimensional superconductor in a thin slab of finite thickness with superconducting transition temperature of 200 milli-Kelvin [45-48].

Maevasana et al. [49] have found a 2DEG with electron density as large as $8x10^{13}$ cm^{-2} formed at the bare $SrTiO_3$ surface of a cleaved crystal and its density can be controlled through exposure of the surface to intense ultraviolet light. The results show that the chemical potential can be tuned above the bottom of the second sub-band due to quantum size effects. These findings have been confirmed by a systematic study using angle-resolved photoemission spectroscopy (ARPES) providing new insights into the electronic structure of the 2DEG [50]. They shed light on previous observations in $SrTiO_3$-based heterostructures and suggest that different forms of electron confinement at the surface of $SrTiO_3$ lead to essentially the same 2DEG. There is therefore compelling evidence for an exotic 2DEG that forms at oxide interfaces based on $SrTiO_3$, but its precise nature remains elusive.

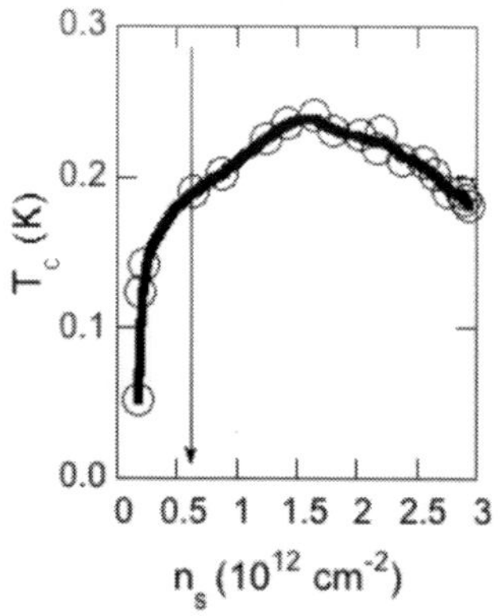

Figure 2. The experimental variation of the critical temperature as a function of superfluid density in the 2DEG at the $LaTiO_3/SrTiO_3$ interface measured by Bert et al. [52]. The superfluid density in the superconducting phase of the 2DEG was controlled by the electrostatic back gate voltage.

Recently Biscaras et al. [51] and Bert et al. [52] have investigated the superconducting phase by controlling the chemical potential in the 2DEG by electrostatic back gate voltage. Biscaras et al. have shown that the superconducting 2DEG is confined into a slab of about 5 nanometers at the $LaTiO_3$ side of the (Mott insulator)/$SrTiO_3$ (band insulator) interface forming subbands due to quantum size effect like it is shown in Figure 1.

Figure 1 shows the 2D metallic slab hosting the 2DEG of thickness L. The total density of states N(E) for the two dimensional subbands made of sharp steps at the band edge of each subband controlled by quantum size effects. By using the gate voltage the chemical potential is tuned to the bottom of the upper subband at about 100 meV above the bottom of the lowest subband. The subbands show a dispersion in the $k_{//}(k_x,k_y)$ plane. The electronic occupation of the upper subband controls the superconducting properties via the small Fermi surface produced by electrons in the upper n-th subband.

The chemical potential in the 2DEG is tuned by the gate voltage to the point where it crosses the bottom of the upper n-th subband, i.e., a 2.5 Lifshitz transition for appearing of a new Fermi surface spot near $k_{//}(k_x,k_y)=0$. Changing the gate voltage field the superconducting transition temperature shows a sharp increase as soon as the small Fermi surface spot due to finite Fermi energy in the n-th subband appears. The critical temperature increases to a maximum value followed by a smooth decrease as a function of the gate voltage [51]. Bert et al. [52] has reported magnetization and susceptibility measurements made by using a scanning SQUID (Superconducting Quantum Interference Device) [52], with a 3 μm diameter pick-up loop and a concentric field coil for applying a local AC magnetic field. They have been able to measure the superfluid density of the superconducting state and to obtain the curve of the critical temperature as a function of the superfluid density shown in Figure 2.

2. Results and discussion

We have calculated the superconducting gaps and the critical temperature in a metallic slab of 5 nanometers using our theoretical approach described in our previous papers [10,11] which is able to describe the superconducing phase near a band edge in a multiband system. The electrons are confined by a very high barrier of 5 eV within the slab and the Schrödinger equation is solved to get the wave-function of electrons in different Fermi surface spots. The electronic structure of the normal phase is made of multiple subbands as shown in Figure 1. The chemical potential is tuned around the bottom of the 6th subband in an energy range of meV. The BCS multi gaps equations for multiband superconductivity are calculated together with the density equation. The chemical potential μ is tuned around the bottom of the upper subband E of n-th subband (n=6). In the band edge region of the n-subband there is a strong renormalization of the chemical potential going from the normal to the superconducting phase which should be included in any theoretical approach.

The superconducting gaps in the subbands of the 2DEG in the slab of 5 nanometer thickness at the $LaTiO_3/SrTiO_3$ interface are shown in Figure 3. The chemical potential is tuned by changing the charge density form the bottom of the first subband. At each 2.5 Lifshitz transition for appearing of a new Fermi surface spot at the band edge of the n-th subband a new condensate is formed and its gap energy increases from zero up to reach a maximum. The superconducting gaps in the n-1 subbands with m<n are very similar. The figure shows three cases a) intraband coupling in the m-th subbands (m<n) in the very weak coupling range c_{mm}=0.1. b) intraband coupling in the n-th subband in the strong coupling range c_{nn}=0.5, with an interband $c_{n,i}$=0.1 c) the intraband coupling is the same as in panel b but now there is a strong exchange pair transfer term c_{ni}=0.5

The position of the chemical potential is measured by the Lifshitz parameter $z=(\mu-E_{n,L})/\omega_o$ where $E_{n,L}$ is the energy of the bottom of the n-th electronic subband and ω_o is the energy cut off of the pairing interaction. When the Lifshitz parameter z is in the range $-1<z<0$ the n-th subband is empty in the normal state but a Bose-like condensate is formed there by bosons created by the strong pairing interaction in the upper n-th subband and the pair transfer from the lower m-th subbands. For $0<z<1$ the Fermi liquid in the n-th subband forms a small Fermi surface where the Fermi energy is smaller than the pairing interaction energy. Therefore the electrons in the n-th subband are in the anti-adiabatic regime where polaron formation is expected and in the superconducting phase bipolarons could form and condense. In this regime the wave-vector of electrons in the n-th Fermi surface k_{Fn} could be smaller than the inverse of the superconducting coherence length $1/\xi_o$ where $(k_{Fn}\ \xi_o)<2\pi$ and T_c versus superfluid density follows the Uemura plot. In cuprates this regime is called the "underdoped regime" or "pseudogap regime". In the range $1<z<2$ the superconducting phase is in a crossover regime approaching the BCS regime for $z>2$.

We report in Figure 3 the calculation of the superconducting gaps in the upper subband n-th and in the (n-1)-th subband and the critical temperature. We assume a weak coupling regime in the lower subbands taking for the intraband coupling strength the value c=0.1.

The green region in Figure 3 corresponds to values of the Lifshitz parameter z, ($-1<z<0$) therefore the n-th subband is empty with no associated Fermi surface in the normal phase. On the other hand, the effective pairing interaction centered around the chemical potential is already able to make available for pair formation states in the n-th subband, thanks to its extension in energy of the order of ω_0 induced by the retardation effects of the electron-phonon coupling. In this situation the condensate component of the n-th subband starts to form well before the appearing of a new Fermi surface in the normal state. This condensate has a Bose-like character, being the underlying Fermi surface not yet formed and its coherence factor $v(k)^2$ quite delocalized as a function k, implying Cooper pairs with a local character in real space.

The yellow region in Figure 3 corresponds to positive values of the Lifshitz parameter z, $0<z<1$: the chemical potential is above the bottom of the n-th subband and a new (small) Fermi surface appears in the electronic spectrum. The superconducting gap is of the order of the energy difference μ-$E_{n,L}$ (which is the Fermi energy for the n-th subband) and the system is in the crossover regime between BCS and BEC.

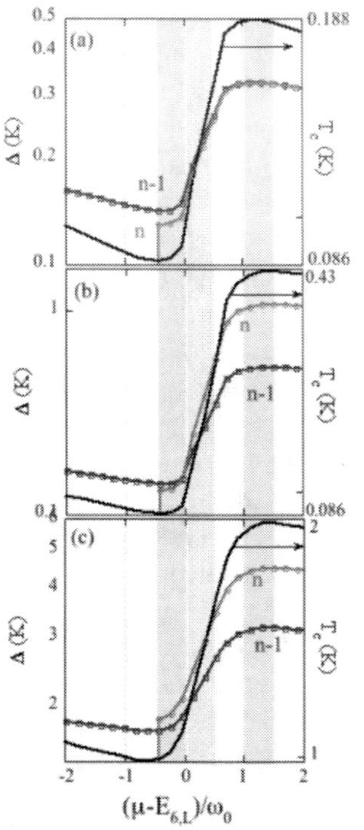

Figure 3. The superconducting gaps and the critical temperature as a function of the Lifshitz parameter z=(μ-$E_{6,L}$)/ω_o. Panel (a): intraband and interband coupling in each subband $c_{nn}=c_{mm}=c_{nm}=0.1$; panel (b) the case of strong intraband coupling in the upper n=6 subband, $c_{nn}=0.5$, while $c_{mm}=c_{nm}=0.1$; c) with strong exchange pair transfer interband term $c_{nm}=c_{nn}=0.5$ while the intraband coupling in the lower subbands is $c_{nm}=0.1$.

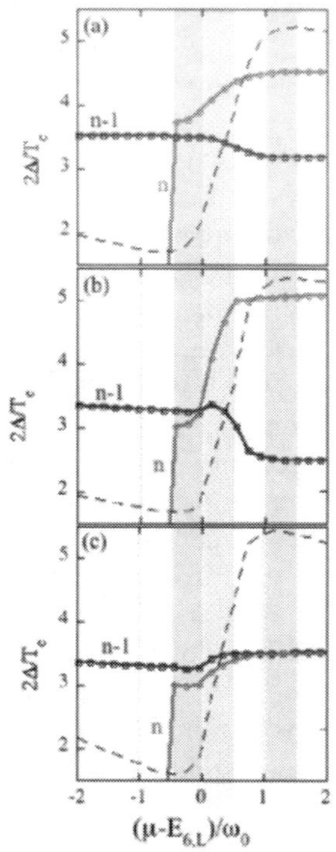

Figure 4. The gap ratio $2\Delta_n/T_c$ in the upper subband and the gap ratio on the lower subbands $2\Delta_m/T_c$, $m<n$ as a function of the Lifshitz parameter z=(μ-$E_{n,L}$)/ω_o, n=6, where m is the chemical potential and $E_{n,L}$ the energy of the bottom of the n-th subband. Panels (a), (b) and (c) show the results of calculations for the same corresponding cases as in Figure 3.

The precise extension of the crossover region depends from the couplings regime, being wider in the Lifshitz parameter z for stronger couplings, leading to larger gap values in the n-th subband, as in panel (c) of Figure 3. This crossover regime is highly non trivial, because the full superconducting system has a condensate which is a coherent mixture of BCS-like (forming in (n-1)-th subband) and Bose-like pairs (in the n-th subband) and the local Fermi energy $E_{Fn,L}$ = μ-$E_{n,L}$ is less or of the same order of the energy cutoff of the interaction ω_0, which leads to non adiabatic effects in the superconducting properties in which the electron-phonon coupling play a role. For this reason this yellow region in Figure 3 can be also addressed as a crossover-antiadiabatic regime.

The blue region in Figure 3, with 1<z<1.5, corresponds to a more conventional crossover to BCS regime for multi-band superconductivity. The frontier between the crossover regime and the BCS regime has the optimal characteristics for the amplification of superconductivity in 2DEG. Anti-adiabatic effects, which in general suppress superconductivity for conventional electron-phonon couplings, start to be negligible, while superconducting fluctuations can be lowered in their detrimental effects by the coexistence of small and large well formed Fermi surface in the spectrum, with the large Cooper pairs forming in the large Fermi surface being able to transfer their large superconducting stiffness to the smaller Copper pairs of the small Fermi surface by means of the interband Josephson-like exchange coupling. Finally, for z>2, the 2DEG superconducting system is in a BCS regime of conventional multi-band superconductivity.

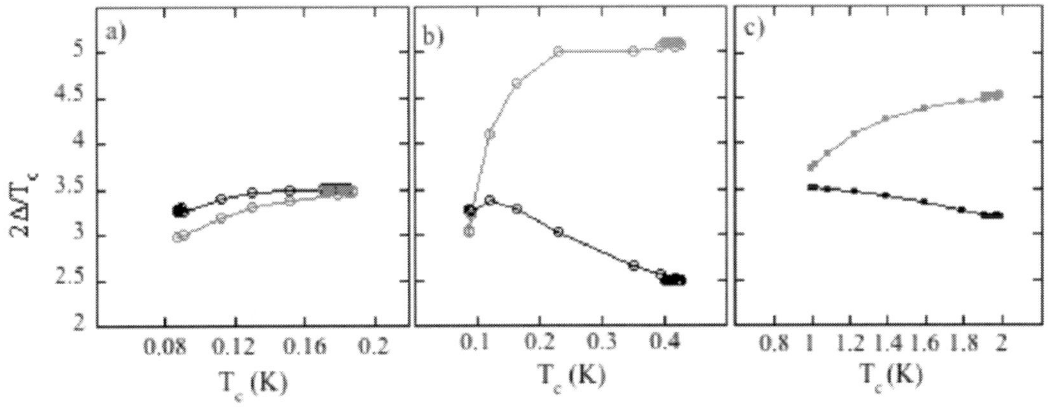

Figure 5: The gap ratio in the n-th (red dots) and m-th subbands (black dots) versus the critical temperature tuning the chemical potential in the 3 considered cases of figures 3 and 4. The figures show three cases a) intraband and interband coupling in all subbands are the weak coupling range 0.1; b) the intraband coupling in the upper n-th subband in the strong coupling range 0.5; c) intraband coupling in the n-th (n=6) and m-th subbands are like in panel (b) but the exchange pair transfer term is much larger, c_{nm}=0.5.

In Figure 5 we report the variation of the gap ratio $2\Delta/T_c$ as a function of the superconducting critical temperature. We show that where the intraband pairing terms are weak as well as the interband pair transfer interaction in panel (a) the gap ratio $2\Delta/T_c$ is similar in different bands, and it is close to the BCS standard value 3.5 for a single large Fermi surface. In panel 2 the BCS ratio in the upper subband, which is in the strong electron-phonon coupling limit, is much larger than 3.5 while for the other subbands, which are in the weak coupling limit c_{mm}=0.1, is much smaller. Panel (c) shows that by increasing the interband pair transfer term c_{nm}=0.5 the difference of the gap ratio in different bands decreases and it approaches again the BCS ratio 3.5.

In Figure 6 we plot the critical temperature as a function of the superfluid density for the three cases we have investigated. The results clearly show that in the mixed Bose-Fermi regime the critical temperature increases as a function superfluid density like in the Uemura plot and by reaching the Fermi regime and the multiband BCS condensation regime the critical temperature saturates and decreases increasing the superfluid density for all investgated cases as in the low superfluid density regime in Figure 2. Further work is needed to clarify the strength of the pair transfer term between different electronic compontens at the Fermi level in this 2DEG.

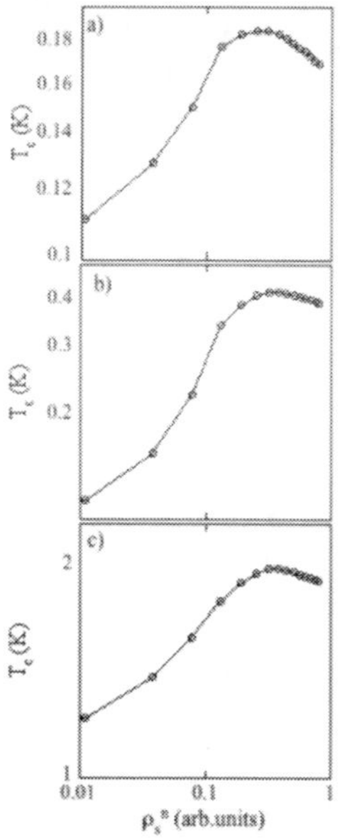

Figure 6: The critical temperature versus the superfluid density in the new appearing subband in the log-log plot. The figure shows three cases (panel a) intraband coupling in the n-th subband in the very weak coupling range 0.1; (panel b) intraband coupling in the n-th subband in the strong coupling range 0.5, (panel c) where the intraband coupling parameters are like in the panel (b) but with a much stronger exchange pair transfer term c_{nm}=0.5.

3. Conclusions

In conclusion one of our most important results: the superconducting shape resonance leading to the optimal amplification of the gap energies and of the critical temperature is located in the range $1<z<2$, and the critical temperature and the gaps in the lower m subbands have minima in the rang $-1<z<0$. It is indeed necessary for the system to go over the Lifshitz critical point to stabilize at the same time lattice and superconducting fluctuations. Optimal superconductivity therefore needs two formed and underlying Fermi surfaces: one large with weak coupling and another smaller one with strong coupling, with the local Fermi level close to the band bottom of the upper band, at an energy distance of the order of the characteristic energy scale of the pairing interaction, together with the interband exchange Josephson-like coupling allowing for (constructive) interference effects between the condensate of the bands.

References

[1] Bianconi A 1994 On the possibility of new high T_c superconductors by producing metal heterostructures as in the cuprate perovskites *Sol. State Commun.* **89** 933
[2] Bianconi A, Saini NL, Rossetti T, Lanzara A, Perali A, Missori M 1996 Stripe structure in the CuO_2 plane of perovskite superconductors *Phys. Rev. B* **54** 12018
[3] Perali A, Bianconi A, Lanzara A, Saini NL 1996 The gap amplification at a shape resonance in a superlattice of quantum stripes: A mechanism for high T_c *Sol. State Commun.* **100**

[4] Bianconi A, Valletta A, Perali A and Saini NL 1997 High T_c superconductivity in a superlattice of quantum stripes *Solid State Commun.* **102** 369

[5] Valletta A, Bianconi A, Perali A, Saini NL 1997 Electronic and superconducting properties of a superlattice of quantum stripes at the atomic limit *Zeitschrift für Physik B Condensed Matter* **104** 707

[6] Bianconi A, Di Castro D, Agrestini S, Campi G, Saini N L, Saccone A, De Negri S, Giovannini M 2001 A superconductor made by a metal heterostructure at the atomic limit tuned at the `shape resonance': MgB_2 *Journal of Physics: Condensed Matter* **13** 7383-7390

[7] Bussmann-Holder A, Bianconi A 2003 Raising the diboride superconductor transition temperature using quantum interference effects *Physical Review B* **67** 132509

[8] Bianconi A, Agrestini S, Bussmann-Holder A 2004 The T_c amplification by shape resonance in doped MgB_2 *Journal of Superconductivity* **17** 205

[9] Bianconi A 2005 Feshbach Shape Resonance in Multiband Superconductivity in Heterostructures *J. Supercond. Novel Magn.* **18** 625

[10] Innocenti D, Poccia N, Ricci A, Valletta A, Caprara S, Perali A, Bianconi A 2010 Resonant and crossover phenomena in a multiband superconductor: Tuning the chemical potential near a band edge *Physical Review B* **82** 184528

[11] Innocenti D, Valletta A, Bianconi A 2011 Shape resonance at a lifshitz transition for high temperature superconductivity in multiband superconductors *Journal of Superconductivity and Novel Magnetism* **24** 1137

[12] Bussmann-Holder A, Keller H, Khasanov R, Simon A, Bianconi A, Bishop AR 2011 Isotope and interband effects in a multi-band model of superconductivity *New Journal of Physics* **13** 093009

[13] Perali A, Innocenti D, Valletta A, Bianconi A 2012 Anomalous isotope effect near a 2.5 lifshitz transition in a multi-band multi-condensate superconductor made of a superlattice of stripes *Superconductor Science and Technology* **25** 124002

[14] Fano U 1935 Sullo spettro di assorbimento dei gas nobili presso il limite dello spettro d'arco *Il Nuovo Cimento (1924-1942)* **12** 154

[15] Fano U 1961 Effects of Configuration Interaction on Intensities and Phase Shifts *Phys. Rev.* 124 1866 URL

[16] Bianconi A 2003 Ugo Fano and shape resonances *AIP Conference Proceedings* 652,13-18.

[17] Vittorini-Orgeas A, Bianconi A 2009 "From Majorana Theory of Atomic Autoionization to Feshbach Resonances in High Temperature Superconductors Su*percond Nov Magn* **52** 215

[18] Blatt J, Thompson C 1963 Shape Resonances in Superconducting Thin Films *Physical Review Letters* **10** 332

[19] Blatt JM 1964 *Theory of Superconductivity* Academic Press New York (for shape resonances see pag. 362 and 215)

[20] Leggett A J 1966 *Prog. Theor. Phys.* **36** 901; Leggett AJ 1966 *Prog. Theor. Phys.* **36** 931

[22] Leggett A J 1980 *Modern Trends in the Theory of Condensed Matter*, edited by A. Pekalski and R. Przystawa, Lecture Notes in Physics, Springer-Verlag, Berlin, vol 115 p 13

[23] Sá de Melo CAR, Randeria M, Engelbrecht JR 1993 Crossover from BCS to Bose superconductivity: Transition temperature and time-dependent Ginzburg–Landau theory *Phys. Rev. Lett.* **71** 3202

[24] Shanenko A A, Croitoru M D, Vagov A V, Axt V M, Perali A, and Peeters F M 2012 Atypical BCS-BEC crossover induced by quantum-size effects *Phys. Rev. A* **86** 033612

[25] de Llano M, Tolmachev VV 2003 Multiple phases in a new statistical boson–fermion model of superconductivity *Physica A: Statistical Mechanics and its Applications* **317** 546 ; Malik G P, Chávez I, de Llano M 2013 *Journal of Modern Physics* **4** 474

[26] Bao XY, Zhang YF, Wang Y, Jia JF, Xue QK, Xie XC, Zhao Z X, 2005 Quantum size effects on the perpendicular upper critical field in ultrathin lead films *Physical Review Letters* **95** 247005

[27] Qin S, Kim J, Niu Q, Shih CK 2009 Superconductivity at the Two-Dimensional limit *Science* **324** 1314

[28] Bose S, Garcia Garcia A M, Ugeda M M, Urbina J D, Michaelis C H, Brihuega I, Kern K 2010 *Nat. Mat.* **9** 550

[29] Zhang T, Cheng P, Li W-J, Sun Y-J, Wang G, Zhu X.-G, He K, Wang L, MaX, Chen X, Wang Y, Liu Y, Lin H-Q, Jia J-F, Xue Q-K 2010 Superconductivity in one-atomic-layer metal films grown on Si(111) *Nature Physics* **6** 104

[30] Shanenko AA, Croitoru MD, and Peeters FM 2006 *EPL* **76** 498

[31] Shanenko AA, Croitoru M D, Zgirski M, Peeters F M, Arutyunov K 2006 *Phys. Rev. B* **74** 052502

[32] García-García AM, Urbina JD, Yuzbashyan EA, Richter K, Altshuler BL 2011 BCS superconductivity in metallic nanograins: Finite-size corrections, low-energy excitations, and robustness of shell effects *Phys. Rev. B* 83, 014510

[33] Campi G, Ricci A, Poccia N, Barba L, Arrighetti G, Burghammer M, Caporale A S, Bianconi A 2013 Scanning micro-x-ray diffraction unveils the distribution of oxygen chain nanoscale puddles in YBa2Cu3O6.33 *Physical Review B* **87** 014517

[34] Ricci A, Poccia N, Campi G, Coneri F, Caporale AS, Innocenti D, Burghammer M, Zimmermann M, Bianconi A 2013 Multiscale distribution of oxygen puddles in 1/8 doped YBa2Cu3O6.67 *Scientific Reports* **3** 2383

[35] Poccia N, Ricci A, Campi G, Fratini M, Puri A, Di Gioacchino D, Marcelli A, Reynolds M, Burghammer M, Saini N L, Aeppli G, Bianconi A 2012 Optimum inhomogeneity of local lattice distortions in La_2CuO_{4+y} *Proceedings of the National Academy of Sciences* **109** 15685

[36] Poccia N, Fratini M, Ricci A, Campi G, Barba L, Vittorini-Orgeas A, Bianconi G, Aeppli G, Bianconi A 2011 Evolution and control of oxygen order in a cuprate superconductor *Nat Mater* **10** 733

[37] Takahashi M, Maeda T, Uemura K, Yao J, Tokuda Y, Yoko T, Kaji H, Marcelli A 2007 Photoinduced Formation of Wrinkled Microstructures with Long−Range Order in Thin Oxide Films *Advanced Materials* **19**

[38] Campi G, Di Castro D, Bianconi G, Agrestini S, Saini NL, Oyanagi H, Bianconi A 2002 Photo-Induced phase transition to a striped polaron crystal in cuprates *Phase Transitions* **75** 927

[39] Di Castro D, Bianconi G, Colapietro M, Pifferi A, Saini NL, Agrestini S, Bianconi A 2000 Evidence for the strain critical point in high T_c superconductors *The European Physical Journal B - Condensed Matter and Complex Systems* **18** 617

[40] Bianconi A, Saini N L, Agrestini S, Di Castro D, Bianconi G 2000 The strain quantum critical point for superstripes in the phase diagram of all cuprate perovskites International *Journal of Modern Physics B* **14** 3342

[41] Bianconi A, Di Castro D, Bianconi G, Pifferi A, Saini N. L, Chou FC, Johnston DC, Colapietro M 2000 Coexistence of stripes and superconductivity: Tc amplification in a superlattice of superconducting stripes *Physica C: Superconductivity* **341** 1719

[42] Bianconi A, Missori M, Oyanagi H, Yamaguchi H, Nishiara Y, Della Longa S 1995 The measurement of the polaron size in the metallic phase of cuprate superconductors *EPL (Europhysics Letters)* **31** 411

[43] Müller KA and Bednorz J G 1987 A Road Towards High-Temperature Superconductivity *Physica Scripta* **1987** 23

[44] Mannhart J, Schlom D G 2010 Oxide Interfaces-An Opportunity for Electronics *Science* **327** 1607

[45] Hwang H Y, Iwasa Y, Kawasaki M, Keimer B, Nagaosa N and Tokura Y 2012 Emergent phenomena at oxide interfaces *Nature Materials* **11** 103

[46] Reyren N, Thiel S, Caviglia AD, Kourkoutis LF, Hammerl G, Richter C, et al. 2007 Superconducting Interfaces Between Insulating Oxides *Science* **317** 1196

[47] Biscaras J, Bergeal N, Kushwaha A, Wolf T, Rastogi A, Budhani R C, Lesueur J 2010 Two-dimensional superconductivity at a Mott insulator/band insulator interface LaTiO3/SrTiO3 *Nature Communications* **1** 89

[48] Bert J A, Kalisky B, Bell C, Kim M, Hikita Y, Hwang HY, Moler K A 2011 Direct imaging of the coexistence of ferromagnetism and superconductivity at the LaAlO3/SrTiO3 interface *Nature Phys* **7** 767

[49] Basletic M, Maurice J-L, Carretero C, Herranz G, Copie O, Bibes M, Jacquet E, Bouzehouane K, Fusil S, Barthelemy A 2008 Mapping the spatial distribution of charge carriers in LaAlO3/SrTiO3 heterostructures *Nature Materials* **7** 621

[49] Caviglia A D, Gariglio S, Reyren N, Jaccard D, Schneider T, Gabay M, Thiel S, Hammerl G, Mannhart J, Triscone J -M 2008 Electric field control of the LaAlO3/SrTiO3 interface ground state *Nature* **456** 624

[50] Meevasana W, King P D C, He R H, Mo S K, Hashimoto M, Tamai A, Songsiriritthigul P, Baumberger F and Shen Z X 2011 Creation and control of a two-dimensional electron liquid at the bare SrTiO3 surface *Nature Materials* **10** 114

[51] Santander-Syro AF, Copie O, Kondo T, Fortuna F, Pailhès S, Weht R, Qiu X G, Bertran F, Nicolaou A, Taleb-Ibrahimi A, et al, 2011 *Nature* **469** 189

[52] Biscaras J, Bergeal N, Hurand S, Grossetête C, Rastogi A, Budhani RC, et al. 2012 Two-Dimensional Superconducting Phase in LaTiO_3/SrTiO_3 Heterostructures Induced by High-Mobility Carrier Doping *Phys Rev Lett* **108** 247004

[53] Bert JA, Nowack KC, Kalisky B, Noad H, Kirtley JR, Bell C, et al. 2012 Gate-tuned superfluid density at the superconducting LaAlO3/SrTiO3 interface *Physical Review* **B** 86 060503

A frustrated spin-1 J_1–J_2 Heisenberg antiferromagnet: An anisotropic planar pyrochlore model

P H Y Li[1], R F Bishop[1], and C E Campbell[2]

[1] School of Physics and Astronomy, Schuster Building, The University of Manchester, Manchester, M13 9PL, UK

[2] School of Physics and Astronomy, University of Minnesota, 116 Church Street SE, Minneapolis, Minnesota 55455, USA

E-mail: raymond.bishop@manchester.ac.uk; peggyhyli@gmail.com

Abstract. The zero-temperature ground-state (GS) properties and phase diagram of a frustrated spin-1 J_1–J_2 Heisenberg model on the checkerboard square lattice are studied, using the coupled cluster method. We consider the case where the nearest-neighbour exchange bonds have strength $J_1 > 0$ and the next-nearest-neighbour exchange bonds present (viz., in the checkerboard pattern of the planar pyrochlore) have strength $J_2 \equiv \kappa J_1 > 0$. We find significant differences from both the spin-1/2 and classical versions of the model. We find that the spin-1 model has a first phase transition at $\kappa_{c_1} \approx 1.00 \pm 0.01$ (as does the classical model at $\kappa_{\rm cl} = 1$) between two antiferromagnetic phases, viz., a quasiclassical Néel phase (for $\kappa < \kappa_{c_1}$) and one of the infinitely degenerate family of quasiclassical phases (for $\kappa > \kappa_{c_1}$) that exists in the classical model for $\kappa > \kappa_{\rm cl}$, which is now chosen by the *order by disorder* mechanism as (probably) the "doubled Néel" (or Néel*) state. By contrast, none of this family survives quantum fluctuations to form a stable GS phase in the spin-1/2 case. We also find evidence for a second quantum critical point at $\kappa_{c_2} \approx 2.0 \pm 0.5$ in the spin-1 model, such that for $\kappa > \kappa_{c_2}$ the quasiclassical (Néel*) ordering melts and a nonclassical phase appears, which, on the basis of preliminary evidence, appears unlikely to have crossed-dimer valence-bond crystalline (CDVBC) ordering, as in the spin-1/2 case. Unlike in the spin-1/2 case, where the Néel and CDVBC phases are separated by a phase with plaquette valence-bond crystalline (PVBC) ordering, we find very preliminary evidence for such a PVBC state in the spin-1 model for all $\kappa > \kappa_{c_2}$.

1. Introduction

Low-dimensional spin-lattice models of magnetic systems, particularly those pertaining to frustrated Heisenberg antiferromagnets (HAFMs) with competing interactions, have been extensively studied from both the theoretical and experimental viewpoints in recent years. Although such spin-lattice models are themselves conceptually simple and easy to write down, these strongly correlated systems often exhibit rich and interesting zero-temperature ($T = 0$) ground-state (GS) phase diagrams as the interaction coupling strengths are varied, due to the strong interplay between quantum fluctuations and frustration. The strength of the quantum fluctuations can itself also be tuned by a variety of methods. These include changing the spin quantum number s of the particles residing on the given lattice sites, while keeping the interaction Hamiltonian unchanged. One expects that, in general, quantum fluctuations will be greatest for the case $s = \frac{1}{2}$, and that they will reduce to zero as the classical limit ($s \to \infty$) is approached.

For this reason the greatest attention has been paid to spin-1/2 magnets. Nevertheless, there has also been an upsurge of interest in spin-1 magnets in recent years. On the theoretical side, although quantum fluctuations will generally be reduced for a given model for the case $s = 1$ compared with its counterpart for the case $s = \frac{1}{2}$, totally new physical effects can also sometimes enter. In one-dimensional (1D) systems these include the by now well-known existence of the gapped Haldane phase [1] with an exponential decay with separation of spin-spin correlations for $s = 1$ (or, more generally, for integral values of s), compared to the gapless phase with a corresponding power-law decay of spin-spin correlations for $s = \frac{1}{2}$ (or, more generally, for half-odd-integral values of s). Also, in any number of dimensions, the possible inclusion in the $s = 1$ case of such additional terms in the Hamiltonian as biquadratic exchange and single-site anisotropy, which are absent for $s = \frac{1}{2}$ systems, can lead both to novel quantum phase transitions and to novel phases with, for example, quadrupolar nematic long-range order (LRO) but with zero magnetic order parameter (taken as the average local on-site magnetization).

On the experimental side many magnetic compounds containing spin-1 ions are now established as being well described by various $s = 1$ spin-lattice models. For the 1D case there are many good experimental realizations of (quasi-)linear chain systems with $s = 1$. These include $CsNiCl_3$ [2] and Y_2BaNiO_5 [3], both with a weak easy-axis single-ion anisotropy, and $CsFeBr_3$ [4] with a strong easy-plane single-ion anisotropy, as well as such complex organo-metallic compounds as NENP ($Ni(C_2H_8N_2)_2(NO_2)(ClO_4)$) [5] with a weak planar anisotropy, and both NENC ($Ni(C_2H_8N_2)_2Ni(CN_4)$) [6] and DTN ($NiCl_2$-$4SC(NH_2)_2$) [7] with a strong planar anisotropy. The spin gaps seen in $CsNiCl_3$, Y_2BaNiO_5 and NENP are now believed to be experimental realizations of the integer-spin gap behaviour predicted by Haldane [1]. For the two-dimensional (2D) case several experimental realizations of spin-1 antiferromagnets exist. For example, K_2NiF_4 [8] provides a realization of an $s = 1$ HAFM on a square lattice. Similarly, $NiGa_2S_4$ [9] is well described as a 2D antiferromagnet on a triangular lattice, for which the GS phase has been argued to have ferro-spin nematic order [10–12].

Of particular relevance to the present study has been the large additional impetus to the study of 2D spin-1 antiferromagnets that was provided by the discovery of superconductivity with a transition temperature $T_c \approx 26\,\text{K}$ in the layered iron-based compound LaOFeAs, when doped by partial substitution of the oxygen atoms by fluorine atoms [13], $La[O_{1-x}F_x]FeAs$, with $x \approx$ 5–11%. That discovery was rapidly followed by finding superconductivity at even higher transition temperatures ($T_c \gtrsim 50\,\text{K}$) in a wide class of similarly doped quaternary oxypnictide materials. First-principles calculations [14] ensued, which showed that the original undoped parent precursor material LaOFeAs is well described by the spin-1 J_1–J_2 HAFM on a square lattice with nearest-neighbour (NN) and next-nearest-neighbour (NNN) Heisenberg exchange couplings, J_1 and J_2 respectively, with $J_1 > 0$, $J_2 > 0$, and with $J_2/J_1 \approx 2$. Other authors have also reached similar conclusions (see, e.g., Ref. [15]).

The J_1–J_2 model on a square lattice has itself received huge theoretical attention over the last 25 or so years, since it provides an archetypal model of a strongly correlated and highly frustrated spin-lattice system. Most attention has naturally been devoted to the spin-1/2 case (see, e.g., Refs. [16–48] and references therein). The consensual view for this model now is that its ($T = 0$) GS phase diagram exhibits two phases with quasiclassical LRO, both with antiferromagnetic (AFM) order, namely, a Néel-ordered phase (with a wavevector $\mathbf{Q} = (\pi, \pi)$) at small values of the frustration parameter ($J_2/J_1 \lesssim 0.4$) and a collinear stripe-ordered phase (with a wavevector $\mathbf{Q} = (\pi, 0)$ or $\mathbf{Q} = (0, \pi)$) at large values ($J_2/J_1 \gtrsim 0.6$). These two magnetically ordered phases are separated by an intermediate quantum paramagnetic (QP) phase without magnetic LRO for $0.4 \lesssim J_2/J_1 \lesssim 0.6$. What makes the system of continuing interest is that the nature of the intermediate QP phase and the order and nature of the two phase transitions bounding it are still not fully resolved and understood.

The classical ($s \rightarrow \infty$) version of the J_1–J_2 model on the square lattice (with a number

$N \to \infty$ of spins) exhibits a unique GS Néel-ordered AFM phase for $J_2/J_1 < \frac{1}{2}$, with an energy per spin $E^{\rm cl}/N = -2s^2(J_1 - J_2)$, but has an infinitely degenerate set of GS phases for $J_2/J_1 > \frac{1}{2}$, all with $E^{\rm cl}/N = -2s^2 J_2$. The latter set comprises two interpenetrating Néel-ordered $\sqrt{2} \times \sqrt{2}$ square lattices, with the relative ordering angle between them completely arbitrary. Quantum fluctuations then act, via the well known *order by disorder* mechanism [49, 50], to lift this (accidental) degeneracy in the quasiclassical ($s \gg 1$) limit where one works to leading order in $1/s$, in favour of collinear ordering, which leads to the two (row or column) stripe-ordered states, with wavevectors $\mathbf{Q} = (0, \pi)$ and $\mathbf{Q} = (\pi, 0)$, discussed above.

This feature of macroscopic classical GS degeneracy in *any* spin-lattice model always makes such models of particular theoretical interest since they are, *a priori*, prime candidates for exhibiting novel quantum GS phases. It also makes them particularly susceptible to small perturbations in the form, for example, of additional spin-orbit interactions, spin-lattice couplings, neglected exchange terms, and anisotropies in the exchange interactions. Hence, considerable attention has also been placed on various such mechanisms, or extra parameters that can be included, to extend the J_1–J_2 model on the square lattice, apart from changing the spin quantum number, both to learn more about the model itself and to enquire how robust are its various properties against any such perturbations.

Naturally, for the purpose of making such detailed comparisons, it is important to use an accurate theoretical technique with controlled approximation hierarchies. One such is the coupled cluster method (CCM) [51–54] that we shall employ here, and which we discuss in more detail in Sec. 3. The CCM has been very successfully applied over the last 20 or more years to a wide variety of quantum spin-lattice systems (see, e.g., Refs. [28, 37, 38, 41, 43, 53–88] and references cited therein). These include applications both to the square-lattice J_1–J_2 model itself [28, 38, 43, 73] as well as to various extensions and refinements of it along the lines discussed above.

Such extensions include, *inter alia*: (a) the J_1–J_2–$J_\perp$ model of a stacked square lattice [37], in which a number of 2D J_1–J_2 square-lattice layers are coupled via a NN inter-layer exchange interaction of strength $J_\perp$; (b) putting a square-plaquette structure on the model [43] by having differing inter- and intra-plaquette NN couplings; (c) the corresponding J_1–J_2–J_3 model [41], which includes next-next-nearest-neighbour Heisenberg couplings of strength J_3; (d) the J_1–J_1'–J_2 model [68], in which a spatial anisotropy between NN bonds along the two perpendicular square-lattice directions is introduced; and (e) the J_1^{XXZ}–J_2^{XXZ} model [67], in which an XXZ-type anisotropy is introduced on both the NN and NNN Heisenberg exchange bonds. The corresponding $s = 1$ cases have also been studied within the CCM framework for the latter two cases of the J_1–J_1'–J_2 model [69] and the J_1^{XXZ}–J_2^{XXZ} model [70] on the square lattice.

Of particular importance for, and relevance to, the present paper, we note that the CCM has also been applied to study the ($T = 0$) GS phase diagrams of several members of the so-called half-depleted spin-1/2 J_1–J_2 models on the square lattice, all of which share the feature that half of the J_2 bonds of the original model are removed. They differ only in the arrangements of the remaining J_2 bonds. When each basic square plaquette (formed from 4 NN J_1 bonds) has a single J_2 bond, they include the three cases of: (a) the interpolating square-triangle model [72], in which the J_2 bonds have the same orientation in each square plaquette; (b) the Union Jack model [74], in which the J_2 bonds have alternating orientations on neighbouring square plaquettes; and (c) the chevron-decorated square-lattice model [87], in which the J_2 bonds alternate in orientation in one direction (say, along rows), but are parallel to each other in the perpendicular direction (say, along columns).

The corresponding $s = \frac{1}{2}$ model on the checkerboard lattice, in which alternating basic square plaquettes have either both or zero J_2 bonds present, has also been studied within the CCM [83]. Furthermore, CCM studies have also been carried out for $s \geq 1$ cases of both the interpolating square-triangle lattice model [79] and the Union Jack model [76], which both show interesting

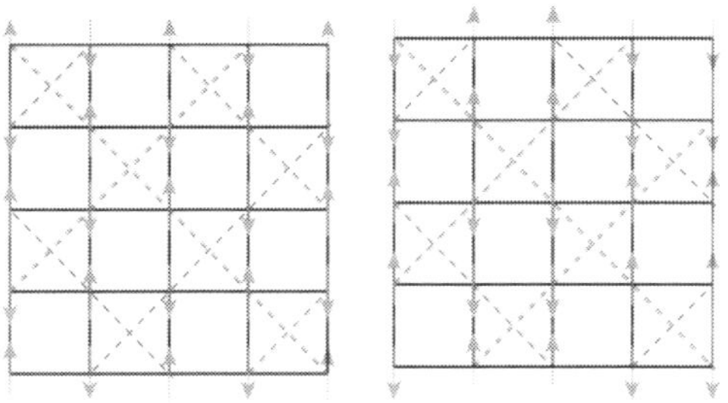

(a) Néel (b) Néel*

Figure 1. The J_1–J_2 checkerboard model, showing (a) the Néel state, and (b) one of the two Néel* states. The NN J_1 bonds are shown as solid (black) lines and the NNN (i.e., also NN) J_2 bonds are shown as dashed (magenta) lines. The arrows represent the orientations of the spins on each lattice site for each of the three states shown.

differences to their $s = \frac{1}{2}$ counterparts. The aim of the present work is to perform a similar CCM analysis of the $s = 1$ version of the J_1–J_2 model on the checkerboard lattice, and to compare it with its $s = \frac{1}{2}$ counterpart studied previously by us [83].

The J_1–J_2 model on the checkerboard lattice, shown schematically in Fig. 1, is also known as the anisotropic planar pyrochlore (APP) model (or, sometimes, the crossed chain model). It may be regarded as a 2D analogue of a three-dimensional (3D) anisotropic pyrochlore model of corner-sharing tetrahedra. The model itself, as we elaborate in Sec. 2, falls into the same interesting class as the full J_1–J_2 model on the square lattice, demonstrating macroscopic classical GS degeneracy above a certain critical value of the anisotropy parameter, $\kappa \equiv J_2/J_1$. For that reason the ($T = 0$) GS phase diagram of the $s = \frac{1}{2}$ version of the model has been much studied earlier by many authors [83, 89–106]. More recently, on the basis of our CCM results for this model [83], a more accurate and, hopefully, more consensual description of its GS phase structure is emerging. The time now seems ripe, therefore, to compare and contrast the model with its $s = 1$ counterpart, which we study here.

In Sec. 2 the model itself is discussed, before we give a brief description of the CCM formalism in Sec. 3. Our main results are then presented in Sec. 4, and we end with a discussion and summary in Sec. 5.

2. The model

The Hamiltonian of the J_1–J_2 model on the checkerboard lattice is given by

$$H = J_1 \sum_{\langle i,j \rangle} \mathbf{s}_i \cdot \mathbf{s}_j + J_2 \sum_{\langle\langle i,k \rangle\rangle'} \mathbf{s}_i \cdot \mathbf{s}_k \, , \tag{1}$$

where the indices runs over all sites of a 2D square lattice, such that the sum over $\langle i,j \rangle$ counts every NN pair (once and once only) and the sum over $\langle\langle i,k \rangle\rangle'$ counts each NNN pair in the checkerboard pattern (once and once only) shown in Fig. 1, such that alternate basic square plaquettes have either two diagonal bonds or none. Each site i of the lattice now carries a particle with spin $s = 1$ described by a spin operator $\mathbf{s}_i = (s_i^x, s_i^y, s_i^z)$.

The lattice and exchange bonds of the anisotropic planar pyrochlore model are thus shown in Fig. 1, from which one sees clearly how it may alternatively be construed as comprising crossed (diagonal) sets of chains, along which the intrachain exchange coupling constant is J_2, coupled by both vertical and horizontal interchain exchange bonds of strength J_1. Both bonds here are assumed to be AFM in nature (i.e., $J_1 > 0$ and $J_2 > 0$) and hence to act to frustrate one another. Clearly, the model interpolates smoothly between the HAFM on the square lattice (when $\kappa \equiv J_2/J_1 = 0$) and decoupled 1D isotropic HAFM chains (when $\kappa \to \infty$). When $\kappa = 1$, in between these limiting cases, we have the isotropic HAFM on the checkerboard lattice, alias the isotropic planar pyrochlore. With no loss of generality, henceforth we choose $J_1 \equiv 1$ to set the overall energy scale.

Classically (i.e., when $s \to \infty$) the model has a single ($T = 0$) GS phase transition at $\kappa = \kappa_{\rm cl} = 1$. For $\kappa < 1$ the GS phase is the Néel state illustrated in Fig. 1(a), for which the ordering along the diagonal chains is hence ferromagnetic (FM) in nature. The GS energy per spin of this classical Néel state is thus $E^{\rm cl}/N = s^2(-2J_1 + J_2)$. For $\kappa > 1$ the corresponding GS phase is now infinitely degenerate. It comprises a set of collinear states in which every checkerboard diagonal chain of sites connected by J_2 bonds has Néel AFM ordering, but where each chain can slide along its own length without changing the overall energy. In other words, the infinite degeneracy in this set is such that the spins along any one row or column can be arbitrarily assigned. All such states then have the same classical energy per spin, $E^{\rm cl}/N = -s^2 J_2$, independent of the value of J_1. Among this infinite family of states is the so-called Néel* state shown in Fig. 1(b), which has pairwise or doubled AFM ordering of the type $\cdots\uparrow\uparrow\downarrow\downarrow\uparrow\uparrow\downarrow\downarrow\cdots$, along every row and column. Thus, the single spin $\uparrow$ or $\downarrow$ of the Néel state is replaced by the two-site unit $\uparrow\uparrow$ or $\downarrow\downarrow$ in the Néel* state. The Néel* state retains a double degeneracy. This can easily be seen, for example, from Fig. 1(b), where the Néel* state exhibits two forms of empty plaquettes (namely, those with four parallel spins and those with two pairs of antiparallel spins), the roles of which can be interchanged.

In our previous [83] CCM analysis of the $s = \frac{1}{2}$ version of the model we found a GS phase diagram with marked differences to its classical ($s \to \infty$) counterpart. Although the quasiclassical state with Néel AFM ordering remains the GS phase for low enough values of κ (viz., $\kappa < \kappa_c^1 \approx 0.80 \pm 0.01$), we found that *none* of the infinitely degenerate set of AFM states in the classical model (which form the GS phase in that case for $\kappa > 1$) can survive the quantum fluctuations present in the $s = \frac{1}{2}$ case to form a stable GS phase. We found instead two different forms of valence-bond crystalline (VBC) ordering for $\kappa > \kappa_c^1$. Firstly, in the region $\kappa_c^1 < \kappa < \kappa_c^2 \approx 1.22 \pm 0.02$ we found the stable GS phase to exhibit plaquette VBC (PVBC) order, while for *all* $\kappa > \kappa_c^2$ the ordering changes to a crossed-dimer VBC (CDVBC) variety. We found that both transitions are probably direct ones, although, as usual, we could not entirely rule out very narrow coexistence regions confined, respectively, to $0.79 \lesssim \kappa \lesssim 0.81$ and $1.20 \lesssim \kappa \lesssim 1.22$.

In view of our earlier discussion in Sec. 1, it is now of great interest to perform a comparable analysis of the $s = 1$ version of the model, the results of which we present in Sec. 4 after a brief discussion of the CCM formalism used to obtain them.

3. The coupled cluster method

Since the CCM is well documented elsewhere (see, e.g., Refs. [51–54]) we give only a brief outline here. We note that it is particularly well suited for the study of such highly frustrated magnets as we consider here, for which alternative methods, such as quantum Monte Carlo (QMC) or exact diagonalization (ED) techniques, run into severe problems. Thus, QMC methods suffer in such cases from the infamous "minus-sign problem," while ED methods are often restricted to too small lattices to be able to sample with sufficient accuracy the details of the often very subtle ordering that is present, even when state-of-the-art calculations are performed with the largest computational resources available. While both QMC and ED calculations are performed on lattices with a finite number N of spins, and hence require finite-size scaling to obtain the $N \to \infty$ limit required, the CCM, as we described below, is a size-extensive method that automatically works in the (infinite-lattice) thermodynamic limit from the outset, at every level of approximation. Since such approximations, as we will see below, can be defined in rigorous hierarchical schemes, the only final extrapolation needed is to the (exact) limit in any such scheme. Furthermore, at the highest levels of approximation feasible with available computational resources, results for physical quantities are often already well converged, as our specific results in Sec. 4 will show.

The CCM starts with the choice of a suitable model state (or reference state), $|\Phi\rangle$, on top of which the quantum correlations present in the exact GS phase under study can be systematically incorporated later, as we described below. For the present model obvious choices are the Néel and Néel* states shown in Fig. 1. At the quasiclassical level we expect these might prove good candidate CCM model states for the regions $\kappa \lesssim 1$ and $\kappa \gtrsim 1$, respectively. Of course, for the latter regime there is an infinite family of classically degenerate states from which to choose. We note, in this context, however, that at the $O(1/s)$ level in a quasiclassical expansion in powers of $1/s$, a fourfold set of states is selected [97] by the *order by disorder* mechanism [49, 50] to lie lowest in energy. These include the (doubly degenerate) Néel* states as well as the two (row and column) striped AFM states, in which alternating rows or columns have spins aligned $\uparrow$ or $\downarrow$.

Once a model state $|\Phi\rangle$ is chosen, the exact GS ket- and bra-state wave functions that satisfy the corresponding Schrödinger equations,

$$H|\Psi\rangle = E|\Psi\rangle\,; \quad \langle\tilde{\Psi}|H = E\langle\tilde{\Psi}|\,, \tag{2}$$

are parametrized as

$$|\Psi\rangle = \mathrm{e}^{S}|\Phi\rangle\,; \quad \langle\tilde{\Psi}| = \langle\Phi|\tilde{S}\mathrm{e}^{-S}\,, \tag{3}$$

where we use the intermediate normalization scheme for $|\Psi\rangle$, such that $\langle\Phi|\Psi\rangle = \langle\Phi|\Phi\rangle \equiv 1$, and then for $\langle\tilde{\Psi}|$ choose its normalization such that $\langle\tilde{\Psi}|\Psi\rangle = 1$. The correlation operators S and $\tilde{S}$ are decomposed in terms of exact sets of multiparticle, multiconfigurational creation and destruction operators, C_I^+ and $C_I^- \equiv (C_I^+)^\dagger$, respectively, as

$$S = \sum_{I\neq 0} \mathcal{S}_I C_I^+\,; \quad \tilde{S} = 1 + \sum_{I\neq 0} \tilde{\mathcal{S}}_I C_I^-\,, \tag{4}$$

where $C_0^+ \equiv 1$, the identity operator, and I is a set index describing a complete set of single-particle configurations for all of the particles. The reference state $|\Phi\rangle$ thus acts as a fiducial (or cyclic) vector, or generalized vacuum state, with respect to the complete set of creation operators $\{C_I^+\}$, which are hence required to satisfy the conditions $\langle\Phi|C_I^+ = 0 = C_I^-|\Phi\rangle\,, \forall I \neq 0$.

In order to consider each site on the spin lattice to be equivalent to all others, whatever the choice of state $|\Phi\rangle$, it is convenient to form a passive rotation of each spin so that in its own local spin-coordinate frame it points in the downward, (i.e., negative z) direction. Clearly, such choices of local spin-coordinate frames leave the basic SU(2) spin commutation relations unchanged, but have the nice effect that the C_I^+ operators can be expressed as products of single-spin raising operators $s_k^+ \equiv s_k^x + is_k^y$, such that $C_I^+ \equiv s_{k_1}^+ s_{k_2}^+ \cdots s_{k_n}^+$; $n = 1, 2, \cdots, 2sN$.

The complete set of multiparticle correlation coefficients $\{\mathcal{S}_I, \tilde{\mathcal{S}}_I\}$ may now be evaluated by extremizing the energy expectation value $\bar{H} \equiv \langle\tilde{\Psi}|H|\Psi\rangle = \langle\Phi|\tilde{S}\mathrm{e}^{-S}H\mathrm{e}^{S}|\Phi\rangle$, with respect to each of them, $\forall I \neq 0$. Variation with respect to each coefficient $\tilde{\mathcal{S}}_I$ yields the coupled set of nonlinear equations,

$$\langle\Phi|C_I^-\mathrm{e}^{-S}H\mathrm{e}^{S}|\Phi\rangle = 0\,, \quad \forall I \neq 0\,, \tag{5}$$

for the coefficients $\{\mathcal{S}_I\}$, while variation with respect to each coefficient $\mathcal{S}_I$ yields the corresponding set of linear equations,

$$\langle\Phi|\tilde{S}(\mathrm{e}^{-S}H\mathrm{e}^{S} - E)C_I^+|\Phi\rangle = 0\,, \quad \forall I \neq 0\,, \tag{6}$$

for the coefficients $\{\tilde{\mathcal{S}}_I\}$, once the coefficients $\{\mathcal{S}_I\}$ have been calculated from Eq. (5), and where in Eq. (6) we have used Eqs. (3) and (4) to introduce the GS energy E.

Up till now everything has been exact. In practice, of course, approximations need to be made, and these are made within the CCM by restricting the set of indices $\{I\}$ retained in the expansions of Eq. (4) for the otherwise exact correlation operators S and $\tilde{S}$. Some specific such hierarchical scheme are described below. It is important to realize, however, that no further approximations are made. In particular, the method is guaranteed by the use of the exponential parametrizations in Eq. (3) to be size-extensive at every level of truncation, and hence we work from the outset in the $N \to \infty$ limit. Similarly, the important Hellmann-Feynman theorem is similarly exactly obeyed at every level of truncation. Lastly, when the similarity-transformed Hamiltonian $\mathrm{e}^{-S}H\mathrm{e}^{S}$ in Eqs. (5) and (6) is expanded in powers of S using the well-known nested commutator expansion, the fact that S contains only spin-raising operators guarantees that the otherwise infinite expansion actually terminates at a finite order, so that no further approximations are needed.

Once an approximation has been chosen and the retained coefficients $\{\mathcal{S}_I, \tilde{\mathcal{S}}_I\}$ calculated from Eqs. (5) and (6), any GS quantity can, in principle, be calculated. For example, the GS energy E can be calculated in terms of the coefficients $\{\mathcal{S}_I\}$ alone, as $E = \langle\Phi|\mathrm{e}^{-S}H\mathrm{e}^{S}|\Phi\rangle$, while the average on-site GS magnetization (or magnetic order parameter) M needs both sets $\{\mathcal{S}_I\}$ and $\{\tilde{\mathcal{S}}_I\}$ for its evaluation as $M = -\frac{1}{N}\langle\Phi|\tilde{S}\mathrm{e}^{-S}\sum_{k=1}^{N} s_k^z \mathrm{e}^{S}|\Phi\rangle$, in terms of the rotated local spin-coordinate frames defined above.

In our previous work for the $s = \frac{1}{2}$ model [83] we employed the well-known and well-tested localized LSUBm CCM approximation scheme (see Refs. [53, 54]). At the mth level of approximation it includes all spin clusters described by multispin configurations in the index set $\{I\}$ that may be defined over any possible lattice animal (or polyomino) of size m on the lattice. Such a lattice animal is defined in the usual graph-theoretic sense to be a configured set of contiguous sites on the lattice, in which every site in the configuration is adjacent (in the NN sense) to at least one other site. Clearly, as $m \to \infty$ the LSUBm approximation becomes exact. The definition of contiguity employed above depends itself on the choice of "geometry" of the lattice, i.e., on the definition of what is meant by a NN pair. Just as in our previous treatment [83] of the $s = \frac{1}{2}$ version of the present $s = 1$ model, we assume the fundamental checkerboard geometry to define the retained configurations, in which pairs of sites connected either by a J_1 bond or by a J_2 bond are defined to be contiguous (or as NN pairs for the sake of defining a lattice animal of a given size). Although the number of retained configurations at a given mth level of approximation is larger in the checkerboard geometry than in the corresponding square-lattice geometry (for which pairs connected by J_2 bonds would be NNN pairs), the advantage is that the former choice retains many of the symmetries of the checkerboard-lattice model at all levels of approximation that would be lost in the latter choice.

At a given mth level of LSUBm approximation (with *any* fixed choice of underlying geometry to define contiguity) the number, N_f, of such distinct (i.e., under the symmetries of the lattice and specified model state) fundamental spin configurations is lowest for $s = \frac{1}{2}$ and rises steeply as s increases. This is because each downward-pointing (in the rotated local frame) spin on each site k may be operated upon by the spin-raising operator s_k^+ up to $2s$ times. Thus each site index k_i in the operators $C_I^+ \equiv s_{k_1}^+ s_{k_2}^+ \cdots s_{k_n}^+$ may be repeated up to a maximum of $2s$ times. For such $s > \frac{1}{2}$ cases, where individual indices may be repeated, an alternative, so-called SUBn–m, CCM scheme has been used. This scheme doubly restricts the configured clusters included to contain no more than n spin-flips (where each spin-flip requires the action of an s_k^+ operator acting once) spanning a range of no more than m contiguous sites on the lattice. We then set $m = n$ and employ here the SUBn–n scheme. Clearly, the LSUBm scheme is equivalent to the SUBn–m scheme when $n = 2sm$ for particles of spin s. For the $s = \frac{1}{2}$ case only, LSUBm = SUBm–m, whereas for the $s = 1$ case LSUB$m \equiv$ SUB$2m$–m. We note that the corresponding numbers, N_f, of fundamental configurations at a given SUBn–n level are higher for the $s = 1$ case than for the $s = \frac{1}{2}$ case. Thus, whereas for the $s = \frac{1}{2}$ case we were able to perform LSUBm calculations with $m \leq 10$ previously [83], with similar supercomputer resources available we are now only able to perform SUBn–n calculations for the $s = 1$ case that are restricted to $n \leq 8$. As before [83] we similarly use massively parallel computation [107] to derive and solve the corresponding coupled sets of CCM equations (5) and (6).

As a last step we need to extrapolate the approximate SUBn–n results thus obtained to the exact $n \to \infty$ limit. Just as for the $s = \frac{1}{2}$ version of the model [83] we use for the $s = 1$ version the very well tested and very robust approximation scheme,

$$\frac{E(n)}{N} = a_0 + a_1 n^{-2} + a_2 n^{-4}, \tag{7}$$

for the GS energy per spin [37, 38, 41, 61–63, 65–70, 72–87]. Not surprisingly, the GS expectation values of other physical observables both tend to converge more slowly than the GS energy, and with leading exponents that can also depend on the model and regime under study. More specifically, the amount of frustration present can then often determine the scaling.

For example, for most systems with no or moderate amounts of frustration present, the magnetic order parameter M has been widely found [61–63, 65, 72, 74–76] to obey a scaling law with leading power $1/n$ (rather than with $1/n^2$ as for the GS energy). In such cases an extrapolation scheme of the form

$$M(n) = b_0 + b_1 n^{-1} + b_2 n^{-2}, \tag{8}$$

works well. However, for systems which are close to a quantum critical point (QCP) or for which the magnetic order parameter for the phase being studied is zero or close to zero, the extrapolation scheme of Eq. (8) tends always to overestimate the (extrapolated value of the) order parameter and hence also to predict a somewhat too large value for the critical strength of the frustrating interaction that drives the transition under study. In such cases much evidence has by now been accumulated that a scaling law with leading power $1/n^{1/2}$ works much better to fit the SUBn–n data. In such cases we then use the alternative well-studied extrapolation scheme [37, 38, 41, 67–70, 73, 77–87],

$$M(n) = c_0 + c_1 n^{-1/2} + c_2 n^{-3/2}. \tag{9}$$

For any physical observable P of any spin-lattice model being studied by the CCM, we may obviously always test for the correct scaling by first fitting the SUBn–n results to a form

$$P(n) = p_0 + p_1 n^{-\nu}, \tag{10}$$

where the leading exponent ν is also a fitting parameter. For the GS energy E such fits generally yield a fitted value of ν very close to 2 for a wide variety of both unfrustrated and (even highly) frustrated systems in different phases, as is also the case here. Such a preliminary fit then justifies the use of Eq. (7) to find the extrapolated value of E/N. A similar preliminary analysis for the order parameter M can be done in specific cases to justify the use of either Eq. (8) or Eq. (9) to find the extrapolated value of M.

For the model at hand we have performed extrapolations for all of the physical observables calculated using each of the SUBn–n data sets with $n = \{2, 4, 6, 8\}$ and $n = \{4, 6, 8\}$ (and also $n = \{2, 4, 6\}$, even though this set is clearly not a preferred set), as a further test of the robustness of the schemes used. In each case we find very similar results for the extrapolated values, thereby lending credence to the extrapolation schemes used to find them.

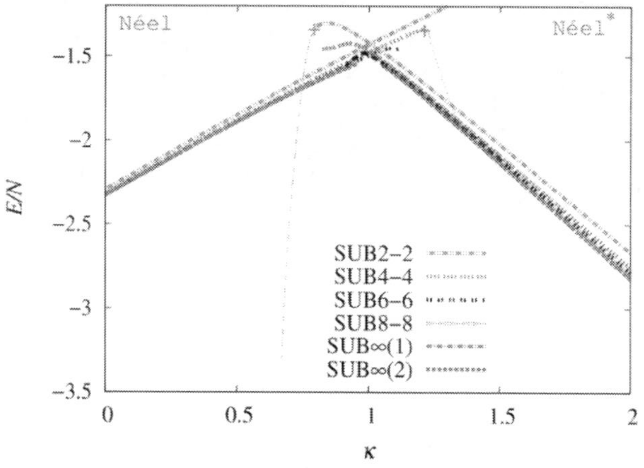

Figure 2. CCM results for the GS energy per spin, E/N, as a function of the frustration parameter, $\kappa \equiv J_2/J_1$, for the spin-1 J_1–J_2 Heisenberg antiferromagnet (with $J_1 \equiv 1$) on the checkerboard lattice. The results for the SUBn–n approximations with $n = \{2, 4, 6, 8\}$ based on the Néel state (left curves) and Néel* state (right curves) as the respective CCM model state are shown, together with the corresponding SUB$\infty(k)$ extrapolations obtained from using Eq. (7) together with the respective SUBn–n data sets: $k = 1$, $n = \{2, 4, 6, 8\}$; and $k = 2$, $n = \{4, 6, 8\}$. All SUBn–n solutions are shown out to their respective (approximately determined) mathematical termination points; and the plus (+) symbols mark those points where the corresponding solutions have vanishing order parameter $M = 0$ (and see Fig. 3(a)). Those portions of the curves beyond the plus (+) symbols, shown with thinner curves, indicate unphysical regions, where $M < 0$ for these approximations (and see text for further details).

4. Results

We now present our CCM results for the spin-1 J_1–J_2 model on the checkerboard lattice, using both the Néel and Néel* states shown in Fig. 1 as model states, and employing the SUBn–n truncation scheme with $n \leq 8$. Figure 2 firstly shows the results for the GS energy per spin, E/N. We display both "raw" SUBn–n results and extrapolations to the $n \to \infty$ limit based on the use of Eq. (7) with either the data set $n = \{2, 4, 6, 8\}$ or $n = \{4, 6, 8\}$. The first thing to notice is how well converged the results are for both sets of results based on the Néel state (left curves) and Néel* state (right curves). Secondly, we note that, exactly as for the $s = \frac{1}{2}$ case studied previously [83], both sets of curves display termination points, at particular values of κ, namely, an upper one for the Néel curves and a lower one for the Néel* curves, beyond which no real solutions exist. The termination points themselves depend on the particular SUBn–n approximation used. What is generally observed is that as the truncation index n is increased the range of values of the frustration parameter κ over which the corresponding SUBn–n approximations have real solutions decreases, as may clearly be seen from Fig. 2. Such CCM termination points are by now well understood [54, 74]. Indeed, they provide a clear first signal of the corresponding QCPs that exist in the system under study. However, we note that it is computationally expensive to obtain the actual termination point with great accuracy, since the CCM SUBn–n solutions require increasingly more computing power the nearer one approaches a termination point. Thus, particularly for the higher values of the truncation index n, it is almost certain that real solutions exist for slightly larger ranges of κ than those shown.

In the vicinity of any such mathematical CCM SUBn–n (or LSUBm) solution termination points it is commonly found, as is the case here too as we shall see explicitly below, that the corresponding solutions themselves become unphysical in the sense that the respective values of the magnetic order parameter (namely, the local on-site magnetization, M) become negative. These values where M changes sign in this way, determined as discussed below, are denoted by plus sign (+) symbols in Fig. 2, and the unphysical regions beyond them, out to the mathematical termination points, where $M < 0$ are shown by thinner curves than the corresponding physical regions where $M > 0$ that are marked by thicker curves.

We note from Fig. 2 that the corresponding SUBn–n curves for E/N based on the Néel and the Néel* states, intersect one another for the same level n of truncation, at least for $n \leq 6$. It is almost certain that this is true too for the $n = 8$ curves, although it is computationally expensive to find the solutions in this region, as indicated above. Furthermore, we note that as the truncation index n increases, the angle of intersection of the corresponding SUBn–n Néel and Néel* curves decreases. Indeed, the corresponding extrapolated curves seem highly likely to join smoothly. Thus, there are clear preliminary indications that the first-order phase transition at $\kappa_{\rm cl} = 1$ in the classical ($s \to \infty$) version of the model might even become a continuous second-order transition in its $s = 1$ counterpart. The energy crossing point of the Néel and Néel* results for the $s = 1$ model is estimated to be $\kappa_{c_1} \approx 1.00 \pm 0.01$ based on our CCM SUBn–n results for $n = \{2, 4, 6\}$ and a corresponding extrapolation based on these results.

Thus, based on the energy results alone, the $s = 1$ model seems to share this transition point at which Néel order melts with its classical ($s \to \infty$) counterpart. Nevertheless, there are first indictions that the classical first-order transition at $\kappa_{\rm cl} = 1$ might become of second-order type for the $s = 1$ model. By contrast, as was shown earlier [83], in the $s = \frac{1}{2}$ model Néel order melts at a lower QCP, namely $\kappa_c^1 \approx 0.80 \pm 0.01$, at which point it gives way not to the quasiclassical Néel* state (or any other of the infinitely degenerate family of classical states) but to a PVBC-ordered state.

In order to investigate the overall level of accuracy of our results it is worthwhile to examine the two special limiting cases of $\kappa = 0$ (square-lattice HAFM) and $\kappa \to \infty$ (decoupled 1D HAFM chains). Thus, firstly, based on the Néel state as the CCM model state, our extrapolated results for the GS energy per spin for the spin-1 square-lattice HAFM are $E(\kappa = 0)/N \approx -2.3287$ based on the SUBn–n results using the data set $n = \{4, 6, 8\}$ and $E(\kappa = 0)/N \approx -2.3292$ using the corresponding data set $n = \{2, 4, 6, 8\}$. These are in very good agreement with the alternative results $E(\kappa = 0)/N \approx -2.3282$ based on third-order spin-wave theory (SWT-3) [108] and $E(\kappa = 0)/N = -2.3279(2)$ based on a linked-cluster series expansion (SE) technique [109]. Our results are also in essentially exact agreement for this $\kappa = 0$ case with previous CCM estimates that are similar limiting cases of spin-1 J_1–J_2 models on the Union Jack lattice [76] and an anisotropic triangular lattice [79], both of which reduce to the square-lattice HAFM when $J_2 = 0$. Secondly, based on the Néel* state as CCM model state our extrapolated results for the GS energy per spin for the spin-1 1D HAFM chain are $E(\kappa \to \infty)/N = -1.3954\kappa$ based on the SUBn–n results using the data set $n = \{4, 6, 8\}$ and $E(\kappa \to \infty)/N = -1.3917\kappa$ using the corresponding set $n = \{2, 4, 6, 8\}$. In this special 1D limiting case essentially exact results are available from density-matrix renormalization group (DMRG) calculations [110], which yield $E(\kappa \to \infty)/N = -1.4015\kappa$. Once again, in this particularly challenging limit for the current model, our CCM results are in very good agreement with the DMRG result.

In Fig. 3 we now show our corresponding results for the GS magnetic order parameter, M, to those shown in Fig. 2 for the GS energy per spin, E/N. Firstly, in Fig. 3(a) we show the "raw" SUBn–n results based on both the Néel state (left curves) and Néel* state (right curves) used as the CCM model state. We note that the plus sign (+) symbols shown in Fig. 2 for the corresponding energy results correspond precisely to the respective points in Fig. 3(a) at which M passes through zero. One sees clearly that in the vicinity of $\kappa = 1$, the SUBn–n curves for M become increasingly steep as n increases in value, thereby also demonstrating why we find increased difficulty in obtaining solutions in this region, which is also close to the corresponding

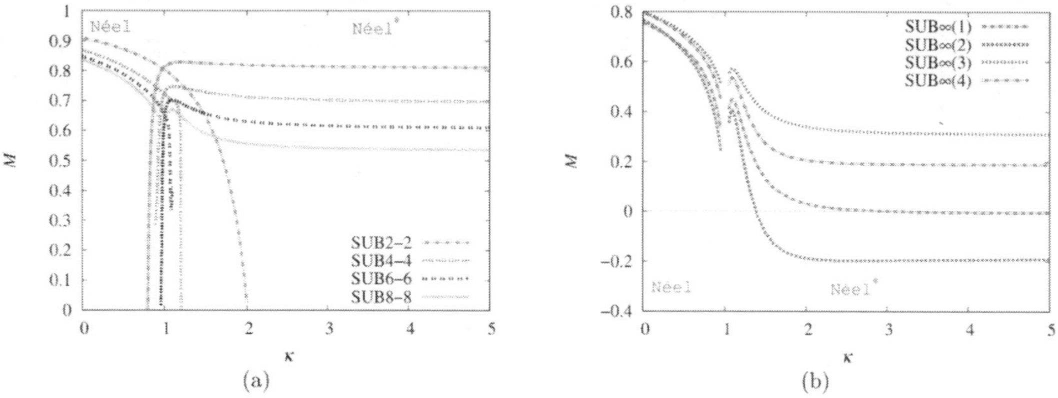

Figure 3. CCM results for the GS magnetic order parameter, M, as a function of the frustration parameter, $\kappa \equiv J_2/J_1$, for the spin-1 J_1–J_2 Heisenberg antiferromagnet (with $J_1 > 0$) on the checkerboard lattice. (a) The results for the SUBn–n approximations with $n = \{2, 4, 6, 8\}$ based on the Néel state (left curves) and Néel* state (right curves) as the respective CCM model state. (b) The corresponding SUB$\infty(k)$ extrapolations: those with $k = \{1, 2\}$ use the extrapolation scheme of Eq. (9), while those with $k = \{3, 4\}$ use the extrapolation scheme of Eq. (8), in both cases with the respective SUBn–n data sets: $k = \{1, 3\}$, $n = \{2, 4, 6, 8\}$; and $k = \{2, 4\}$, $n = \{4, 6, 8\}$.

termination points discussed previously. The raw SUBn–n results show very clearly the existence of a phase transition with a QCP at or very close to $\kappa = 1$, just as did the corresponding results for the GS energy shown in Fig. 2.

In Fig. 3(b) we also show the extrapolations of the results in Fig. 3(a), based on both the extrapolation schemes of Eqs. (8) and (9), and using each of the SUBn–n data sets $n = \{2, 4, 6, 8\}$ and $n = \{4, 6, 8\}$ in the two cases, for completeness and the sake of comparison. As we noted earlier, fits to each data set for M for both model states used, of the form of Eq. (10), can also be performed to estimate the leading exponent and hence to determine which (if either) of the two forms of Eqs. (8) or (9) is appropriate in particular cases. We refer the interested reader to Ref. [88], for example, for a fuller description of the procedure. Based on such fits and much prior experience, the extrapolation scheme of Eq. (8) is now clearly indicated for our Néel-state results over most of the range of values of κ for which solutions exist, except very near the point $\kappa = 1$ where the extrapolation scheme of Eq. (9) becomes validated.

In particular, the use of Eq. (8) is clearly indicated at the point $\kappa = 0$, corresponding to the special limiting case of the square-lattice HAFM. Our corresponding extrapolated results for the GS magnetic order parameter for the spin-1 square-lattice HAFM, using Eq. (8), are $M(\kappa = 0) \approx 0.796$ based on SUBn–n results with $n = \{4, 6, 8\}$, and $M(\kappa = 0) \approx 0.798$ with $n = \{2, 4, 6, 8\}$. Once again, these are in very good agreement with such independent results as $M(\kappa = 0) \approx 0.804$ based on SWT-3 calculations [108] and $M(\kappa = 0) = 0.8039(4)$ from linked-cluster SE calculations [109]. Our current results are also in complete accord with previous CCM estimates [76, 79] discussed above in connection with the corresponding results for the GS energy.

We note from Fig. 3(a) that the SUBn–n results based on the Néel state as CCM model state converge (with increasing values of n) much faster than those for the Néel* state, which is fully consistent with a smaller value of the leading exponent ν in the scaling law for the latter results than for the former, over most of their respective ranges of existence. In this case, as explained before, the extrapolation scheme of Eq. (9), which has been validated by much prior experience, is found to be appropriate here too. The corresponding extrapolations based on the two data sets $n = \{2, 4, 6, 8\}$ and $n = \{4, 6, 8\}$, shown in Fig. 3(b), now differ somewhat from each other. There are two reasons that come immediately to mind as possible explanations for this difference. The first is that the inclusion of the results with $n = 2$ might bias the results somewhat, in view of them being rather far from the $n \to \infty$ limit. For this reason we usually prefer to exclude data with $n = 2$. Another argument for doing so in general for square lattices is that since the basic square plaquette is an important structural element of models on the square lattice, approximations with $n \geq 4$ are preferred *a priori*. However, since we employ here SUBn–n approximations based on the checkerboard geometry rather than the square-lattice geometry, such an argument loses much of its validity. By contrast, a second reason for the difference in extrapolations based on the two data sets shown is that 3-parameter fits based on only 3 inputs are inherently less stable than those based on 4 or more inputs, and for that reason we generally argue for fits of the latter type. Clearly, the above two arguments are in conflict over whether or not it is better to include or exclude the SUB2-2 results in our fits. From both Figs. 2 and 3(b) we observe that the fits are, nonetheless, remarkably similar in both cases for each of the GS quantities E/N and M, with the single exception of results for M in the region $\kappa \gtrsim 1$.

Despite the above caveat, it is abundantly clear that all of the results for M in Figs. 3(a) and 3(b) point very strongly to a quantum phase transition at the same value $\kappa_{c_1} \approx 1.00 \pm 0.01$ as indicated by the GS energy results in Fig. 2. It is clear too that the transition there is a very sharp one. The more likely scenario, based on these results, is of a continuous (but very steep or sharp) second-order transition at which $M \to 0$ on both sides of the transition. Clearly, however, we cannot exclude a weak first-order transition either in which M approaches the same small (but nonzero) value on both sides of the transition or in which $M \to 0$ continuously on one side (likely the Néel side) and then discontinuously jumps to a small (but nonzero) value on the other (most likely, Néel*) side.

We turn finally to the results of M in Fig. 3(b) based on the Néel* state in the region $\kappa > 1$. As indicated above the appropriate extrapolations in this region should be based on Eq. (9), and hence the relevant curves are those labelled SUB∞(1) and SUB∞(2). Conflicting reasons have already been discussed with regard to which of these two extrapolations to prefer, and it is hard to make an *a priori* decision on this basis. Nevertheless, we may also appeal to the known result that the spin-1 1D Haldane chain has $M = 0$. Since this is precisely our limiting case $\kappa \to \infty$, on this basis the extrapolation curve SUB∞(1) is clearly preferred, since it gives this result within very small errors.

What is clear from Fig. 3(b), however, is that whether we use the SUB∞(1) or SUB∞(2) result, there is clear evidence that Néel* order is present only over a range of values of $\kappa_{c_1} < \kappa < \kappa_{c_2}$ of the frustration parameter. The SUB∞(2) curve shows $M < 0$ in this case for $\kappa > \kappa_{c_2} \approx 1.4$ whereas the SUB∞(1) curve shows the more physical result that $M = 0$ (within very small errors) for $\kappa > \kappa_{c_2} \approx 2.5$. The existence of this finite region of stable Néel* order is quite different both from the $s = \frac{1}{2}$ version of the model (for which Néel* order exists nowhere as a stable GS phase) and the classical ($s \to \infty$) version (for which it co-exists with an infinite family of states with AFM ordering along crossed J_2-coupled chains as the stable GS phase for *all* values $\kappa > \kappa_{\rm cl} = 1$).

In Sec. 5 we summarize our results and compare further the present $s = 1$ model with its $s = \frac{1}{2}$ and classical ($s \to \infty$) counterparts.

5. Discussion and summary

In this paper we have investigated the GS properties and GS ($T = 0$) phase structure of the frustrated spin-1 AFM J_1–J_2 model (with $J_1 > 0, J_2 \equiv \kappa J_1 > 0$) on the checkerboard lattice. To do so we have employed the CCM in the hierarchical SUBn–n approximation scheme carried out to orders $n \leq 8$. As CCM model states we have employed two quasiclassical states, namely the Néel state and the Néel* state. The former is the unique GS phase for $\kappa < \kappa_{\rm cl} = 1$ for the classical version of the model, while the latter is one of an infinitely degenerate family of classical GS phases for $\kappa > \kappa_{\rm cl} = 1$.

We find that for the $s = 1$ model the GS phase is an AFM Néel-ordered state for $\kappa < \kappa_{c_1}$, at which point the staggered Néel magnetization vanishes. Our best estimate for this lower QCP is $\kappa_{c_1} \approx 1.00 \pm 0.01$. On the one hand this value seems to concur with the classical value. On the other hand it is quite different from the $s = \frac{1}{2}$ value of $\kappa_c^1 \approx 0.80 \pm 0.01$, found previously [83].

In the classical version of the model the transition at $\kappa_{\rm cl} = 1$ is a direct first-order one (with a discontinuity in the slope, $dE/d\kappa$, of the GS energy there). We find for the $s = 1$ model that the corresponding transition at κ_{c_1} is considerably "softened," with the most likely scenario being that the transition is now a continuous (second-order) one, although we cannot on the available evidence rule out a weak first-order one. All of our evidence is that, as in the classical model, the transition at κ_{c_1} is a direct one to a GS phase with another quasiclassical form of AFM ordering. We find zero evidence of any intermediate phase, and we can positively exclude such a phase except for a very narrow region around κ_{c_1}. If any such intermediate phase does exist (which we doubt on the present evidence), it can do so only over a tiny region confined to the range $0.99 \lesssim \kappa \lesssim 1.01$. Similarly, if present at all, any region of coexistence of Néel and Néel* ordering is restricted by our results to a correspondingly narrow range.

In the classical checkerboard model the Néel ordering that exists for $\kappa < \kappa_{\rm cl} = 1$ gives way for all $\kappa > \kappa_{\rm cl}$ to an infinitely degenerate family of GS phases characterized by AFM ordering along the (crossed) J_2 chains. In the quasiclassical limit (where one works to leading order in $1/s$) it has been shown [97] that quantum fluctuations select, by the *order by disorder* mechanism [49, 50], a fourfold family of collinear states from among all other classically degenerate states. These comprise the stripe-ordered and Néel* states, both of which are doubly degenerate. In a previous CCM study of the $s = \frac{1}{2}$ version of the present checkerboard model, the Néel* state was found to have lower energy than the stripe-ordered state, and other authors [102] have also found evidence in favour of the Néel* state for this model. Consequently in the present analysis we have also been motivated to use the Néel* state as a CCM model state to investigate the phase structure of the $s = 1$ checkerboard model for $\kappa > \kappa_{c_1}$. Nevertheless, further work should be done in the future to investigate, for example, whether a quasiclassical striped state might lie lower in energy than the Néel* state for the $s = 1$ model.

We have found clear evidence that a Néel* state with nonzero values of the order parameter exists for the $s = 1$ case for values of the frustration parameter $\kappa_{c_1} < \kappa < \kappa_{c_2} \approx 2.0 \pm 0.5$. This finding, perhaps the most striking of this study, differs from both the $s = \frac{1}{2}$ and $s \to \infty$ (classical) versions of the model. Firstly, for the $s = \frac{1}{2}$ model, the previous CCM study [83] found that the Néel* state could not survive quantum fluctuations to form a stable GS phase for *any* values of κ. When used as a CCM model state, although solutions could be found for $\kappa > \kappa_c^1$, the calculated (i.e., extrapolated) value of its order parameter M was found to be zero (or negative) everywhere. By contrast, for the $s = 1$ case we find that when the Néel* state is used as the CCM model state, the calculated value of M is zero (or negative) only for $\kappa > \kappa_{c_2}$ ($> \kappa_{c_1}$). Secondly, by contrast with the classical checkerboard model, where stable nonzero Néel* ordering exists for *all* values $\kappa > \kappa_{\rm cl}$, its $s = 1$ counterpart exists only over a finite range of values of κ.

The question thus remains as to what is the stable GS phase of the $s = 1$ checkerboard model for $\kappa > \kappa_{c_2}$. We note that for its $s = \frac{1}{2}$ counterpart, the previous CCM study [83] found a PVBC-ordered phase for $\kappa_c^1 < \kappa < \kappa_c^2 \approx 1.22 \pm 0.02$, which then gave way to a CDVBC-ordered phase for all $\kappa > \kappa_c^2$. We have also performed very preliminary calculations (which we will report on, more fully, elsewhere) for these phases for the $s = 1$ model, using the same CCM technique as reported previously [83] for the $s = \frac{1}{2}$ case. Very interestingly, the evidence so far is that for all values $\kappa > \kappa_{c_2}$ the GS phase of the $s = 1$ model seems to take PVBC ordering. Again, if this result stands up to further scrutiny, the $s = 1$ model will again show distinct differences to both its $s = \frac{1}{2}$ and $s \to \infty$ counterparts.

Finally, we make some brief remarks regarding the order of the various quantum phase transitions exhibited by the checkerboard model, and in particular whether specific transitions are allowed to be continuous. Although a general renormalization group theory approach to continuous critical phenomena itself places only rather weak constraints on the existence of a continuous phase transition between any two quantum phases, the traditional (and conventional) Landau-Ginzburg-Wilson (LGW) approach [111, 112] places stricter criteria. In a nutshell LGW theory places as a necessary condition on a continuous transition from a phase $\mathcal{P}_1$ to a phase $\mathcal{P}_2$ that the symmetry group of phase $\mathcal{P}_2$ is a subgroup of that $\mathcal{P}_1$ [111]. Physically, the order parameter of some mode of phase $\mathcal{P}_1$ goes to zero at the transition as the mode becomes soft and macroscopic condensation into it hence occurs, with a consequent symmetry reduction.

A detailed analysis and description of the various LGW-allowed continuous phase transitions for the checkerboard model has been given elsewhere [102]. In particular it is shown explicitly that the symmetry group of the Néel* state is a subgroup of that of the PVBC state. Hence, our tentative identification of the QCP at κ_{c_2} in the $s = 1$ checkerboard model as being between states with Néel* and PVBC order, would be LGW-allowed as a continuous transition, and it will be interesting to examine the order of the transition in more detail. By contrast, it is not difficult to show (and see, e.g., Ref. [102] for explicit details) that the Néel* and CDVBC states break *different* symmetries of the checkerboard lattice (and hence of the Hamiltonian), so that the symmetry group of neither state is a subgroup of the other. Hence, any continuous phase transition between states with Néel* and CDVBC ordering would be LGW-forbidden. In such a case the most likely scenarios would be either a direct first-order transition or one that involves an intermediate coexistence phase with intermediate ordering and bond modulation. Reference [102] describes the possible properties of such a coexistence phase and the nature of the symmetry breakings at the two transition points from it into each of the pure phases.

The nature of the QCP at κ_c^2 in the $s = \frac{1}{2}$ model is also more open than that at κ_{c_2} in the $s = 1$ model. Thus, again, while both the CDVBC and PVBC phases are doubly degenerate (and can thus be described via Ising order parameters), it is easy to see that they have distinct symmetries [102]. Thus, again, neither state has a symmetry group which is a subgroup of that of the other, and a continuous transition between them is LGW-forbidden.

Lastly, we note too that the transitions at κ_c^1 in the $s = \frac{1}{2}$ model and at κ_{c_1} in the $s = 1$ model, from the Néel phase to, respectively, the PVBC phase (for $s = \frac{1}{2}$) and the Néel* phase (for $s = 1$) are both also LGW-forbidden as continuous transitions. Thus, for example, the Néel* (and PVBC) phases both have rotations of $\frac{1}{2}\pi$ about the centre of any square plaquette containing four parallel spins as symmetries, which are not shared with the Néel phase. Hence, if the transition at κ_{c_1} for the present $s = 1$ model is indeed continuous, as seems possible from our results, its nature has to be sought outside the conventional LGW paradigm. One such possibility, which takes us too far afield to study further here however, is via the deconfinement scenario [113].

In conclusion, the $s = 1$ checkerboard model has been shown to exhibit some very interesting features of its GS ($T = 0$) phase diagram that are qualitatively different to those of both its $s = \frac{1}{2}$ and $s \to \infty$ (classical) counterparts. In future work we intend both to investigate the relative stabilities of the striped and Néel* GS phases in the intermediate regime $\kappa_{c_1} < \kappa < \kappa_{c_2}$, and to perform a rigorous study of the stability of possible phases with VBC order, particularly those with PVBC and CDVBC ordering, in the regime $\kappa > \kappa_{c_2}$.

Acknowledgment
We thank the University of Minnesota Supercomputing Institute for Digital Simulation and Advanced Computation for the grant of supercomputing facilities, on which we relied heavily for the numerical calculations reported here.

References

[1] Haldane F D M 1983 *Phys. Rev. Lett.* **50** 1153; Haldane F D M 1983 *Phys. Lett.* A **93** 464
[2] Steiner M, Kakurai K, Kjems J K, Petitgrand D and Pynn R 1987 *J. Appl. Phys.* **61** 3953
[3] Darriet J and Regnault L P 1993 *Solid State Commun.* **86** 409
[4] Dorner B, Visser D, Steigenberger U, Kakurai K and Steiner M 1988 *Z. Phys.* B **72** 487
[5] Renard J P, Verdaguer M, Regnault L P, Erkelens W A C and Rossat-Mignod J 1988 *J. Appl. Phys.* **63** 3538
[6] Orendáč M, Orendáčová A, Černák J, Feher A, Signore P J C, Meisel M W, Merah S and Verdaguer M 1995 *Phys. Rev.* B **52** 3435
[7] Zapf V S, Zocco D, Hansen B R, Jaime M, Harrison N, Batista C D, Kenzelmann M, Niedermayer C, Lacerda A and Paduan-Filho A 2006 *Phys. Rev. Lett.* **96** 077204
[8] Birgeneau R J, Skalyo J Jr. and Shirane G 1970 *J. Appl. Phys.* **41** 1303
[9] Nakatsuji S, Nambu Y, Tonomura H, Sakai O, Jonas S, Broholm C, Tsunetsugu H, Qiu Y M and Maeno Y 2005 *Science* **309** 1697
[10] Läuchli A, Mila F and Penc K 2006 *Phys. Rev. Lett.* **97** 087205
[11] Tsunetsugu H and Arikawa M 2006 *J. Phys. Soc. Japan* **75** 083701
[12] Bhattacharjee S, Shenoy V B and Senthil T 2006, *Phys. Rev.* B **74** 092406
[13] Kamihara Y, Watanabe T, Hirano M and Hosono H 2008 *J. Am. Chem. Soc.* **130** 3296
[14] Ma F, Lu Z-Y and Xiang T 2008 *Phys. Rev.* B **78** 224517
[15] Si Q and Abrahams E 2008 *Phys. Rev. Lett.* **101** 076401
[16] Chandra P and Doucot B 1988 *Phys. Rev.* B **38** 9335
[17] Gelfand M P, Singh R R P and Huse D A 1989 *Phys. Rev.* B **40** 10801
[18] Dagotto E and Moreo A 1989 *Phys. Rev. Lett.* **63** 2148
[19] Sachdev S and Bhatt R 1990 *Phys. Rev.* B **41** 9323
[20] Chubukov A V and Jolicoeur Th 1991 *Phys. Rev.* B **44** 12050
[21] Read N and Sachdev S 1991 *Phys. Rev. Lett.* **66** 1773
[22] Schulz H J and Ziman T A L 1992 *Europhys. Lett.* **18** 355
[23] Richter J 1993 *Phys. Rev.* B **47** 5794
[24] Richter J, Ivanov N B and Retzlaff K 1994 *Europhys. Lett.* **25** 545
[25] Oitmaa J and Weihong Z 1996 *Phys. Rev.* B **54** 3022
[26] Schulz H J, Ziman T A L and Poilblanc D 1996 *J. Physique* I **6** 675
[27] Zhitomirsky M E and Ueda K 1996 *Phys. Rev.* B **54** 9007
[28] Bishop R F, Farnell D J J and Parkinson J B 1998 *Phys. Rev.* B **58** 6394
[29] Singh R R P, Weihong Z, Hamer C J and Oitmaa J 1999 *Phys. Rev.* B **60** 7278
[30] Kotov V N, Oitmaa J, Sushkov O P and Weihong Z 1999 *Phys. Rev.* B **60** 14613
[31] Capriotti L and Sorella S 2000 *Phys. Rev. Lett.* **84** 3173
[32] Capriotti L, Becca F, Parola A and Sorella S 2001 *Phys. Rev. Lett.* **87** 097201
[33] Takano K, Kito Y, Ono Y and Sano K 2003 *Phys. Rev. Lett.* **91** 197202
[34] Roscilde T, Feiguin A, Chernyshev A L, Liu S and Haas S 2004 *Phys. Rev. Lett.* **93** 017203
[35] Sirker J, Weihong Z, Sushkov O P and Oitmaa J 2006 *Phys. Rev.* B **73** 184420
[36] Mambrini M, Läuchli A, Poiblanc D and Mila F 2006 *Phys. Rev.* B **74** 144422
[37] Schmalfuß D, Darradi R, Richter J, Schulenburg J and Ihle D 2006 *Phys. Rev. Lett.* **97** 157201
[38] Darradi R, Derzhko O, Zinke R, Schulenburg J, Krüger S E and Richter J 2008 *Phys. Rev.* B **78** 214415
[39] Ralko A, Mambrini M and Poiblanc D 2009 *Phys. Rev.* B **80** 184427
[40] Richter J and Schulenburg J 2010 *Eur. Phys. J.* B **73** 117
[41] Reuther J, Wölfle P, Darradi R, Brenig W, Arlego M and Richter J 2011 *Phys. Rev.* B **83** 064416
[42] Yu J-F and Kao Y-J 2012 *Phys. Rev.* B **85** 094407
[43] Götze O, Krüger S E, Fleck F, Schulenburg J and Richter J 2012 *Phys. Rev.* B **85** 224424
[44] Jiang H-C, Yao H and Balents L 2012 *Phys. Rev.* B **86** 024424
[45] Mezzacapo F 2012 *Phys. Rev.* B **86** 045115
[46] Li T, Becca F, Hu W and Sorella S 2012 *Phys. Rev.* B **86** 075111
[47] Wang L, Poilblanc D, Gu Z-C, Wen X-G and Verstraete F 2013 *Phys. Rev. Lett.* **111** 037202
[48] Hu W-J, Becca F, Parola A and Sorella S 2013 *Phys. Rev.* B **88** 060402(R)

[49] Villain J 1977 *J. Physique* **38** 385; Villain J, Bidaux R, Carton J P and Conte R 1980 *J. Physique* **41** 1263
[50] Shender E 1982 *Sov. Phys.-JETP* **56** 178
[51] Bishop R F 1991 *Theor. Chim. Acta* **80** 95
[52] Bishop R F 1998 *Microscopic Quantum Many-Body Theories and Their Applications* (*Springer Lecture Notes in Physics* vol 510) ed J Navarro and A Polls (Berlin: Springer) p 1
[53] Zeng C, Farnell D J J and Bishop R F 1998 *J. Stat. Phys.* **90** 327
[54] Farnell D J J and Bishop R F 2004 *Quantum Magnetism* (*Springer Lecture Notes in Physics* vol 645) ed U Schollwöck, J Richter, D J J Farnell and R F Bishop (Berlin: Springer) p 307
[55] Roger M and Hetherington J H 1990 *Phys. Rev.* B **41** 200
[56] Bishop R F, Parkinson J B and Xian Y 1991 *Phys. Rev.* B **43** 13782(R)
[57] Bishop R F, Parkinson J B and Xian Y 1992 *Phys. Rev.* B **46** 880
[58] Bishop R F, Parkinson J B and Xian Y 1992 *J. Phys.: Condens. Matter* **4** 5783
[59] Bishop R F, Parkinson J B and Xian Y 1993 *J. Phys.: Condens. Matter* **5** 9169
[60] Zeng C, Staples I and Bishop R F 1996 *Phys. Rev.* B **53** 9168
[61] Bishop R F, Farnell D J J, Krüger S E, Parkinson J B, Richter J and Zeng C 2000 *J. Phys.: Condens. Matter* **12** 6887
[62] Krüger S E, Richter J, Schulenburg J, Farnell D J J and Bishop R F 2000 *Phys. Rev.* B **61** 14607
[63] Farnell D J J, Gernoth K A and Bishop R F 2001 *Phys. Rev.* B **64** 172409
[64] Farnell D J J, Bishop R F and Gernoth K A 2002 *J. Stat. Phys.* **108** 401
[65] Darradi R, Richter J and Farnell D J J 2005 *Phys. Rev.* B **72** 104425
[66] Farnell D J J and Bishop R F 2008 *Int. J. Mod. Phys.* B **22** 3369
[67] Bishop R F, Li P H Y, Darradi R, Schulenburg J and Richter J 2008 *Phys. Rev.* B **78** 054412
[68] Bishop R F, Li P H Y, Darradi R and Richter J 2008 *J. Phys.: Condens. Matter* **20** 255251
[69] Bishop R F, Li P H Y, Darradi R and Richter J 2008 *Europhys. Lett.* **83** 47004
[70] Bishop R F, Li P H Y, Darradi R, Richter J and Campbell C E 2008 *J. Phys.: Condens. Matter* **20** 415213
[71] Farnell D J J, Richter J, Zinke R and Bishop R F 2009 *J. Stat. Phys.* **135** 175
[72] Bishop R F, Li P H Y, Farnell D J J and Campbell C E 2009 *Phys. Rev.* B **79** 174405
[73] Richter J, Darradi R, Schulenburg J, Farnell D J J and Rosner H 2010 *Phys. Rev.* B **81** 174429
[74] Bishop R F, Li P H Y, Farnell D J J and Campbell C E 2010 *Phys. Rev.* B **82** 024416
[75] Bishop R F, Li P H Y, Farnell D J J and Campbell C E 2010 *Phys. Rev.* B **82** 104406
[76] Bishop R F and Li P H Y 2011 *Eur. Phys. J.* B **81** 37
[77] Farnell D J J, Bishop R F, Li P H Y, Richter J and Campbell C E 2011 *Phys. Rev.* B **84** 012403
[78] Götze O, Farnell D J J, Bishop R F, Li P H Y and Richter J 2011 *Phys. Rev.* B **84** 224428
[79] Bishop R F and Li P H Y 2012 *Eur. Phys. J.* B **85** 25
[80] Li P H Y, Bishop R F, Farnell D J J, Richter J and Campbell C E 2012 *Phys. Rev.* B **85** 085115
[81] Bishop R F, Li P H Y, Farnell D J J and Campbell C E 2012 *J. Phys.: Condens. Matter* **24** 236002
[82] Bishop R F and Li P H Y 2012 *Phys. Rev.* B **85** 155135
[83] Bishop R F, Li P H Y, Farnell D J J, Richter J and Campbell C E 2012 *Phys. Rev.* B **85** 205122
[84] Li P H Y, Bishop R F, Farnell D J J and Campbell C E 2012 *Phys. Rev.* B **86** 144404
[85] Li P H Y, Bishop R F, Campbell C E, Farnell D J J, Götze O and Richter J 2012 *Phys. Rev.* B **86** 214403
[86] Bishop R F, Li P H Y and Campbell C E 2013 *J. Phys.: Condens. Matter* **25** 306002
[87] Li P H Y, Bishop R F and Campbell C E 2013 *Phys. Rev.* B **88** 144423
[88] Bishop R F, Li P H Y and Campbell C E 2013 *arXiv:1308.4573v2* [cond-mat.str-el]
[89] Singh R R P, Starykh O A and Freitas P J 1998 *J. Appl. Phys.* **83** 7387
[90] Palmer S E and Chalker J T 2001 *Phys. Rev.* B **64** 094412
[91] Brenig W and Honecker A 2002 *Phys. Rev.* B **65** 140407(R)
[92] Canals B 2002 *Phys. Rev.* B **65** 184408
[93] Starykh O A, Singh R R P and Levine G C 2002 *Phys. Rev. Lett.* **88** 167203
[94] Sindzingre P, Fouet J-B and Lhuillier C 2002 *Phys. Rev.* B **66** 174424
[95] Fouet J-B, Mambrini M, Sindzingre P and Lhuillier C 2003 *Phys. Rev.* B **67** 054411
[96] Berg E, Altman E and Auerbach A 2003 *Phys. Rev. Lett.* **90** 147204
[97] Tchernyshyov O, Starykh O A, Moessner R and Abanov A G 2003 *Phys. Rev.* B **68** 144422
[98] Moessner R, Tchernyshyov O and Sondhi S L 2004 *J. Stat. Phys.* **116** 755
[99] Hermele M, Fisher M P A and Balents L 2004 *Phys. Rev.* B **69** 064404
[100] Brenig W and Grzeschik M 2004 *Phys. Rev.* B **69** 064420
[101] Bernier J S, Chung C H, Kim Y B and Sachdev S 2004 *Phys. Rev.* B **69** 214427
[102] Starykh O A, Furusaki A and Balents L 2005 *Phys. Rev.* B **72** 094416
[103] Schmidt H-J, Richter J and Moessner R 2006 *J. Phys. A: Math. Gen.* **39** 10673
[104] Arlego M and Brenig W 2007 *Phys. Rev.* B **75** 024409; Arlego M and Brenig W 2009 *Phys. Rev.* B **80**

099902(E)
[105] Moukouri S 2008 *Phys. Rev.* B **77** 052408
[106] Chan Y-H, Han Y-J and Duan L-M 2011 *Phys. Rev.* B **84** 224407
[107] We use the program package CCCM of Farnell D J J and Schulenburg J, see http://www-e.uni-magdeburg.de/jschulen/ccm/index.html
[108] Hamer C J, Zheng Weihong and Arndt P 1992 *Phys. Rev.* B **46** 6276
[109] Zheng Weihong, Oitmaa J and Hamer C J 1991 *Phys. Rev.* B **43** 8321
[110] White S R and Huse D A 1993 *Phys. Rev.* B **48** 3844
[111] Landau L D, Lifshitz E M and Pitaevskii L P 1980 *Statistical Physics* (3rd ed) (Oxford: Butterworth-Heinemann)
[112] Wilson K G and Kogut J 1974 *Phys. Rep.* **12** 75
[113] Senthil T, Vishwanath A, Balents L, Sachdev S and Fisher M P A 2004 *Science* **303** 1490

Journal of Physics: Conference Series **529** (2014) 012009 doi:10.1088/1742-6596/529/1/012009

Dynamic structure function of a cold Fermi gas at unitarity

G E Astrakharchik[‡], **J Boronat**[‡], **E Krotscheck**[+†] **and T Lichtenegger**[+†]

[‡]Departament de Física i Enginyeria Nuclear, Campus Nord B4-B5, Universitat Politècnica de Catalunya, E-08034 Barcelona, Spain

[+]Department of Physics, University at Buffalo, SUNY Buffalo NY 14260

[†]Institut für Theoretische Physik, Johannes Kepler Universität, A 4040 Linz, Austria

E-mail: eckhard.krotscheck@jku.at

Abstract. We present a theoretical study of the dynamic structure function of a resonantly interacting two-component Fermi gas at zero temperature. Our approach is based on dynamic many-body theory able to describe excitations in strongly correlated Fermi systems. The fixed-node diffusion Monte Carlo method is used to produce the ground-state correlation functions which are used as an input for the excitation theory. Our approach reproduces recent Bragg scattering data in both the density and the spin channel. In the BCS regime, the response is close to that of the ideal Fermi gas. On the BEC side, the Bose peak associated with the formation of dimers dominates the density channel of the dynamic response. When the fraction of dimers is large our theory departs from the experimental data, mainly in the spin channel.

1. Introduction

An impressive advance in realizing and controlling ultracold Fermi gases has permitted to study physical phenomena whose appearance was previously only a matter of speculation [1]. One of the major advances has been the first clean observation of a crossover between two limits of Fermi matter, the Bose-Einstein condensate (BEC) regime composed by Fermi molecules and the Bardeen-Cooper-Schrieffer (BCS) superfluid gas where Cooper pairs are formed [2, 3, 4, 5].

In the dilute limit, interactions between atoms can be described by a single parameter, namely the s-wave scattering length a. The system evolves from a BCS regime with no bound state ($a < 0$) to a molecular one (BEC) with a two-body bound state ($a > 0$), crossing a singular point where $|a| \to \infty$ corresponding to a Fano-Feshbach resonance [6, 7, 8]. This special point is referred to as the unitary limit and is expected to show a number of universal properties. Indeed, in this regime the only relevant length and energy scales are the inverse Fermi momentum k_F^{-1} and the Fermi energy $E_{\rm F}$, respectively.

Nowadays, the unitary Fermi gas is routinely realized in experiments. An accurate theoretical description of the unitary limit requires the use of advanced methods of many-body theory that can deal with strongly interacting systems. Quantum Monte Carlo methods have proven to be very useful for the calculation of the finite- and zero-temperature equation of state [6, 7, 8, 9] which is found to be in good overall agreement with experimental data [10, 11]. Also, it was possible to predict the effect of the unbalanced population [12] and different masses [13]. More recently, static responses such as the spin susceptibility and spin diffusivity [14, 15] have been measured and also studied theoretically using quantum Monte Carlo (QMC) methods [16] and a diagrammatic approach [17].

Recently, Hoinka *et al.* [18, 19] have measured the dynamic spin and density responses of a Fermi gas at unitarity and in the BEC regime. In order to measure both responses, they used two-photon Bragg scattering, acting on ^{6}Li at very low temperature ($T \sim 0.06\,E_{\rm F}$ at unitarity), and a proper choice of the laser detuning. The density response at unitarity turns out to be significantly different from that of an ideal Fermi gas, showing a clear peak which corresponds to the formation of dimers composed of spin-up and spin-down particles. This signature of dimer formation is not visible in the spin channel, nevertheless the experimental results are quite different from the ideal Fermi gas response. This provides evidence for the importance of correlations in the unitary gas despite its diluteness.

Some of the features shown by the dynamic response function at unitarity were previously determined in theoretical work: at $T = 0$ using a self-consistent mean-field approach based on BCS theory [20], and at $T > 0$ using virial expansions [21]. However, the dynamic properties along the BCS-BEC crossover are not completely understood due to the need of a full many-body theory and the impossibility to use QMC methods for dynamic properties directly. The goal of our study is to utilize successful methods stemming from the many-body theory of strongly interacting systems, combining the virtues of Monte Carlo methods and modern diagrammatic many-body theory, in order to study the dynamic structure function of a low–density gas. For pedagogical material on the techniques to be employed here, see Ref. [22].

2. Dynamic Many-Body Theory

In the present work, we study the density and spin responses of the two-component Fermi gas at unitarity and in characteristic points on both the BEC and BCS sides of the crossover. We use a fully microscopic approach utilizing correlated basis function (CBF) theory [22]. The input to the theory are the ground-state structure functions $S_{\uparrow\uparrow}(k)$ and $S_{\uparrow\downarrow}(k)$ obtained from diffusion Monte Carlo simulations within the fixed-node (FN-DMC) approximation [6]. In these calculations, an attractive square-well potential was used for the interaction. In its most advanced form [23], the dynamic many-body theory (DMBT) used here is the fermion version of the CBF-Brillouin-Wigner perturbation theory previously developed for bosons [24, 25, 26]. The power of DMBT for fermions has been demonstrated by the prediction of a stable roton excitation in two-dimensional ^{3}He, which has independently been confirmed by inelastic neutron scattering experiments [27].

The only further inputs for the calculation of the dynamic response is the fraction of dimers along the BCS-BEC crossover. With that, we obtain a complete description of the density and spin dynamic responses in the momentum-energy plane. In the unitary limit, both theoretical responses are in satisfactory agreement with the recent experimental Bragg scattering measurements [18].

In weakly interacting systems, time-dependent Hartree-Fock (TDHF) theory [28] is a well-established method for capturing the dynamics. The system is subjected to a perturbing Hamiltonian

$$\delta \hat{H}_{\rm ext}(t) = \int d^3r \hat{\rho}^{(\rho/\sigma)}(\mathbf{r}) h_{\rm ext}(\mathbf{r}, t) ; \tag{1}$$

the superscripts ρ and σ stand for density and spin excitations, respectively. $\hat{\rho}^{(\rho/\sigma)}(\mathbf{r})$ is the (spin-)density operator, and $h_{\rm ext}(\mathbf{r}, t)$ is a weak, local external field. The wave function resulting from such a perturbation is assumed to be of the form

$$|\psi(t)\rangle = e^{1/2 \sum_{ph} c_{ph}(t) a_p^\dagger a_h} |\phi_0\rangle \ , \tag{2}$$

where $|\phi_0\rangle$ is the ground state of a non-interacting system with the same density. The particle ("p") and hole ("h") labels run over spatial quantum numbers and spin degrees of freedom.

The amplitudes $c_{ph}(t)$ are Fourier decomposed

$$c_{ph}(t) = c_{ph}^{(+)}(\omega) e^{-\imath(\omega + \imath\eta/\hbar)t} + c_{ph}^{(-)*}(\omega) e^{\imath(\omega + \imath\eta/\hbar)t} \ . \tag{3}$$

Equations of motion are then derived from a least action principle [29, 30]. They are given by

$$(\hbar\omega + \imath\eta - \epsilon_{ph}) c_{ph}^{(+)}(\omega) - \sum_{p'h'} V_{ph;p'h'} c_{p'h'}^{(+)}(\omega) - \sum_{p'h'} V_{pp'hh';0} c_{p'h'}^{(-)}(\omega) = 2 \int d^3r \rho_{ph;0}^{(\rho/\sigma)}(\mathbf{r}) h_{\rm ext}(\mathbf{r}, \omega) \tag{4}$$

$$(-\hbar\omega - \imath\eta - \epsilon_{ph}) c_{ph}^{(-)}(\omega) - \sum_{p'h'} V_{p'h';ph} c_{p'h'}^{(-)}(\omega) - \sum_{p'h'} V_{0;pp'hh'} c_{p'h'}^{(+)}(\omega) = 2 \int d^3r \rho_{0;ph}^{(\rho/\sigma)}(\mathbf{r}) h_{\rm ext}(\mathbf{r}, \omega), \tag{5}$$

where $h_{\rm ext}(\mathbf{r}, \omega)$ is the Fourier component of the external field. To maintain causality, the real frequency ω is supplemented with a small imaginary contribution corresponding to the adiabatic swichting-on of the perturbation. While usually the limit $\eta \to 0^+$ is performed, we use for η a finite, monotonically increasing function of momentum transfer to phenomenologically incorporate the possibility of excitations to decay. Outside the regime of Landau-damping, this effect is due to multi-pair excitations [23].

From the solutions $c_{ph}^{(\pm)}(\omega)$ we can calculate the (spin-) density response function

$$\chi^{(\rho/\sigma)}(\mathbf{r}, \mathbf{r}', \omega) \equiv \frac{1}{\rho(\mathbf{r})\rho(\mathbf{r}')} \frac{\delta \langle\psi| \hat{\rho}^{(\rho/\sigma)} |\psi\rangle (\mathbf{r}, \omega)}{\delta h_{\rm ext}(\mathbf{r}', \omega)} \ , \tag{6}$$

and the (spin-)density dynamic structure function

$$S^{(\rho/\sigma)}(\mathbf{r},\mathbf{r}',\omega) \equiv -\frac{1}{\pi}\Im m\chi^{(\rho/\sigma)}(\mathbf{r},\mathbf{r}',\omega) \tag{7}$$

in coordinate space. In a translationally invariant geometry, their momentum space representations, $\chi^{(\rho/\sigma)}(k,\omega)$ and $S^{(\rho/\sigma)}(k,\omega)$, are functions of the momentum transfer k.

The key ingredients of the equations of motion are the matrix elements of the interparticle interaction

$$V_{ph;p'h'} = \left\langle ph'\right| V_{\rm d} \left|hp'\right\rangle - \left\langle ph'\right| V_{\rm ex} \left|p'h\right\rangle \tag{8}$$

$$V_{pp'hh';0} = \left\langle pp'\right| V_{\rm d} \left|hh'\right\rangle - \left\langle pp'\right| V_{\rm ex} \left|h'h\right\rangle \tag{9}$$

and the single-particle spectrum $\epsilon_{ph} = \epsilon_p - \epsilon_h$. In TDHF theory, the potentials $V_{\rm d}$ and $V_{\rm ex}$ are the same and just the bare interaction between the particles, and the ϵ_p, ϵ_h are the Hartree-Fock single-particle energies. However, close to unitarity, our system is strongly interacting and the TDHF approximation is, as we shall see, vastly inappropriate. For such strongly interacting systems like the helium liquids or strongly correlated electrons, CBF theory has proven to be a suitable tool for making quantitative predictions of the dynamic structure function [23], and there is no reason that the method should not be applicable to cold gases.

3. Calculation of the effective interactions

The effective interactions $V_{\rm d}$ and $V_{\rm ex}$ can be derived in basically two different ways: The first, and original derivation [31] was within correlated basis functions theory: CBF theory starts with a correlated ground state

$$\left|\psi_0\right\rangle = F\left|\phi_0\right\rangle \tag{10}$$

where F is a local correlation operator, most frequently taken in the Jastrow-Feenberg form including two- and three- body correlations. This is among others realized in the fixed-node Monte Carlo method for the non-dimerized component, which basically determines the best local correlation operator. Excitations in strongly correlated systems are then treated by generalizing the *ansatz* (2) to

$$\left|\psi(t)\right\rangle = Fe^{1/2\sum_{ph} c_{ph}(t)a_p^\dagger a_h}\left|\phi_0\right\rangle . \tag{11}$$

Equations of motion are then derived along the same lines as the TDHF equations (4,5). In a rather intricate diagrammatic analysis [22, 32, 33] one arrives at the same equations of motion (4,5) for a strongly interacting system, the basic difference being that the matrix elements of the bare potential are replaced by correlated matrix elements of the form

$$\left\langle \phi_0 a_{h'}^\dagger a_{p'} \middle| F^\dagger(H-E_0)F \middle| a_{p'}^\dagger a_{h'}\phi_0\right\rangle \quad\text{and}\quad \left\langle \phi_0 \middle| F^\dagger(H-E_0)F \middle| a_p^\dagger a_{p'}^\dagger a_{h'} a_h \phi_0\right\rangle . \tag{12}$$

These matrix elements can obviously be written as matrix-elements of a two-body operator. A diagrammatic analysis of these interactions [31, 33, 34] leads to the working formulas for their local parts and, thus, to the replacement of the bare interaction by *effective interactions* $V_{\rm d}$ and $V_{\rm ex}$:

$$\tilde{V}_{\rm d}(k) = \frac{t(k)}{2}\left(\frac{1}{S^{(\rho)}(k)^2} - \frac{1}{S_{\rm F}(k)^2}\right), \tag{13}$$

and

$$\tilde{V}_{\rm ex}(k) = -t(k)\,\frac{S^{(\rho)}(k) - S_{\rm F}(k)}{S_{\rm F}(k)^3}. \tag{14}$$

where $S_{\rm F}(k)$ is the static structure factor of the free Fermi gas, and $t(k) = \frac{\hbar^2 k^2}{2m}$.

The procedure has been described in pedagogical literature [22] and review papers [34] and does not need to be repeated here. The fixed-node approximation is, for the non-dimerized component, equivalent to the optimized Jastrow-Feenberg method, these are the correct expressions to use.

One can obtain the same quantities from the diagrammatic methods of Feynman-diagram based perturbation theory and the connections between that method and CBF theory unveiled by the parquet-diagram analysis [35, 36, 37]. From that analysis, and by comparing the structure of ring-diagrams in perturbation theory and CBF theory [33], it is known that the Jastrow-Feenberg wave function and, equivalently, the fixed-node approximation, amounts to approximating the Lindhard Function $\chi_0(q,\omega)$ by a "collective" or "mean spherical" approximation (MSA):

$$\chi_0^{\rm MSA}(k,\omega) \equiv \frac{2t(k)}{(\omega+i\eta)^2 - \left(\frac{t(k)}{S_{\rm F}(k)}\right)^2}. \tag{15}$$

This approximation for $\chi_0(k,\omega)$ replaces the particle–hole band by an effective collective mode whose energy is chosen such that the first two energy weighted sumrules are fulfilled:

$$\begin{aligned}
\Im m\int_0^\infty d\omega\,\chi_0^{\rm MSA}(k,\omega) &= \Im m\int_0^\infty d\omega\,\chi_0(k,\omega)\,,\\
\Im m\int_0^\infty d\omega\,\omega\chi_0^{\rm MSA}(k,\omega) &= \Im m\int_0^\infty d\omega\,\omega\chi_0(k,\omega)\,.
\end{aligned} \tag{16}$$

Given a local, effective interaction $\tilde{V}_{\rm d}(q)$, one can construct the static structure function by

$$\begin{aligned}
S^{\rm RPA}(k) &= -\int_0^\infty \frac{d(\hbar\omega)}{\pi}\Im m\,\chi^{\rm RPA}(k,\omega)\,,\\
\chi^{\rm RPA}(k,\omega) &= \frac{\chi_0(k,\omega)}{1-\tilde{V}_{\rm d}(k)\chi_0(k,\omega)}.
\end{aligned} \tag{17}$$

When the MSA (15) is used, the frequency integration can be carried out exactly and leads to

$$S^{\rm MSA}(k) = \frac{S_{\rm F}(k)}{\sqrt{1+\frac{2S_{\rm F}^2(k)}{t(k)}\tilde{V}_{\rm d}(k)}} \tag{18}$$

Journal of Physics: Conference Series **529** (2014) 012009 doi:10.1088/1742-6596/529/1/012009

which is seen to be exactly the relationship (13).

The effective interaction in the "direct" channel is particle-hole irreducible, in the exchange channel we must include all particle-hole reducible diagrams. This leads *a priori* to an energy dependent exchange interaction,

$$\tilde{V}_{\rm ex}^{\rm RPA}(k,\omega) = \frac{\tilde{V}_{\rm d}(k)}{1-\tilde{V}_{\rm d}(k)\chi_0(k,\omega)} \tag{19}$$

where in the Jastrow-Feenberg or fixed-node approximation, the Lindhard function $\chi_0(k,\omega)$ is replaced by $\chi_0^{\rm MSA}(k,\omega)$.

The prescription from parquet-theory to make this energy dependent interaction local is as follows: Construct the RPA static structure function by summing the ring diagrams

$$S^{\rm RPA}(k) = -\Im m \int \frac{d(\hbar\omega)}{2\pi}\left[\chi_0(k,\omega) + \chi_0^2(k,\omega)\tilde{V}_{\rm ex}^{\rm RPA}(k,\omega)\right] \tag{20}$$

which leads, when the MSA is used, to Eq. (18). Also, construct the *ladder approximation* for the same quantity in terms of a different and yet unspecified local and energy independent effective interaction, say $\tilde{V}_{\rm L}(q)$

$$\begin{aligned} S^{\rm ladder}(k) &= -\Im m \int \frac{d(\hbar\omega)}{2\pi}\left[\chi_0(k,\omega) + \chi_0^2(k,\omega)\,\tilde{V}_{\rm L}(k)\right] \\ &= 1 - \frac{S_F^3(k)}{t(k)}\,\tilde{V}_{\rm L}(k)\,. \end{aligned} \tag{21}$$

Now choose $\tilde{V}_{\rm L}(k)$ such that $S^{\rm ladder}(k) = S^{\rm RPA}(k)$. This *defines* $\tilde{V}_{\rm L}(k)$. Through the same relationship, we can identify $\tilde{V}_{\rm L}(k)$ with a local, energy independent approximation to $\tilde{V}_{\rm ex}(k,\omega)$. This leads immediately to Eq. (14).

The effective exchange interaction defines the single-particle spectrum

$$\epsilon_k = t(k) - \frac{1}{\nu N}\sum_h V_{\rm ex}(|\mathbf{k}-\mathbf{h}|)\,, \tag{22}$$

where ν is the spin degeneracy. A consistent treatment of both the exchange potential and the single particle spectrum is needed to satisfy the ω^1 sum rule.

We see again, that the analysis of parquet-diagram theory [35, 36] for how to construct the energy independent interactions of CBF theory from the energy dependent vertices of Feynman-diagram based perturbation theory lead to the same results, provided the Lindhard function is replaced by its collective approximation. The above analysis has been carried out only for a very specific class of diagrams, but, since the result is as expected, there is no reason that the analysis should not carry over to other, more complicated structures. Such an analysis is, however, subject to a study in formal many-body theory and beyond the scope of this work.

Our analysis also provides a method to relate methods of diagrammatic many-body theory to Monte Carlo simulation methods: If one neglects, for example, exchange effects, one can construct $\tilde{V}_{\rm d}(k)$ from the static structure function by demanding that $\tilde{V}_{\rm d}(k)$ reproduces $S^{(\rho)}(k)$ through the ω^0 sum rule

$$S^{(\rho)}(k) = \int_0^\infty d(\hbar\omega) S^{(\rho)}(k, \omega) \,. \tag{23}$$

The prescription provides the *best local* potential that reproduces the known $S^{(\rho)}(k)$. If the fixed-node approximation is used, the relationship (13) is the correct one to use; if the $S^{(\rho)}(k)$ is taken from an improved calculation or from experiment, the full Lindhard function should be used to determine $\tilde{V}_{\rm d}(k)$. We note, however, that the resultant interactions differ only by a few percent.

4. Results

For spin-independent interactions, the spin-density fluctuations depend on $V_{\rm ex}(k)$ but not on $V_{\rm d}(k)$. Thus one might be tempted, as an alternative to using Eq. (14), to construct $V_{\rm ex}(k)$ from $S^{(\sigma)}(k)$ in a similar fashion as $V_{\rm d}(k)$ is derived from $S^{(\rho)}(k)$. Unfortunately, this procedure has numerical difficulties because the resulting problem is very poorly conditioned. We have therefore utilized the local exchange interaction (14). The calculation of the potentials $V_{\rm d}(k)$ (13) and $V_{\rm ex}(k)$ (14) then requires knowledge of no more than the density structure function of the interacting system, $S^{(\rho)}(k)$. For that, we use ground-state results obtained using the FN-DMC method [6], which has proven its accuracy in the description of the BCS-BEC crossover.

The resulting $V_{\rm d}(k)$ and $V_{\rm ex}(k)$ are shown in Fig. 1. The linear behavior at large momenta is related to Tan's relations [38], specifically

$$S^{(\rho)}(k \to \infty) = 1 + \frac{\mathcal{I}}{N}\frac{1}{4k}, \tag{24}$$

where $\mathcal{I}$ is Tan's universal contact parameter and N the number of particles. These lead to the properties

$$\tilde{V}_{\rm d}(k) \sim \frac{\mathcal{I}}{N}\frac{\hbar^2 k}{8m} \qquad \text{and} \qquad \tilde{V}_{\rm ex}(k) \sim \frac{\mathcal{I}}{N}\frac{\hbar^2 k}{8m} \tag{25}$$

which are the cause for the linar increase of the interactions as a function of momentum transfer seen in Fig. 1

For comparison we also display the bare interaction, which corresponds to an attractive square-well potential: $V(r) = -V_0$ for $r < R_0$, and $V(r) = 0$ otherwise. This potential acts only between particles of different spin since s-wave collisions between equal-spin particles are forbidden by the Pauli principle. We use $nR_0^3 = 10^{-6}$ to ensure that the mean interparticle distance is much larger than the range of the potential. The value of the s-wave scattering length is easily changed by modifying the depth V_0 of the potential. Evidently, the bare interaction is much weaker than the effective interaction that is needed to explain the static structure function. This is the compelling evidence alluded to above that the system is in fact very strongly interacting. Note that this is rather different from what one finds in the helium liquids: There, the effective interactions are, in coordinate space, rather similar to the bare interaction except that

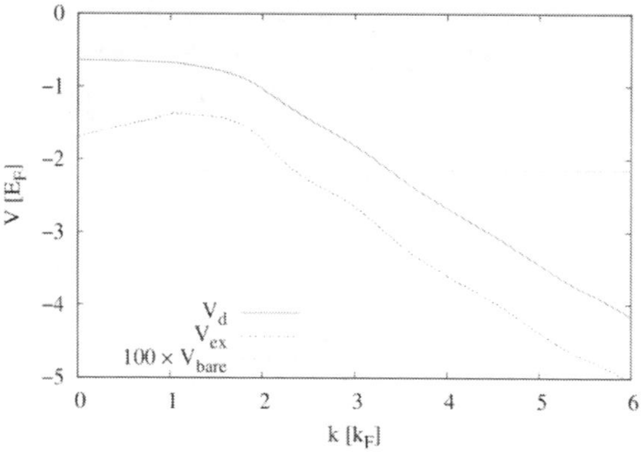

Figure 1. (Color online) Direct and exchange interaction at unitarity ($1/(k_F a) = 0$). The linear behavior at large momenta is a consequence of Tan's relations (24) [38]. The bare potential is much weaker than the effective ones, underlining the strong correlations within the system.

the repulsive hard core is screened off. For a description of the physical effects that determine the properties of these interactions, see Ref. [39].

We have calculated the dynamic structure function in the density and spin channels for $1/(k_\mathrm{F} a) = -1$ (BCS like), $1/(k_\mathrm{F} a) = 0$ (unitarity) and $1/(k_\mathrm{F} a) = 1$ (BEC like) including exchange terms and non-local, energy dependent CBF corrections that have not spelled out explicitly here, see Ref. [33]. Since the system favors, with decreasing $1/(k_\mathrm{F} a)$, the formation of bound dimers, we consider here a mixture of interacting fermions and non-interacting dimers ("bosons" with $2m$). The relative concentrations by which we weight the two contributions to the dynamic structure function in the density channel are obtained from our FN-DMC ground-state calculations. In particular, the concentration of dimers as a function of the scattering length is estimated assuming that all pairs are in the zero-momentum state. According to this criterion, the concentration of dimers is 100% deep inside the BEC regime and decreases to zero quickly after crossing the unitary limit where it amounts $\sim 50\%$ [40]. Our model cannot predict the width of the "Bose" peak, hence it has been fitted to reproduce the width of the experimental data subject to the constraint that its contribution to the ω^0 sum rule is constant. The spin channel dynamics on the other hand is not affected by the presence of dimers because they do not contribute to spin-density fluctuations.

In Fig. 2, we show results of $S^{(\rho)}(k, \omega)$ and $S^{(\sigma)}(k, \omega)$ for the wave vector $k = 4.2\, k_\mathrm{F}$. We have selected this particular k value because it is the only momentum measured by Bragg scattering [18]. (This wave vector deviates, due to recalibration of the experimental values [41], slightly from the wave number $k = 4.5\, k_\mathrm{F}$ reported in Ref. [18].) In agreement with the experimental data, the density channel of the dynamic structure function shows a clear "Bose" peak centered at energy $\hbar^2 k^2/4m$ coming from the scattering off the ↑-↓ pairs. At higher energies, contributions from single Fermi

Journal of Physics: Conference Series **529** (2014) 012009 doi:10.1088/1742-6596/529/1/012009

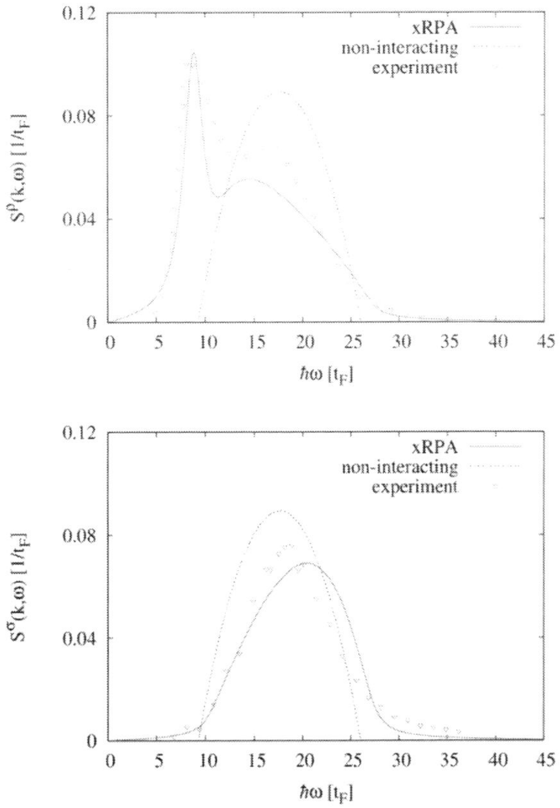

Figure 2. (Color online) The dynamic structure function in the density (*top*) and spin (*bottom*) channels at unitarity $(1/(k_F a) = 0)$ and at a momentum $k = 4.2k_F$ (solid line). The damping constant for this momentum transfer was chosen $\eta \approx t_F$. The points are the experimental data of Ref. [18]. Dashed line stands for the non-interacting response.

atoms appear as a broader response. The importance of correlations at unitarity is best demonstrated by comparing the response coming from the full theory with the non-interacting Lindhard function, also shown in Fig. 2. Our result for the spin channel at unitarity and the same k value as in the experiment is also shown in Fig. 1 (bottom panel). As mentioned above, dimers do not contribute to the spin channel and only atomic contributions appear. As in the density channel, the use of many-body theory is crucial to describe the experimental data, note that the response of a non-interacting system has a completely different structure and differs substantially from what was observed in the experiment.

The full energy-momentum dependence of the density and spin response functions is shown in the form of stacked line plots in Fig. 3. At low momenta, $k \lesssim 4k_F$, the dimer peak lies inside the particle-hole $(p-h)$ band and has a large overlap with the single-atom

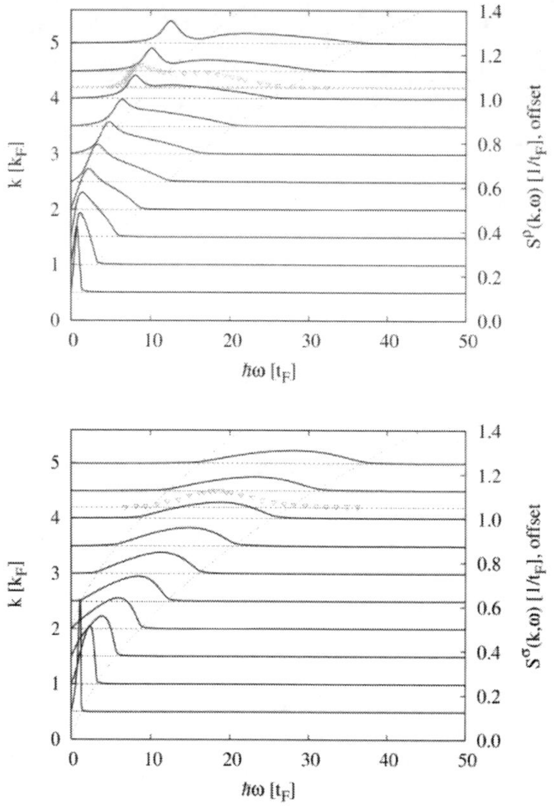

Figure 3. (Color online) The figure shows line plots of the dynamic structure function in the density (*top*) and spin (*bottom*) channels at unitarity ($1/(k_Fa) = 0$) as a function of the momentum transfer k and energy transfer $\hbar\omega$ in units of the Fermi energy $E_{\rm F}$. The left vertical scale shows the momentum offset, the right scale the value of $S^{\rho,\sigma}(k,\omega)$. The dashed lines show the limits of the particle-hole band. The triangles correspond to the momentum transfer at which the experiment of Ref. [18, 41] was performed.

scattering contribution. At higher momenta, the "Bose" peak is progressively decoupled of Fermi excitations with a position given by half the recoil energy. The strength of the spin response is localized inside the $p-h$ band but interactions make the peak slightly asymmetric and shifted to higher energy compared with the non-interacting response; this effect is washed out at momenta $k \gtrsim 4k_F$.

Bragg scattering [18, 19] was also applied to a point in the phase diagram located inside the BEC side, in particular at $1/(k_Fa) = 1$, whereas no experimental data exists on the BCS side. We have applied dynamic many-body theory to the experimental BEC point and to the symmetric one in the BCS side, with $1/(k_Fa) = -1$, and at a momentum transfer $k = 4.2k_F$. The response in the BCS regime is rather close to that of

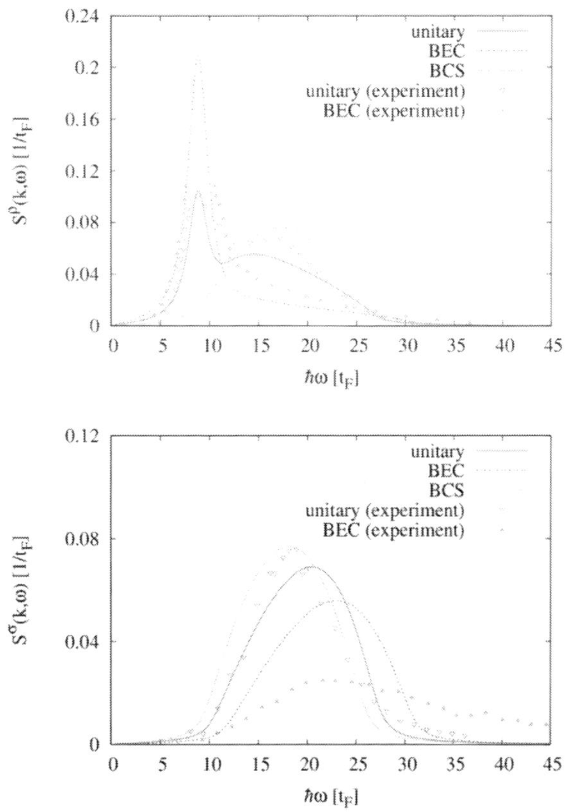

Figure 4. (Color online) The dynamic structure function in the density (*top*) and spin (*bottom*) channels at unitarity ($1/(k_F a) = 0$), BEC ($1/(k_F a) = 1$), and BCS ($1/(k_F a) = -1$). The lines are results from dynamic many-body theory and points are experimental data of Ref. [18]. All data corresponds to $k = 4.2k_F$.

the free Fermi gas, with only an incipient Bose peak from the dimer contribution due to its very low concentration (< 10 %). On the contrary, the dimer contribution dominates the density channel in the BEC side since the concentration of dimers is ~ 90 %. Our theory, which produces results in good agreement with experiment in the unitary limit, seems to worsen in the BEC regime. This is seen most clearly in the spin response at $1/(k_F a) = 1$ where our result does not show the broad response obtained from Bragg scattering.

5. Conclusion

To conclude, motivated by the first measurements of the dynamic structure function of a two-component resonantly interacting Fermi mixture, we have employed a well-tested theory able to predict its dynamic response. Dynamic many-body theory has been used together with a FN-DMC input for the static structure factor. From a direct comparison with experimental data [18], we show that the ideal Fermi gas model is not sufficient to reproduce correctly the dynamic response functions. We have formulated a description in terms of a mixture of dimers (bosons) and atoms (fermions). This simple model permits to capture correctly the complicated structure of the density response at the unitary limit. Our theory departs from the experimental data, mainly in the spin channel, when the fraction of dimers is large. Finally, we provide information on the dynamic response for different (previously not measured) values of the momentum in the unitary limit and on the BCS (not measured) side.

We acknowledge partial financial support from the DGI (Spain) Grant No. FIS2011-25275 and Generalitat de Catalunya Grant No. 2009SGR-1003 as well as from the Austrian Science Fund FWF under project I-602. G. E. A. acknowledges support from the Spanish MEC through the Ramon y Cajal fellowship program. Additional support was provided by a grant from the Qatar National Research Fund # NPRP NPRP 5 - 674 - 1 - 114.

[1] Giorgini S, Pitaevskii L P and Stringari S 2008 *Rev. Mod. Phys.* **80** 1215
[2] Zwierlein M W, Schirotzek A, Schunck C H and Ketterle W 2006 *Science* **311** 492
[3] Partridge G B, Li W, Kamar R I, Liao Y and Hulet R G 2006 *Science* **311** 503
[4] Shin Y, Schunck C H, Schirotzek A and Ketterle W 2008 *Nature* **451** 689
[5] Navon N, Nascimbene S, Chevy F and Salomon C 2010 *Science* **328** 729–732
[6] Astrakharchik G E, Boronat J, Casulleras J and Giorgini S 2004 *Phys. Rev. Lett.* **93** 200404
[7] Carlson J and Reddy S 2005 *Phys. Rev. Lett.* **95** 060401
[8] McNeil Forbes M, Gandolfi S and Gezerlis A 2011 *Phys. Rev. Lett.* **106** 235303
[9] Drut J E, Lähde T A, Wlazłowski G and Magierski P 2012 *Phys. Rev. A* **85** 051601
[10] Nascimbène S, Navon N, Jiang K J, Chevy F and Salomon C 2010 *Nature* **463** 1057
[11] Ku M J H, Sommer A T, Cheuk L W and Zwierlein M W 2012 *Science* **335** 563
[12] Pilati S and Giorgini S 2008 *Phys. Rev. Lett.* **100** 030401
[13] Astrakharchik G E, Giorgini S and Boronat J 2012 *Phys. Rev. B* **86** 174518
[14] Sanner C, Su E J, Keshet A, Huang W, Gillen J, Gommers R and Ketterle W 2011 *Phys. Rev. Lett.* **106** 010402
[15] Sommer A, Ku M, Roati G and Zwierlein M W 2011 *Nature* **472** 201
[16] Wlazłowski G, Magierski P, Drut J E, Bulgac A and Roche K J 2013 *Phys. Rev. Lett.* **110** 090401
[17] Palestini F, Pieri P and Strinati G C 2012 *Phys. Rev. Lett.* **108** 080401
[18] Hoinka S, Lingham M, Delehaye M and Vale C J 2012 *Phys. Rev. Lett.* **109**(5) 050403
[19] Hoinka S, Lingham M, Fenech K, Hu H, Vale C J, Drut J E and Gandolfi S 2013 *Phys. Rev. Lett.* **110** 055305
[20] Combescot R, Giorgini S and Stringari S 2006 *Europhys. Lett.* **75** 695
[21] Hu H and Liu X J 2012 *Phys. Rev. A* **85** 023612
[22] Fabrocini A, Fantoni S and Krotscheck E 2002 *Introduction to Modern Methods of Quantum Many-Body Theory and their Applications* (*Advances in Quantum Many-Body Theory* vol 7) (Singapore: World Scientific)
[23] Böhm H M, Holler R, Krotscheck E and Panholzer M 2010 *Phys. Rev. B* **82** 224505/1–28
[24] Jackson H W and Feenberg E 1961 *Ann. Phys. (NY)* **15** 266–295
[25] Jackson H W and Feenberg E 1962 *Rev. Mod. Phys.* **34** 686–693
[26] Chang C C and Campbell C E 1976 *Phys. Rev. B* **13** 3779–3782
[27] Godfrin H, Meschke M, Lauter H J, Sultan A, Böhm H M, Krotscheck E and Panholzer M 2012 *Nature* **483** 576579
[28] Thouless D J 1972 *The quantum mechanics of many-body systems* 2nd ed (New York: Academic Press)
[29] Kerman A K and Koonin S E 1976 *Ann. Phys. (NY)* **100** 332–358
[30] Kramer P and Saraceno M 1981 *Geometry of the time-dependent variational principle in quantum*

mechanics (*Lecture Notes in Physics* vol 140) (Berlin, Heidelberg, and New York: Springer)
[31] Krotscheck E and Clark J W 1979 *Nucl. Phys. A* **328** 73–103
[32] Chen J M C, Clark J W and Sandler D G 1982 *Z. Physik A* **305** 223–229
[33] Krotscheck E 1982 *Phys. Rev. A* **26** 3536–3556
[34] Krotscheck E 2000 *J. Low Temp. Phys.* **119** 103–145
[35] Jackson A D, Lande A and Smith R A 1982 *Physics Reports* **86** 55–111
[36] Jackson A D, Lande A and Smith R A 1985 *Phys. Rev. Lett.* **54** 1469–1471
[37] Krotscheck E, Smith R A and Jackson A D 1986 *Phys. Rev. A* **33** 3535–3536
[38] Tan W C and Inkson J C 1999 *Phys. Rev. B* **60** 5626
[39] Aldrich C H and Pines D 1976 *J. Low Temp. Phys.* **25** 677–690
[40] Astrakharchik G E, Boronat J, Casulleras J and Giorgini S 2005 *Phys. Rev. Lett.* **95** 230405
[41] Hoinka S 2013 (private communication)

Kohn anomaly in phonon driven superconductors

M P Das[1] and R Chaudhury[2]

[1]Research School of Physics and Engineering, The Australian National University, Canberra ACT 0200, Australia

[2]S.N. Bose National Centre For Basic Sciences, Block-JD, Sector-3, Salt Lake, Kolkata 700098, India

E-mail: mukunda.das@anu.edu.au

Abstract. Anomalies often occur in the physical world. Sometimes quite unexpectedly anomalies may give rise to new insight to an unrecognized phenomenon. In this paper we shall discuss about Kohn anomaly in a conventional phonon-driven superconductor by using a microscopic approach. Recently Aynajian et al.'s experiment showed a striking feature; the energy of phonon at a particular wave-vector is almost exactly equal to twice the energy of the superconducting gap. Although the phonon mechanism of superconductivity is well known for many conventional superconductors, as has been noted by Scalapino, the new experimental results reveal a genuine puzzle. In our recent work we have presented a detailed theoretical analysis with the help of microscopic calculations to unravel this mystery. We probe this aspect of phonon behaviour from the properties of electronic polarizability function in the super-conducting phase of a Fermi liquid metal, leading to the appearance of a Kohn singularity. We show the crossover to the standard Kohn anomaly of the normal phase for temperatures above the transition temperature. Our analysis provides a nearly complete explanation of this new experimentally discovered phenomenon. This report is a shorter version of our recent work in JPCM.

Occurrence of Kohn anomaly in normal conventional metals predicted by W. Kohn [1] and soon after observed in experiments [2,3], is well understood over the past five decades [4]. This anomalous feature originates from the discontinuity of electron occupation number at the Fermi surface and the dielectric screening of the bare phonon modes by the electrons of the Fermi sea. In contrast, the apparent signature of the occurrence of the similar phenomenon even in elemental conventional superconductors, as observed experimentally in recent times, still look rather mysterious [5]. Based on an extension of our recent works [6,7], we present in this paper some quantitative aspect of our theoretical and microscopic analysis on the background of recent observations. The experiments involve resonant spin-echo spectroscopy with neutrons in real superconducting materials. Moreover, we also discuss the possible implications of these studies towards a full understanding of results from another important experimental technique viz. ultrasonic attenuation even for conventional superconductors, considering the various electron-phonon processes [8-10].

We begin with the results of our investigation mostly on the phonon dispersion due to the emergence of the superconducting gap in conventional superconductors. Recently Johnston et al. studied theoretically and computationally the consequences for the phonon line-widths under this situation [11]. They performed a standard BCS calculation of the phonon line-width and showed the ratio of twice the energy gap at T=0 to the Kohn energy ($2\Delta(0)/\hbar\omega_{qKA}$) is close to unity in comparison with the experiments of Aynajian et al. [5].

We calculate the frequency and wave-vector dependent total electronic polarizability function ($P_s(q, \omega)$) in the superconducting phase. All the details can be found in ref. [7]. Making use of the retarded and advanced fermionic Green functions corresponding to Bogoliubov quasi-particles and quasi-holes respectively, we obtain the polarization bubble in the random phase approximation (RPA).

In the static limit ($\omega \to 0$) the q-space derivative of the polarization function is

$$dP_s(q)/dq = (2\pi N(0)m/\hbar^2 k_F)\,[(-1/q^2)\ln(A/B) + (1/q)\,\{(1/A)(dA/dq) - (1/B)(dB/dq)\}], \qquad (1)$$

where $A = 2[(\{\hbar^2 q/(2m)(q-2k_F)\}^2 + \Delta^2]^{1/2} + (\hbar^2 q/m)(q-2k_F)$ and B is same as A excepting that in the expression $(q-2k_F)$ is replaced by $(q+2k_F)$. The quantities N(0), m, k_F and Δ occurring in the above equation represent electronic density of states at the Fermi level for one kind of spin (in the normal phase), conduction electron mass, Fermi wave-vector and the superconducting gap respectively.

The q-space derivative of the electronic polarizability function as shown above, in the superconducting phase at T=0, evaluated at $q = 2k_F$ possesses the following property. When it is scaled down by the quantity $P_s(q=2k_F)/2k_F$, it exhibits a logarithmic divergence in the limiting situation of the superconducting coherence length ξ becoming either (i) extremely large or (ii) vanishingly small, in comparison to the lattice spacing or more correctly the average inter-electron distance. This spectacular feature in fact, is an analogue of Kohn singularity in the BCS superconducting phase and the very large coherence length limit is expected to be reached in the conventional phonon- driven superconductors implying that Kohn singularity can occur quite naturally in these systems [7].

By taking the limits $\Delta \to 0$ and $q \to 2k_F$ it can be seen easily from the previous expressions (see eqn. (1)) that in this limit $dP_s(q)/dq$ reducing to $dP_N(q)/dq$ diverges. Thus, we recover the usual Kohn singularity for normal metals, as expected from microscopic considerations.

Next we focus on the nature of the q-space derivative of the static electronic polarizability function for any general wave-vector and explore its relation with the renormalized phonon dispersion function as well. This is for a microscopic investigation into the possible existence of Kohn anomaly in real BCS superconductors with phonon based pairing mechanism.

For a superconductor at very low temperature, where the electron-phonon interaction is crucial for BCS superconductivity, the phonons get renormalized because of creation and destruction of quasi-particles and quasi-hole pairs in the electron-phonon scattering processes. Furthermore, in the case of elemental conventional superconductors, neglecting the vertex corrections in the spirit of Migdal's approximation the Dyson Equation for the phonon self-energy in the superconducting phase leads to the following approximate equation in the region $\omega \ll \omega_D$ [12, 7].

$$\omega^2(q) = \omega_0^2(q)\,[1 + |g(q)|^2 P_s(q)], \qquad (2)$$

where $\omega(q)$ and $\omega_0(q)$ are the renormalized and bare phonon dispersions respectively and $\omega_0(q)$ is taken according to the Debye model (relevant for an acoustic phonon). The quantity P_s is the polarizability function in the superconducting phase. Therefore, any anomaly in the dressed phonon spectrum has its origin in the functional property of this polarizability function. This nearly static approximation ($\omega \to 0$) would not lead to any signature of damping of the phonon modes. In fact, to study the complete effect on the line-width of the phonon modes, one has to go beyond the RPA for the polarizability function adopted here and to consider the vertex corrections in the finite frequency regime.

The above eqn. (2) leads to the following important relation between the derivative of the dispersion function and that of the polarizability function :

$$2\omega(q)\,d\omega(q)/dq = \omega_0^2(q)\,|g(q)|^2\,dP_s(q)/dq, \qquad (3)$$

assuming g(q) to be a rather slowly varying function of q or nearly a constant and hence we neglect its derivative with respect to q.

Now considering the experimental observation of Aynajian et al to be a genuine manifestation of Kohn anomaly in the superconducting phase [5], represented by the occurrence of the point of inflexion (not discontinuity or divergence) in the observed i.e. the renormalized phonon dispersion function at a wave-vector near $q = 2k_F$, we must have (making use of equation (3)),

$$[dP_s(q) / dq]_{q\sim 2kF} = 0. \tag{4}$$

Thorough examination and inspection involving the combination of the equations (1) and (4), lead to the conclusion that the equation (4) can indeed have solution quite close to $q = 2k_F$. In fact, the solution occurs at $q = 2k_F \pm \delta$, where $\delta << 2k_F$ and is +ve. Furthermore it turns out that δ is of the order of $\sim \Delta/\hbar v_F$, for $\Delta << E_F$ (which holds easily for weak coupling phonon mediated superconducting pairing) and v_F being the Fermi velocity.

Quite interestingly it is also found that the parameter δ is material dependent and is essentially of the order of the inverse of the superconducting coherence length. Therefore, the measurement of the latter would give an estimate for δ. Again it can be shown that in the limit $\Delta \to 0$, we have $\delta \to 0$ leading to the Kohn anomaly exactly at $q = 2k_F$. Thus, we recover the normal state scenario exactly.

Next we consider the pairing in the superconducting phase to be mediated by the phonons, as is the case with the elemental s-wave superconductors Pb and Nb, the likely candidates for exhibiting Kohn anomaly in the superconducting phase. This enables us to further rewrite the BCS equation for the superconducting gap at zero temperature by considering the mediator phonons to be predominantly the dressed phonons which themselves exhibit the Kohn anomaly at the onset of the superconducting transition from the normal state. Purely isotropic weak coupling BCS gap function (Δ_0) at T=0K would then satisfy the following approximate equation under such condition [assuming a jellium model and keeping in mind that at Kohn anomaly $\hbar\omega(q) = 2\Delta_0$] [7],

$$\Delta_0 \sim 3.5\, \Delta_0 \exp[-\hbar^2 (3\pi^2)^{(2/3)}/3n^{(1/3)} m|g_{eff}|^2], \tag{5}$$

where $|g_{eff}|^2$ is the Fermi surface averaged effective electron-electron (more correctly the Landau quasiparticle-quasiparticle residual) attractive interaction in the normal phase at the onset of superconducting transition, mediated primarily by the phonons undergoing Kohn anomaly at the same time and 'n' is the conduction electron density.

For simplicity, making use of the jellium model in calculating the usual BCS attractive coupling constant 'g' and incorporating the standard values for the parameters corresponding to the above elemental superconducting materials, we find that indeed the above equation (5) can be obeyed quite realistically. Therefore, our proposal that the observations of Aynajian et al indicate a true occurrence of Kohn anomaly in the phonon driven BCS superconducting phase of a 3d Fermi Liquid metal has a very strong microscopic support.

Our work presented here has an interesting and a very natural connection with the acoustic attenuation problem studied both theoretically and experimentally earlier for some of the conventional BCS superconductors [8-10, 12]. In particular, the notable feature of ultrasonic waves exhibiting a discontinuity in attenuation at energy of the phonon equal to twice the superconducting gap, even at finite temperature as observed in various experimental phonon mechanism based 3-dimensional superconducting systems like Nb and Nb_3Sn, are directly related to our investigation reported here [8]. In fact abrupt changes in phonon life times have also been detected near T_C in Nb_3Sn. These results are very much consistent with the observed peak in the phonon line-width occurring at the above mentioned value of phonon energy seen in the very high resolution measurements involving resonant spin echo spectroscopy with neutrons [5]. The magnitude of the jump or discontinuity in the ratio of attenuation in the superconducting state to that in the normal state was predicted to be the highest (about 1.57) for energy of the ultrasonic phonon equal to twice the superconducting gap [9]. This feature is unlike that seen for the usual phonon attenuation in the pure acoustic regime ($\hbar\omega/2\Pi << 2\Delta$) both experimentally and theoretically [14]. Moreover, there is a substantial phonon softening (about 3%) observed below T_C in Nb_3Sn [8, 10].

The detailed quantitative calculation of the line-width or attenuation for these ultrasonic phonons around Kohn anomaly point can be carried out by retaining the frequency dependence in the self-energy diagrams corresponding to our calculations involving quasi-particle-quasi-hole excitations and preferably with inclusion of vertex correction as well. Johnston et al. have done such a calculation (neglecting vertex corrections) taking into account also the electronic structure for Pb and Nb [11]. However, the very important change in the phonon dispersion function from the bare linear one near q_{KA} is completely neglected in their calculation. Alternatively, it is easier to understand the substantial enhancement of attenuation at Kohn anomaly point in terms of simple 'Fermi's golden rule' type of calculations by considering the increase in the number of possible scattering channels in this region.

The possibilities of occurrences of Kohn singularity and Kohn anomaly with non-conventional pairing schemes in exotic systems are rather hypothetical ones , since no materials belonging to this class have been found to exhibit these effects clearly either in the normal or in the superconducting phases. Some of these questions have been discussed by Flatte [13].

In summary, in this short report we have tried to present a more realistic analysis of the occurrences of both Kohn singularity and Kohn anomaly in the superconducting phase, keeping in mind the real condensed matter systems. Our calculations strongly indicate that the acoustic dispersion can indeed display anomaly at the phonon energy equal to 2Δ in the real phonon driven BCS superconductors. We also highlight that our microscopic approach is very different from that proposed in [5, 14-16] for an explanation of Kohn anomaly in superconductors. Besides, our analysis and calculational results establish a genuine and strong microscopic justification for the observations of Aynajian et al. regarding the simultaneous occurrences of both singularity in phonon dispersion function and a sharp peak in phonon line-width, to be unified and identified with a Kohn anomaly in the BCS superconducting phase based on phonon mechanism for pairing. This is in sharp contrast to the proposal of Johnston et al. [11] that these two phenomena are different and the strong similarity between them is just a coincidence.

Acknowledgements

One of us (MPD) is grateful to Professor Gerd Röpke, Dr Heidi Reinholz and the organising team of MBT-17 for their hospitality. We thank Professor Doug Scalapino for helpful discussions.

References

[1] Kohn W 1959 *Phys. Rev. Lett.* **2** 393
[2] Brockhouse B N, Rao K R and Woods A D B 1961 *Phys. Rev. Lett.* **7** 93
[3] Brockhouse B N, Arase T, Caglioti G, Rao K R and Woods A D B 1962 *Phys. Rev.* **128** 93
[4] Sugai S 1985 *Phys. Stat. Sol (b)* **129** 13
[5] Aynajian P, Keller T, Boeri L, Shapiro S M, Habicht K and Keimer B 2008 *Science* **319** 1509
[6] Chaudhury R and Das M P 2010 *Int. J. Mod. Phys.* B **24** 5172
[7] Chaudhury R and Das M P 2013 *J. Phys. Condens Matter* **25**, 12202
[8] Shapiro S, Shirane G and Axe J D 1975 *Phys. Rev.* **12** 4899
[9] Bobetic V M 1964 *Phys. Rev.* A**136** 1535
Zeyher R and Zwicknagl G 1990 *Z. Phys.* B **78** 175
Privorotskii A 1963 *Sov. Phys. JETP* **16** 945

[10] Axe J D and Shirane G 1973 *Phys. Rev.* B **8** 1965
[11] Johnston S , Sorini A P, Moritz B, Devereaux T and D J Scalapino 2011 *Phys. Rev.* B **84** 174523 (2011)
[12] Schrieffer J R 1999 *Theory of Superconductivity* (Perseus Books Publishing, Massachusetts) Chapter 6 156 ; Chapter 3 62
Abrikosov A A, Gorkov L P and Dzyaloshinski I E 1961 *Methods of Quantum Field Theory In Statistical Physics* (Dover Publications INC. New York) Chapter 4 181
Migdal A B 1958 *Sov. Phys. JETP* **7** 996
[13] Flatte M E 1994 *Phys. Rev.* B **50** 1190
[14] Moncton D E , Axe J D and DiSalvo F J 1977 *Phys. Rev.* B **16** 801
[15] Kiss T, Yokoya T, Chainani A, Shin S, Hanaguri T, Nohara M and Takagi H 2007 *Nat. Phys.* **3** 721
[16] Terashima K, Matsui H, Sato T, Takahashi T, Kofu M and Hirota K 2007 *Phys. Rev. Lett.* **99** 017003
Lee W S , Vishik M, Tanaka K, Lu D H, Sasagawa T, Nagaosa N, Devereaux T P, Hussain Z and Shen Z X 2007 *Nature* **450** 81

Quantum Monte Carlo study of strongly interacting Fermi gases

S Gandolfi

Theoretical Division, Los Alamos National Laboratory Los Alamos, NM 87545, USA

E-mail: stefano@lanl.gov

Abstract. In recent years Quantum Monte Carlo techniques provided to be a valuable tool to study strongly interacting Fermi gases at zero temperature. We have used QMC methods to investigate several properties of the two-components Fermi gas at unitarity and in the BCS-BEC crossover, both with equal and unequal masses corresponding to the $Li - K$ Fermi mixture. In this paper we present several recent QMC results, including the energy at zero and finite effective range, the contact parameter and the static structure factor, which, at low momentum, depends strongly on the phonons in the unitary Fermi gas.

1. Introduction

During the last years important efforts have been devoted to study ultra-cold Fermi gases, both experimentally and theoretically (for a review see for example Ref. [1]). By means of Feshbach resonance it is possible experimentally to adjust the interaction between atoms. The system can be tuned to a BCS state where Fermions are weakly interacting, to the so called unitary regime where the two-body scattering length is infinite, or to form a BEC condensate of bosonic dimers. Theoretically, ultra-cold Fermi gases have become an important testing ground for a host of many-body methods, as well as an avenue to studying new physics.

The study of ultra-cold Fermi gases is very intriguing, and the unitary limit became a very interesting many-body problem to solve for several reasons. These systems can be experimentally studied at very low temperatures of the order of 0.1 T_F, essentially zero temperature. They are very dilute, and their properties are independent of the form of the Hamiltonian used to describe the system, the results given by any model satisfying the limits of large scattering length and small effective range must be identical. Moreover these systems are very strongly interacting, methods based on perturbation theory, mean-field, or expansions of small parameters are not obviously convergent, and the calculation of the properties of the Unitary Fermi gas is very challenging. The properties scale with powers of the Fermi momentum $k_F = (3\pi^2\rho)^{1/3}$. The interparticle separation

r_0 is approximately given by $r_0 = (9\pi/4)^{1/3}/k_F$, and the dilute regime is guarantee by imposing $r_0 \gg r_e$, where r_e is the effective range of the two-body interaction. The universal regime is given when $r_e \ll r_0 \ll a$, where a is the scattering length, and the unitary limit correspond to the case of $|a| = \infty$. In the limit of $r_e = 0$ the only scale of the system is r_0 or more conveniently the Fermi momentum k_F. Thus a natural unit for the energy if given by the Fermi-gas energy $E_{FG} = 3\hbar^2 k_F^2/10\,m$, and there are no other free parameters. For this reason, these systems are called universal.

Ultra-cold Fermi gases can also be realized with population imbalance, and the system can eventually exhibit different phases [2, 3] where superfluid and normal phases may coexist or phase separate [4, 5], or other intriguing more exotic phases like LOFF or p-wave superfluidity could appear [6, 7]. In the last few years a new direction of research has been undertaken: the realization of trapped two–component Fermi gases with the mixture of Fermions with different masses, in particular the $^6Li - ^{40}K$ mixture has been addressed [8–10]. The properties of systems with mass imbalance can be very different from the equal masses case; for example, for majority light population the Chandrasekhar–Clogston limit is very small and close to zero [11]. Then, at unitarity, the two–components Fermi gas with different masses may exhibit very different properties with respect to the equal mass case [12, 13], even in the BCS limit [14].

Several predictions for equal mass Fermi gases were made possible through Quantum Monte Carlo techniques (QMC), later confirmed by experiments [4, 11, 15–21]. The Variational Monte Carlo (VMC) and the Diffusion Monte Carlo (DMC) enable accurate calculations of the ground state of strongly interacting Fermions at unitarity, and to study its properties [20]. In addition, the unbalanced two-component Fermi gases with equal masses do not exhibit a sign problem using the Auxiliary Field Monte Carlo (AFMC) [21], and this permits a very accurate calculation to benchmark other many-body calculations. In the latter case the sign problem is removed at the cost of introducing a discrete lattice and extrapolating to infinite volume and zero lattice spacing. Experiments are possible at very low temperatures of the order of fractions of T_F, and then the zero temperature properties can be directly obtained from experiments.

In this paper we will review several properties calculated using QMC methods, including the exact calculation of the energy, the role of the effective range, the BCS-BEC crossover, and other properties including the contact parameter and the static structure factor.

2. The model and QMC methods

The AFMC performed on a lattice through the use of auxiliary fields does not suffer of the Fermion sign problem and permits to calculate exactly the energy of the unpolarized Fermi gas [21]. The AFMC method uses a BCS wave function as a trial wave function. In this section we describe the variational wave function and the DMC method in the continuum.

In our study we model the ground state of the system in the continuum using the Hamiltonian

$$H = \sum_{i=1}^{N_l} \frac{-\hbar^2}{2m_l}\nabla_i^2 + \sum_{j=1}^{N_h} \frac{-\hbar^2}{2m_h}\nabla_j^2 + \sum_{i,j} v(r_{ij}), \tag{1}$$

where N_l and N_h is the number of particles with mass m_l and m_h. Since we want to model a system that is dilute and cold, particles interact only in s-wave, and the potential v_{ij} is non-zero only between particles with opposite spin state or with different mass. Several forms of the potential v_{ij} has been explored, and provided to be equivalent in the limit of small s-wave effective range [22]. In most of our studies we have used the Pöschl-Teller already employed in several previous QMC calculations [11, 15, 17, 19, 20, 22–25]:

$$v(r) = -v_0 \frac{\hbar^2}{m_r} \frac{\mu^2}{\cosh^2(\mu r)}, \tag{2}$$

where m_r is the reduced mass, and the parameters μ and v_0 tune respectively the s-wave effective range r_e and the scattering length a of the interaction. For this potential, the unitary limit corresponding to the zero-energy ground state between two particles is with $v_0 = 1$ and the s-wave effective range is $r_e = 2/\mu$ [15]. Most of the results presented in this work have been obtained by fixing $r_e k_F \approx 0.03$ but in several cases we considered different values to check effects due to the effective range of the potential. It is also possible to use other forms of the potential $v(r)$, and the results are universal in the limit of small effective range as we shall discuss in the next sections.

The QMC calculations are performed using the Jastrow-BCS form of the wave function of Refs. [15, 20]:

$$\Psi_v(\mathbf{r}_1 \ldots \mathbf{r}_N) = F(\mathbf{r}_1 \ldots \mathbf{r}_N)\, \Phi_{\rm BCS}\,. \tag{3}$$

The antisymmetric part $\Phi_{\rm BCS}$ is a particle-projected BCS wave function including pairing correlations. It is given by

$$\Phi_{\rm BCS} = \mathcal{A}[\phi(r_{11'})\phi(r_{22'})...\phi(r_{nn'})]\,. \tag{4}$$

The operator $\mathcal{A}$ is an anti-symmetrization operator, the unprimed coordinates are for spin-up or heavy particles and the primed are for spin-down or light particles, and $n = N/2$ for the unpolarized case. The pairing function is parametrized as

$$\begin{aligned} \phi(\mathbf{r}) &= \sum_n \alpha_{k_n^2} \exp[i\mathbf{k}_n \cdot \mathbf{r}] + \beta(r)\,, \\ \beta(r) &= \tilde{\beta}(r) + \tilde{\beta}(L-r) - 2\tilde{\beta}(L/2)\,, \\ \tilde{\beta}(r) &= [1 + cbr]\,[1 - \exp(-dbr)]\frac{\exp(-br)}{dbr}\,. \end{aligned} \tag{5}$$

In the infinite volume limit $\phi(\mathbf{r})$ is a function of one scalar variable, the product $k_F r$. The simulations use periodic boundary conditions in a finite volume. In this finite volume, the function $\beta(r)$ has a range of $L/2$, L is the size of the simulation box, the value of c is chosen such that $\beta(r)$ has zero slope at the origin, and b, d and $\alpha_{k_m^2}$ are variational parameters. Note that if $\beta(r) = 0$, and $\alpha_{k_m^2} = 0$ for $k > k_F$, $\Phi_{\rm BCS} = \Phi_{FG}$, where the latter is the wave function describing the non interacting Fermi gas in the normal phase. As Φ_{FG} and $\Phi_{\rm BCS}$ are orthogonal, we can study the Fermi gases both in the superfluid and in the normal phase [4, 5, 11]. The pairing wave function used here contains several free parameters that have been optimized by minimizing the energy of the system using VMC and following the strategy of Ref. [26]. In the BEC side of the transition, when $1/a\, k_F \geq 2$, we found that using the pairing function as described in Ref. [16], i.e. the two-body wave function instead of the function of Eq. 5, gives lower energies. Instead, in the BCS case for $1/a\, k_F \leq -1$, the BCS wave function Φ_{BCS} gives almost the same energy of Φ_{FG} as the pairing becomes less important to the equation of state. The polarized systems can be simulated by extending the wave function to include single-particle states for the unpaired particles [15, 23]. Note that the wave function previously described cannot reproduce more exotic phases like LOFF or p-wave pairing.

The Jastrow $F(R)$ includes a short- and a long-range part. The short-range term is given by

$$F_{sr}(R) = \prod_{i<j} f_{sr}(r_{ij}), \tag{6}$$

where the function $f_{sr}(r)$ acts only between particles with different spin or mass, and it is obtained by solving the equation

$$-\frac{\hbar^2}{2m_r}\nabla^2 f_{sr}(r) + \alpha v(r) f_{sr}(r) = \lambda f_{sr}(r). \tag{7}$$

The parameter lambda is obtained by imposing the boundary condition $f(r > d_0) = 1$, where the healing distance d_0 is a variational parameter. The quenching parameter α is adjusted to be less than one only in the BEC regime in order to define a Jastrow function $F(R)$ that satisfy the above boundary condition, and reduce the variance in the calculation. The correct boundary conditions to the wave function are guaranteed by constraining $d_0 \leq L/2$ where L is the size of the simulation box, but we found that a typical good choice is given by taking $d_0 \approx L/10$. By solving Eq. 7 we also assure the correct behavior of the wave function at small distances so that the Jastrow function f is defined to have $\partial f/\partial r$=0 at the origin. The calculation in the deep BEC can be also improved by adding a short-range function $f_{sr}(r)$ between particles with same spin or mass, and solving the equation above with a repulsive interaction.

In order to precisely calculate the static structure function at low momenta, it is important to include the effect of phonons as long-distance correlations. In our calculation, we include a term

$$F_{lr}(R) = \prod_{i<j} \exp\left[\gamma \sum_n \frac{\exp(-\beta|\mathbf{q_n}|)}{|\mathbf{q_n}|}\exp(-i\mathbf{q_n}\cdot\mathbf{r_{ij}})\right], \tag{8}$$

where γ and β are new variational parameters [27]. The VMC algorithm is then used to optimize over wave functions of this form. Diffusion Monte Carlo (DMC) is then used to project out the ground state from:

$$\Phi_0 = \lim_{\tau\to\infty} \Psi(\tau), \tag{9}$$

where

$$\Psi(\tau) = e^{-(H-E_T)\tau}\Psi(0) = e^{-(H-E_T)\tau}\Psi_v. \tag{10}$$

The factor E_T is a constant used to control the normalization of the ground state. The propagation is achieved using a many-body propagator defined as

$$G(R', R, \delta\tau) = \langle R'|e^{-H\delta\tau}|R\rangle, \tag{11}$$

and

$$\Psi(R', \tau+\delta\tau) = \int dR G(R', R, \delta\tau)\Psi(R, \tau). \tag{12}$$

A good approximation of G can be obtained using a Trotter expansion, and iterating the above equation for small values of $\delta\tau$. Note that for a zero-range interaction the exact two-body propagator could also be used to sample the many-body Green's function [28]. In the case of Fermions there is a sign problem that needs to be taken care. A common approach combined with the DMC method is the fixed-node approximation. The sampling of paths is restricted to regions where the trial wave function is positive, and the problem is then recasted into a Bosonic problem (without sign problem) in a restricted subspace. This approximation results in providing an upper bound to the exact energy of the system, whose accuracy depends to the quality of the variational wave function. More details on the implementation of the VMC and DMC algorithms can be found in Ref. [29].

3. The BCS-BEC crossover and the unitary limit

The BCS-BEC crossover has been calculated extensively over the past ten years. The unitary limit is very interesting because perturbative methods cannot be applied. In addition, there are no small parameters to expand, and thus using numerical calculations is the only way to accurately explore the properties of the system. The energy of these systems is given by

$$E = \xi E_{FG} , \tag{13}$$

where ξ is called Bertsch parameter. There have been several calculations of ξ based on QMC techniques, and recently we used the AFMC method to calculate exactly the energy of the system [21], the results are summarized in Fig. 1. Since the lattice space provides a natural regularization of the Hamiltonian, it is possible to test different models of kinetic energy or dispersion relations. In the figure, we show the results obtained using different regularization of the kinetic energy:

$$\begin{aligned}
\epsilon_k^{(2)} &= \frac{\hbar^2 k^2}{2m} , \\
\epsilon_k^{(4)} &= \frac{\hbar^2 k^2}{2m} \left[1 - \beta^2 k^2 \alpha^2\right] , \\
\epsilon_k^{(h)} &= \frac{\hbar^2}{m\alpha^2} \left[3 - \cos(k_x\alpha) - \cos(k_y\alpha) - \cos(k_z\alpha)\right] ,
\end{aligned} \tag{14}$$

where α is the lattice spacing, and β is fixed to have a s-wave interaction with zero effective range, see Ref. [21] for details. All the results correspond the case of infinite scattering length, and in the case of $\epsilon_k^{(h)}$ and $\epsilon_k^{(2)}$, the results are obtained at finite values of effective range and need to be extrapolated. Results obtained using different number of particles are included in the figure. We note that finite size effects are basically absent, as using 38, 48 or 66 Fermions yields to the same energy. This fact is because the interaction is very short-range, and because pairing of the system is very strong. Particles behave like Bosons, and finite size effects associated to the kinetic energy vanish. The absence of sizeable finite size effects for more than ~40 particles has been also shown with a DMC calculation [25]. By extrapolating the results to zero s-wave effective range gives the result of $\xi = 0.372(5)$ [28] that is in excellent agreement with a subsequent experiment performed at MIT [30]. The best upper bound estimate using the fixed-node DMC method is $\xi = 0.389(1)$ [22].

Another quantity of interest is the dependence of the energy to the s-wave effective range:

$$\xi = \xi_0 + s\, k_F r_e + \ldots . \tag{15}$$

At unitarity the parameter s has been calculated using DMC and AFMC, and the different calculations yield $s = 0.11(3)$ and $s = 0.12(2)$ [21, 22]. In Fig. 2 we show the results obtained using DMC with various potentials for the equal mass case, and for a mass ratio $m_h/m_l = 6.5$ corresponding to the $Li - K$ mixture. It is pretty interesting that the value of s does not depend to the mass ratio of the two species.

The knowledge of the energy dependence to the effective range is important to compare properties of neutron matter and cold atoms. At low-densities and zero temperature, neutron matter has similar properties to cold atoms: the neutron-neutron interaction is mainly s-wave, the effective

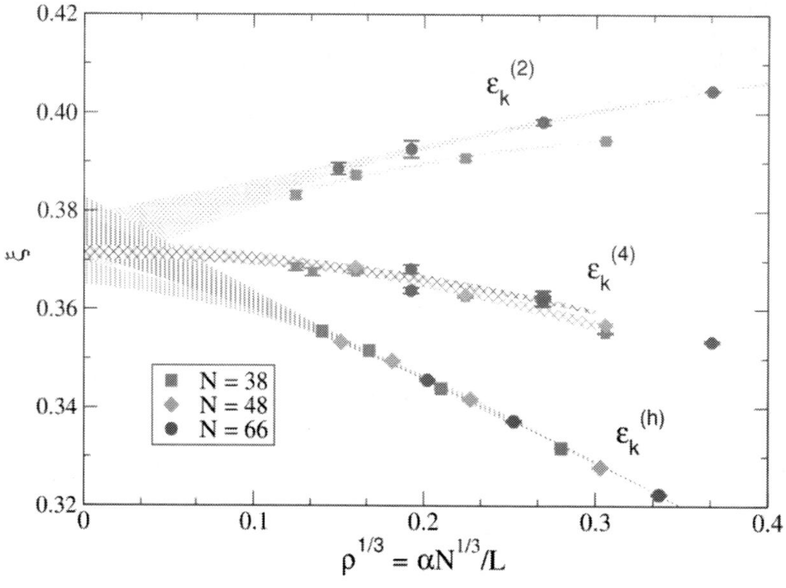

Figure 1. Ground state energy of the unitary Fermi gas calculated using the AFMC method. We show the value of ξ as a function of the lattice size for various particle numbers N and Hamiltonians, α is the lattice spacing. See Ref. [21] for more details.

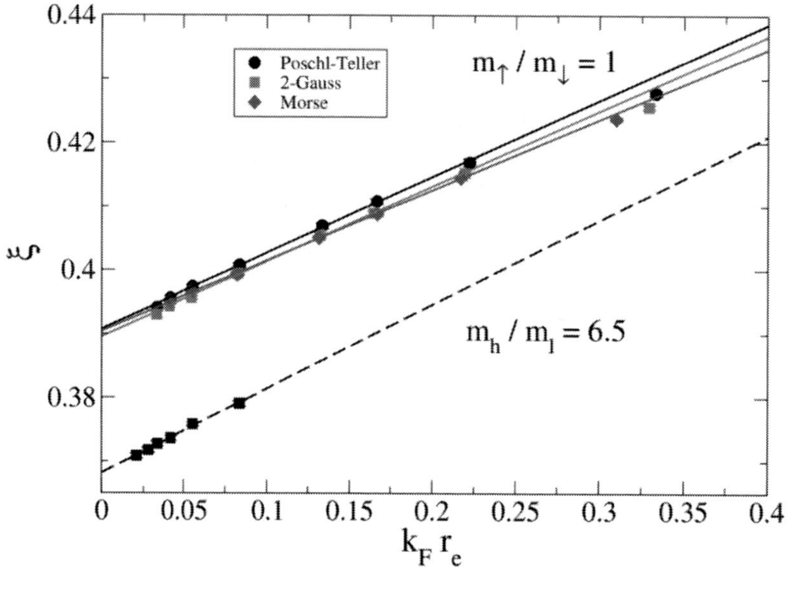

Figure 2. The value of ξ as a function of $k_F r_e$ calculated using DMC for the two-components with equal and unequal masses unitary Fermi gas. In the case of equal masses, the results obtained using different two-body potentials are presented. See Ref. [22] for more details.

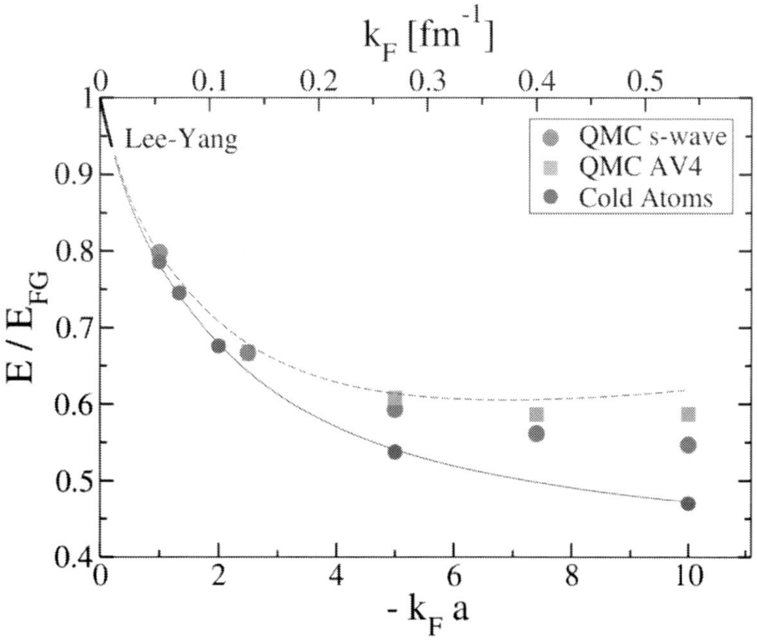

Figure 3. The equation of state of cold atoms and low-density neutron matter. For the latter, we show results obtained from the s-wave part of a realistic neutron-neutron potential, and those obtained by extrapolating the cold-atoms results to large effective ranges. See the text for details. This figure is taken from Ref. [28].

range is small, about 2.7 fm, and the scattering length is large, -18.5 fm [31]. In order to make qualitative comparisons between cold atoms and neutron matter it is important to take into account the fact that the neutron-neutron interaction has a small but finite r_e. Using the estimate of s as a function of $a\,k_F$ opens the possibility to compare these two different systems. An example of this comparison is reported in Fig. 3. In the figure we compare the cold atoms results corresponding to $r_e = 0$, the neutron matter equation of state calculated using the s-wave component of realistic nuclear forces [31, 32], and the results obtained by extrapolating the cold-atoms results to the effective range of neutron-neutron potential using Eq. 15.

In the BCS-BEC crossover, the two limits correspond to the BCS where Fermions are weakly attractive and to the BEC regime with very strong coupling. In the weak coupling limit, the energy of the interacting Fermi gas is given by [33]

$$\frac{E}{E_{FG}} = 1 + \frac{10}{9\pi} a\,k_F + \frac{4(11 - 2\log 2)}{21\pi^2}(a\,k_F)^2 + \dots , \tag{16}$$

where a is the scattering length and k_F is the Fermi momentum of the system. In the very strong coupling limit, when $a\,k_F \to 0^+$, the system becomes made of Bosonic molecules, formed by one

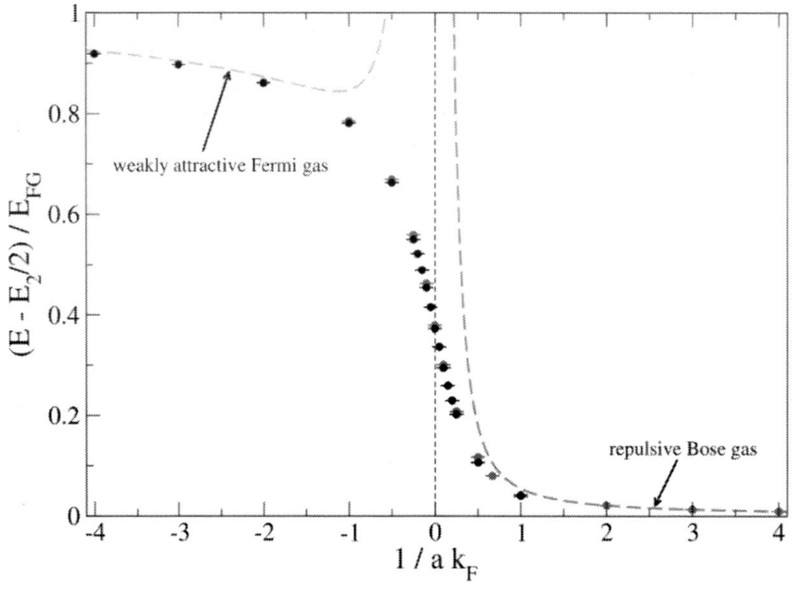

Figure 4. The BCS-BEC crossover calculated using DMC for the two-component unequal masses case corresponding to the $Li-K$ mixture. The two different set of points correspond to the results obtained using different values of r_e. The dashed lines correspond to the perturbative calculation of weakly attractive Fermi gas in the BCS, and of the weakly repulsive Bose gas in the BEC.

spin-up or heavy and one spin-down or light Fermion, that are weakly repulsive. Their energy is given by [34]

$$\frac{E/N - E_2/2}{E_{FG}} = \frac{10}{9\pi} a_{dd} k_F \frac{m_h m_l}{(m_h + m_l)^2} \times \left[1 + \frac{128}{15\sqrt{6\pi^3}} (a_{dd} k_F)^{3/2} + \ldots\right], \tag{17}$$

where a_{dd} is the Boson-Boson scattering length that can be obtained from a few-body calculation [35], or even by fitting QMC results [16]. The BCS-BEC crossover for a two-component $Li-K$ Fermi-Fermi mixture is shown in Fig. 4. In the figure, the blue and black points correspond to the QMC results obtained using two different values of r_e in the two-body interaction. In the unitary limit, the energy can be expanded as

$$\frac{E}{E_{FG}} = \xi - \frac{\zeta}{a\,k_F} - \frac{5\nu}{3(a\,k_F)^2} + \ldots . \tag{18}$$

The fit to QMC results, extrapolated to $r_e = 0$, gives $\xi = 0.3726(6)$, $\zeta = 0.900(2)$ and $\nu = 0.46(1)$. The values of ζ and ν are very similar to the case of equal masses reported in Ref. [20]. In Fig. 4 we also show the two limits given by Eq. 16 and 17 that agree with the QMC results. The crossover corresponding to the $Li-K$ mixture is qualitatively similar to the equal masses case [16, 23].

4. Contact parameter and the static structure factor

One of the most intriguing properties of strongly interacting Fermi gases is the contact parameter. Shina Tan showed that several quantities, including the pair distribution function, momentum distribution, and the static structure factor in the limit of small distances or high momenta are fully described by the contact parameter [36–38]. Also the equation of state is related to the contact parameter. This particular property is very interesting because the contact can be calculated in different ways, and, more important, can be measured in different experiments (see for example Refs. [39–41]).

In Ref. [20] several observables related to the contact have been calculated using QMC methods. We note that, apart the fixed-node approximation, any other observable depends to the choice of Ψ_v. Within DMC, we have computed observables with

$$\langle O\rangle = 2O_m - O_V\,, \tag{19}$$

where

$$O_m = \langle\Phi_0|O|\Psi_v\rangle\,, \tag{20}$$

and

$$O_V = \langle\Psi_v|O|\Psi_v\rangle\,. \tag{21}$$

Note that the energy is a special case, because the operator H commutes with the propagator G, and in this case it's easy to see that $\langle H\rangle = H_m$. The fact that several operators basically describe the same contact parameter is a proof of Tan's relations but also of the very good quality of the trial wave function employed in our calculations.

We have calculated the contact parameter in the BCS-BEC crossover using the adiabatic relation that relates the contact to the equation of state:

$$\frac{C}{Nk_F} = -\frac{6\pi}{5}\frac{\partial\xi}{\partial(k_Fa)^{-1}}\,. \tag{22}$$

The contact parameter is shown in Fig. 5, and compared with recent measurements and with the result obtained by Enss *et al.* [42]. In the inset we show the equation of state. The experimental measurements based on the Bragg spectroscopy are in a very good agreement with our calculations, both at unitarity and in the BEC limit [43].

One of the quantities related to the contact is the static structure factor $S(q)$, that corresponds to the sum-rule of the density-response function:

$$S(q) = \int d\omega\, S(\omega, q)\,. \tag{23}$$

The density-response function $S(\omega, q)$ has been measured by means of the Bragg spectroscopy [43, 45,46]. At high momenta, this quantity is related to the contact through the relation

$$\frac{C}{Nk_F} = \frac{4q}{k_F}\left[\frac{S(q)-1}{1-4/(\pi\, q\, a)}\right]. \tag{24}$$

At low momenta $S(q)$ is related to the long-wavelength phonons in the system. The quantity $S(q)$ has been calculated using QMC for different values of $a\,k_F$ in Ref. [43], and the agreement with

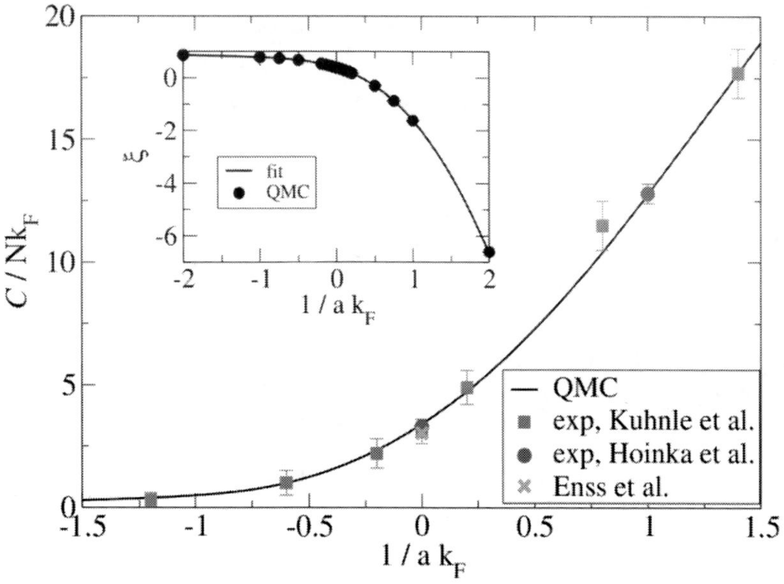

Figure 5. The contact parameter as a function of $1/a\,k_F$ obtained from the equation of state calculated using DMC (shown in the inset). The experimental points of Refs. [43,44] and the result of Ref. [42] are also shown.

experimental data is within 1%. The calculation of $S(q)$ at small q requires extra care. Due to the phonon dispersion, we found it very important to include long-range correlations in the variational wave function as discussed in Sec. 2. In Fig 6 we show the VMC results for $S(q)$ obtained by including (or not) the long range correlations in the Jastrow. We note that when the DMC algorithm is also used, the expectation value of $S(q)$ is less dependent to the presence of long-range correlations though even in this case the accuracy of the calculated S(q) is much greater with the improved trial function. It would be very interesting to measure the behavior of $S(q)$ at small momenta, where collective modes are expected to be important.

5. Conclusions

In this paper we present recent results obtained using QMC methods for two-component strongly interacting Fermi gases. The energy and its dependence on the product of the effective range and Fermi momentum have been precisely determined theoretically and compared to experiment across the BCS-BEC crossover. Other properties related to the contact have also been computed, in particular the static structure factor, and the results are in very good agreement with experimental data. Future directions include dynamic response of the unitary Fermi gas, inhomogeneous systems and the transitions from three- to two-dimensions, and multi-component Fermi gases.

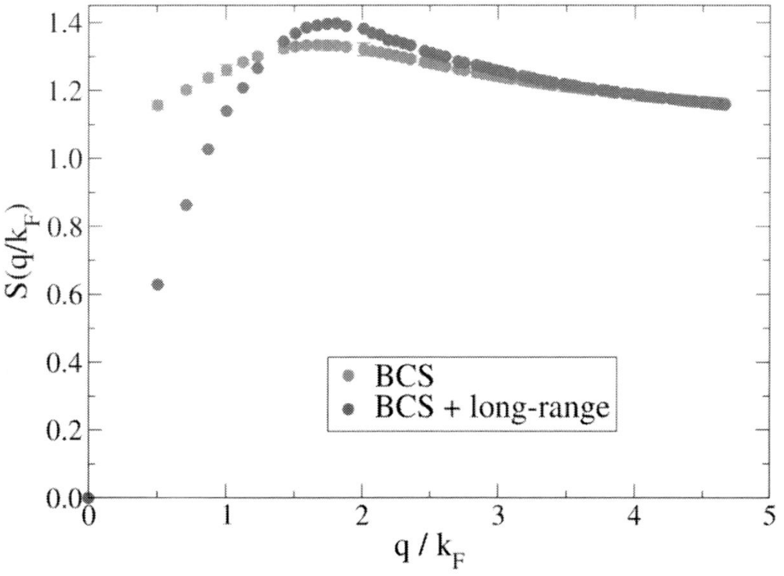

Figure 6. The static structure factor (density-response sum rule) as a function of q. The results have been obtained with the VMC, with and without long-range correlations in the Jastrow.

Acknowledgments

The author is grateful to J. Carlson for very useful discussions and critical comments on the manuscript. The research presented in this paper has been supported by the U.S. Department of Energy, Office of Nuclear Physics, the NUCLEI SciDAC program and by the LANL LDRD program. Computing time time were made available by Los Alamos Open Supercomputing. This research used also resources of the National Energy Research Scientific Computing Center, which is supported by the Office of Science of the U.S. Department of Energy under Contract No. DE-AC02-05CH11231.

References

[1] Giorgini S, Pitaevskii L P and Stringari S 2008 *Rev. Mod. Phys.* **80** 1215–1274
[2] Zwierlein M W, Schirotzek A, Schunck C H and Ketterle W 2006 *Science* **311** 492
[3] Partridge G B, Li W, Kamar R I, an Liao Y and Hulet R G 2006 *Science* **311** 503
[4] Pilati S and Giorgini S 2008 *Phys. Rev. Lett.* **100** 030401
[5] Lobo C, Recati A, Giorgini S and Stringari S 2006 *Phys. Rev. Lett.* **97** 200403
[6] Bulgac A, Forbes M M and Schwenk A 2006 *Phys. Rev. Lett.* **97** 020402
[7] Bulgac A and Forbes M M 2008 *Phys. Rev. Lett.* **101** 215301
[8] Wille E, Spiegelhalder F M, Kerner G, Naik D, Trenkwalder A, Hendl G, Schreck F, Grimm R, Tiecke T G, Walraven J T M, Kokkelmans S J J M F, Tiesinga E and Julienne P S 2008 *Phys. Rev. Lett.* **100** 053201
[9] Spiegelhalder F M, Trenkwalder A, Naik D, Kerner G, Wille E, Hendl G, Schreck F and Grimm R 2010 *Phys. Rev.* **A 81** 043637

[10] Kohstall C, Zaccanti M, Jag M, Trenkwalder A, Massignan P, Bruun G M, Schreck F and Grimm R 2012 *Nature* **485** 615–618
[11] Gezerlis A, Gandolfi S, Schmidt K E and Carlson J 2009 *Phys. Rev. Lett.* **103** 060403
[12] Gubbels K B, Baarsma J E and Stoof H T C 2009 *Phys. Rev. Lett.* **103** 195301
[13] Baarsma J E, Gubbels K B and Stoof H T C 2010 *Phys. Rev.* **A 82** 013624
[14] Baranov M A, Lobo C and Shlyapnikov G V 2008 *Phys. Rev.* **A 78** 033620
[15] Carlson J, Chang S Y, Pandharipande V R and Schmidt K E 2003 *Phys. Rev. Lett.* **91** 050401
[16] Astrakharchik G E, Boronat J, Casulleras J and Giorgini S 2004 *Phys. Rev. Lett.* **93** 200404
[17] Carlson J and Reddy S 2005 *Phys. Rev. Lett.* **95** 060401
[18] Bulgac A, Drut J E and Magierski P 2006 *Phys. Rev. Lett.* **96**(9) 090404
[19] Carlson J and Reddy S 2008 *Phys. Rev. Lett.* **100** 150403
[20] Gandolfi S, Schmidt K E and Carlson J 2011 *Phys. Rev.* **A 83** 041601
[21] Carlson J, Gandolfi S, Schmidt K E and Zhang S 2011 *Phys. Rev.* **A 84**(6) 061602
[22] Forbes M M, Gandolfi S and Gezerlis A 2012 *Phys. Rev.* **A 86**(5) 053603
[23] Chang S Y, Pandharipande V R, Carlson J and Schmidt K E 2004 *Phys. Rev.* **A 70** 043602
[24] Morris A J, López Ríos P and Needs R J 2010 *Phys. Rev.* **A 81** 033619
[25] Forbes M M, Gandolfi S and Gezerlis A 2011 *Phys. Rev. Lett.* **106** 235303
[26] Sorella S 2001 *Phys. Rev.* **B 64** 024512
[27] Vitali E, Arrighetti P, Rossi M and Galli D E 2011 *Molecular Physics* **109** 2855–2862
[28] Carlson J, Gandolfi S and Gezerlis A 2012 *Progress of Theoretical and Experimental Physics* **2012** 01A209
[29] Foulkes W M C, Mitas L, Needs R J and Rajagopal G 2001 *Rev. Mod. Phys.* **73**(1) 33–83
[30] Ku M J H, Sommer A T, Cheuk L W and Zwierlein M W 2012 *Science* **335** 563
[31] Gezerlis A and Carlson J 2008 *Phys. Rev.* **C 77**(3) 032801
[32] Gezerlis A and Carlson J 2010 *Phys. Rev.* **C 81**(2) 025803
[33] Huang K and Yang C N 1957 *Phys. Rev.* **105**(3) 767–775
[34] Lee T D, Huang K and Yang C N 1957 *Phys. Rev.* **106** 1135–1145
[35] Petrov D S, Salomon C and Shlyapnikov G V 2005 *J. Phys. B: At. Mol. Opt. Phys.* **38** S645
[36] Tan S 2008 *Ann. Phys.* **323** 2952–2970
[37] Tan S 2008 *Ann. Phys.* **323** 2971–2986
[38] Tan S 2008 *Ann. Phys.* **323** 2987–2990
[39] Navon N, Nascimbène S, Chevy F and Salomon C 2010 *Science* **328** 729–732
[40] Kuhnle E D, Hu H, Liu X J, Dyke P, Mark M, Drummond P D, Hannaford P and Vale C J 2010 *Phys. Rev. Lett.* **105** 070402
[41] Stewart J T, Gaebler J P, Drake T E and Jin D S 2010 *Phys. Rev. Lett.* **104** 235301
[42] Enss T, Haussmann R and Zwerger W 2011 *Annals of Physics* **326** 770 – 796
[43] Hoinka S, Lingham M, Fenech K, Hu H, Vale C J, Drut J E and Gandolfi S 2013 *Phys. Rev. Lett.* **110**(5) 055305
[44] Kuhnle E D, Hoinka S, Hu H, Dyke P, Hannaford P and Vale C J 2011 *New Journal of Physics* **13** 055010
[45] Veeravalli G, Kuhnle E, Dyke P and Vale C J 2008 *Phys. Rev. Lett.* **101**(25) 250403
[46] Hoinka S, Lingham M, Delehaye M and Vale C J 2012 *Phys. Rev. Lett.* **109**(5) 050403

From hypernuclei to the Inner Core of Neutron Stars: A Quantum Monte Carlo Study

D Lonardoni[1], **F Pederiva**[2,3], **S Gandolfi**[4]

[1]Physics Division, Argonne National Laboratory, Argonne, IL 60439, USA
[2]Dipartimento di Fisica, Università di Trento, via Sommarive 14, I-38123 Trento, Italy
[3]I.N.F.N. - T.I.F.P.A, Trento Institute for Fundamental Physics and Applications, I-38123 Trento, Italy
[4]Theoretical Division, Los Alamos National Laboratory, Los Alamos, New Mexico 87545, USA

E-mail: lonardoni@science.unitn.it, pederiva@science.unitn.it, stefano@lanl.gov

Abstract. Auxiliary Field Diffusion Monte Carlo (AFDMC) calculations have been employed to revise the interaction beween Λ-hyperons and nucleons in hypernuclei. The scheme used to describe the interaction, inspired by the phenomenological Argonne-Urbana forces, is the $\Lambda N + \Lambda NN$ potential firstly introduced by Bodmer, Usmani *et al.*. Within this framework, we performed calculations on light and medium mass hypernuclei in order to assess the extent of the repulsive contribution of the three-body part. By tuning this contribution in order to reproduce the Λ separation energy in ${}^5_\Lambda$He and ${}^{17}_{\Lambda}$O, experimental findings are reproduced over a wide range of masses. Calculations have then been extended to Λ-neutron matter in order to derive an analogous of the symmetry energy to be used in determining the equation of state of matter in the typical conditions found in the inner core of neutron stars.

1. Introduction

The composition of the inner core of neutron stars (NS) still remains a largely unsolved question. The fairly recent observation of neutron stars with masses of order $2M_\odot$ [1, 2] has set a rather strong constraint on the stiffness of the equation of state (EoS) of hadronic matter at high densities. On the other hand simple physical arguments can be made, that introduce mechanisms for softening the EoS at high densities. One of them is based on the idea that if the chemical potential of matter at β-equilibrium reaches a large enough value, particles with non-zero strangeness can be stabilized in the system. In particular one can expect by charge neutrality arguments that Σ^- and Λ hyperons could be the first species to appear. The larger mass of these hadrons ($m_\Sigma \simeq 1193$ MeV and $m_\Lambda \simeq 1116$ MeV), together with the fact that they become distinguishable with respect to nucleons, lowers the energy of the system, making the EoS softer. Several calculations performed in the past using the available hyperon-nucleon interactions all confirmed this picture (see for example Refs. [3, 4, 5, 6]). However, the softening of the EoS is such that the predicted maximum mass of a NS is not compatible with the current observational data.

This apparent puzzle could be solved if the interaction between hyperons and nucleons becomes so repulsive at large baryon density, that the system does not find energetically convenient to push the creation of hyperons beyond a certain limit, thereby preventing an excessive softening of the EoS. In this spirit, the aim of this work was to perform a deeper analysis of the hyperon-nucleon interaction starting from the data available for Λ hypernuclei. Our scheme is based on the use of Quantum Monte Carlo methods (more specifically the Auxiliary Field Diffusion Monte Carlo method [7, 8, 9]) in order to solve for the ground state of an Hamiltonian including a realistic local potential which contains hyperon-nucleon and hyperon-nucleon-nucleon terms. The solutions allow us for discussing the systematics of the binding energy of the hyperon in hypernuclei, and consequently fine tuning the interaction itself. The final outcome of these calculations should be a realistic potential to be used in determining the properties of hyperon-nucleon matter, and in this paper some of the progress along this way will be presented.

2. ΛN Potential

In this paper we will consider only systems including neutral Λ hyperons. So far, experimental binding energies and excitation energies are available only for a limited set of Λ hypernuclei [10, 11, 12, 13, 14], which is hopefully going to be extended by ongoing measurements at several facilities worldwide. The number of available $p\Lambda$ scattering events is also relatively small [15, 16, 17], but enough to constrain the main contributions of the hyperon-nucleon interaction. Bodmer, Usmani *et al.* proposed a phenomenological potential inspired by the Argonne-Urbana forces (see [18, 19, 20, 21] and references therein), that is sufficient to capture the physical information that is presently at disposal.

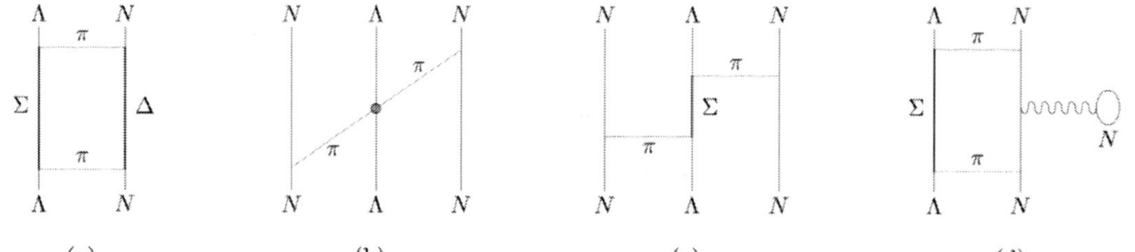

Figure 1. A schematic representation of the phenomenological Λ-nucleon interaction. In this scheme 2π-exchange processes between nucleons and the Λ particle appear as two-body (a) and three-body ((b), (c) and (d)) contributions. The last three will be labelled in the following as S-wave, P-wave and Dispersive terms respectively.

It should be noticed that isospin conservation prevents the occurrence of vertices $\Lambda\pi\Lambda$. As a consequence, the Λ-nucleon interaction requires at least the exchange of two pions. In Fig. 1 we list all the possible occurring processes of this kind. As it can be seen, only one of this contributions involve two particles. All the others are three-body forces of the ΛNN kind. It should be stressed out that this specific subdivision in two and three-body terms is somewhat depending on the scheme used. In the ΛN channel we should also include a process involving the exchange of kaons, which would give rise to a permutation of Λ and nucleons. For technical reasons we will include this contribution effectively in the coefficients of the interaction. In Refs. [20, 21] Usmani *et al.* reported this term to contribute order 10% and 30% of the central interaction.

The two-body terms has the form:

$$v_{\lambda i} = v_0(r_{\lambda i})(1 - \varepsilon + \varepsilon \mathcal{P}_x) + \frac{1}{4} v_\sigma T_\pi^2(r_{\lambda i})\, \boldsymbol{\sigma}_\lambda \cdot \boldsymbol{\sigma}_i \,, \tag{1}$$

where $v_0(r) = v_c(r) - \bar{v}\,T_\pi^2(r)$ is a central term. The coefficients $\bar{v} = (v_s + 3v_t)/4$ and $v_\sigma = v_s - v_t$ are the spin-average and spin-dependent strengths, where v_s and v_t denote singlet- and triplet-state strengths, respectively. Both the spin-dependent and the central radial terms contain the usual regularized OPE tensor operator $T_\pi(r)$

$$T_\pi(r) = \left[1 + \frac{3}{\mu_\pi r} + \frac{3}{(\mu_\pi r)^2}\right] \frac{e^{-\mu_\pi r}}{\mu_\pi r} \left(1 - e^{-cr^2}\right)^2 , \tag{2}$$

where μ_π is the pion reduced mass

$$\mu_\pi = \frac{1}{\hbar} \frac{m_{\pi^0} + 2\, m_{\pi^\pm}}{3} \qquad \frac{1}{\mu_\pi} \simeq 1.4 \text{ fm} . \tag{3}$$

The three-body terms have the following expressions:

$$v_{\lambda ij}^S = C_S\, Z\left(r_{\lambda i}\right) Z\left(r_{\lambda j}\right) \boldsymbol{\sigma}_i \cdot \hat{\boldsymbol{r}}_{i\lambda}\, \boldsymbol{\sigma}_j \cdot \hat{\boldsymbol{r}}_{j\lambda}\, \boldsymbol{\tau}_i \cdot \boldsymbol{\tau}_j , \tag{4}$$

$$v_{\lambda ij}^P = -\frac{C_P}{6} \Big\{ X_{i\lambda}, X_{\lambda j} \Big\}\, \boldsymbol{\tau}_i \cdot \boldsymbol{\tau}_j , \tag{5}$$

$$v_{\lambda ij}^D = W_D\, T_\pi^2\left(r_{\lambda i}\right) T_\pi^2\left(r_{\lambda j}\right) \left[1 + \frac{1}{6} \boldsymbol{\sigma}_\lambda \cdot (\boldsymbol{\sigma}_i + \boldsymbol{\sigma}_j)\right] . \tag{6}$$

The functions $X_{\lambda i}$ and $Z(r)$ are defined by

$$X_{\lambda i} = Y_\pi(r_{\lambda i})\, \boldsymbol{\sigma}_\lambda \cdot \boldsymbol{\sigma}_i + T_\pi(r_{\lambda i})\, S_{\lambda i} , \tag{7}$$

$$Z(r) = \frac{\mu_\pi r}{3} \Big[Y_\pi(r) - T_\pi(r) \Big] , \tag{8}$$

where the function $Y_\pi(r)$ is the usual regularized Yukawa potential

$$Y_\pi(r) = \frac{e^{-\mu_\pi r}}{\mu_\pi r} \left(1 - e^{-cr^2}\right) , \tag{9}$$

and $S_{\lambda i}$ is the tensor operator.

The parameters appearing in Eq. (1) have been determined by fitting the available scattering data, and have never been varied in our studies. Details on the parametrization can be found in Refs. [19, 20, 21]. For a complete review see Ref. [22].

3. Computational Methods

We model hypernuclei made up of $A-1$ nucleons and one Λ by the non-relativistic Hamiltonian

$$\hat{H} = \sum_{i=1}^{A-1} \frac{\hat{p}_i^{\,2}}{2m_N} + \frac{\hat{p}_\Lambda^2}{2m_\Lambda} + \sum_{i<j} v_{ij}(r_{ij}) + \sum_{i=1}^{A-1} v_{\lambda i}(r_{\Lambda i}) + \sum_{i<j} v_{\lambda ij}(r_{\Lambda i}, r_{\Lambda j}) , \tag{10}$$

where v_{ij} is the two-nucleons interaction. In order to obtain information about the ground state properties we use a stochastic projection algorithm, namely the Auxiliary Field Diffusion Monte Carlo (AFDMC) method [7, 8, 9]. AFDMC is based on propagating a number of points in the extended configuration space (including the coordinates, spin and isospin degrees of freedom), by sampling an approximation of the Green's function of the imaginary time Hamiltonian operator

$$\Psi(R, S, T; \tau + d\tau) = \sum_{S', T'} \int dR'\, G(R, S, T; R', S', T'; d\tau)\, \Psi(R', S', T'; \tau) , \tag{11}$$

where $R = \{\boldsymbol{r}_1 \ldots \boldsymbol{r}_A\}$, $S = \{\boldsymbol{\sigma}_1 \ldots \boldsymbol{\sigma}_A\}$, $T = \{\boldsymbol{\tau}_1 \ldots \boldsymbol{\tau}_A\}$ are the coordinates, spin and isospin of the baryons respectively. The sum over S' and T' has to be intended as the sum over all the possible spin/isospin states of the baryons. The approximated Green's function for a small step $d\tau$ in imaginary time is then written as:

$$G(R,S,T;R',S',T';d\tau) \simeq \left(\frac{m}{2\pi\hbar^2 d\tau}\right)^{\frac{3A}{2}} \mathrm{e}^{-\frac{m(R-R')}{2\hbar^2 d\tau}} \mathrm{e}^{-\left[\hat{V}(R,S,T)-E\right]d\tau} \,. \tag{12}$$

In general the last factor contains terms that are operatorial and/or non-local. In standard Quantum Monte Carlo calculations one is usually limited to use in the Green's function interactions that are local and at most quadratic in the spin/isospin operators, with the only exception of spin-orbit terms [23]. Therefore, the sum over spin and isospin states requires the use of a wave function that has a number of components exponentially growing with A. The AFDMC method makes use of the Hubbard-Stratonovich transformation to make only single particle spin/isospin operators in the Green's function. The advantage is the possibility to work on a single particle representation of the wave function, thereby reducing the computational cost of the calculation from exponential to cubic in A. On the other hand, this method makes more challenging the use of accurate wave functions including quantum correlations in operatorial channels. This has the consequence of reducing the accuracy that is achievable when imposing the approximate sampling constraints necessary to avoid the exponential growth of the variance due to the Fermion Sign Problem. Because of such technical reasons we limited ourselves to a nucleon-nucleon interaction that can be considered under control in AFDMC calculations, namely the Argonne V4' potential [24]. This force is known to give a reasonable binding energy for small nuclei, but to strongly overbind medium/heavy nuclei. However, we are interested in making a connection with experiments only through the Λ separation energy defined as

$$B_\Lambda = B_{\text{nuc}}(A-1) - B_{\text{hyp}}(A-1,\Lambda) \,, \tag{13}$$

where B_{nuc} and B_{hyp} are the total binding energies of the the nucleus and the Λ-hypernucleus with $A-1$ nucleons respectively. The hypothesis we make, rather well justified a posteriori, is that in the difference the contributions due to the nucleon-nucleon interaction mostly cancel out. Besides the reasonable agreement we obtain for the available experimental results for medium mass-hypernuclei, this assumption is supported by the fact that changing the interaction from AV4' to AV6' [24] (which gives a systematic underbinding of nuclei) does not seem to give a strong influence on the results [25]. The presence of the Λ hyperon also requires a modification of the wave function and a special treatment of the kinetic energy, in order to correctly subtract the contributions of the center of mass motion. Details can be found in Refs. [25, 26, 22, 27].

4. Results

4.1. Hypernuclei

As previously mentioned, the key point of this work is to begin the journey towards an accurate phenomenological hyperon-nucleon interaction that can be employed in many-body Quantum Monte Carlo calculations. Given that at present the number and quality of available scattering data do not allow for a substantial improvement of the ΛN contribution, we focused on refitting the coefficients appearing in the three-body ΛNN term. It emerges form the calculations that the binding energies are substantially insensitive to the value of the coefficient of the S-wave term in the 2π-exchange channel C_S, that was fixed to the value 1.5 MeV [21]. We are therefore left with two parameters to determine. This can be done by reproducing the experimental results in two hypernuclei. We chose two closed-shell systems, which we believe to be much less prone to suffer from the necessary approximations related to the use of AFDMC, namely ${}^{5}_{\Lambda}$He and ${}^{17}_{\Lambda}$O.

In Fig. 2 the dependence of B_Λ in ${}^{5}_{\Lambda}$He as a function of the parameters W_D and C_P is shown. The parametrizations compatible with the experimental result $B_\Lambda = 3.12(2)$ MeV [10]

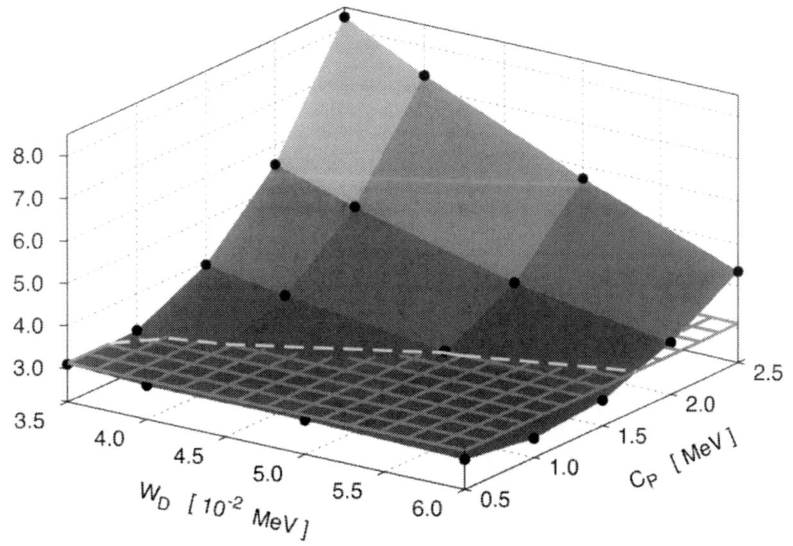

Figure 2. The Λ separation energy B_Λ as a function of the coefficients W_D and C_P of the ΛNN force, relative to the the P-wave component of the 2π-exchange term and the dispersive term respectively. The red plane is the experimental value from Ref. [10].

(represented by the red plane) are those found at the intersection of the two surfaces. Among them, we picked the one that better reproduces the (extrapolated) B_Λ in ${}^{17}_{\Lambda}$O [19]. The resulting values are $W_D = 0.035$ MeV and $C_P = 1.00$ MeV [27]. With this parametrization, we proceeded to compute the Λ separation energy for a number of other light and medium-heavy hypernuclei. In particular, we were interested in testing the saturation property of B_Λ, which is of extreme importance in view of extending the calculations to infinite matter.

Given the available computational resources we were able to simulate closed-shell nucleus/hypernucleus pairs up to ${}^{91}_{\Lambda}$Zr [25, 27]. The results are summarized in Fig. 3, where B_Λ is plotted as a function of $A^{-2/3}$. As it can be seen, the effect of including the ΛNN term in the Hamiltonian is very strong. It provides the repulsion necessary to realistically reproduce the limiting value of B_Λ. One important test of our calculation also comes from the fact that the new parametrization of the ΛNN term, that was fixed only using data for two hypernuclei, is very good in reproducing the overall behavior of the experimental data. Some discrepancies appear at very low masses ($A < 5$), where other effects should be expected, such as a rather strong contribution from charge symmetry breaking components of the interaction.

In Fig. 4 we report the results for the single particle densities for ^{4}He and ${}^5_\Lambda$He. Densities here are computed without taking into account the effect of the displacement of the center of mass due to the difference between m_Λ and m_N. The green curves are the densities of nucleons in the nucleus, while the red and blue curves are, respectively, the density of nucleons and of the lambda particle in the hypernucleus. In the left panel the results are obtained using AV4' for the nuclear part and the two-body ΛN interaction alone for the hypernuclear component. In the right panel the densities are calculated with the full two- plus three-body hyperon-nucleon interaction in the new parametrization.

The correct estimators for the single particle densities are obtained starting from the mixed

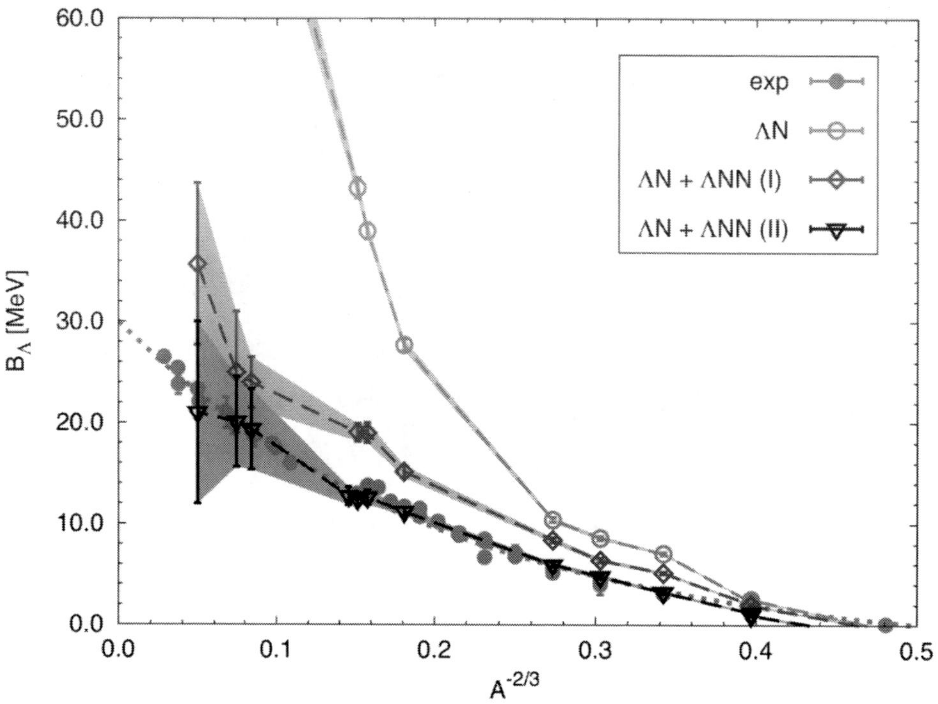

Figure 3. The Λ separation energy B_Λ as a function of $A^{-2/3}$. The red dots are the results obtained with an Hamiltonian containing a two-body ΛN force only. Blue diamonds are AFDMC calculations with the original 2- and 3-body ΛN potential by Usmani [28]. Black triangles are the current AFDMC results with the refitted 3-body interaction [27]. Green dots are experimental results. Lines and bands are drawn as a guid to the eye.

DMC results and the variational ones via the positive defined relation [23]:

$$\langle \mathcal{O} \rangle_{real} = \frac{\langle \psi_0 | \mathcal{O} | \psi_0 \rangle}{\langle \psi_0 | \psi_0 \rangle} = \frac{\left(\frac{\langle \psi_T | \mathcal{O} | \psi_0 \rangle}{\langle \psi_T | \psi_0 \rangle} \right)^2}{\frac{\langle \psi_T | \mathcal{O} | \psi_T \rangle}{\langle \psi_T | \psi_T \rangle}} = \frac{\langle \mathcal{O} \rangle^2_{DMC}}{\langle \mathcal{O} \rangle_{VMC}} , \tag{14}$$

where $\mathcal{O}$ is the density operator $\hat{\rho} = \sum_i \delta(r - r_i)$, ψ_T is the trial wave function and ψ_0 is the projected ground state wave function. The addition of the Λ particle to the nuclear core of ^{4}He has the effect to reduce the nucleon density in the center. The Λ particle tries to localize close to $r = 0$, enlarging therefore the nucleon distribution. When the three-body ΛNN interaction is turned on (right panel of Fig. 4), the repulsion moves the nucleons to large distances but the main effect is that the hyperon is pushed away from the center of the system.

As can be seen from Fig. 5, this effect is much more evident for large A. When the hypernucleus is described by the ΛN interaction alone, the Λ particle is localized near the center, in the range $r < 2$ fm (top panel of Fig. 5). The inclusion of the three-body ΛNN potential forces the hyperon to move from the center, in a region that roughly correspond to the skin of nucleons. It should be noticed that the nuclear densities given by the AV4' are widely overestimated. However we want to point out a couple of interesting facts. First of all, when using the AV6' potential we found the same effects on the Λ particle, confirming the importance of the three-body hyperon-nucleon

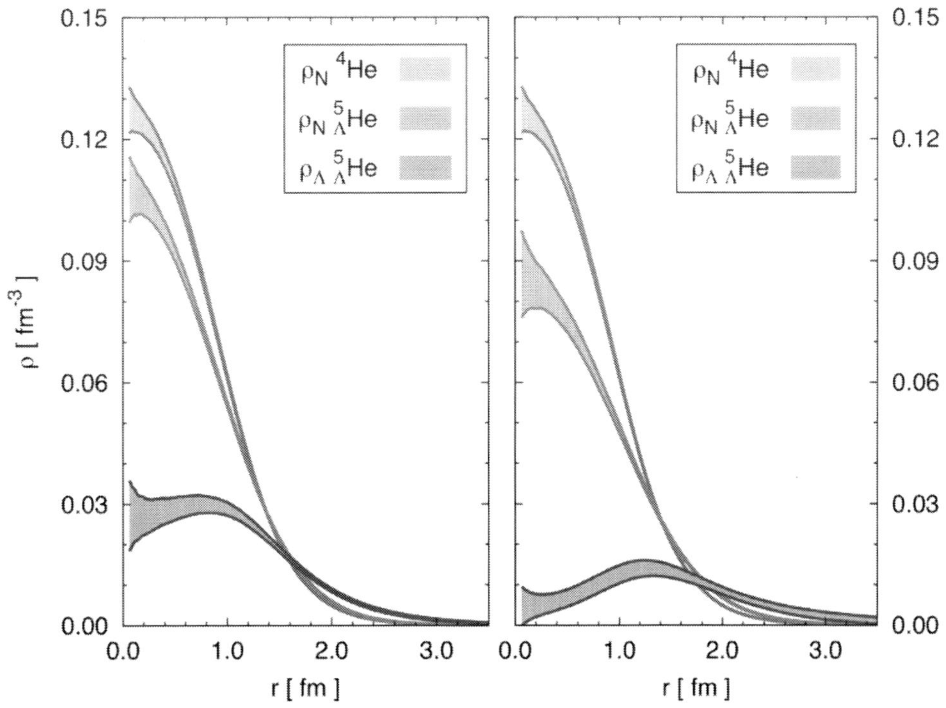

Figure 4. Single particle densities for nucleons in ^{4}He [green, upper banded curve] and for nucleons [red, middle banded curve] and the lambda particle [blue, lower banded curve] in $^5_\Lambda$He. In the left panel the results for the two-body ΛN interaction alone. In the right panel the results with the inclusion also of the three-body hyperon-nucleon force with the new set of parameters.

interaction and its strongly repulsive nature. Second, it can be noticed that the position of the peak of the Λ density roughly correspond to the distance from the center at which the nuclear density is about the saturation value. This might indicate that when using a correct NN force the hyperon could be found more towards the center. This might also reconcile the observed shrinkage effect of the core nucleus [13, 29], as obtained for example in Ref. [30].

4.2. Infinite Matter

The prediction of the structure of a NS requires the knowledge of the EoS of an infinite homogeneous matter. In the NS core weak interactions always sustain the presence of a finite fraction of protons, electrons and muons. Chemical equilibrium among these species determines the composition of the NS core. As already mentioned, when the density increases hyperons can appear. In general, it is required that the number of baryons and the charge neutrality are preserved.

In order to simplify the description, as a first approximation, the interior of a NS can be assumed to be made of neutrons only. In this case the chemical equilibrium between hyperons and neutrons is simply given by the condition $\mu_\Lambda = \mu_n$. In principle it is possible to compute

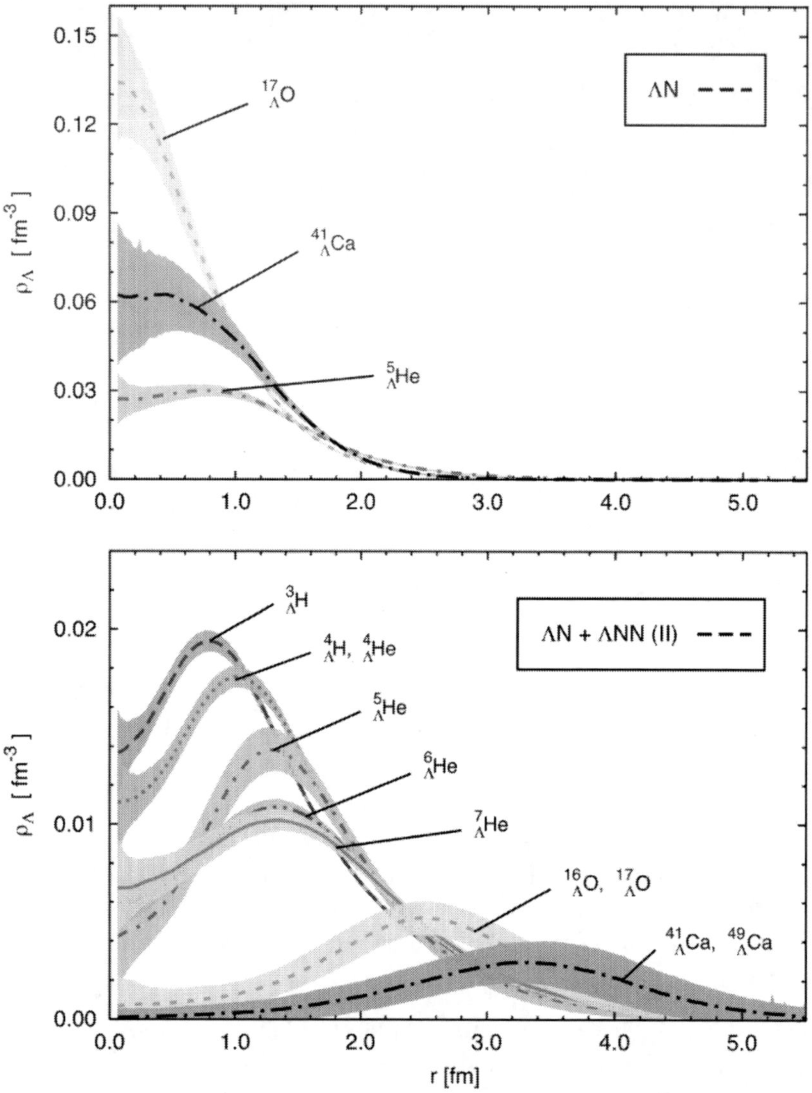

Figure 5. Single particle densities for the Λ particle in different hypernuclei. Top panel reports the results for the two-body ΛN interaction alone. Bottom panel shows the results when the three-body hyperon-nucleon interaction with the new set of parameters is also included.

the chemical potential of the species κ as:

$$\mu_\kappa = \left.\frac{\partial \mathcal{E}\left(\{\rho_\kappa\}\right)}{\partial \rho_\kappa}\right|_V , \tag{15}$$

where $\mathcal{E}$ is the total energy density (time component of the relativistic energy-momentum density 4-vector) of the system. Besides of the masses of the baryons, $\mathcal{E}$ must include the interaction energy. In the limit of a small fraction of hyperons present in the neutron medium, it is sensible to express the ΛN interaction energy per baryon as a function of the total baryon density ρ_b in

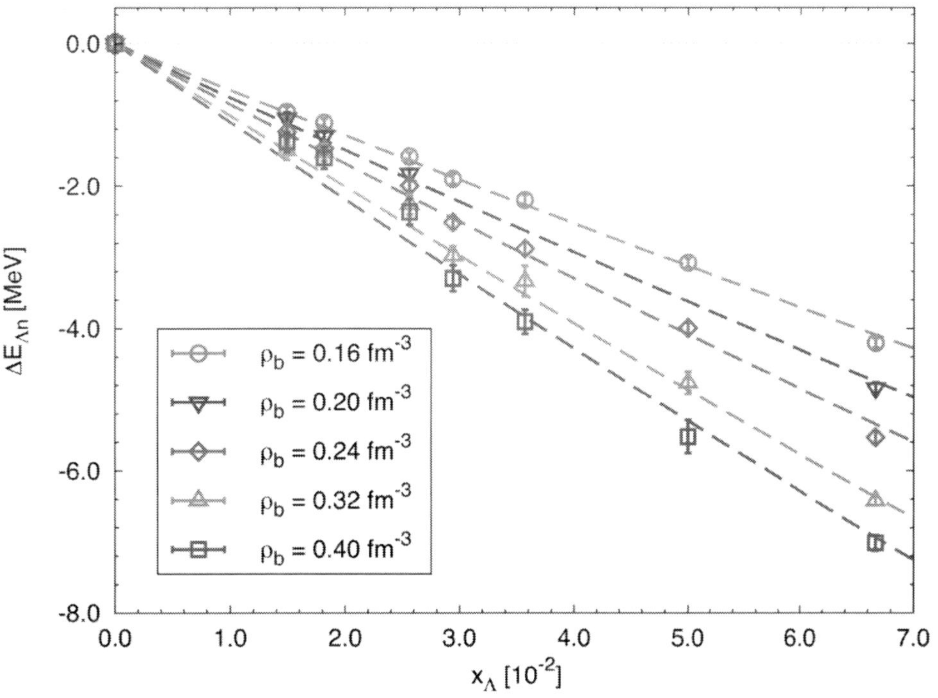

Figure 6. The energy difference $\Delta E_{\Lambda n} = E_{\Lambda n} - E_{PNM}$ per baryon as a function of the Λ fraction x_Λ in a mixed Λ-neutron matter for different values of the total baryon density ρ_b. The points are AFDMC results, while the corresponding lines are linear fits according to Eq. (16).

the form:

$$E_{\Lambda n}(\rho_b, x_\Lambda) = E_{PNM}(\rho_b) - S_{\Lambda n}(\rho_b)\, x_\Lambda \,, \tag{16}$$

where E_{PNM} is the energy per baryon of pure neutron matter (PNM), x_Λ is the hyperon fraction $x_\Lambda = \rho_\Lambda/\rho_b$ and $S_{\Lambda N}(\rho_b)$ could be considered as an analog of the symmetry energy. In the previous equation we neglect terms order x_Λ^2 which might be in principle non negligible. We should remember that being $m_\Lambda \neq m_n$, there is no reason for linear terms in the hyperon asymmetry not to be present in the EoS. In Fig. 6 we report the difference between the energy of Λ-neutron matter for finite hyperon fractions and the energy of PNM as a function of x_Λ, and for different total baryon densities. These results are obtained by simulating a system in which we add one or two hyperons in a periodic cubic box containing a given number of neutrons at a given density. Particular care must be paid here to correct for finite size effects. As it can be seen from the figure, the behavior of the energy difference can be reproduced by a linear fit.

From the slopes of the linear fits it is possible to obtain the behavior of $S_{\Lambda n}(\rho_b)$, which is reported in Fig. 7. By knowing the $S_{\Lambda n}$ function, the EoS of the Λ-neutron matter is given by Eq. (16) as a function of both ρ_b and x_Λ. It is then possible to impose the chemical equilibrium condition between hyperons and neutrons and derive the threshold density for the appearance of hyperons and the Λ fraction as a function of the total baryon density. These are key ingredients in the description of the Λ-neutron medium that affect the derivation of the mass-radius relation and the maximum mass of a NS. First steps in this direction have been taken and preliminary results are reported in Ref. [22].

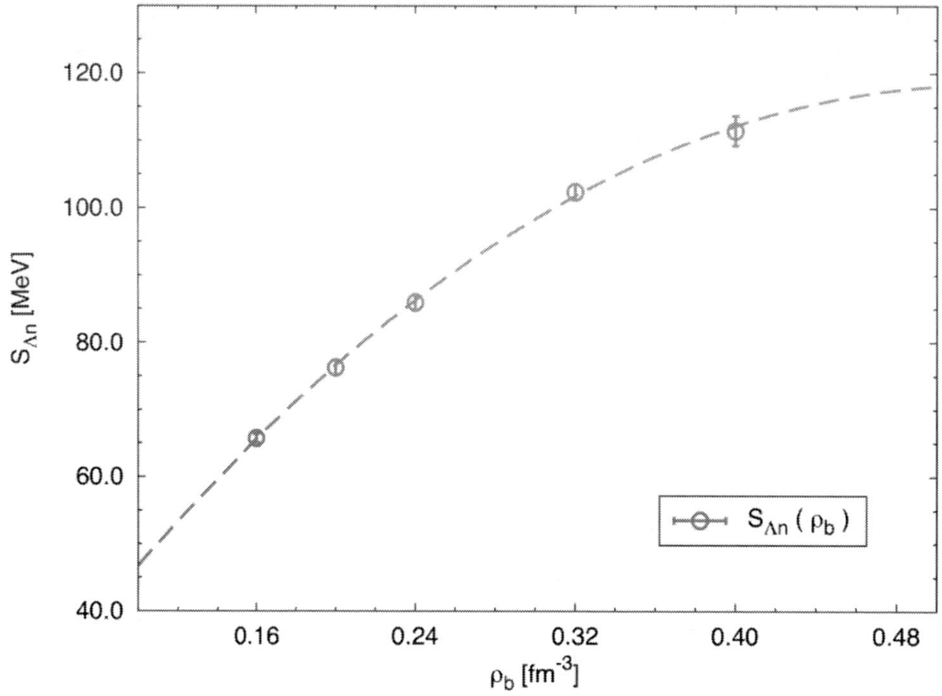

Figure 7. The coefficient $S_{\Lambda n}$ (analog of the symmetry energy in a mixed Λ-neutron matter) as a function of the total baryon density ρ_b. The dots are the values obtained by fitting the AFDMC results for the energy difference defined from Eq. (16). The dashed line is a fit to the data.

5. Conclusions

Ground state properties of hypernuclei over a wide range of masses were studied by means of the Auxiliary Field Diffusion Monte Carlo method. The main outcome of this is the determination of a realistic three-body local ΛNN interaction that reproduces with high accuracy the saturation properties of the hyperon separation energy. We also showed how by this method a good insight on the properties of homogeneous matter can be gained. In particular, we determined the behavior of $S_{\Lambda n}$, an analog of the symmetry energy for a mixed hyperon-neutron matter, which is the necessary ingredient to determine the equation of state of Λ-neutron matter, of great interest for the determination of the inner structure of neutron stars.

Acknowledgments

This work has been partially performed at LISC, Interdisciplinary Laboratory for Computational Science, a joint venture of the University of Trento and Bruno Kessler Foundation. Support and computer time were partly made available by the AuroraScience project (funded by the Autonomous Province of Trento and INFN), and by Los Alamos Open Supercomputing. This research used also resources of the National Energy Research Scientific Computing Center, which is supported by the Office of Science of the U.S. Department of Energy under Contract No. DE-AC02-05CH11231. The work of S. G. was supported by the Department of Energy Nuclear Physics Office, by the NUCLEI SciDAC program, and by a Los Alamos LDRD early career grant.

References

[1] Demorest P B, Pennucci T, Ransom S M, Roberts M S E and Hessels J W T 2010 *Nature* **467** 1081–1083
[2] Antoniadis J *et al* 2013 *Science* **340** 1233232
[3] Đapo H, Schaefer B J and Wambach J 2010 *Phys. Rev.* C **81**(3) 035803
[4] Schulze H J and Rijken T 2011 *Phys. Rev.* C **84**(3) 035801
[5] Vidaña I, Logoteta D, Providência C, Polls A and Bombaci I 2011 *EuroPhys. Lett.* **94** 11002
[6] Massot E, Margueron J and Chanfray G 2012 *EuroPhys. Lett.* **97** 39002
[7] Schmidt K E and Fantoni S 1999 *Phys. Lett.* B **446** 99–103
[8] Gandolfi S, Pederiva F, Fantoni S and Schmidt K E 2007 *Phys. Rev. Lett.* **99**(2) 022507
[9] Gandolfi S, Illarionov A Y, Schmidt K E, Pederiva F and Fantoni S 2009 *Phys. Rev.* C **79**(5) 054005
[10] Jurič M *et al* 1973 *Nucl. Phys.* B **52** 1–30
[11] Cantwell T *et al* 1974 *Nucl. Phys.* A **236** 445–456
[12] Pile P *et al* 1991 *Phys. Rev. Lett.* **66**(20) 2585–2588
[13] Hashimoto O and Tamura H 2006 *Progr. Part. Nucl. Phys.* **57** 564–653
[14] Nakamura S *et al* (HKS (JLab E01-011) Collaboration) 2013 *Phys. Rev. Lett.* **110**(1) 012502
[15] de Swart J J, Nagels M M, Rijken T A and Verhoeven P A 1971 *Springer Tracts in Modern Physics* **60** 138
[16] Kadyk J, Alexander G, Chan J, Gaposchkin P and Trilling G 1971 *Nucl. Phys.* B **27** 13–22
[17] Ahn J K *et al* 2005 *Nucl. Phys.* A **761** 41–66
[18] Bodmer A and Usmani Q 1988 *Nucl. Phys.* A **477** 621–651
[19] Usmani A A, Pieper S C and Usmani Q N 1995 *Phys. Rev.* C **51**(5) 2347
[20] Usmani A A 2006 *Phys. Rev.* C **73** 1–5
[21] Usmani A A and Khanna F C 2008 *J. Phys. G: Nucl. Part. Phys.* **35** 025105
[22] Lonardoni D 2013 *arXiv e-prints* Ph.D. thesis (*Preprint* nucl-th/1311.6672)
[23] Pieper S C 2008 *Nuovo Cimento Rivista Serie* **31** 709–740
[24] Wiringa R and Pieper S 2002 *Phys. Rev. Lett.* **89** 18–21
[25] Lonardoni D, Gandolfi S and Pederiva F 2013 *Phys. Rev.* C **87**(4) 041303
[26] Lonardoni D, Pederiva F and Gandolfi S 2013 *Nucl. Phys.* A **914** 243–247
[27] Lonardoni D, Pederiva F and Gandolfi S 2013 Phys. Rev.C **89**(1) 014314
[28] Usmani A A 1995 *Phys. Rev.* C **52**(4) 1773–1777
[29] Tanida K *et al* 2001 *Phys. Rev. Lett.* **86**(10) 1982–1985
[30] Sinha R, Usmani Q N and Taib B M 2002 *Phys. Rev.* C **66**(2) 024006

Superfluid instability in ultra-cold gas of fermionic atoms with attractive potential in a one-dimensional trap

P Schlottmann[1] **and A A Zvyagin**[2]

[1] Department of Physics, Florida State University, Tallahassee, Florida 32306, USA
[2] B.I. Verkin Institute for Low Temperature Physics and Engineering, Ukrainian National Academy of Sciences, 47 Lenin Avenue, Kharkov, 61103, Ukraine, and Max-Planck-Institut für Physik komplexer Systeme, D-01187, Dresden, Germany

E-mail: `schlottmann@physics.fsu.edu`, `zvyagin@pks.mpg.de`

Abstract. In the context of ultracold fermionic atoms with effective spin S confined to an elongated trap we study the one-dimensional gas interacting via an attractive δ-function potential using the Bethe *ansatz* solution. There are $N = 2S + 1$ fundamental states: The particles can either be unpaired or clustered in bound states of 2, 3, $\cdots$, $2S$ and $2S + 1$ atoms. In a magnetic field, the rich ground state phase diagram consists of these N states and various mixed phases in which combinations of the fundamental states coexist. The phase diagram simplifies considerably in zero-field, where only bound states of N atoms can exist. Due to the harmonic confinement and within the local density approximation, the density profile of bound states decreases along the tube from the center of the trap to its boundaries. In an array of tubes with weak Josephson tunneling superfluid order may arise. In zero-field the response functions determining the superfluid and density wave order are calculated using conformal field theory and the exact Bethe *ansatz* solution. The response function for superfluidity consists of a power law with distance, while the correlation function for density waves is a power law of distance times a sinusoidal factor oscillating with distance with a period given by two times the Fermi momentum. For $S = 1/2$ superfluidity is a possibility for all densities and density waves can be excluded. For $S \geq 3/2$ superfluidity may occur at low densities but at high densities it gives way to density waves. We discuss the scenario of phase separation where for $S \geq 3/2$ the system has superfluid long-range order toward the trap boundaries and density waves at its center.

1. Introduction

Spin-imbalanced ultracold gases of fermionic atoms confined to one-dimensional traps have been the subject of several recent studies [1, 2, 3]. Confinement to nearly one-dimensional tubes can be achieved if the ultracold cloud of atoms is subjected to a two-dimensional optical lattice, which defines a two-dimensional array of tubes [1]. The tubes can be regarded as isolated if the confinement by the laser beams is strong enough to suppress tunneling between tubes. The scattering between atoms under transverse harmonic confinement is subject to a confinement-induced resonance [4]. Fine-tuning this Feshbach-type resonance, the interaction between particles can be made attractive and its strength can be varied [5]. The interaction is local and can be approximated by a δ-function potential in space. The weak confinement along the tube is roughly harmonic and can be locally incorporated into the chemical potential. Consequently, these systems are only locally homogeneous and within the local density approximation display phase separation with the variation of the chemical potential along the tube [2, 3].

For fermions with a spin S an attractive potential may lead to bound states of up to $N = 2S+1$ atoms, extending the concept of preformed Cooper pairs to larger clusters. The ground state phase diagram will have large number of possible pure (consisting of only one kind of clusters) and mixed phases (coexistence of two or more basic states) [6, 7]. Interesting physics arises from the transitions between these phases as a function of the chemical potential and the external magnetic field. In this paper we consider an attractive δ-function potential between the atoms of spin S and assume the model has SU(N) symmetry, i.e. the scattering is independent of the spin components.

One-dimensional spinless bosons with δ-function interaction were first studied by Lieb and Liniger [8] using Bethe's *ansatz*. These results where extended to spin-1/2 gases by M. Gaudin [9] and C.N. Yang [10]. For an attractive interaction the $S = 1/2$ ground state has two classes of solutions of the discrete Bethe *ansatz* equations, namely, real charge rapidities and paired complex conjugated rapidities [9, 11, 12], representing spin polarized particles and bound states of the Cooper type, respectively. The preformed Cooper pairs are gapped in a magnetic field smaller than a critical one (energy required to break up the pairs) and do not display long-range order. Similar results were obtained for the Hubbard model with attractive U [13, 14]. Sutherland [15] generalized C.N. Yang's solution for spin 1/2 [10] to an arbitrary number of colors $N = 2S + 1$ [SU(N)-symmetry]. Takahashi [16] derived the integral equations for the ground state density functions for bound states up to $N = 2S+1$ particles. The classification of states, the thermodynamics, the ground state equations and elementary excitations of the gas with arbitrary number of colors have been derived by Schlottmann [6, 17, 18] for both attractive and repulsive potentials. Several of these results have been recently rederived in the context of ultracold fermion gases [19, 20, 21, 22].

So far experimental efforts were mostly focussed on ^{6}Li, i.e. a $S = 1/2$ system. Tubes with ultracold gases of fermionic atoms provide the unique possibility to study systems with a spin larger than 1/2, e.g., ^{40}K (spin 9/2), ^{43}Ca (spin 7/2), ^{87}Sr (spin 9/2) or ^{173}Yb (spin 5/2) atoms. In a recent publication [7] we obtained the μ versus H ground state phase diagrams for Fermi gases with spins $S = 3/2$, 5/2, 7/2 and 9/2, where μ is the chemical potential and H the Zeeman splitting, extending this way Orso's [2] results for $S = 1/2$. Similar results for $S = 3/2$ have been presented previously for a fixed number of particles [20].

For an isolated tube there is no long-range order of the bound states. A weakly interacting array of tubes, e.g. via tunneling of bound states between tubes, however, increases the effective dimension of the system so that long-range order can arise [23, 24]. For a spin-imbalanced system this could lead to the realization of the Fulde-Ferrell-Larkin-Ovchinnikov (FFLO) [25] phase for superfluidity in an ultracold gas of atoms [1, 26]. The system remains strongly anisotropic and pure (there are almost no impurities) and is hence favorable for inhomogeneities like modulations of the order parameter of the FFLO type in the presence of an external magnetic field. As for unconventional superconductors, the instability to superconductivity is determined by the properties of the normal phase. In this case the normal phase is a Luttinger liquid. Evaluating the superfluid response functions for the Luttinger liquid, one can then determine which one is the first superfluid instability reached from the normal phase. There is a large number of possible superfluid order parameters. The correlation functions have been studied in detail for $S = 3/2$ and $S = 5/2$ using conformal field theory and the Bethe *ansatz* solution [27, 28]. It was concluded that in a strong magnetic field the dominant phase is the pairing of atoms with spin components $S_z = S$ and $S_z = S - 1$, one atom being a forward mover and the other one a backward mover. Since the Fermi gas is spin-imbalanced in a magnetic field, the Fermi momenta of the different spin components are different yielding sinusoidal oscillations in space, characteristic of an FFLO state. In the one-dimensional case the periods of oscillation can be extracted from the corresponding correlation functions. The correlation functions are the product of a power-law dependence of the distance and a cosine term with the desired periodicity.

In this paper we investigate the possibility of superfluidity in zero magnetic field. In zero-field only bound states of N atoms, each with a different spin-component, exist. Using the Bethe *ansatz* equations we show that the critical exponent of the superfluid response function increases with the particle density. The superfluid order parameter will be homogeneous and not of the FFLO type. On the other hand, the exponent of the particle density wave response function decreases with the particle density and for $S \geq 3/2$ density waves are the dominating form order at high densities. As a consequence of the longitudinal confining harmonic potential, phase separation is expected, with the trap boundaries having a higher chance of being superfluid for $S \geq 3/2$ than the center of the trap, which is more prone to density waves.

The remainder of the paper is organized as follows. In Sect. II we restate the model, the classification of states of the Bethe *ansatz* equations and the integral equations satisfied by the dressed energies and the densities for bound states of N particles in zero magnetic field. We also discuss the density profile due to the varying chemical potential along the trap in zero magnetic field. In Sect. III we study the response functions for superfluidity and density waves in zero magnetic field using the conformal invariance and the Bethe *ansatz* solution of the system. Conclusions are presented in Sect. IV.

2. Model and Bethe *ansatz* equations

2.1. The model

The Hamiltonian for a gas of nonrelativistic particles with $(2S+1)$ colors (spin S) interacting via an attractive δ-function potential is

$$\mathcal{H} = -\sum_{i=1}^{N_t} \frac{\partial^2}{\partial x_i^2} - 2|c| \sum_{i<j} \delta(x_i - x_j) \ , \tag{1}$$

where x_i are the coordinates, N_t is the total number of particles and c is the interaction strength. By fine-tuning the confinement-induced resonance [4] the interaction can become attractive and its strength can be varied. Here $\hbar^2/2m$, where m is the mass of the particles, has been equated to 1, or alternatively it has been scaled into $\mathcal{H}$ and c.

2.2. Classification of Bethe states

The states of the coordinate Bethe *ansatz* are plane waves constructed from the two-particle scattering matrix. This scattering matrix satisfies the so-called Yang-Baxter triangular relation, which is a necessary condition for integrability. As a consequence of the triangular relation many-particle scattering processes can be factorized into two-particle processes and the order in which the individual scattering processes take place can be interchanged (the order becomes arbitrary). A solution can be constructed by iteratively applying (nesting) $N = 2S+1$ Bethe *ansätze*, such that one color is eliminated at each step [15]. Each Bethe *ansatz* gives rise to a new set of rapidities, $\{k_j\}$, $j = 1, \cdots, N_t$ for the charges (coordinate Bethe *ansatz*) and $\{\Lambda_\alpha^{(l)}\}$, $l = 1, \cdots, N-1$, with $\alpha = 1, \cdots, M^{(l)}$ for the internal degrees of freedom (spin). Here $M^{(l)}$ is the number of rapidities in the lth set and α is the running index within each set. All rapidities within a given set have to be different to ensure linearly independent solutions.

For an attractive interaction and large length of the trap, L, the solutions of the discrete Bethe equations can be classified according to (i) real charge rapidities, belonging to the set $\{k_j\}$, associated with unpaired propagating spin-polarized particles, (ii) complex spin and charge rapidities, which correspond to bound states of particles with different spin components, and

(iii) strings of complex spin rapidities, which represent bound spin states [17, 18]. States in class (iii) are not represented in the ground state: these states correspond to excited states and are not considered here. The classification of states is completely determined by the SU(N)-invariance of the model and the attractive effective potential. The classification of states is therefore completely analogous to that of the Anderson impurity of arbitrary spin in the $U \to \infty$ limit [29, 30, 31, 32, 33] and the one-dimensional degenerate supersymmetric $t-J$ model [34, 35]. The interaction strength in the case of the Anderson impurity model is determined by the hybridization, while the supersymmetric $t-J$ model has no energy scale and the interaction strength is equal to one.

Since only particles with different spin components experience an effective attractive interaction, we may build bound states of up to $(2S+1)$ particles. A bound state of n particles ($n \leq N = 2S+1$) is characterized by one real $\xi^{(n-1)}$ rapidity and in general complex $\Lambda^{(l)}$ rapidities, $l < n-1$, given by [16]

$$\Lambda_p^{(l)} = \xi^{(n-1)} + ip|c|/2 \ , \quad l \leq n-1 \leq 2S \ ,$$
$$p = -(n-l-1), -(n-l-3), \cdots, (n-l-1) \ . \tag{2}$$

These spin and charge strings form classes (i) and (ii), which are present in the ground state [16, 17]. The real rapidities $\xi^{(n-1)}$ have all to be different and satisfy the Fermi-Dirac statistics, i.e. the states are either occupied or empty. In the ground state the rapidities are densely distributed in the interval $[-B_l, B_l]$ and we denote with $\varepsilon^{(l)}(\xi)$, $l = 0, 1, \cdots, 2S$, the dressed energy potentials (entering the Fermi-Dirac distribution).

2.3. Dressed energy potential in zero magnetic field

In zero magnetic field and in the thermodynamic limit the number of particles of each spin-component (color) is the same, i.e. there is no spin imbalance. It is then energetically favorable to bind all the particles into clusters of N atoms. Each cluster has one atom of each color. These clusters are the generalization of preformed Cooper pairs to N colors. Hence, $B_l = 0$ for $l < 2S$ and only $B_{2S} \geq 0$. Only one energy potential $\varepsilon^{(2S)}(\xi)$ needs to be considered, which satisfies the integral equation [6, 18]

$$\varepsilon^{(2S)}(\xi) = D_{2S}(\xi) - \int_{-B_{2S}}^{B_{2S}} d\xi' K_{2S,2S}(\xi - \xi')\varepsilon^{(2S)}(\xi'), \tag{3}$$

where the $D_{2S}(\xi)$ is the driving term given by [6, 18]

$$D_{2S}(\xi) = (2S+1)\Big[\xi^2 - \frac{S(S+1)}{3}c^2 - \mu\Big] \ , \tag{4}$$

and $K_{2S,2S}(\xi)$ is the integration kernel, which can be written in a compact form [6]

$$K_{2S,2S}(\xi) = \int \frac{d\omega}{2\pi} \exp[i\xi\omega - (2S+1)|\omega c|/2] \frac{\sinh[S\omega c]}{\sinh(\omega c/2)}. \tag{5}$$

Here μ is the chemical potential, playing the role of the Lagrange parameter for the conservation of the total number of particles. μ determines the integration limit B_{2S} through the condition that $\varepsilon^{(2S)}(\pm B_{2S}) = 0$. Here B_{2S} determines the two Fermi points of the system, since occupied states correspond to $\varepsilon^{(2S)}(\xi) < 0$ and for empty states $\varepsilon^{(2S)}(\xi)$ is positive. Since bound states of N particles have one particle with each spin component, these clusters do not have magnetic response and Eq. (3) does not depend on the magnetic field (assumed to be zero or less than the binding energy of the bound states).

If $\varepsilon^{(2S)}$ is rescaled to $\varepsilon^{(2S)}/c^2$, μ to μ/c^2, B_{2S} to $B_{2S}/|c|$ and ξ to $\xi/|c|$, the equation (3) is universal, i.e., independent of the magnitude of $|c|$. Hence, within the framework of the grand canonical ensemble, without loss of generality, it is sufficient to present the results for $|c| = 1$.

2.4. Distribution density of rapidities

The density function of the rapidities is obtained from the dressed energy $\varepsilon^{(2S)}(\xi)$ by differentiation with respect to μ, i.e., [6, 18]

$$\rho_h^{(2S)}(\xi) + \rho^{(2S)}(\xi) = -\frac{1}{2\pi}\frac{\partial \varepsilon^{(2S)}(\xi)}{\partial \mu}, \tag{6}$$

where $\rho^{(2S)}(\xi)$ is the particle density and $\rho_h^{(2S)}(\xi)$ the corresponding hole density for bound states involving $2S+1$ particles. The integral equation satisfied by the density function is

$$\rho_h^{(2S)}(\xi) + \rho^{(2S)}(\xi) = \frac{2S+1}{2\pi} - \int_{-B_{2S}}^{B_{2S}} d\xi' K_{2S,2S}(\xi - \xi')\rho^{(2S)}(\xi'), \tag{7}$$

where the integration kernel and the integration limits are the same ones as for the dressed energy potential, but the driving term is different [16]. The number of bound states per unit length is given by

$$n_{2S} = \int_{-B_{2S}}^{B_{2S}} d\xi \rho^{(2S)}(\xi)\ . \tag{8}$$

If ξ and B_{2S} are rescaled to $\xi/|c|$ and $B_{2S}/|c|$, respectively, as discussed above, then $\rho^{(2S)}$ remains invariant and n_{2S} rescales to $|c|n_{2S}$.

The density of the total number of particles is given by $N_t/L = (2S+1)n_{2S}$ and the ground state energy of the system by

$$E_{GS} = L\epsilon_\infty = (2S+1)\int_{-B_{2S}}^{B_{2S}} d\xi \Big[\xi^2 - \frac{S(S+1)}{3}c^2\Big]\rho^{(2S)}(\xi)\ , \tag{9}$$

where ϵ_∞ is the energy density.

If the band is partially filled, the group velocity for bound states of N particles is defined as [6, 18]

$$v_{2S} = \left(\frac{d\varepsilon^{(2S)}(\xi)}{d\xi}\Big|_{\xi=B_{2S}}\right) \Big/ \left(2\pi\rho^{(2S)}(B_{2S})\right)\ . \tag{10}$$

The low lying excitation states form a simple Dirac sea with two Fermi points at $\xi = \pm B_{2S}$. The Fermi momentum is $p_{F,2S} = \pi n_{2S}$, i.e., it is completely determined by the occupation of the band.

2.5. Local density approximation

An ultra-cold atom system is inherently inhomogeneous since the diameter of the tube gradually changes with position from the center of the trap to its boundaries. As a consequence of the changing diameter of the tube, the quantization in the plane transversal to the tube gradually changes the zero of energy. This change can be represented by a harmonic potential of frequency ω_{ho}, so that the actual local chemical potential μ is a function of x given by

$$\mu(x) + \tfrac{1}{2}m\omega_{ho}^2 x^2 = \mu(0) = const\ . \tag{11}$$

Within the local density approximation, it is $\mu(x)$ that enters the Bethe equations, (3)-(5). The Bethe *ansatz* solution is then exact for the one-dimensional system, but approximate for the trap. This approximation [1, 2, 3] is expected to be good if the variation of μ with x is slow, i.e. it is the largest length scale in the system other than the length of the trap. The approximation neglects the quantization of the harmonic confinement, which is treated classically and is locally incorporated into the chemical potential. Assuming the trap is symmetric and has length L, then given $\mu(0)$ and $\mu(L/2)$, i.e. the chemical potential at the center and boundary of the trap, the position along the trap can be written as (from Eq. (11))

$$x/(L/2) = \sqrt{[\mu(x) - \mu(0)]/[\mu(L/2) - \mu(0)]} \,. \tag{12}$$

This parametrization eliminates the constants m and ω_{ho} from Eq. (11), and the local density profile can be expressed as a function of the position along the trap, x.

2.6. *Numerical results*

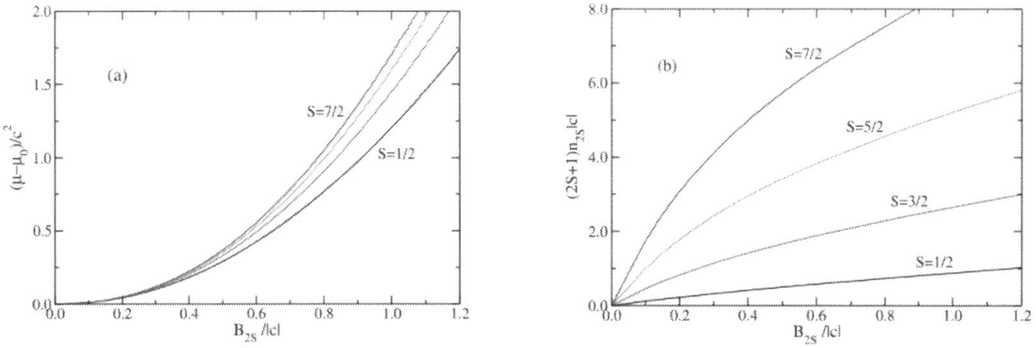

Figure 1. (a) Chemical potential $(\mu-\mu_0)/c^2$ with $\mu_0 = -S(S+1)c^2/3$ and (b) particle density, $N_t|c|/L$, as a function of the integration limit $B_{2S}/|c|$ for $S = 1/2$ (black), $S = 3/2$ (red), $S = 5/2$ (green) and $S = 7/2$ (blue). The quantities have been scaled with $|c|$ so that the plot is universal.

The occupation of the band of bound states of N particles grows monotonically with the chemical potential. For $\mu \leq -S(S+1)c^2/3$ the band is empty, i.e. there are no particles in the system. The integration limit B_{2S} grows with μ, being zero for $\mu_0 = -S(S+1)c^2/3$. The variation of μ and $(2S+1)n_{2S}$ as a function of B_{2S} is shown in Fig. 1 for various S. The quantities have been scaled with $|c|$ so that the results are universal for all interaction strengths. Note that $\mu - \mu_0$ grows approximately quadratically with B_{2S} as a consequence of the parabolic dispersion of the Hamiltonian. For small B_{2S} the dependence of n_{2S} on B_{2S} is roughly linear.

In Fig. 2 we show the density profile of bound states of N atoms, n_{2S}, as a function of the position along the trap x, $S = 1/2, \cdots, 7/2$, $\mu(L/2) = -S(S+1)c^2/3$ and $\mu(0) = c^2$. Note that for the parameters chosen the density of particles decreases rapidly when the boundary of the trap is approached.

3. Conformal towers and correlation functions

In an array of tubes (optical lattice) interactions between particles in different tubes and/or Josephson tunneling between tubes may give rise to a dimensional crossover from one dimension to three dimensions and opens the possibility for superfluid long-range order [23, 26]. As for high temperature superconductors, where the instability to superconductivity and properties of the superconducting phase are determined by the properties of the normal phase, in the present case superfluidity is determined by the Luttinger liquid. To investigate the possibility of superfluidity we need to calculate the superfluid response function in the disordered phase, i.e., for the one-dimensional gas. The asymptotic behavior of the equal time correlation function at large distances can be determined using conformal field theory.

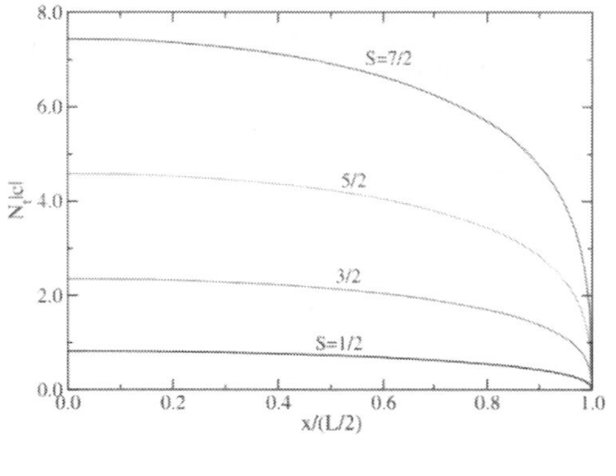

Figure 2. Zero-field density profile N_t/L within the local particle density approximation as a function of the position along the trap x for $S = 1/2$ (black), $S = 3/2$ (red), $S = 5/2$ (green) and $S = 7/2$ (blue). N_t has been scaled with $|c|$ so that the plot is universal. We have chosen $\mu(L/2) = -S(S+1)c^2/3$ and $\mu(0) = 1.0c^2$.

3.1. Conformal towers

The low-lying excitations of the system are described by the finite size corrections to the ground state energy [36, 37, 38]. While the ground state energy for the bulk, E_{GS} given by Eq. (9), is an extensive quantity, the mesoscopic corrections are of order $1/L$, where L is the length of the system. In zero magnetic field, we need to consider only the contribution of the populated Dirac sea of bound states of N atoms to the finite size corrections. For periodic boundary conditions, four quantum numbers determine the excitations of this branch, namely, ΔN_{2S} corresponds to the added or removed number of bound states, D_{2S} is the parity variable, i.e. $2D_{2S}$ is the difference between forward and backward movers, and the $n_{2S}^{\pm}$ count the number of particle and hole excitations about each Fermi point (+ for forward movers and − for backward movers). The cases $S = 1/2$ [27, 39, 40], $S = 3/2$ [27] and $S = 5/2$ [28] in a magnetic field have been extensively investigated previously.

The finite size corrections to the ground state energy are given by [27, 36, 37, 38, 39, 40]

$$E = E_{GS} + \frac{\pi v_{2S}}{2L}\left[\Delta N_{2S}/z_{2S,2S}\right]^2 + \frac{2\pi v_{2S}}{L}\left\{\left[z_{2S,2S}D_{2S}\right]^2 + n_{2S}^{+} + n_{2S}^{-} - \frac{1}{12}\right\} \qquad (13)$$

and the corresponding change in momentum for the excitations is

$$\Delta P = \frac{2\pi}{L}\left[D_{2S}\Delta N_{2S} + n_{2S}^{+} - n_{2S}^{-}\right] + p_{F,2S}D_{2S}\ . \qquad (14)$$

In Eq. (13) v_{2S} is the group velocity and $z_{2S,2S}$ is the generalized dressed charge for the band. The generalized dressed charge determines how the two Fermi points interact with each other, i.e. the way a change of a quantum number ΔN_{2S} or D_{2S} affects the contribution to the energy. For elementary excitations from the ground state, the values of the D_{2S} quantum number are constrained by the discrete Bethe *ansatz* equations for periodic boundary conditions. A change in the population, ΔN_{2S}, changes the backscattering quantum number by

$$D_{2S} = \tfrac{1}{2}(2S+1)\Delta N_{2S} \quad (\text{mod } 1)\ . \qquad (15)$$

Note that D_{2S} is only determined modulo 1.

The dressed generalized charge is obtained as $z_{2S,2S} = \Xi_{2S,2S}(B_{2S})$, where the $\Xi_{2S,2S}$ is the solution of the integral equation

$$\Xi_{2S,2S}(\xi) = 1 - \int_{-B_{2S}}^{B_{2S}} d\xi' K_{2S,2S}(\xi-\xi')\Xi_{2S,2S}(\xi'). \qquad (16)$$

Here the integration kernel is the same as for the integral equation for $\varepsilon^{(2S)}(\xi)$, i.e. Eq. (5). Comparing Eqs. (7) and (16) it follows that $\Xi_{2S,2S}(\xi) = [2\pi/(2S+1)][\rho_h^{(2S)}(\xi) + \rho^{(2S)}(\xi)]$, so that $z_{2S,2S} = [2\pi/(2S+1)]\rho^{(2S)}(B_{2S})$. Note that $z_{2S,2S}$ is always nonnegative. $z_{2S,2S}$ is a monotonically decreasing function of $B_{2S}/|c|$ and is shown in Fig. 3(a) for various spin values. Note that the dependence of $z_{2S,2S}$ with the integration limit increases dramatically with increasing S.

For very large B_{2S} the Fredholm integral equation (16) can be reduced to a Wiener-Hopf integral equation, by shifting the variable ξ by the amount B_{2S}. The Wiener-Hopf equation can be solved analytically with standard methods (see e.g. Appendix B of [33]). This way we obtain asymptotically $z_{2S,2S}(B_{2S} \to \infty) = 1/\sqrt{2S+1}$.

3.2. Asymptotic behavior of correlation functions

The asymptotic long-time and large distance behavior of the correlation function of a given conformal field operator $\mathcal{O}$ can be obtained using conformal field theory. The asymptote is determined by the low energy excitations of the system as a function of the set of quantum numbers ΔN_{2S}, D_{2S}, n^+_{2S} and n^-_{2S}. At $T=0$ the space and time dependent correlation function for the operator $\mathcal{O}$ is given by [37, 38, 39, 41, 42]

$$\langle \mathcal{O}^\dagger(x,t)\mathcal{O}(0,0)\rangle = \frac{\exp[-2i(D_{2S}\,p_{F,2S})x]}{(x-iv_{2S}t)^{2\Delta^+_{2S}}(x+iv_{2S}t)^{2\Delta^-_{2S}}} , \tag{17}$$

where $p_{F,2S} = \pi n_{2S}$ is the Fermi momentum defined previously. Here $\Delta^\pm_{2S}$ are the conformal dimensions that are obtained from the finite size corrections to the ground state energy, Eq. (13), and the momentum, Eq. (14), for the excitations [37, 38, 39, 41, 42]

$$2\Delta^\pm_{2S} = 2n^\pm_{2S} + \left[z_{2S,2S}D_{2S} \pm \tfrac{1}{2}\Delta N_{2S}/z_{2S,2S}\right]^2 . \tag{18}$$

There are two factors in the denominator of Eq. (17), one for forward moving excitations and one for backward movers. The smallest conformal dimensions are obtained for $n^\pm_{2S} = 0$, which correspond to the smallest exponents in the correlation functions. Hence, the correlations with the longest range correspond to $n^\pm_{2S} = 0$. The above results are easily extended to finite temperature T by replacing each factor $(x \pm ivt)^{-2\Delta^\pm}$ in Eq. (17) by

$$\left(\frac{\pi T/v}{\sinh[\pi T(x \pm ivt)/v]}\right)^{2\Delta^\pm} , \tag{19}$$

but this extension is not needed here.

The leading term of the equal time correlation function for the operator $\mathcal{O}$ is then of the form

$$\langle \mathcal{O}^\dagger(x,0)\mathcal{O}(0,0)\rangle = Ax^{-\theta_{2S}}\cos(2D_{2S}p_{F,2S}x) , \tag{20}$$

where the amplitude A cannot be determined from conformal field theory. The exponent θ_{2S} is given by

$$\theta_{2S} = 2(\Delta^+_{2S} + \Delta^-_{2S}) , \quad n^\pm_{2S} = 0 . \tag{21}$$

The factor $\cos(2D_{2S}p_{F,2S}x)$ reflects the space modulation of the order parameter. For superfluidity in a finite magnetic field it arises from the spin-imbalance in the system and is reminiscent of the FFLO phase, where $\lambda = 1/(2\pi D_{2S}n_{2S})$ is the distance between nodes. For a particle density wave response function, on the other hand, it reflects the modulation of the density wave.

3.3. Response function for superfluidity

Weak Josephson tunneling between tubes and interactions between particles in different tubes in an array of tubes [23, 26], may give rise to a dimensional crossover from one-dimension to a higher dimension, yielding superfluid long-range order. The superfluid order parameter in zero magnetic field is a generalization of Cooper pairs for $S = 1/2$ to higher spin, in this case the bound states of $N = 2S + 1$ atoms. In the presence of a magnetic field there are many possible order parameters, which have been discussed in [27, 28]. The correlation function for the bound states determines if long-range order is a possibility. A small exponent θ_{2S} favors superfluidity, in particular, the exponent θ_{2S} should be smaller than 2, which is the critical scaling dimension in one-dimension. This criterion is simple for the zero-field case, but has to be complemented by other criteria if the system is spin-imbalanced.

Correlation functions for pure phases have been studied previously for other models [42, 43]. The operator $\mathcal{O}$ corresponds then to adding or removing one cluster, i.e. $\Delta N_{2S} = \pm 1$. Consequently, from Eq. (15) for a half-integer spin D_{2S} is an integer. The smallest exponent corresponds to $D_{2S} = 0$; it is easily seen that $D_{2S} = \pm 1$ yields a larger θ_{2S}. From Eqs. (18) and (21) we obtain that $\theta_{2S} = 1/(2z^2_{2S,2S})$. The exponent θ_{2S} as a function of $B_{2S}/|c|$ for various spin values is shown in Fig. 3(b).

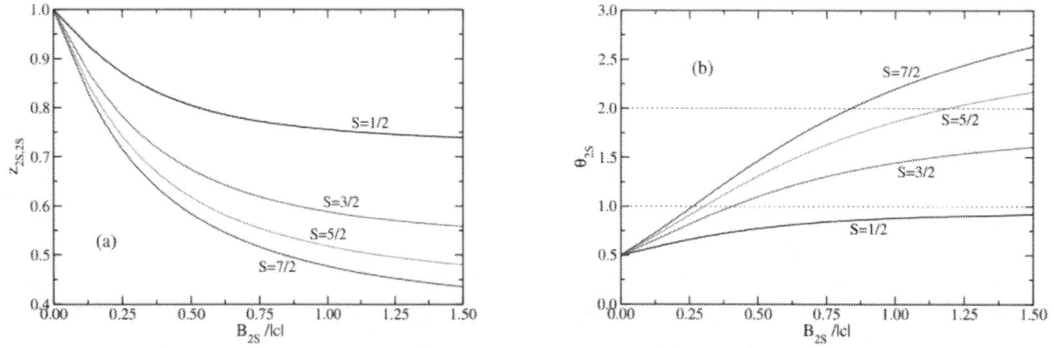

Figure 3. (a) Dressed generalized charge $z_{2S,2S}$ and (b) critical exponent θ_{2S} for superfluidity as a function of the integration limit $B_{2S}/|c|$ for $S = 1/2$ (black), $S = 3/2$ (red), $S = 5/2$ (green) and $S = 7/2$ (blue). The integration limit has been scaled with $|c|$ so that the plot is universal. The dashed line at $\theta = 2$ corresponds to the critical scaling dimension in 1D. The dotted line, $\theta = 1$, represents equal exponents for superfluidity and density waves and is discussed in the text.

In the limit $B_{2S} \to 0$ the exponent θ_{2S} tends to the value 0.5 for all values of S. With increasing atom density in the tube (increasing B_{2S}) the exponent increases. In the limit $B_{2S} \to \infty$ we obtain $\theta_{2S} = (2S + 1)/2$. θ_{2S} increases monotonically with B_{2S} from 0.5 to $(2S + 1)/2$. Hence, for $S = 1/2$ and $S = 3/2$ the exponent θ_{2S} is always less than the critical scaling dimension and consequently the situation is favorable for superfluidity for all densities. The critical scaling dimension in 1D is 2, so that for $\theta > 2$ the correlation function falls off too fast to yield long-range order. For $S > 3/2$ the exponent θ_{2S} is larger than 2 for larger densities. Hence, if $S > 3/2$ only low atom densities (or strong interactions) are favorable for superfluid long-range order in an array of tubes with Josephson tunneling. $\theta = 2$ is indicated by the faint dashed line in Fig. 3(b).

3.4. Response function for density waves
Particle density waves oscillate in space with a wave number $2p_{F,2S}$. The density wave operator does not change the number of particles, so that $\Delta N_{2S} = 0$ and consequently D_{2S} is an integer. We choose $D_{2S} = \pm 1$, which corresponds to oscillations with $2p_{F,2S}$, and $n^{\pm}_{2S} = 0$. The equal time correlation function is then proportional to

$$\cos(2p_{F,2S}x)x^{-\theta_{DW}} \quad , \quad \theta_{DW} = 2z^2_{2S,2S} \quad , \tag{22}$$

i.e. the exponent is just the inverse of the one of superfluidity. Hence, θ_{DW} decreases from the value 2 as B_{2S} increases. Density waves are then, in principle, favorable for all spins and all band fillings.

We now have to compare θ_{2S} with θ_{DW}. The simplest criterion is that the smallest exponent yields the dominant long-range order. The exponents are equal if $\theta_{2S} = \theta_{DW} = 1$ or $z_{2S,2S} = 1/\sqrt{2}$. The corresponding line is shown as dotted in Fig. 3(b). We may conclude that for $S = 1/2$ superfluidity is more favorable than density waves for all particle densities and interaction strengths. For $S \geq 3/2$ there is a crossover from superfluidity at low densities to density waves at large $B_{2S}/|c|$. The transition is where the exponents are equal one.

4. Conclusions

We studied an ultracold gas of fermionic atoms with spin S interacting via an attractive contact potential in zero magnetic field by solving the corresponding Bethe *ansatz* equations. The atoms form bound states of $(2S+1)$ particles, one of each spin component. These bound states are the generalization of preformed Cooper pairs to larger spins. Possible applications are to ultracold Fermi gases of ^{40}K (spin 9/2), ^{43}Ca (spin 7/2), ^{87}Sr (spin 9/2), ^{173}Yb (spin 5/2), ^{9}Be (spin 3/2), ^{135}Ba (spin 3/2), ^{137}Ba (spin 3/2), ^{53}Cr (spin 3/2), and ^{6}Li (spin 1/2) atoms [7].

The focus of our previous papers [7, 27, 28] was the possibility of finding inhomogeneous phases of two types in the gas in a magnetic field: (a) We considered the scenario of phase separation along the tube, and (b) modulations of the order parameters of the FFLO type.

In case (a), the confining harmonic potential varies with the position x along the tube. Within the local density approximation, which absorbs this variation into the chemical potential, μ is a function of x and hence different phases for a given magnetic field appear along the trap giving rise to phase separation [1, 2, 3, 7, 28]. In zero-field there is only one phase, so that in the present study, such a phase separation cannot occur. The density profile, however, changes dramatically along the tube, being larger at the center of the tube than at its boundaries.

In case (b), inhomogeneities like modulations of the order parameter of the FFLO type, may arise due to the spin-imbalance of the atoms in a magnetic field. In general this requires an array of tubes with interactions between particles in different tubes or Josephson tunneling between tubes [26, 23, 27] giving rise to a dimensional crossover from one-dimension to a three dimensions. The dimensional crossover opens the door to the possibility of superfluid long-range order at finite temperature. There is a very large number of order parameters for superfluidity for the spin-imbalanced gas, but only one in zero magnetic field.

The response functions in the disordered phase are determined by the Luttinger properties of the one-dimensional gas. They determine the instability toward superfluidity from the normal phase. A combination of three criteria determine the dominating order in the case of a spin imbalanced gas: (i) the smallest exponent θ corresponds to the longest range and is hence favorable to order if it is less than 2, the critical scaling dimension, (ii) a large distance between nodes, λ, favors order because it is energetically less unfavorable, and (iii) the preformed bound states should carry a small momentum, since a large momentum of the bound states is energetically unfavorable to a condensate. Cooper-*pairing* of atoms with spin-component S and $S-1$ (one particle being a forward mover and the other a backward mover) was found to be the first superfluid phase accessed from the normal state as the temperature is lowered in a sufficiently large field.

Note that the two main conditions for realization of the FFLO phase are satisfied in cold atom tubes: (1) the system is very pure (no impurities) and (2) it has a low effective dimension (extreme anisotropy). In a sufficiently strong magnetic field there are in general several bands of bound states represented in the ground state. Each band contributes additively to the critical exponent of a correlation function. Hence, critical exponents tend to be larger in a magnetic field and superfluidity is less favorable than in zero-field.

In zero-field there is only one operator for superfluidity, namely the creation (annihilation) operator for one preformed bound state of $(2S+1)$ atoms. The superfluid phase is homogeneous and not of the FFLO type. For $S = 1/2$ superfluidity is favorable because the critical exponent is smaller than one and smaller than θ_{DW} for all densities and coupling strengths. For $S \geq 3/2$, on the other hand, superfluidity is only possible at sufficiently low atom densities and/or strong interactions $|c|$, while for large $B_{2S}/|c|$ density waves are the likely candidate for long-range order. The critical exponent grows quite rapidly with increasing spin size and particle density, and eventually θ_{2S} crosses the value one (dotted line in Fig. 3(b)) and beyond that point $\theta_{DW} = 1/\theta_{2S}$ will be less than one and density waves are the dominating feature.

It should be mentioned that a related SU(4) invariant Hubbard model (attractive U) has been studied on a lattice [44]. The methods employed are the density-matrix renormalization group technique and a phenomenological Luttinger liquid, where the charge sector is represented by a free massless bosonic field. Their phase diagram ($S = 3/2$) displays two regions, a superfluid quartet phase at low density, which gives way to a charge density wave phase at larger densities. This crossover and the trend that superfluidity and DW exclude each other is confirmed by our exact calculation for $N = 4$. A quantitative comparison is more difficult because the models are different.

A possible application of a SU(4) system with attractive interactions is the alpha-particle condensation in nuclear matter. The four degrees of freedom are made up of the spin and the pseudospin. The alpha particles are strongly bound and exist in a Bose-Einstein condensation phase, which implies a low density. Of course a three-dimensional system has to be considered here. An excellent review on the status of the subject can be found in [45].

Acknowledgments

P.S. is supported by the U.S. Department of Energy under grant DE-FG02-98ER45707. A.A.Z. acknowledges the support from the Institute for Chemistry of V.N. Karazin Kharkov National University.

References

[1] Liao Y A, Rittner A S C, Paprotta T, Li W H, Partridge G B, Hulet R G, Baur S K, and Mueller E J 2010 *Nature* (London) **467** 567
[2] Orso G 2007 *Phys. Rev. Lett.* **98** 070402
[3] Hu H, Liu X-J, and Drummond P D 2007 *Phys. Rev. Lett.* **98** 070403
[4] Olshanii M 1998 *Phys. Rev. Lett.* **81** 938
[5] Bergeman T, Moore M G, and Olshanii M 2003 *Phys. Rev. Lett.* **91** 163201
[6] Schlottmann P 1994 *J. Phys.: Condens. Matter* **6** 1359
[7] Schlottmann P and Zvyagin A A 2012 *Phys. Rev. B* **85** 024535
[8] Lieb E H and Liniger W 1963 *Phys. Rev.* **130** 1605
[9] Gaudin M 1967 *Phys. Lett. A* **24** 55
[10] Yang C N 1967 *Phys. Rev. Lett.* **19** 1312
[11] Takahashi M 1971 *Prog. Theor. Phys.* **46** 1388
[12] Lai C K 1971 *Phys. Rev. Lett.* **26** 1472; Lai C K 1973 *Phys. Rev. A* **8** 2567
[13] Bahder T B and Woynarovich F 1986 *Phys. Rev. B* **33** 2114
[14] Lee K J B and Schlottmann P 1989 *Phys. Rev. B* **40** 9104

[15] Sutherland B 1968 *Phys. Rev. Lett.* **20** 98
[16] Takahashi M 1970 *Prog. Theor. Phys.* **44** 899
[17] Schlottmann P 1993 *J. Phys.: Condens. Matter* **5** 5869
[18] Schlottmann P 1997 *Int. J. Mod. Phys. B* **11** 355
[19] Guan X W, Batchelor M T, Lee C, and Zhou H-Q 2008 *Phys. Rev. Lett.* **100** 200401
[20] Guan X W, Batchelor M T, Lee C, and Lee J Y 2009 *Europhysics Letters* **86** 50003
[21] Guan X W, Lee J-K, Batchelor M T, Yin X-G, and Chen S 2010 *Phys. Rev. A* **82** 021606(R)
[22] He P, Yin X-G, Guan X W, Batchelor M T, and Wang Y 2010 *Phys. Rev. A* **82** 053633
[23] Parish M M, Baur S K, Mueller E J, and Huse D A 2007 *Phys. Rev. Lett.* **99** 250403
[24] Rizzi M, Polini M, Cazalilla M A, Bakhtiari M R, Tosi M P, and Fazio R 2008 *Phys. Rev. B* **77** 245105
[25] Fulde P and Ferrell A 1964 *Phys. Rev.* **135** A550; Larkin A and Ovchinnikov Y N 1964 *Zh. Eksp. Teor. Fiz.* **47** 1136 [1965 *Sov. Phys. JETP* **20** 762]
[26] Yang K 2001 *Phys. Rev. B* **63** 140511(R)
[27] Schlottmann P and Zvyagin A A 2012 *Phys. Rev. B* **85** 205129
[28] Schlottmann P and Zvyagin A A 2012 *Modern Physics Letters B* **147** 1230009
[29] Schlottmann P 1984 *Z. Phys.* **54** 207
[30] Schlottmann P 1982 *Z. Phys.* **49** 109
[31] Tsvelick A M 1984 *J. Phys. C* **17** 2299
[32] Kawakami N, Tokuono S, and Okiji A 1984 *J. Phys. Soc. Jpn.* **53** 51
[33] Schlottmann P 1989 *Phys. Rep.* **181** 1
[34] Schlottmann P 1987 *Phys. Rev. B* **36** 5177
[35] Schlottmann P 1992 *J. Phys.: Condens. Matter* **4** 7565
[36] Woynarovich F 1989 *J. Phys. A* **22** 4243
[37] Izergin A G, Korepin V E, and Reshetikhin N Yu 1989 *J. Phys. A* **22** 2615
[38] Frahm H and Korepin V E 1990 *Phys. Rev. B* **42** 10553
[39] Essler F H L 2010 *Phys. Rev. B* **81** 205120
[40] Lee J Y and Guan X-W 2011 *Nucl. Phys. B* **853** [FS] 125
[41] Kawakami N 1993 *Phys. Rev. B* **47** 2928
[42] Schlottmann P 2004 *Phys. Rev. B* **69** 035110
[43] Bariev R Z, Klümper A, Schadschneider A, and Zittartz J 1995 *Z. Phys. B* **96** 395
[44] Capponi S, Roux G, Lecheminant P, Azaria P, Boulat E, and White S R 2008 *Phys. Rev. A* **77** 013624
[45] Schuck P 2013 arXiv: 1303.2943v1

Theory for Quartet Condensation in Fermi Systems with Applications to Nuclei and Nuclear Matter

P Schuck[1 2 3],Y Funaki[4], H Horiuchi[5], G Röpke[6], A Tohsaki[5], T Yamada[7]

[1]Institut de Physique Nucléaire, CNRS, UMR8608,
Orsay, F-91406, France
[2]Université Paris-Sud, Orsay, F-91505, France
[3]Laboratoire de Physique et de Modélisation des Milieux Condensés, CNRS et Université Joseph Fourier, UMR5493, 25 Av. des Martyrs, BP 166, F-38042 Grenoble Cedex 9, France
[4]The Institute of Physical and Chemical Research (RIKEN), Wako,
Saitama, 351-0198, Japan
[5]Research Center for Nuclear Physics (RCNP), Osaka University,
Ibaraki, Osaka 567-0047, Japan
[6]Institut für Physik, University of Rostock, Universitätsplatz 1,
18051 Rostock, Germany
[7]Laboratory of Physics, Kanto Gakuin University,
Yokohama 236-8501, Japan

Abstract. The theory of quartet condensation is further developed. The onset of quartetting in homogeneous fermionic matter is studied with the help of an in-medium modified four fermion equation. It is found that at very low density quartetting wins over pairing. At zero temperature, in analogy to pairing, a set of equations for the quartet order parameter is given. Contrary to pairing, quartetting only exists for strong coupling and breaks down for weak coupling. Reasons for this finding are detailed. In an application to nuclear matter, the critical temperature for α particle condensation can reach values up to around 8 MeV. The disappearance of α-particles with increasing density, i.e. the Mott transition, is investigated. In finite nuclei the Hoyle state, that is the $0_2{}^+$ of ^{12}C is identified as an 'α-particle condensate' state. It is conjectured that such states also exist in heavier $n\alpha$-nuclei, like ^{16}O, ^{20}Ne, etc. The sixth 0^+ state in ^{16}O is proposed as an analogue to the Hoyle state. The Gross-Pitaevski equation is employed to make an estimate of the maximum number of α particles a condensate state can contain. Possible quartet condensation in other systems is discussed briefly.

1. Introduction

One of the most amazing phenomena in quantum many-particle systems is the formation of quantum condensates. At present, the formation of condensates is of particular interest in strongly coupled fermion systems in which the crossover from Bardeen-Cooper-Schrieffer (BCS) pairing to Bose-Einstein condensation (BEC) may be investigated. Among very different quantum systems such as the electron-hole exciton and bi-exciton systems in excited semiconductors, atoms in traps at extremely low temperatures, etc., nuclear matter is especially well suited for the study of correlation effects in a quantum liquid. However, more exotic systems like the dense hydrogen gas may also feature condensation phenomena of electron-electron, proton-proton, and electron-proton pairs [1] where the strongly bound hydrogen molecule can be thought of as a candidate for condensation in the same context. Coming to nuclear systems, neutron matter, nuclear matter, but also finite nuclei are superfluid. However, at low density, nuclear matter will not cluster into pairs, i.e. deuterons, but rather into α -particles which are much more stable. Also heavier clusters, starting with Carbon, may be of importance but are presently not considered for condensation phenomena. Therefore, one may ask the question whether there exists quartetting, i.e. α-particle condensation, in nuclei, analogous to nuclear pairing. The only nucleus which in its ground state has a pronounced α -cluster structure is ^{8}Be. In section 5 we will show a figure of ^{8}Be in the laboratory frame and in the intrinsic deformed frame. We will see that ^{8}Be is formed out of two almost free α - particles roughly 4 fm apart, only weakly overlapping with their surface tails. Actually ^{8}Be is slightly unstable and the two α's only hold together via the Coulomb barrier. Because of the large distance of the two α-particles, the 0^+ ground state of ^{8}Be has, in the laboratory frame, a spherical density distribution whose average is very low: about 1/3 of ordinary saturation density ρ_0. ^{8}Be is, therefore, a very large object with an rms radius of about 3.7 fm to be compared with the nuclear systematics of $R = r_0 A^{1/3} = 2.44$ fm. Definitely ^{8}Be is a rather unusual and, in its kind, unique nucleus. One may ask the question what happens when one brings a third α-particle alongside of ^{8}Be. We know the answer: the 3α system collapses to the ground state of ^{12}C which is much denser than ^{8}Be and can not accommodate, with its small radius of 2.4 fm, three more or less free α-particles barely touching one another. One nevertheless may ask the question whether the dilute three α configuration ^{8}Be- α, or $\alpha - \alpha - \alpha$, may not form an isomeric or excited state of ^{12}C. That such a state indeed exists will be one of the main subjects of our considerations. Once one accepts the idea of the existence of an α-gas state in ^{12}C, there is no reason why equivalent states at low density should not also exist in heavier $n\alpha$-nuclei, like ^{16}O, ^{20}Ne, etc. In a mean field picture, i.e. all α's being ideal bosons (in this context remember that the first excited state of an α-particle is at $\sim$ 20 MeV, by factors higher than in all other nuclei), all α's will occupy the lowest 0S-state, i.e. they will condense into this state. This forms, of course, not a macroscopic condensate but it can be understood in the same sense as we know that nuclei are superfluid because of the presence of a finite number of Cooper pairs. These α condensate states are generally close to the α disintegration threshold and, therefore, at higher and higher excitation energy as the number of α's increases. One may think that, as a consequence, these states decay very fast but due to their unusual structure, they couple little to ordinary excited states and will, thus, have an unusual long life time. On the other hand, for example during the cooling process of compact stars [2], where one predicts the presence of α-particles [3], a real macroscopic phase of condensed α's may be formed. In the present contribution we will mainly concentrate on nuclear systems but we also can think about the possibility of quartetting in other Fermi-systems, as mentioned above. One should, however, keep in mind that a pre-requisite for its existence is, as in nuclear physics, that there exists a bound quartet in free space. This is facilitated with the existence of four different types of fermions, all attracting one another. For example to form quartets with cold atoms one could try to trap fermions in four different magnetic substates, a task which eventually seems possible [4]. In the next section we will outline the general theory for quartet condensation and in section 3 we investigate how the binding energy of various nuclear clusters changes with density as well as the critical temperature of α-particle condensation in infinite matter via an in- medium four-nucleon equation (Thouless criterion). In section 4, we give some results concerning α condensation in infinite nuclear matter, and in section 5 we treat α-particle condensation in finite nuclei. Finally in section 6 we conclude with an outlook and further discussions.

2. General Theory for Quartetting

The aim will be to develope a theory for quartet condensation which in many aspects is analogous to the BCS approach for the condensation of pairs. It is, however, evident that a microscopic theory for quartet condensation, involving a highly correlated four fermion cluster, is at least by an order of magnitude more complicated than the pairing case. It also will turn out that the physics is quite different, that is, we will find that quartet condensation only exists in the BEC phase and no quartet condensation in weak coupling with a very long coherence length, the analogue to the BCS phase of pairs, is possible. We will dwell on this aspect in quite some detail. In order to set the frame of our approach, let us start to repeat the BCS approach for pairing. It is well known that the BCS ground state wave function can be written as a coherent state [5]

$$|\mathrm{BCS}\rangle = e^{\sum_{k<k'} z_{kk'} c_k^+ c_{k'}^+}|\mathrm{vac}\rangle \tag{1}$$

with c^+, c fermion creation and annihilation operators and indices k implying momentum vector, spin, and, eventually other quantum numbers. Standard singlet pairing considers $k' = \bar{k}$, i.e. the pair is at rest and the two fermions occupy time reversed states k and $\bar{k}$. The BCS ground state is the vacuum to the quasiparticle destructors $\beta_k = u_k c_k - v_k c_{\bar{k}}^+$ (with $u_k^2 + v_k^2 = 1$), that is

$$\beta_k|\mathrm{BCS}\rangle = 0 \tag{2}$$

where the standard relation $z_{k\bar{k}} = v_k/u_k$ is to be used. The BCS equations can then be obtained by minimising the following mean single particle energy (which can be identified with an energy weighted sum rule [5])

$$e_k = \frac{\langle \mathrm{BCS}|\{\beta_k, [H - \mu\hat{N}, \beta_k^+]\}\mathrm{BCS}\rangle}{\langle \mathrm{BCS}|\{\beta_k, \beta_k^+\}|\mathrm{BCS}\rangle} \tag{3}$$

where $\{.,.\}$ and $[.,.]$ are anticommutators and commutators, respectively and $\hat{N}$ and μ are particle number operator and chemical potential. The Hamiltonian is given by

$$H = \sum \varepsilon_k^0 c_k^+ c_k + \frac{1}{4}\sum_{k_1k_2k_3k_4} \bar{v}_{k_1k_2k_3k_4} c_{k_1}^+ c_{k_2}^+ c_{k_4} c_{k_3} \tag{4}$$

with $\varepsilon_k^0 = k^2/(2m)$ the kinetic energy and $\bar{v}_{k_1k_2k_3k_4}$ the antisymmetrised matrix element of the two body interaction. Since the quasiparticle operators β^+, β are obtained from a canonical transformation among the c^+, c operators, the transformation can be inverted and the c^+, c expressed in terms of β^+, β. Using the killing condition, all expectation values contained in the variational equations $\delta e_k = 0$ can then be expressed in terms of the u, v amplitudes leading directly to the usual coupled BCS equations for u, v. For later reasons of comparison with the quartet case, let us give the BCS in a particular form eliminating the v-amplitudes

$$\xi_k u_k + \frac{|\Delta_k|^2}{E_k + \xi_k} u_k = E_k u_k \tag{5}$$

where $E_k = \sqrt{\xi_k^2 + \Delta_k^2}$ is the quasi-particle energy, $\Delta_k = \sum_{k'} V_{kk'} u_{k'} v_{k'}$ the usual gap function with $V_{kk'}$ the angle averaged two body force in momentum space, and $\xi_k = \varepsilon_k - \mu$ with

$$\varepsilon_k = \varepsilon_k^0 + \sum_{k'} \bar{v}_{kk'kk'} n_{k'} \tag{6}$$

being the Hartree-Fock (HF) single particle energies where $n_k = v_k^2 = 1 - u_k^2$ are the single particle occupation numbers. We also want to give the equation for the pairing order parameter $\kappa_k \equiv \langle c_{\bar{k}} c_k \rangle = u_k v_k$ which is equivalent to the standard gap equation

$$2\varepsilon_k \kappa_k - (1 - 2n_k) \sum_{k'} V_{kk'} \kappa_{k'} = 2\mu \kappa_k \tag{7}$$

It is seen that (7) looks similar to a two-body Schrödinger equation in momentum space, with eigenvalue 2μ and the interaction modified by the Pauli blocking factor $(1 - 2n_k)$. We will see that in the case of quartetting, we will get two coupled equations, analogous to (5) and (7).

So far for the BCS approach concerning pairs. Let us try to set up an analogous procedure for quartets. Obviously we should write for the wave function

$$|Z\rangle = e^{\frac{1}{4!} \sum_{k_1 k_2 k_3 k_4} Z_{k_1 k_2 k_3 k_4} c_{k_1}^+ c_{k_2}^+ c_{k_3}^+ c_{k_4}^+} |\text{vac}\rangle \tag{8}$$

where the quartet amplitudes Z are fully antisymmetric (symmetric) with respect to an odd (even) permutation of the indices. The task will now be to find a killing operator for this quartet condensate state. Whereas in the pairing case the partitioning of the pair operator into a linear combination of a fermion creator and a fermion destructor is unambiguous, in the quartet case there exist two ways to partition the quartet operator, that is into a single plus a triple or into two doubles. Let us start with the superposition of a single and a triple. As a matter of fact it is easy to show that (in the following, we always will assume that all amplitudes are real)

$$q_\nu = u_{k_1}^\nu c_{k_1} - \frac{1}{3!} \sum v_{k_2 k_3 k_4}^\nu c_{k_1}^+ c_{k_2}^+ c_{k_3}^+ \tag{9}$$

kills the quartet state under the condition

$$Z_{k_1 k_2 k_3 k_4} = \sum_\nu (u^{-1})_{k_1}^\nu v_{k_2 k_3 k_4}^\nu \tag{10}$$

However, so far, we barely have gained anything, since above quartet destructor contains a non-linear fermion transformation which, a priory, cannot be handled. Therefore, let us try with a superposition of two fermion pair operators which is, in a way, the natural extension of the Bogoliubov transformation in the pairing case, i.e. with $Q = \sum[XP - YP^+]$ where $P^+ = c^+ c^+$ is a fermion pair creator. We will, however, find out that such an operator cannot kill the quartet state of Eq. (8). In analogy to the so-called Self-Consistent RPA (SCRPA) approach [6], we will introduce a slightly more general operator, that is

$$Q_\nu = \sum_{k<k'} [X_{kk'}^\nu c_k c_{k'} - Y_{kk'}^\nu c_{k'}^+ c_k^+] - \sum_{k_1<k_2<k_3 k_4} \eta_{k_1 k_2 k_3 k_4}^\nu c_{k_1}^+ c_{k_2}^+ c_{k_3}^+ c_{k_4} \tag{11}$$

with X, Y antisymmetric in k, k'. Applying this operator on our quartet state, we find $Q_\nu |Z\rangle = 0$ where the relations between the various amplitudes turn out to be

$$\sum_{k<k'} X_{kk'}^\nu Z_{kk'll'} = Y_{ll'}^\nu \qquad \text{and} \qquad \eta_{l_2 l_3 l_4;k'}^\nu = \sum_k X_{kk'}^\nu Z_{k l_2 l_3 l_4} \tag{12}$$

These relations are quite analogous to the ones which hold in the case of the SCRPA approach [6]. One also notices that the relation between X, Y, Z amplitudes is similar in structure to the one of BCS theory for pairing. As with SCRPA, in order to proceed, we have to approximate the additional η-term. The quite suggestive recipe is to replace in the η-term of Eq. (11) the density operator $c_{k'}^+ c_k$ by its mean value $\langle Z|c_{k'}^+ c_k|Z\rangle / \langle Z|Z\rangle \equiv \langle c_{k'}^+ c_k \rangle = \delta_{kk'} n_k$, i.e. $c^+_{k_1}c^+_{k_2}c^+_{k_3}c_{k_4} \to c^+_{k_1}c^+_{k_2}n_{k_3}\delta_{k_3k_4}$ where we supposed that we work in the basis where the single particle density matrix is diagonal, that is, it is given by the occupation probabilities n_k. This approximation, of course, violates the Pauli principle but, as it was found in applications of SCRPA [6], we suppose that also here this violation will be quite mild (of the order of a couple of percent). With this approximation, we see that the η-term only renormalises the Y amplitudes and, thus, the killing operator boils down to a linear super position of a fermion pair destructor with a pair creator. This can then be seen as a Hartree-Fock-Bogoliubov (HFB) transformation of fermion pair operators, i.e., pairing of 'pairs'. Replacing the pair operators by ideal bosons as done in RPA, would lead to a standard bosonic HFB approach [7] (see also, [5],ch.9). Here, however, we will stay with the fermionic description and elaborate an HFB theory for fermion pairs. For this, we will suppose that we can use the killing property $Q_\nu|Z\rangle = 0$ even with the approximate Q-operator. As with our experience from SCRPA, we assume that this violation of consistency is weak.

Let us continue with elaborating our just defined frame. We will then use for the pair-killing operator

$$Q_\nu = \sum_{k<k'} [X^\nu_{kk'}c_kc_{k'} - Y^\nu_{kk'}c^+_{k'}c^+_k]/N^{1/2}_{kk'} \tag{13}$$

with (the approximate) property $Q|Z\rangle = 0$ and the first relation in (12). The normalisation factor $N_{kk'} = |1 - n_k - n_{k'}|$ has been introduced so that $< [Q,Q^+] >= \frac{1}{2}\sum(X^2 - Y^2) = 1$, i.e., the quasi-pair state $Q^+|Z\rangle$ and the X,Y amplitudes being normalised to one. In analogy with (3), we now will minimise the following energy weighted sum rule

$$\Omega_\nu = \frac{\langle Z|[Q_\nu,[H - 2\mu\hat{N}, Q^+_\nu]]|Z\rangle}{\langle Z|[Q_\nu, Q^+_\nu]|Z\rangle} \tag{14}$$

The minimisation with respect to X,Y amplitudes leads to

$$\begin{pmatrix} \mathbf{H} & \mathbf{\Delta}^{(22)} \\ -\mathbf{\Delta}^{(22)+} & -\mathbf{H}^* \end{pmatrix} \begin{pmatrix} X^\nu \\ Y^\nu \end{pmatrix} = \Omega_\nu \begin{pmatrix} X^\nu \\ Y^\nu \end{pmatrix} \tag{15}$$

with (we eventually will consider a symmetrized double commutator in $\mathbf{H}$)

$$\begin{aligned} \mathbf{H}_{k_1k_2,k'_1k'_2} &= \langle [c_{k_2}c_{k_1}, [H - 2\mu\hat{N}, c^+_{k'_1}c^+_{k'_2}]]\rangle/(N^{1/2}_{k_1k_2}N^{1/2}_{k'_1k'_2}) \\ &= (\xi_{k_1} + \xi_{k_2})\delta k_1k_2, k'_1k'_2 + N^{-1/2}_{k_1k_2}N^{-1/2}_{k'_1k'_2}\{N_{k_1k_2}\bar{v}_{k_1k_2k'_1k'_2}N_{k'_1k'_2} \\ &+ [(\frac{1}{2}\delta_{k_1k'_1}\bar{v}_{l_1k_2l_3l_4}C_{l_3l_4k'_2l_1} + \bar{v}_{l_1k_2l_4k'_2}C_{l_4k_1l_1k'_1}) - (k_1 \leftrightarrow k_2)] - [k'_1 \leftrightarrow k'_2]\} \end{aligned} \tag{16}$$

where

$$C_{k_1k_2k'_1k'_2} = \langle c^+_{k'_1}c^+_{k'_2}c_{k_2}c_{k_1}\rangle - n_{k_1}n_{k_2}\delta_{k_1k_2k'_1k'_2} \tag{17}$$

is the two body correlation function and

$$\begin{aligned} \mathbf{\Delta}^{(22)}_{k_1k_2,k'_1k'_2} &= -\langle [c_{k_2}c_{k_1}, [H - 2\mu\hat{N}, c_{k'_1}c_{k'_2}]]\rangle/(N^{1/2}_{k_1k_2}N^{1/2}_{k'_1k'_2}) \\ &= N^{-1/2}_{k_1k_2}[(\Delta_{k_1k'_2;k'_1k_2} - k_1 \leftrightarrow k_2) - (k'_1 \leftrightarrow k'_2)]N^{-1/2}_{k'_1k'_2} \end{aligned} \tag{18}$$

with

$$\Delta_{k_1k_2';k_1'k_2} = \sum_{l<l'} \bar{v}_{k_1k_2'll'} \langle c_{k_1'} c_{k_2} c_{l'} c_l \rangle \tag{19}$$

In (15) the matrix multiplication is to be understood as $\sum_{k_1'<k_2'}$ for restricted summation (or as $\frac{1}{2}\sum_{k_1'k_2'}$ for unrestricted summation) . We see from (18) and (19) that the bosonic gap $\mathbf{\Delta}^{(22)}$ involves the quartet order parameter quite in analogy to the usual gap field in the BCS case. The $\mathbf{H}$ operator in (15) has already been discussed in [8] in connection with SCRPA in the particle-particle channel. Equation (15) has the typical structure of a bosonic HFB equation but, here, for fermion pairs, instead of bosons. It remains the task to close those HFB equations in expressing all expectation values involved in the $\mathbf{H}$ and $\mathbf{\Delta}^{(22)}$ fields by the X, Y amplitudes. This goes in the following way. Because of the HFB structure of (15), the X, Y amplitudes obey the usual orthonormality relations, see [5]. Therefore, one can invert relation (13) to obtain

$$c_{k'}^+ c_k^+ = N_{kk'}^{1/2} \sum_\nu [X_{kk'}^\nu Q_\nu^+ + Y^\nu Q_\nu] \qquad (k < k') \tag{20}$$

and by conjugation the expression for cc. With this relation, we can calculate all two body correlation functions in (18) and (16) in terms of X, Y amplitudes. This is achieved in commuting the destruction operators Q to the right hand side and use the killing property. For example, the quartet order parameter in the gap-field (19) is obtained as $\langle c_{k_1'} c_{k_2} c_{l'} c_l \rangle = N_{k_1'k_2}^{1/2} \sum_\nu X_{k_2k_1'}^\nu Y_{ll'}^\nu N_{ll'}^{1/2}$. Remains the task how to link the occupation numbers $n_k = \langle c_k^+ c_k \rangle$ to the X, Y amplitudes. Of course, that is where our partitioning of the quartet operator into singles and triples comes into play. Therefore, let us try to work with the operator (9). First, as a side-remark, let us notice that if in (9) we replace $c_{k_1}^+ c_{k_2}^+$ by its expectation value which is the pairing tensor, we are back to the standard Bogoliubov transformation for pairing. Here we want to consider quartetting and, thus, we have to keep the triple operator fully. Minimising, as in (3) an average single particle energy, we arrive at the following equation for the amplitudes u, v in (9)

$$\begin{pmatrix} \xi & \mathbf{\Delta}^{(13)} \\ \mathbf{\Delta}^{(13)+} & -\mathcal{N}\mathcal{H}^* \end{pmatrix} \begin{pmatrix} u \\ v \end{pmatrix} = E \begin{pmatrix} 1 & 0 \\ 0 & \mathcal{N} \end{pmatrix} \begin{pmatrix} u \\ v \end{pmatrix} \tag{21}$$

with (we disregard pairing, i.e., $\langle cc \rangle$ amplitudes)

$$\mathbf{\Delta}_{k;k_1k_2k_3}^{(13)} = \Delta_{kk_3;k_2k_1} - [(k_2 \leftrightarrow k_3) - (k_1 \leftrightarrow k_2)] \tag{22}$$

and

$$(\mathcal{N}\mathcal{H}^*)_{k_1k_2k_3;k_1'k_2'k_3'} = \langle \{ c_{k_3}^+ c_{k_2}^+ c_{k_1}^+, [H - 3\mu\hat{N}, c_{k_1'} c_{k_2'} c_{k_3'}] \} \rangle \tag{23}$$

$$\mathcal{N}_{k_1k_2k_3;k_1'k_2'k_3'} = \langle \{ c_{k_3}^+ c_{k_2}^+ c_{k_1}^+, c_{k_1'} c_{k_2'} c_{k_3'} \} \rangle \tag{24}$$

We will not give $\mathcal{H}$ in full because it is a very complicated expression involving self-consistent determination of three-body densities. To lowest order in the interaction it is given by

$$\mathcal{H}_{k_1k_2k_3;k_1'k_2'k_3'} = (\xi_{k_1} + \xi_{k_2} + \xi_{k_3}) \delta_{k_1k_2k_3,k_1'k_2'k_3'} + [(1 - n_{k_1} - n_{k_2}) \bar{v}_{k_1k_2k_1'k_2'} \delta_{k_3k_3'} + \text{ permutations}] \tag{25}$$

where $\delta_{k_1k_2k_3,k_1'k_2'k_3'}$ is the fully antisymmetrised three fermion Kronecker symbol. Even this operator is still rather complicated for numerical applications and mostly one will replace the

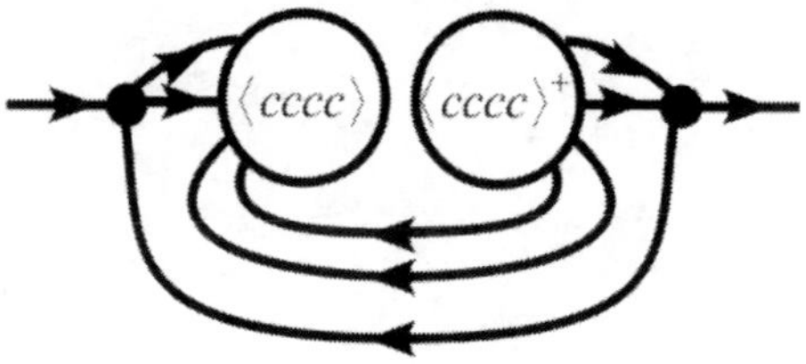

Figure 1. Schematic representation of self-energy for α particle condensation in an uncorrelated nucleon gas.

correlated occupation numbers by their free Fermi- Dirac steps, i.e., $n_k \to n_k^0$. To this order the three body norm in (24) is given by

$$\mathcal{N}_{k_1k_2k_3;k_1'k_2'k_3'} \simeq [\bar{n}_{k_1}^0\bar{n}_{k_2}^0\bar{n}_{k_3}^0 + n_{k_1}^0n_{k_2}^0n_{k_3}^0]\delta_{k_1k_2k_3,k_1'k_2'k_3'} \tag{26}$$

with $\bar{n}^0 = 1 - n^0$. In principle this effective three-body Hamiltonian leads to three-body bound and scattering states. In our application to nuclear matter given below, we will make an even more drastic approximation and completely neglect the interaction term in the three-body Hamiltonian. Eliminating under this condition the v-amplitudes from (21), one can write down the following effective single particle equation

$$\xi_k u_k^{(\nu)} + \sum_{k_1<k_2<k_3k'} \frac{\mathbf{\Delta}^{(13)}_{k,k_1k_2k_3}(\bar{n}_{k_1}^0\bar{n}_{k_2}^0\bar{n}_{k_3}^0 + n_{k_1}^0n_{k_2}^0n_{k_3}^0)\mathbf{\Delta}^{(13)^*}_{k_3k_2k_1k'}}{E_\nu + \xi_{k_1} + \xi_{k_2} + \xi_{k_3}} u_{k'}^{(\nu)} = E_\nu u_k^{(\nu)} \tag{27}$$

The occupation numbers are given by

$$n_k = 1 - \sum_\nu |u_k^{(\nu)}|^2 \tag{28}$$

The effective single particle field in (27) is grapphically interpreted in Fig. 1. The gap-fields in (27) are then to be calculated as in (22) and (19) with (20) and the system of equations is fully closed. This is quite in parallel to the pairing case. In cases, where the quartet consists out of four different fermions and in addition is rather strongly bound, as this will be the case for the α particle in nuclear physics, one still can make a very good but drastic simplification: one writes the quartic order parameter as a translationally invariant product of four times the same single particle wave function in momentum space. We will see, how this goes below when we apply our theory to α particle condensation in nuclear matter. Comparing the effective single particle field in (27) with the one of standard pairing, Eq. (5), we find strong analogies but also several structural differences. The most striking is that in the quartet case Pauli factors figure in the numerator of (27) whereas this is not the case for pairing. In principle in the pairing case, they are also there, but since $\bar{n}_k + n_k = 1$, they drop out. This difference has quite dramatic consequences between the pairing and the quartetting case. Namely when the chemical potential μ changes from negative (binding) to positive, the implicit three hole level density

$$g_{3h}(\omega) = \sum_{k_1k_2k_3} (\bar{n}_{k_1}^0\bar{n}_{k_2}^0\bar{n}_{k_3}^0 + n_{k_1}^0n_{k_2}^0n_{k_3}^0)\delta(\omega + \xi_{k_1} + \xi_{k_2} + \xi_{k_3}) \tag{29}$$

passes through zero at $\omega - 3\mu = 0$ because phase space constraints and energy conservation cannot be fullfilled simultaneously at that point. This is not at all the case for the single particle level density

$$g_{1h}(\omega) = \sum_k (\bar{n}_k^0 + n_k^0)\delta(\omega + \xi_k) \quad = \quad \sum_k \delta(\omega - \xi_k) \tag{30}$$

entering implicitly in the pairing case, since, as indicated $\bar{n}_k^0 + n_k^0 = 1$. Therefore, for positive μ, in the case of quartetting essentially no correlations around the Fermi energy ($\omega - 3\mu = 0$) can build up and, therefore, there is no quartet condensation for positive chemical potential. Quartetting only exists on the BEC side where μ is negative and, thus, $n_k^0 = 0$ and no structural difference exists between the pairing case and the quartetting one for $\mu < 0$. We will show explicit examples of level densities in section 4.

So far, we have discarded possible existence of ordinary pairing mixed with quartetting. Formally, this can easily be achieved in writing everywhere quasiparticle operators instead of particle ones. For example the ground state (8) is then given by

$$|Z\rangle = e^{\frac{1}{4!}\sum Z_{k_1k_2k_3k_4}\beta_{k_1}^+\beta_{k_2}^+\beta_{k_3}^+\beta_{k_4}^+}|\mathrm{BCS}\rangle \tag{31}$$

and the killing operator

$$Q_\nu = \sum_{k<k'} [X^\nu \beta_{k'}\beta_k - Y_{kk'}^\nu \beta_k^+\beta_{k'}^+]/N_{kk'} \tag{32}$$

and analogously for the superposition of singles with triples. Of course, if the amplitudes Z in (31) are zero, we are back to ordinary BCS theory of pairs. All the algebra goes through as before, only we have to deal in the equations of motion with the Hamiltonian in the quasi-particle representation (see [5], App. E). This description of quartetting with quasiparticles shall be worked out in the future. It may help to understand the precise nature of the transition from quartetting to pairing when μ changes from negative to positive values. It is worth noticing that (32) is exactly the ansatz used to derive quasi-particle RPA (QRPA) [5] what implies a linearised equation of motion. Extending to self-consistent QRPA [6] will describe pairing and quartetting consistently within the same formalism.

Of course, the description of quartetting can also be obtained from an equivalent Gorkov-type or Green's function approach. Since this may shed a complementary light on the theory, we very briefly want to sketch how this goes. Let us intruduce two types of matrix Green's functions

$$\mathcal{G}^{(13)} = \begin{pmatrix} G^{(11)} & G^{(13)} \\ G^{(13)+} & G^{(33)} \end{pmatrix} \tag{33}$$

with the usual definition of time ordered Green's functions at zero temperature to be found, e.g., in [9],[5]

$$G_{kk'}^{(11)} = -i\langle Tc_k(t)c_{k'}^+(t')\rangle; \qquad G_{k;k_1'k_2'k_3'}^{(13)} = -i\langle Tc_k(t)(c_{k_1'}c_{k_2'}c_{k_3'})_{t'}\rangle$$

$$G_{k_1k_2k_3;k'}^{(13)} = i\langle T(c_{k_3}^+c_{k_2}^+c_{k_1}^+)_t c_{k'}^+(t')\rangle; \qquad G_{k_1k_2k_3;k_1'k_2'k_3'}^{(33)} = i\langle T(c_{k_3}^+c_{k_2}^+c_{k_1}^+)_t(c_{k_1'}c_{k_2'}c_{k_3'})_{t'}\rangle$$

and

$$\mathcal{G}^{(22)} = \begin{pmatrix} G_{-+}^{(22)} & G_{--}^{(22)} \\ G_{++}^{(22)} & G_{+-}^{(22)} \end{pmatrix} \tag{34}$$

with

$$G_{-+,\ k_1k_2;k_1'k_2'}^{(22)} = -i\langle P_{k_1k_2}^{(-)}(t)P_{k_1'k_2'}^{(+)}(t')\rangle$$

and $P^{(-)}_{k_1k_2} \equiv c_{k_1}c_{k_2}$, and $P^{(+)} = P^{(-)+}$. The other Green's functions in the matrix of (34) are defined analogously. With the equation of motion method for fermion cluster Green's functions, see [8], one then can write down a Dyson equation for these matrix Green's functions

$$\mathcal{G} = \mathcal{G}^{(0)} + \mathcal{G}^{(0)}\Sigma^{(0)}\mathcal{G} \tag{35}$$

where $\mathcal{G}$ is either the matrix Green's function (33) or (34) and $\Sigma^{(0)}$ is a corresponding instantaneous part of the exact self-energy. $\mathcal{G}^{(0)}$ is the free Green's function, as usual. The self-energies $\Sigma^{(0,22)}$ and $\Sigma^{(0,13)}$ can easily be read off from the effective 2x2 Hamiltonians in (15) and (21). The one and two body density matrices entering these Hamiltonians are then given by the residua of the various Green's functions and the system of equations is again closed. This is completely equivalent to the system of equations we established before, using the killing operators. The Green's function formalism has the advantage of easily being generalizable to finite temperature.

This ends the formal aspects of quartetting theory. We will give some further insight when we describe the applications to nuclear systems in the next sections.

3. Nuclear clusters in the medium and critical temperature of quartetting in nuclear matter

As mentioned in the preceding sections, we will apply our theory for quartetting to nuclear matter and also to finite nuclei. In this context it may be useful to study first the behaviour of a single fermion pair, e.g. the deuteron, and a single quartet, i.e., the α particle in a gas of uncorrelated nucleons. For example the in-medium deuteron equation is given by

$$[\varepsilon_{k_1} + \varepsilon_{k_2} - E_{d,P}]\,\psi_{d,P}(12) + \sum_{k_1'<k_2'} [1 - f_{k_1} - f_{k_2}]\,\bar{v}_{k_1k_2k_1'k_2'}\psi_{d,P}(1'2') = 0. \tag{36}$$

with P the total momentum, f_k the Fermi-Dirac distribution at finite temperature (equivalent to the previously defined Fermi step n_k^0 at zero temperature, and ε_k the HF single particle energies defined in (6). This *effective wave equation* describes bound states as well as scattering states. The onset of pair condensation is achieved when the binding energy $E_{d,P=0}$ coincides with 2μ.

Similar equations have been derived from the Green function approach for the case of nucleon numbers $A = 3$ and $A = 4$, describing triton/helion (^{3}He) nuclei as well as α-particles in nuclear matter. The effective wave equation contains in mean field approximation the Hartree-Fock self-energy shift of the single-particle energies as well as the Pauli blocking of the interaction. We give the effective wave equation for the α particle,

$$\begin{aligned}[\varepsilon_{k_1} + \varepsilon_{k_2} + \varepsilon_{k_3} + \varepsilon_{k_4} - E_{\alpha,P}]\psi_{\alpha,P}(1234) \;&+\; \sum_{k_1'<k_2'} [1 - f_{k_1} - f_{k_2}]\bar{v}_{k_1k_2k_1'k_2'}\psi_{\alpha,P}(1'2'34) \\ &+\; \text{permutations} = 0. \end{aligned} \tag{37}$$

A similar equation is obtained for $A = 3$.

The effective wave equation has been solved using separable potentials for $A = 2$ by integration. For $A = 3, 4$ we can use a *Faddeev approach* [10]. The shifts of binding energy can also be calculated approximately via perturbation theory. In Fig. 2 we show the shift of the binding energy of the light clusters ($d, t/h$ and α) in symmetric nuclear matter as a function of density for temperature $T = 10$ MeV. It is found that the cluster binding energy decreases with increasing density. Finally, at the *Mott density* $\rho^{\rm Mott}_{A,n,P}(T)$, the bound state is dissolved. The clusters are not present at higher densities, merging into the nucleonic medium. For a given cluster type characterized by A, n, we can also introduce the Mott momentum $P^{\rm Mott}_{A,n}(\rho, T)$ in terms of the ambient

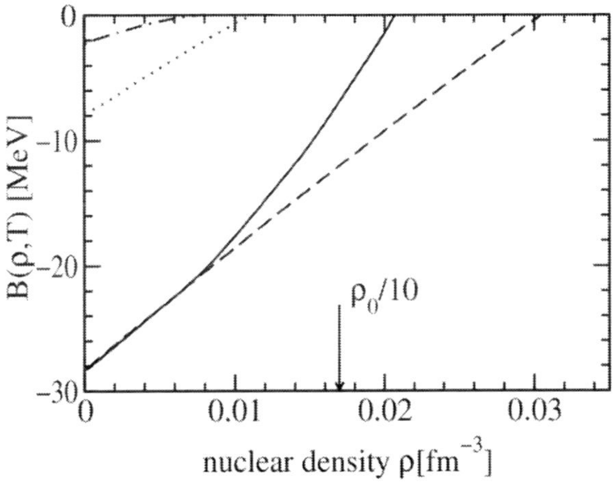

Figure 2. Shift of binding energies of the light clusters (d - dash dotted, t/h - dotted, and α - dashed: perturbation theory, full line : non-perturbative Faddeev-Yakubovski equation) in symmetric nuclear matter as a function of density for given temperature $T = 10$ MeV.

temperature T and nucleon density ρ, such that the bound states exist only for $P \geq P_{A,n}^{\mathrm{Mott}}(\rho, T)$.

In general, it is necessary to take account of *all bosonic clusters* to gain a complete picture of the onset of superfluidity. As is well known, the deuteron is weakly bound compared to other nuclei. Higher A-clusters can arise that are more stable. In this section, we will consider the formation of α-particles, which are of special importance because of their large binding energy per nucleon ($\sim$7 MeV). We will not include tritons or helions, which are fermions and not so tightly bound. Moreover, we will not consider nuclei in the iron region, which have even larger binding energy per nucleon than the α-particle and thus constitute, in principle, the dominant component at low temperatures and densities. However, the latter are complex structures of many particles and are strongly affected by the medium as the density increases, since their excitation spectrum starts at much lower energy than the one of the α particle, for instance. So they are assumed not to be of relevance in the density region considered here. Given the medium-modified bound-state energy $E_{4,P}$, the bound-state contribution to the EOS is

$$\rho_4(\beta, \mu) = \sum_P \left[e^{\beta(E_{4,P} - 2\mu_p - 2\mu_n)} - 1 \right]^{-1} . \tag{38}$$

We will not include the contribution of the excited states or that of scattering states. Because of the large specific binding energy of the α particle, low-density nuclear matter is predominantly composed of α particles. This observation underlies the concept of α matter and its relevance to diverse nuclear phenomena.

As exemplified by Eq. (37), the effect of the medium on the properties of an α particle in mean-field approximation (i.e., for an uncorrelated medium) is produced by the Hartree-Fock self-energy shift and Pauli blocking. The shift of the α-like bound state has been calculated using perturbation theory as well as by solution of the Faddeev-Yakubovski equation [10]. It is found that this bound state merges with the continuum of scattering states at a Mott density $\rho_\alpha^{\mathrm{Mott}} \approx \rho_0/10$, see Fig. 2. The bound states of clusters d, t, and h with $A < 4$ are already dissolved at the density $\rho_\alpha^{\mathrm{Mott}}$. Consequently, if we neglect the contribution of the four-particle scattering phase shifts in the different channels, we can now construct an equation of state $\rho(T, \mu) = \rho^{\mathrm{free}}(T, \mu) + \rho^{\mathrm{bound},d}(T, \mu) + \rho^{\mathrm{bound},\alpha}(T, \mu)$ so that α-particles determine the behavior of symmetric nuclear matter at densities below $\rho_\alpha^{\mathrm{Mott}}$ and temperatures below the binding energy per nucleon of the α-particle. The formation of deuteron clusters alone gives an incorrect description because the deuteron binding energy is small, and the abundance of d-clusters is small compared with that of α-clusters. In the low density region of the phase diagram, α-

matter emerges as an adequate model for describing the nuclear-matter equation of state. With increasing density, the medium modifications – especially Pauli blocking – will lead to a deviation of the critical temperature $T_c(\rho)$ from that of an ideal Bose gas of α-particles (the analogous situation holds for deuteron clusters, i.e., in the isospin-singlet channel). At a critical density which more or less coincides with the point where the chemical potential turns from negative to positive value, the quartet will be quite abruptly dissolved and no α particle survives. This is in line with the arguments given above concerning the level densities.

Symmetric nuclear matter is characterized by the equality of the proton and neutron chemical potentials, i.e., $\mu_p = \mu_n = \mu$. Then an extended Thouless condition based on the relation for the four-body in medium wave function, Eq. (37), at eigenvalue 4μ serves to determine the onset of Bose condensation of α-like clusters, noting that existence of a solution of this relation signals a divergence of the four-particle correlation function. An approximate solution has been obtained by a variational approach, in which the wave function is taken of the following projected mean field form [11] (see also [12] for another variational ansatz):

$$\psi(1234) = \delta(\mathbf{k}_1 + \mathbf{k}_2 + \mathbf{k}_3 + \mathbf{k}_4 - \mathbf{K})\varphi(\mathbf{k}_1)\varphi(\mathbf{k}_2)\varphi(\mathbf{k}_3)\varphi(\mathbf{k}_4)\chi(1234) \tag{39}$$

where the φ's are single particle wave functions in momentum space to be determined variationally out of the in medium four body wave equation and $\chi(1234)$ is the scalar spin-isospin function. The delta function is a projector on total momentum $\mathbf{K}$ which, for condensates at rest, is to be taken at $\mathbf{K} = 0$.

The results for the critical temperature of symmetric and asymmetric nuclear matter with a separable force which reproduces binding energy and radius of a free α particle are presented in Fig. 3. The crosses on the figure indicate an exact solution of the in medium four body equation with a realistic nucleon-nucleon force [11] [13]. The excellent agreement with the approximate solution justifies the choice of our variational wave function. An important feature is that at the lowest temperatures, Bose-Einstein condensation occurs for α particles rather than for deuterons. As the density increases within the low-temperature regime, the chemical potential μ first reaches -7 MeV, where the α's Bose-condense. By contrast, Bose condensation of deuterons would not occur until μ rises to -1.1 MeV. In the asymmetric case, it is interesting to note that at strong asymmetry α particle condensation wins over the deuteron condensation because of the strong binding of the α's.

The *"quartetting"* transition temperature sharply drops as the rising density approaches the critical Mott value ($\mu \simeq 0$) at which the four-body bound states disappear. The deeper reason for this abrupt disappearence is again the same as we discussed above concerning the non-existence of quartet condensation for positive chemical potential, namely the in medium four fermion level density

$$g_4(\omega) = \sum(\bar{n}^0_{k_1}\bar{n}^0_{k_2}\bar{n}^0_{k_3}\bar{n}^0_{k_4} - n^0_{k_1}n^0_{k_2}n^0_{k_3}n^0_{k_4})\delta(\omega - \xi_{k_1} - \xi_{k_2} - \xi_{k_3} - \xi_{k_4}) \tag{40}$$

goes, for positive μ through zero at $\omega - 4\mu = 0$. At that point, pair formation in the isospin-singlet deuteron-like channel comes into play, and a deuteron condensate will exist below the critical temperature for BCS pairing up to densities above the nuclear-matter saturation density ρ_0. The reason why the critical temperature for deuterons does not break down for positive μ can again be given via the level density. Namely the two particle in medium level density, for pairs *at rest*, equals, up to a factor, the one for single particle states. The critical density at which the α condensate disappears is estimated to be around $\rho_0/10 - \rho_0/5$. Therefore, α-particle condensation primarily only exists in the Bose-Einstein-Condensed (BEC) phase and there does not seem to exist a phase where the quartets acquire a large extension as Cooper pairs do in the weak coupling regime. However, our variational approach on which this conclusion is based represents only a first attempt at the description of the transition from quartetting

Journal of Physics: Conference Series **529** (2014) 012014 doi:10.1088/1742-6596/529/1/012014

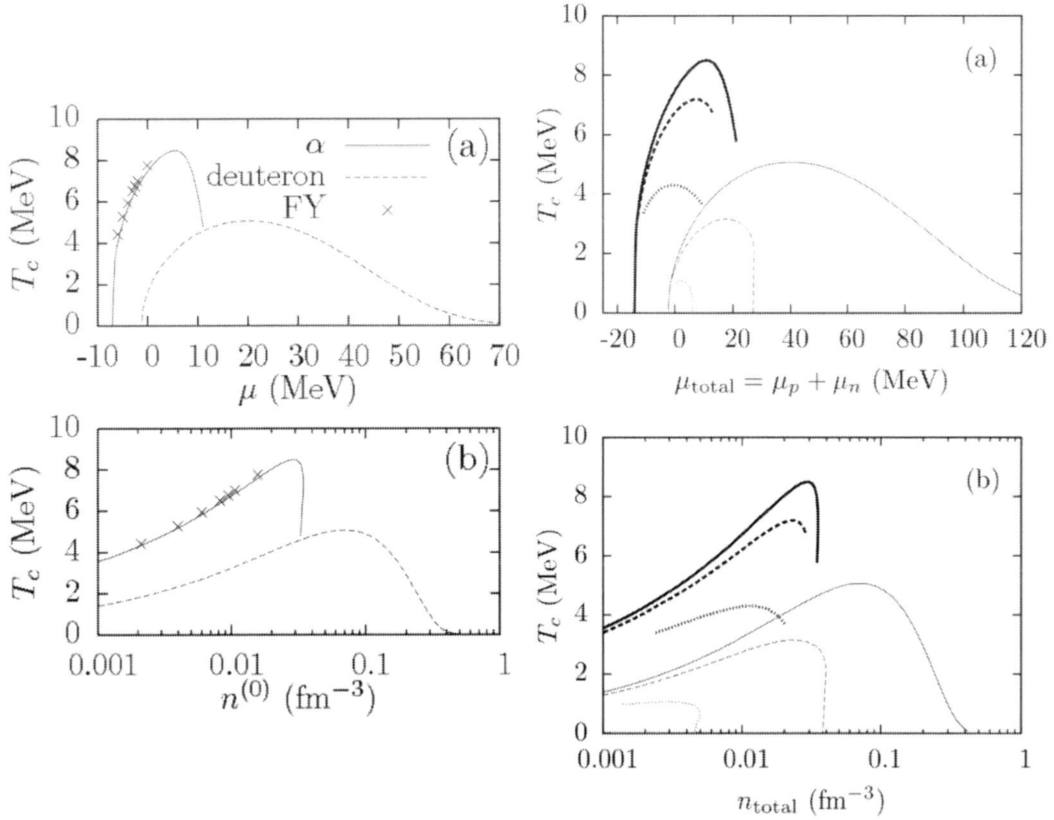

Figure 3. (Left) Critical temperatures for α particle and deuteron condensation in symmetric nuclear matter as a function of μ (a) and density $n^{(0)}$ (b). (Right) Same as in (Left) but for asymmetric nuclear matter for various asymmetry parameters $\delta = (n_n - n_p)/(n_n + n_p)$. $\delta = 0$ (full line); $\delta = 0.5$ (broken line); $\delta = 0.9$ (dotted line).

to pairing. The detailed nature of this fascinating transition remains to be clarified. Further discussions on this phenomenon can be found in [14]. We also should mention that the estimate of our critical temperature for α particle condensation which is based on a generalisation of the Thouless criterion for pairing, is only valid at the upper end of densities for which condensation occurs. For very low densities, the α's should form an ideal Bose gas. For the description of this limit, a theory should be developed which is analogous to the one of Nozières-Scmitt-Rink [15] for pairing. For quartets, this is more involved and has not been worked out so far.

A very intriguing question in relation with non existence of an α condensate at higher densities can be asked: is it possible that in heavy nuclei an α particle condensate exists in the nuclear surface, at least in its fluctuating form? The preformation assumption of α particles in the surface to explain α decay may give a hint to this.

4. Alpha-Particle Condensation in Infinite Nuclear Matter at Zero Temperature

The equation for the four-body order parameter $K_{k_1k_2k_3k_4} \equiv \langle c_{k_1}c_{k_2}c_{k_3}c_{k_4}\rangle$ figuring in (18),(19),(22), obeys, to lowest order in the interaction, formally the same equation as the the one which determines the critical temperature, namely

$$(\varepsilon_{k_1} + \varepsilon_{k_2} + \varepsilon_{k_3} + \varepsilon_{k_4})K_{k_1k_2k_3k_4} - \sum_{k'_1<k'_2} [1 - n_{k_1} - n_{k_2}]\bar{v}_{k_1k_2k'_1k'_2} K_{k'_1k'_2k_3k_4} \tag{41}$$

$$+ \text{ permutations} = 4\mu K_{k_1k_2k_3k_4} \tag{42}$$

where the occupation numbers n_k should be calculated self-consistently from Eq. (28). These two coupled equations are analogous to (7) and (5) in the case of pairing. Of course, with an un-approximated 4-body order parameter K, this would be a tremendously complicated self-consistent system of in medium 4-body equations to be solved. However, as before, for the critical temperature of α particle condensation, we very effectively approximate the order parameter K by our projected mean field ansatz in Eq. (39). Therefore the only unknown is now the single particle 0S wave function $\varphi(k)$. This constitutes a strong simplification of the problem. It has been solved in [14] and we refer the reader to that reference for further details. We want, however, to show the $3h$ level density from that paper, see Fig. 4, to demonstrate its vanishing behavior around the Fermi-energy for positive μ. The explicit solution of the single particle wave function φ, see Fig. 5 . In this latter figure, we also give the values of the occupation number distribution n_k for various (negative) chemical potentials. It is very interesting to see that even for a slightly positive value of μ (i.e., just before the break down of a solution), the n_k distribution is far from saturation, i.e., from one at $k = 0$. Its maximum value is about 0.35 indicating that one is still far in the BEC side when α particle condensation starts to end.

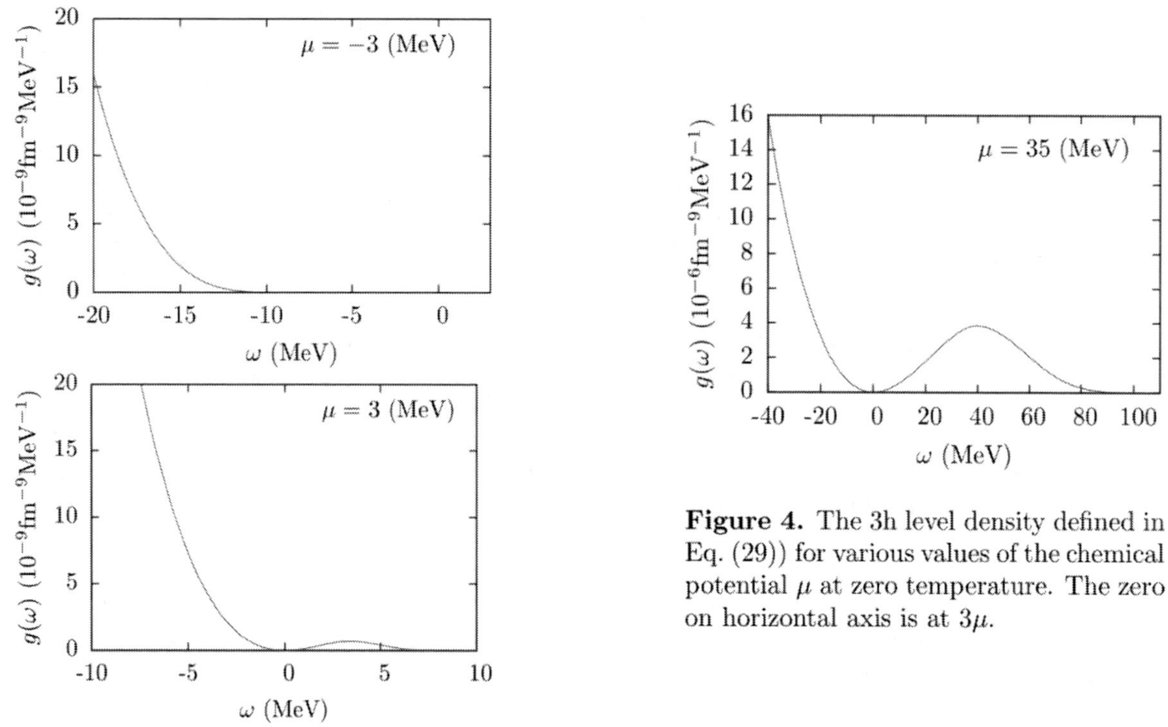

Figure 4. The 3h level density defined in Eq. (29)) for various values of the chemical potential μ at zero temperature. The zero on horizontal axis is at 3μ.

Before we will come to the description of α particle condensation in finite nuclei, let us make some further remarks concerning the approach we have worked out so far. This concerns for instance the approximate treatment of the single particle self energy leading to quartet condensation shown in Fig. 1 . This process implies that α particles are directly formed out

Journal of Physics: Conference Series **529** (2014) 012014 doi:10.1088/1742-6596/529/1/012014

of an uncorrelated nucleon gas. That the scattering between those particles is neglected may not be the worst approximation. On the other hand, at least in nuclear physics, there exist three nucleon bound states like the triton or ^{3}He. In a gas of nucleons, they will be present and then a scattering process of a nucleon on one of those 'trions' may lead to an α particle (or the other way round). In addition, there may be deuterons around and two deuterons may again form an α particle. All these processes, depicted in Fig. 6, are neglected in our present application and should be included in future studies. One will then have a coherent description of a hot mixture of nucleons, Cooper pairs, deuterons, trions, and α-particles which awaits future solution. Another feature which distinguishes the quartet case from pairing is that, as seen in Fig. 1, there are three hole lines and not just one as for pairing. Because one has to sum over the relative momenta of those three holes, the effective single particle field acquires an imaginary part and no sharp quasi-particle pole developes. More on this has been worked out in [14].

5. Alpha-Particle Condensate States in Self-Conjugate 4n Nuclei

Let us discuss the possibility of quartetting in nuclei. The only nucleus having a pronounced α-cluster structure in its ground state is ^{8}Be. In Fig. 7(a), the result of an exact calculation of the density distribution of ^{8}Be in the laboratory frame is shown. In Fig. 7(b), for comparison, the result of the same calculation in the intrinsic, deformed frame is displayed. We see a pronounced two α-cluster structure where the two α's are $\sim$ 4 fm apart, giving rise to a very low average density $\rho \sim \rho_0/3$ as seen in Fig. 7(a). As already discussed in the introduction, ^{8}Be is a rather unusual and unique nucleus. One may be intrigued by the question, already raised earlier, whether loosely bound α- particle configurations may not also exist in heavier nα-nuclei, at least in excited states, naturally close to the nα disintegration threshold. Since α-particles are rather inert bosons (first excited state at $\sim$ 20 MeV), they then would all condense in the lowest S-wavefunction, very much in the same way as do bosonic atoms in magneto-optical traps [16]. This question and exploring related issues of quartetting in finite nuclei will consume most of the rest of the present outline on α particle condensation. In fact, we will be able to offer strong arguments that the 0_2^+ state of ^{12}C at 7.654 MeV is a state of α-condensate nature. First, it should be understood that the 0_2^+ state in ^{12}C is in fact hadronically unstable (as ^{8}Be), being situated about 300 keV above the three α-break up threshold. This state is stabilized only by the Coulomb barrier. It has a width of 8.7 eV and a corresponding lifetime of 7.6×10^{-17} s. As well known, this state is of paramount astrophysical (and biological!) importance due to its role in the creation of ^{12}C in stellar nucleosynthesis. Its existence was predicted in 1953 by the astrophysicist Fred Hoyle [17]; his prediction was confirmed experimentally a few years later by Willy Fowler and coworkers at Caltech [18]. It is also well known that this *Hoyle state*, as it is called, is a notoriously difficult state for any nuclear theory to explain. For example, the most modern no-core shell-model calculations predict the 0_2^+ state in ^{12}C to lie at around 17 MeV above the ground state – more than twice the actual value [19]. This fact alone tells us that the Hoyle state must have a very unusual structure. It is easy to understand that, should it indeed have the proposed loosely bound three α-particle structure, a shell-model type of calculation would have great difficulties in reproducing its properties.

An important development bearing on this issue took place some thirty years ago. Two Japanese physicists, M. Kamimura [20] and K. Uegaki [21], along with their collaborators, almost simultaneously reproduced the Hoyle state from a microscopic theory. They employed a twelve-nucleon wave function together with a Hamiltonian containing an effective nucleon-nucleon interaction. At that time, their work did not attract the attention it deserved; the true importance of their achievement has been appreciated only recently. The two groups started from practically the same ansatz for the ^{12}C wave function, which has the following three α-cluster structure:

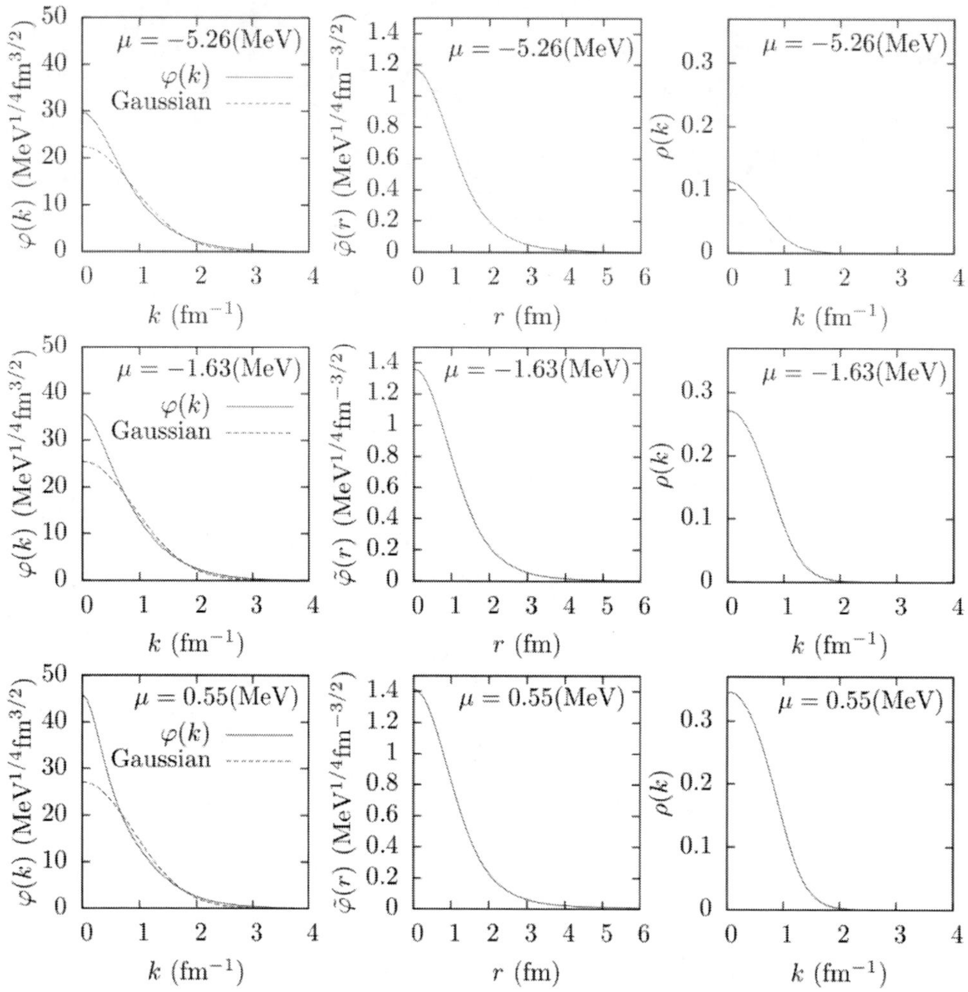

Figure 5. Single particle wave function $\varphi(k)$ for k-space (left), for r-space $\tilde{\varphi}(r)$ (middle), and occupation numbers $\rho(k) \equiv n_k$ (right) at $\mu = -5.26$ (top), -1.63 (middle) and 0.55 (bottom). The r-space wave function $\tilde{\varphi}(r)$ is derived from the Fourier transform of $\varphi(k)$ by $\tilde{\varphi}(r) = \int d^3k e^{i\mathbf{k}\cdot\mathbf{r}} \varphi(k)/(2\pi)^3$. The dashed line in the left figure correspond to the Gaussian approximation with same norm and rms momentum as $\varphi(k)$.

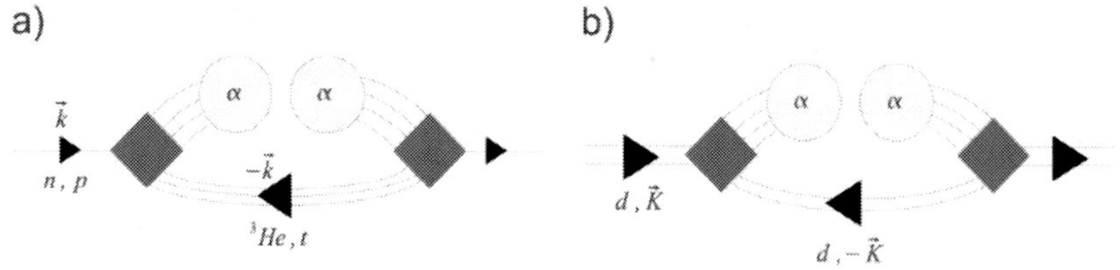

Figure 6. Alpha particle creation by scattering of a nucleon on a triton (^{3}He) (left) or of two deuterons (right).

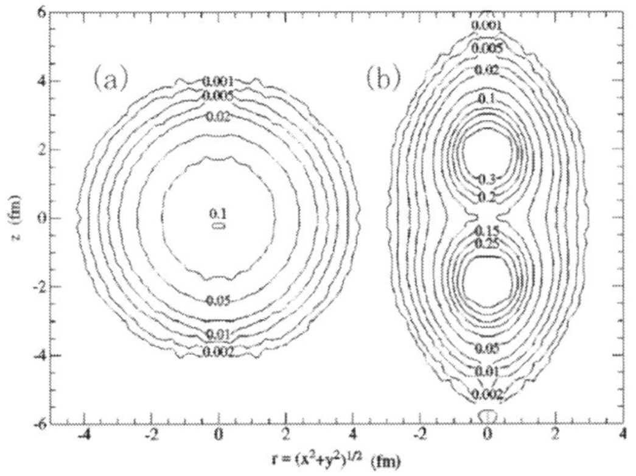

Figure 7. Contours of constant density (taken from Ref. [22]), plotted in cylindrical coordinates, for ^{8}Be(0^+). The left side (a) is in the 'laboratory" frame while the right side (b) is in the intrinsic frame.

$$\langle \mathbf{r}_1 ... \mathbf{r}_{12} | ^{12}\mathrm{C} \rangle = \mathcal{A} \left[\chi(\mathbf{R}, \mathbf{s}) \phi_1 \phi_2 \phi_3 \right] . \tag{43}$$

In this expression, the operator $\mathcal{A}$ imposes antisymmetry in the nucleonic degrees of freedom and ϕ_i, with $i = 1, 2, 3$ for the three α's, is an intrinsic α-particle wave function of prescribed Gaussian form,

$$\phi(\mathbf{r}_1, \mathbf{r}_2, \mathbf{r}_3, \mathbf{r}_4) = \exp \left\{ - \left[(\mathbf{r}_1 - \mathbf{r}_2)^2 + (\mathbf{r}_1 - \mathbf{r}_3)^2 + ... \right] / 2b^2 \right\} , \tag{44}$$

where the size parameter b is adjusted to fit the rms of the free α-particle, and $\chi(\mathbf{R}, \mathbf{s})$ is a yet-to-be determined three-body wave function for the c.o.m. motion of the three α's, their corresponding Jakobi coordinates being denoted by $\mathbf{R}$ and $\mathbf{s}$. The unknown function χ was determined via calculations based on the Generator Coordinate Method [21] (GCM) and the Resonating Group Method [20] (RGM) calculations using the Volkov I and Volkov II effective nucleon-nucleon forces, which fit α-α phase shifts. The precise solution of this complicated three body problem, carried out three decades ago, was truly a pioneering achievement, with results fulfilling expectations. The position of the Hoyle state, as well as other properties including the inelastic form factor and transition probability, successfully reproduced the experimental data. Other states of ^{12}C below and around the energy of the Hoyle state were also successfully described. Moreover, it was already recognized that the three α's in the Hoyle state form sort of a gas-like state. In fact, this feature had already been noted by H. Horiuchi [23] prior to the appearance of Refs. [20, 21], based on results from the orthogonality condition model (OCM) [24]. All three Japanese research groups concluded from their studies that the linear-chain state of three α-particles, postulated by Morinaga many years earlier [25], had to be rejected.

Although the evidence for interpreting the Hoyle state in terms of an α gas was stressed in the cited papers from the late 1970's, two important aspects of the situation were missed at that time. First, because the three α's move in identical S-wave orbits, one is dealing with an α-condensate state what may, in fact, be a general feature in $n\alpha$ nuclei. The second and most important point is that the complicated three-body wave function $\chi(\mathbf{R}, \mathbf{s})$ for the c.o.m. motion of the three α's can be replaced by a structurally and conceptually very simple microscopic three-α wave function of the condensate type, which has practically 100 percent overlap with the previously constructed ones [26] [27] (see also Ref. [28]). We now describe this condensate wave function.

We start by examining the BCS wave function of ordinary fermion pairing, obtained by projecting the familiar BCS ground-state ansatz onto an N-particle subspace of Fock space. In the position representation, this wave function is

$$\langle \mathbf{r}_1 ... \mathbf{r}_N | \mathrm{BCS} \rangle = \mathcal{A} \left[\phi(\mathbf{r}_1, \mathbf{r}_2) \phi(\mathbf{r}_3, \mathbf{r}_4) ... \phi(\mathbf{r}_{N-1} \mathbf{r}_N) \right] , \tag{45}$$

where $\phi(\mathbf{r}_1, \mathbf{r}_2)$ is the Cooper-pair wave function (including spin and isospin), which is to be determined variationally through the familiar BCS equations. The condensate character of the BCS ansatz is born out by the fact that within the antisymmetrizer $\mathcal{A}$, one has a product of $N/2$ times the same pair wave function ϕ, with one such function for each distinct pair in the reference partition of $\{1, 2, \ldots, N-1, N\}$. Formally, it is now a simple matter to generalize (45) to quartet or α-particle condensation. We write

$$\langle \mathbf{r}_1, \ldots, \mathbf{r}_N | \Phi_{n\alpha} \rangle = \mathcal{A} \left[\phi_\alpha(\mathbf{r}_1, \mathbf{r}_2, \mathbf{r}_3, \mathbf{r}_4) \phi_\alpha(\mathbf{r}_5, \ldots, \mathbf{r}_8) \cdots \phi_\alpha(\mathbf{r}_{N-3}, \ldots, \mathbf{r}_N) \right] , \tag{46}$$

where ϕ_α is the wave function common to all condensed α-particles. It may be considered as the number projected quartet wave function introduced in (8). Of course, finding the variational solution for this function is, in general, extraordinarily more complicated than finding the Cooper pair-wave function ϕ of Eq. (45). Even so, in the present case that the α-particle is the four-body cluster involved, and for applications to relatively light nuclei, the complexity of the problem can be reduced dramatically. This possibility stems from the fact that an excellent variational ansatz for the intrinsic wave function of the α-particle is provided [as in Eq. (44)], by a Gaussian form with only the size parameter b to be determined. The new aspect in [26] was that in addition even the center-of-mass motion of the system of α-particles can be described very well by a Gaussian wave function with, this time, a size parameter $B \gg b$ to account for the motion over the whole nuclear space. This is a strong technical simplification and, at the same time, underlines the boson condensate character of the wave function. We therefore write

$$\phi_\alpha(\mathbf{r}_1, \mathbf{r}_2, \mathbf{r}_3, \mathbf{r}_4) = e^{-2\mathbf{R}^2/B^2} \phi(\mathbf{r}_1 - \mathbf{r}_2, \mathbf{r}_1 - \mathbf{r}_3, \cdots) , \tag{47}$$

where $\mathbf{R} = (\mathbf{r}_1 + \mathbf{r}_2 + \mathbf{r}_3 + \mathbf{r}_4)/4$ is the c.o.m. coordinate of one α-particle and $\phi(\mathbf{r}_1 - \mathbf{r}_2, ...)$ is the same intrinsic α-particle wave function of Gaussian form as already used in Refs. [20, 21] and given explicitly in Eq. (44). This wave function has a close relation to the one we used in the infinite matter case. To this end, let us Fourier transform the infinite matter ansatz (39) into real space (with a projection onto a finite $\mathbf{K}$)

$$\phi^\infty_{\alpha,\mathbf{K}}(\mathbf{r}_1, \mathbf{r}_2, \mathbf{r}_3, \mathbf{r}_4) = e^{i\mathbf{K}\mathbf{R}} \int d^3R' e^{-i\mathbf{K}\mathbf{R}'} \tilde{\varphi}(\mathbf{r}_1 - \mathbf{R}') \tilde{\varphi}(\mathbf{r}_2 - \mathbf{R}') \tilde{\varphi}(\mathbf{r}_3 - \mathbf{R}') \tilde{\varphi}(\mathbf{r}_4 - \mathbf{R}') \tag{48}$$

where $\tilde{\varphi}$ is the Fourier transform of φ. Taking for φ a Gaussian, one obtains

$$\phi^\infty_{\alpha,\mathbf{K}}(\mathbf{r}_1, \mathbf{r}_2, \mathbf{r}_3, \mathbf{r}_4) = e^{i\mathbf{K}\mathbf{R}} \exp\left\{ - \left[(\mathbf{r}_1 - \mathbf{r}_2)^2 + (\mathbf{r}_1 - \mathbf{r}_3)^2 + ... \right] / 2b^2 \right\} \tag{49}$$

Comparing with (47), we, therefore, see that our variational ansatz for the α particle condensate wave function is of the same spirit as the one we used already in the homogeneous case, only the plane wave c.o.m. wave function has, naturally, been replaced by a Gaussian. For a small number of α particles, it is, of course, important to work with a wave function with a definite number of particles and not with the coherent state (8). However, for a larger number of α's where the handling of (46) becomes more and more difficult because of the explicit antisymmetrisation of all nucleons, it may be worth to use (47) also in our quartet coherent state (8) as a variational wave function with the two parameters B, b.

Naturally, in Eq. (46) the center of mass $\mathbf{X}_{cm}$ of the three α's, i.e., of the whole nucleus, should be eliminated; this is easily achieved by replacing $\mathbf{R}$ by $\mathbf{R} - \mathbf{X}_{cm}$ in each of the α wave functions in Eq. (46). The α-particle condensate wave function specified by Eqs. (46) and (47), proposed in Ref. [26] and called the THSR wave function, now depends on only two parameters, B and b. The expectation value of an assumed microscopic Hamiltonian H,

$$\mathcal{H}(B, b) = \frac{\langle \Phi_{n\alpha}(B, b)|H|\Phi_{n\alpha}(B, b)\rangle}{\langle \Phi_{n\alpha}|\Phi_{n\alpha}\rangle}, \tag{50}$$

can be evaluated, and the corresponding two-dimensional energy surface can be quantized using the two parameters B and b as Hill-Wheeler coordinates.

Before presenting the results, let us discuss the THSR wave function in somewhat more detail. This innocuous-looking variational ansatz, namely Eq. (46) together with Eq. (47), is actually more subtle than it might at first appear. One should realize that two limits are incorporated exactly. One is obtained by choosing $B = b$, for which Eq. (46) reduces to a standard Slater determinant with harmonic-oscillator single-nucleon wave functions, leaving the oscillator length b as the single adjustable parameter. This holds because the right-hand-side of expression (47), with $B = b$, becomes a product of four identical Gaussians, and the antisymmetrization creates all the necessary P, D, etc. harmonic oscillator wave functions automatically [26]. On the other hand, when $B \gg b$, the density of α-particles is very low, and in the limit $B \to \infty$, the average distance between α's is so large that the antisymmetrisation between them can be neglected, i.e., the operator $\mathcal{A}$ in front of Eq. (46) becomes irrelevant and can be removed. In this limiting case, our wave function then describes an ideal gas of independent, condensed α-particles – it is a pure product state of α's! An elucidating study on this aspect is given in Ref. [29].

Evidently, in realistic cases the antisymmetrizer $\mathcal{A}$ cannot be neglected, and evaluation of the expectation value (50) becomes a nontrivial task. The Hamiltonian in Eq. (50) was taken to be the one used in Ref. [30], which features an effective nucleon-nucleon force of the Gogny type, with parameters fitted to α-α scattering phase shifts as available about fifteen years ago. This force also leads to very reasonable properties of ordinary nuclear matter. Our theory is therefore free of any adjustable parameters. The energy landscapes $\mathcal{H}(B, b)$ for various $n\,\alpha$ nuclei are interesting in themselves [31], but for the sake of brevity they are not shown here.

As we discussed already, the variational wave function constructed from the Hill-Wheeler equation based on Eqs. (46), (47), and (50) has practically 100 percent overlap with the RGM and GCM wave functions constructed in Refs. [20] and [21], once the same Volkov force is used [27]. However, one can take fixed optimised values for b and B parameters and, then, the corresponding *single* THSR wave function still has 98 percent squared overlap with the RGM or GCM solutions. It is, thus, not astonishing that our results are very similar to the RGM and GCM ones. For ^{12}C we obtain two eigenvalues: the ground state and the Hoyle state. Theoretical values for positions, rms values, and transition probabilities are given in Table 1 and compared to the data. Inspecting the rms radii, we see that the Hoyle state has a volume 3 to 4 larger than that of the ground state of ^{12}C. This is the primary aspect of the dilute-gas state we highlighted above. Constructing a pure-state α-particle density matrix $\rho(\mathbf{R}, \mathbf{R}')$ from our wave function, integrating out of the total density matrix all intrinsic α-particle coordinates, and diagonalizing this reduced density matrix, we find that the corresponding $0S$ α-particle orbit is occupied to 70 percent by the three α-particles [29, 32] whereas the occupation of all other states is down by at least a factor of ten, see Fig. 8.

This is a huge percentage, giving vivid support to the view that the Hoyle state is an almost

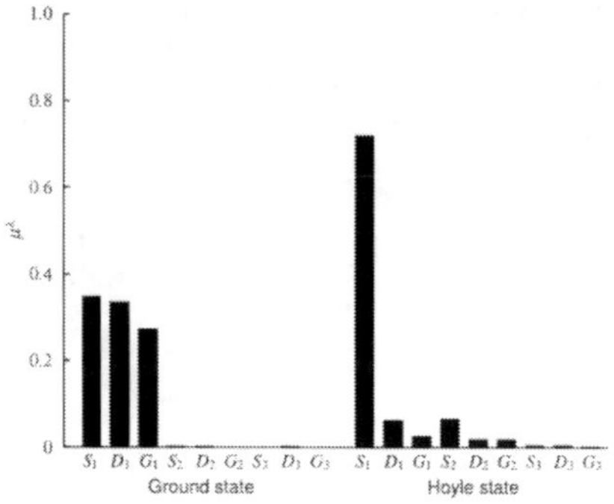

Figure 8. α occupation numbers in the ground state (left) and Hoyle state (right) of ^{12}C.

		condensate w.f. (Hill-Wheeler)	RGM [20]	Exp.
E(MeV)	0_1^+	-89.52	-89.4	-92.2
	0_2^+	-81.79	-81.7	-84.6
$R_{\text{r.m.s.}}$(fm)	0_1^+	2.40	2.40	2.44
	0_2^+	3.83	3.47	
$M(0_2^+ \to 0_1^+)(\text{fm}^2)$		6.45	6.7	5.4

Table 1. Comparison of the binding energies, rms radii ($R_{\text{r.m.s.}}$), and monopole matrix elements ($M(0_2^+ \to 0_1^+)$) for ^{12}C given by solving Hill-Wheeler equation based on Eq. (46) and by Ref. [20]. The effective two-nucleon force Volkov No. 2 was adopted in the two cases for which the 3α threshold energy is calculated to be -82.04 MeV.

ideal α-particle condensate. By way of contrast, we should also mention that in the ground state of ^{12}C, the α-particle occupation is about equally shared between the $0S$, $0D$, and $0G$ orbits, i.e. yielding the shell model limit, clearly invalidating a condensate picture of the ground state, see Fig. 8 (it is important to note that the ground-state energy of ^{12}C is also reasonably reproduced by our theory). We, thus, think that the investigation of the bosonic occupancies is the most adequate way to demonstrate whether a given state can, in nuclei, be qualified as an α condensate or not.

Let us now discuss what to our mind is the most convincing evidence that our description of the Hoyle state is the correct one. Like the authors of Ref. [20], we reproduce very accurately the inelastic form factor $0_1^+ \to 0_2^+$ of ^{12}C, as shown in Fig. 9. As such, the agreement with experiment is already quite impressive in view of the fact that we did not use a single adjustable parameter. Additionally, however, the following study was made, results from which are presented in Fig. 10. We artificially varied the extension of the Hoyle state and examined the influence on the form factor. It was found that the overall shape of the form factor shows little variation, for example in the position of the minimum. On the other hand, we found a strong dependence of the absolute magnitude of the form factor; Fig. 10 illustrates this behavior with a plot showing the variation of the height of the first maximum of the inelastic form factor as a function of the percentage change of the rms radius of the Hoyle state [27]. It can be seen that a 20 percent *increase of the rms radius* produces a remarkable decrease of the maximum – by a factor of two! This strong sensitivity of the magnitude of the form factor to the size of the Hoyle state enhances our firm belief that the agreement with the actual measurement

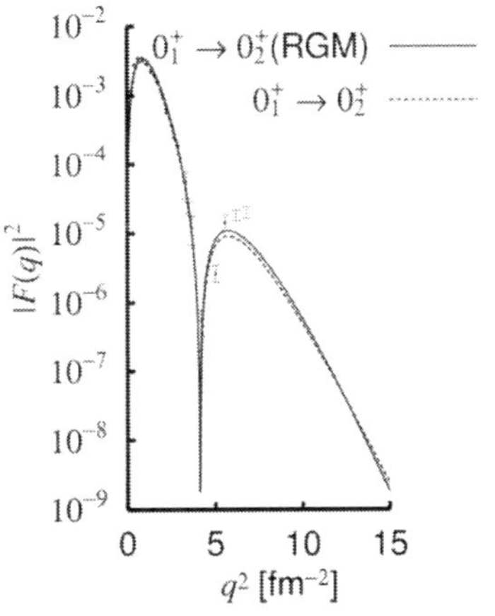

Figure 9. Experimental values of inelastic form factor in ^{12}C to the Hoyle state are compared with our values and those given by Kamimura et al. in Ref. [20] (RGM). In our result, the Hoyle-state wave function is obtained by solving the Hill-Wheeler equation based on Eq. (46).

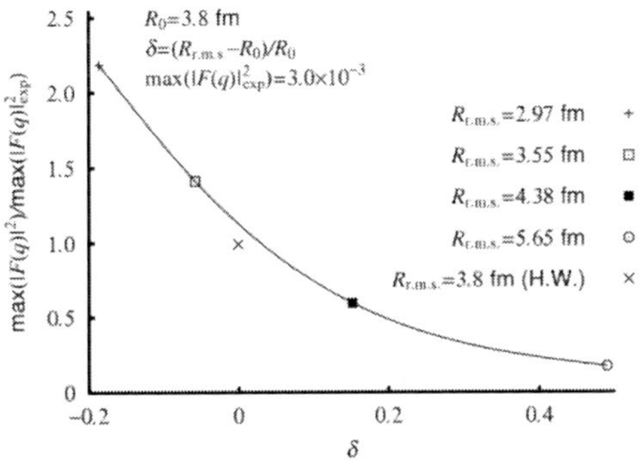

Figure 10. The ratio of the value of the maximum height, theory versus experiment, of the inelastic form factor, i.e. $\max|F(q)|^2/\max|F(q)|^2_{\rm exp}$, is plotted as a function of $\delta = (R_{\rm r.m.s.} - R_0)/R_0$. Here $R_{\rm r.m.s.}$ and R_0 are the rms radii corresponding, respectively, to the wave function of Eq. (46) and that obtained by solving the Hill-Wheeler equation based on Eq. (46).

is tantamount to a proof that the calculated wide extension of the Hoyle state corresponds to reality. We thus advocate and support the view that the Hoyle state can be regarded as the ground state of an α-particle condensate [33]. We should, however, be aware of the fact that it is not an ideal condensate and that the α's occupy the lowest S-state only with 70 percent, as already discussed. The major part of the correlations comes from the Pauli principle and intermediate ^{8}Be formation. It is by the way not clear whether a gas of α particles condenses as such or as 'molecules' of ^{8}Be. The latter are also bosons, of course. We furthermore performed a deformed calculation to investigate the structure of the 2_2^+ state in ^{12}C. The state came at the right energy. Our analysis showed that this state essentially corresponds to exciting one α-particle out of the condensate and putting it into the $0D$ orbit. Without going into details, we also affirm that the width of this state is correctly reproduced [34]. It should also be mentioned that this 2_2^+ state has in our calculations an enormous extension with an rms radius of 4.3 fm what corresponds approximately to eight times the ground state volume of ^{12}C or also to the size of ^{40}Ca. As a matter of fact the properties of this state have been subject of a vivid debate among the experimentalists in the recent past. The situation seems clarified now [35][36][37]. It is for instance, the very nice experiment by Moshe Gai [37] which confirms the properties of the 2_2^+ state beyond any doubt. In that reference it also is given an estimate of the radius which agrees with our value. One can talk about an α- halo state. Further experimental verification of this giant α-gas state would be very welcome.

It is tempting to imagine that the 0_3^+ state which – experimentally – is almost degenerate with the 2_2^+ state, is obtained by lifting one α-particle into the $1S$ orbit. Initial theoretical studies

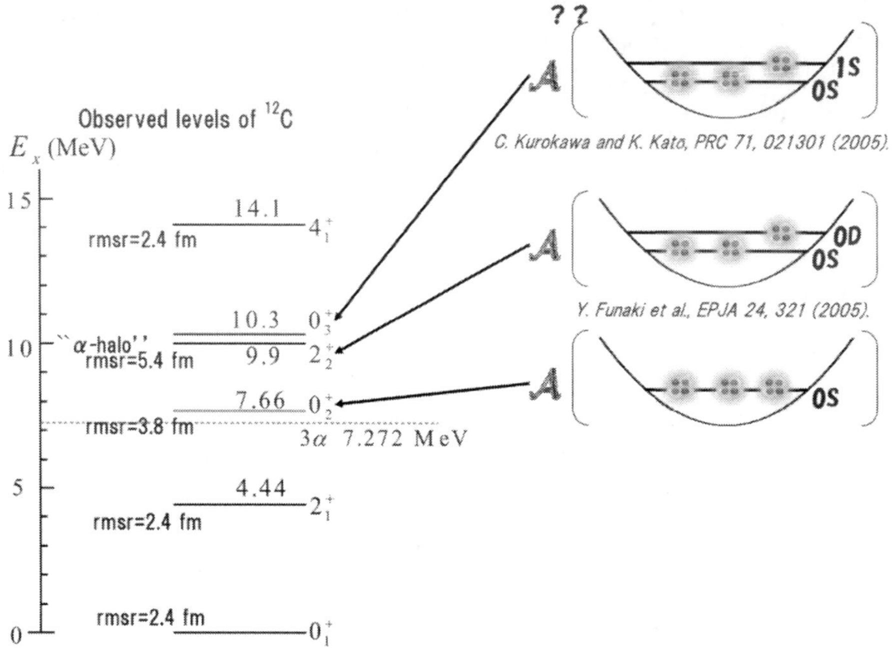

Figure 11. Spectrum of ^{12}C and its interpretation via an analysis with the THSR wave function concerning α cluster states.

[38] indicate that this scenario might indeed apply. However, the width of the 0_3^+ state ($\sim$ 3 MeV) is very broad, rendering a theoretical treatment rather delicate. Further investigations are necessary to validate or reject this picture. At any rate, it would be quite satisfying if the triplet of states ($0_2{}^+$, $2_2{}^+$, $0_3{}^+$) could all be explained from the α-particle perspective, since those three states are *precisely* the ones which cannot be reproduced within a (no core) shell-model approach [19]. In Fig. 11 we represent this scenario.

Summarizing our inquiry into the possible role of α clustering in ^{12}C, we have accumulated enough facts to be convinced that the Hoyle state is, indeed, what one may call to first approximation an α-particle condensate state. At the same time, we acknowledge that referring to only three particles as a "condensate" constitutes a certain abuse of the word. However, in this regard it should be remembered that also in the case of nuclear Cooper pairing, only a few pairs are sufficient to obtain clear signatures of superfluidity in nuclei!

What about α-particle condensation in heavier nuclei? Once one accepts the idea that the Hoyle state is essentially a state of three free α-particles held together only by the Coulomb barrier, it is hard to see why analogous states would not also exist in heavier $n\,\alpha$ nuclei like ^{16}O, ^{20}Ne, ^{24}Mg, etc. In this respect, it is important to recognize the argument that the successively higher excitation energies of α condensate states do not necessarily imply very short life times because of their very unusual structure, having little in common with ordinary nuclear states. Our calculations on such nuclei systematically yield a 0^+-state close to the α-particle disintegration threshold. For example in ^{16}O we obtain three 0^+-states [26]: the ground state at $E_0 = -124.8$ MeV (experimental value: 127.62 MeV), a second state at excitation energy $E_{0_2{}^+} = 8.8$ MeV, and a third one at $E_{0_3{}^+} = 14.1$ MeV. The threshold in ^{16}O is at 14.4 MeV. Unfortunately, the relevant experimental information in ^{16}O is not nearly so complete as in ^{12}C. In particular, no measurements are available for transition probabilities of 0^+-states near the threshold or for inelastic form factors.

In contrast to the situation for ^{12}C, the THSR wave function is certainly not able to describe the structure of all 0^+-states in ^{16}O lying below the disintegration threshold. In ^{12}C knocking loose one α particle, the other two are also loosely bound, since what remains is ^{8}Be. However, in ^{16}O this is not the case. Before reaching a four α particle gas state, there will appear configurations where one α particle orbits around a ^{12}C core in its ground state or in excited states of particle- hole type. A case in point is the first excited state in ^{16}O, i.e., the $0_2{}^+$-state at 6.06 MeV, which is believed to have a structure corresponding to an α-particle orbiting in an S wave around a ^{12}C core in its ground state. Such a configuration is clearly missing from our wave function (46). As a matter of fact a calculation for the first six 0^+ states has been performed in the meanwhile employing a somewhat different, not completely microscopic approach [39]. This method is the so-called 'orthogonality condition model (OCM)' which generally works also quite well for the description of cluster states. The novelty with respect to former calculations of ^{16}O states with this method was that the configuration space was strongly enlarged. In Fig. 12, we show the comparison of the calculated with the experimental sprectrum. In view of the fact that the states are complicated cluster states, the agreement between experiment and theory is very good. The interpretation goes as follows: the first four excited 0^+ states have a ^{12}C + α structure where the α orbits in 0S, 0D, 1S waves around the ground state core of ^{12}C and also in a 0P wave orbiting around the first 1^- of ^{12}C. It is only the first state above the four α particle threshold at 15.1 MeV which is interpreted as an α gas state or a condensate. This state has, indeed, some analogies with the Hoyle state: it is just some hundreds of keV above threshold, it is strongly excited by inelastic electron scattering what means that monopole transition is quite large. Unfortunately the inelastic form factor has not been measured so far.

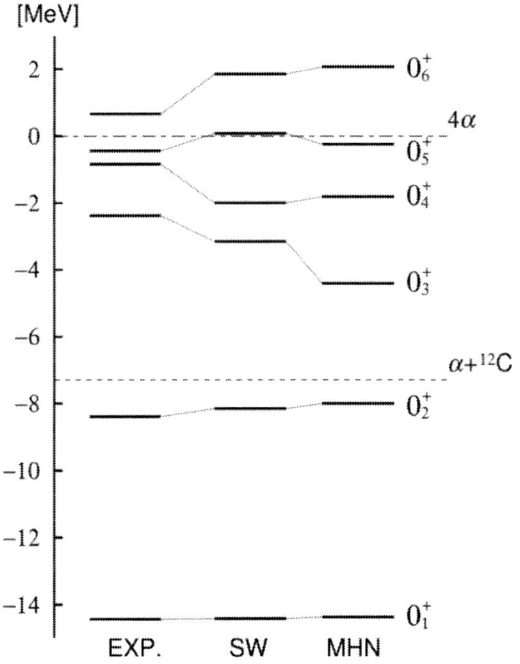

Figure 12. Spectrum of first six 0^+ states in ^{16}O. Left: experiment; middle: theory with SW force; right: theory with MHN force; see [39] for more details.

One interesting question that can be asked at this point is: How many α's can maximally exist in a self-bound α-gas state? Seeking an answer, we performed a schematic investigation using an effective α-α interaction of the Ali-Bodmer form [40] within an α-gas mean-field calculation of the Gross-Pitaevskii type [41]. The parameters of the force were slightly adjusted to reproduce

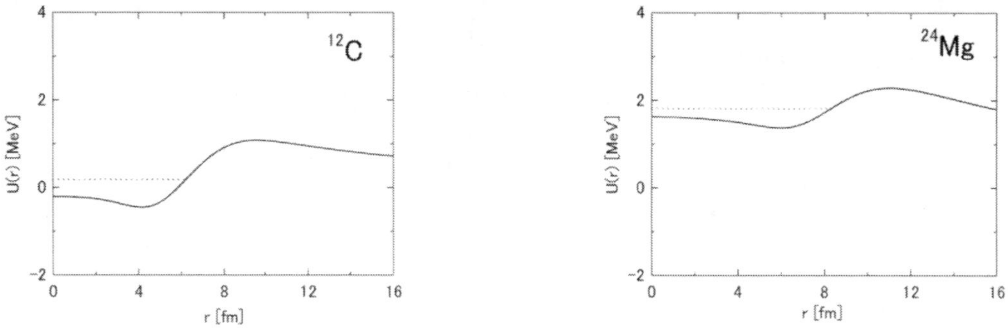

Figure 13. Alpha-particle mean-field potential for three α's in ^{12}C and six α's in ^{24}Mg. Note the lower Coulomb barrier for ^{24}Mg (from Ref. [42]).

our microscopic results for ^{12}C. The corresponding α mean-field potential is shown in Fig. 5. One sees the $0S$-state lying slightly above threshold but below the Coulomb barrier. As more α-particles are added, the Coulomb repulsion drives the loosely bound system of α-particles farther and farther apart, so that the Coulomb barrier fades away. According to our estimate [42], a maximum of eight to ten α-particles can be held together in a condensate. However, there may be ways to lend additional stability to such systems. We know that in the case of ^{8}Be, adding one or two neutrons produces extra binding without seriously disturbing the pronounced α-cluster structure. Therefore, one has reason to speculate that adding a few of neutrons to a many-α state may stabilize the condensate. But again, state-of-the-art microscopic investigations are necessary before anything definite can be said about how extra neutrons will influence an α-particle condensate.

Another interesting idea concerning α-particle condensates was put forward by von Oertzen and collaborators [43, 44]. Adding more and more α-particles to the, e.g., ^{40}Ca core, one will arrive sooner or later at the point of α-particle drip. Therefore minimal further excitation may be sufficient to shake loose some α-particles, so that an $n\,\alpha$-condensate could be created on top of an inert ^{40}Ca core. Similar ideas also have been advanced by Ogloblin [45], who envisions a three-α-particle condensate on top of ^{100}Sn, and earlier by Brenner and Gridnev, who have presented evidence of experimental detection of gaseous α-particles in ^{28}Si and ^{32}S on top of an inert ^{16}O core [46].

Interesting new theoretical developments generalizing the THSR wave function [26] are presently going on, putting nuclear cluster physics on a completely new basis [47]. For example, it was found that the rotational parity doublet ground state bands in ^{20}Ne can be described with a slightly generalised THSR wave function. Also the intriguing question about the intrinsic cluster structure inherent to the THSR wave function is further elucidated in [47]. In this respect it is also interesting that Hartree-Fock-Bogoliubov calculations for expanding $n\alpha$ nuclei show clusterisation into geometrical arrangements of the α's, as, e.g., a tetrahedron for ^{16}O [48].

6. Conclusions, Discussion, Outlook

We have investigated the role that pairing and multiparticle correlations may play in nuclear matter existing in dense astrophysical objects and in finite nuclei. A complete and quantitative description of nuclear matter must allow for the presence of clusters of nucleons, bound or metastable, possibly forming a quantum condensate. In particular, quartetting correlations, responsible for the emergence of α-like clusters, are identified as uniquely important in determining the behavior of nuclear matter in the limiting regime of low density and low temperature. We have calculated the transition temperature for the onset of quantum condensates made up of α-like and deuteron-like bosonic clusters, and considered in considerable detail the intriguing example of Bose-Einstein condensation of α particles. It turns out that contrary to pairing, quartet condensation primarily exists in the BEC phase at low density. In which way quartet condensation is lost by increasing the density is still an open question. It may be similar to a liquid-gas phase transition. Anyway, it is clear that there can not exist a condensate of quartets with a long coherence length for arbitrarily small attraction as this is the case for pairing in the BCS phase. Arguments for this based on many fermion level densities have been presented. It is inevitable that under increasing density or pressure, the bound α, d, or other nuclidic clusters, present at low density, experience significant modification due to the background medium (and eventually merge with it). We have shown how self-energy corrections and Pauli blocking alter the properties of cluster states, and we have formulated a cluster mean-field approximation to provide an initial description of this process. One result of special interest is the suppression of the α-like condensate, which is dominant at lower densities, as the density reaches and exceeds the Mott value, allowing the pairing transition to occur. Even at lower densities α-particle condensation may be influenced by neutron excess, i.e. in the case of asymmetric nuclear matter, see Fig. 3. A genuine theory for the quartet order parameter in homogeneous infinite matter, similar to BCS theory, is demanded. First results in this direction have been published [14]. A theory for quartet condensation on firmer grounds which parallels the one of pairing is presented in this contribution in section 2.

A truly remarkable manifestation of α-particle condensation seems to be present in finite nuclei. Indeed, the so-called Hoyle state ($0_2{}^+$) in ^{12}C at 7.654 MeV is very likely a dilute gas of three α-particles, held together only by the Coulomb barrier. This view is encouraged by the fact that we can explain all the experimental data in terms of a conceptually simple wave function of the quartet-condensate type. Within the same model, we also systematically predict such states in heavier $n\,\alpha$ nuclei, and the search is on for their experimental identification. With the more phenomenological OCM method, we found that in ^{16}O the sixth 0^+ state at 15.1 MeV should be the candidate for an α condensed state. It is quite natural that such states should exist up to some maximum number of α particles inspite of their increasing excitation energy: these α-gas states are almost orhtogonal to the rest of nuclear states so that decay is strongly hindered. We estimate that the phenomenon will terminate at about eight to ten α's as the confining Coulomb barrier fades away. However, there is the possibility that larger condensates could be stabilized by addition of a few neutrons. Indeed, consider ^{9}Be, which, contrary to ^{8}Be, is bound by $\sim$ 1.5 MeV, still showing a pronounced two α-structure similar to the one of Fig. 7 (b). One could imagine ten α's or more, stabilised by two or four extra neutrons in a low density phase. However, even without being stabilised, if a compressed hot nuclear blob as e.g. produced in a central Heavy Ion collision expands and cools, it may turn on its way out, at a certain low density, into an expanding α condensed state where all α's are in relative S-waves. One may also, in one way or the other (photons?) excite, e.g., ^{40}Ca to the 10 α threshold at about 60 MeV where a slow Coulomb explosion of an α gas would then take place. This would be an analogous situation to an expanding Bose condensate of atoms, once the trapping potential has been switched off. Future dedicated experiments with high resolution multiparticle detectors will tell whether such scenarios can be realised. Intriguing news in this respect come from GANIL where one may have achieved the disintegration of ^{56}Ni into 14 α's (seven α's have been detected with high yield, the other seven may not have been seen because of the detectors were not sensitive to very low energy α particles [49][50]). Other possibilities of loose α-gas states may exist on top of particularly stable cores, like ^{16}O or ^{40}Ca. Indeed in adding α's to e.g. ^{40}Ca, one will reach the α-particle drip line. Compound states of heavy N = Z nuclei of this kind may be produced in heavy ion reactions and an enhanced α-decay rate may reveal the existence of an α-particle condensate. Ideas of this type have been promoted by von Oertzen

[44], and also M. Brenner [46], and A. Ogloblin [45]. However, coincidence measurements of multiple α's of decaying lighter nuclei like ^{16}O may also be very useful [51][52] [53] to detect at least one additional α- condensate state beyond the only one that has been identified so far, namely the $0_2{}^+$-state in ^{12}C.

Another issue which may be raised in the context of α-particle condensation is the question, also discussed in condensed matter physics [54], whether α's condense as singles or as doubles, i.e. as ^{8}Be. In microscopic studies of ^{12}C one, indeed, can see that in the $0_2{}^+$-state two of the three α's are slightly more closer to one another than to the third one [55]. The question is definitely very interesting and desserves future investigation, for instance in what concerns the identification of the α structure of the Hoyle state with the THSR wave function in the intrinsic frame. However, quantitatively, this constitutes probably only a slight modification over the present formulation of α-condensation.

Very recent triple α coincidence experiments [56] have shown that there exists a decay channel of the Hoyle state where the three α's share democratically the available energy, that is, each α carries away one third of the initial energy. This very rare three body decay may give further credit to the idea that the three α particles occupy the same 0S-orbit , that is, they are condensed in this state. Same results have been confirmed by a second group [57] and earlier, with more yield, with heavy ion reactions [58].

What about 'ab initio' calculations of the Hoyle state? Presently several groups are on the track [59][60]. The Los Alamos group has achieved to calculate the density of the Hoyle state, though it is not yet converged in the far tail [60]. It agrees very well with the density from the THSR wave function besides in the far tail.

In conclusion, we see that the idea of α-particle condensation in nuclei and nuclear systems has triggered many new ideas , calculations, and experiments, in spite of the fact that, so far, a compelling case for such a state has only been made in ^{12}C. Even so, the possible existence of a completely new nuclear phase in which α-particles play the role of quasi-elementary constituents is surely fascinating. Hopefully, many more α-particle states of nuclei will be detected in the near future, bringing deeper insights into the role of clustering and quantum condensates in systems of strongly interacting fermions.

Let us mention in the end that a more elaborate report on α particle condensation can be found in [61]

7. Acknowledgements

We thank T. Sogo for contributions. Useful discussions and a careful reading of the mauscript by M. Tohyama are greatfully achknowledged. Discussions with B. Zhou and Z. Ren have been appreciated.

References

[1] Moulopoulos K Ashcroft N W 1991 *Phys. Rev. Lett.***66** 2915
[2] Shapiro L S and Teukolsky S A 1983 *Black holes, white Dwarfs and Neutron Stars: The Physics of Compact Objects* (Wiley, N.Y.); Pines D Tamagaki R and Tsuruta S (eds.) 1992 *Neutron Stars* (Addison-Wesley, N.Y.)
[3] Shen H Toki H Oyamatsu K and Sumiyoshi K 1998 *Progr. Theor. Phys.* **100** 1013
[4] Salomon Ch Zwierlein M W 2012 private communications.
[5] Ring P Schuck P 1980 *The Nuclear Many-Body Problem*(Springer)
[6] Jemai M Delion D S Schuck P 2013 *Phys. Rev.* C **88** 044004
[7] Blaizot J-P Ripka G 1985 *Quantum Theory of Finite Systems* (The MIT Press, Cambridge, Massachusetts)
[8] Dukelsky J Roepke G Schuck P 1998 *Nucl. Phys.* A **628** 17
[9] Fetter A L Walecka J D 2003 *Quantum Theory of Many-Particle Systems* (Dover Publications, Mineaola, New York)

[10] Beyer M Sofianos S A Kurths C Röpke G and Schuck P 2000 *Phys. Lett.* B **478** 86
[11] Sogo T Lazauskas R Röpke G Schuck P 2009 *Phys. Rev.* C **79** 051301
[12] Röpke G Schnell A Schuck P and Nozières P 1998 *Phys. Rev. Lett.* **80** 3177
[13] Sogo T Röpke G Schuck P 2010 *Phys. Rev.* C **82** 034322
[14] Sogo T Röpke G Schuck P 2010 *Phys. Rev.* C **81** 064310
[15] Nozières P Schmitt-Rink S 1985 *J. Low Temp. Phys.* **59** 195
[16] Dalfovo L Giorgini S Pitaevskii L D Stringari S 1999 *Rev. Mod. Phys.* **71** 463
[17] Hoyle F Dunbar D N F Wenzel W A Whaling W 1953 *Phys. Rev.* **92** 1095
[18] Cook C W Fowler W A Lauritsen C C Lauritsen T B 1957 *Phys. Rev.* **107** 508
[19] Navrátil P Vary J P and Barrett B R 2000 *Phys. Rev. Lett.* **84** 5728; Navrátil P Vary J P and Barrett B R 2000 *Phys. Rev.* **C 62** 054311; Barrett B R Mihaila B Pieper S C and Wiringa R B 2003 *Nucl. Phys. News* **13** 17
[20] Fukushima Y and Kamimura M 1978 *Proc. Int. Conf. on Nuclear Structure* Ed. Marumori T *Suppl J. Phys. Soc. Japan* **44** 225; Kamimura M 1981 *Nucl. Phys.* A **351** 456
[21] Uegaki E Okabe S Abe Y and Tanaka H 1977 *Prog. Theor. Phys.* **57** 1262; Uegaki E Abe Y Okabe S and Tanaka H 1978 *Prog. Theor. Phys.* **59** 1031; 1979 **62**, 1621
[22] Wiringa R B S. Pieper C Carlson J and Pandharipande V R 2000 *Phys. Rev.* C **62** 014001
[23] Horiuchi H 1974 *Prog. Theor. Phys.* **51** 1266; 1975 **53** 447
[24] Saito S 1968 *Prog. Theor. Phys.* **40**; 1969 **41** 705 1977 *Prog. Theor. Phys. Suppl.* **62** 11
[25] Morinaga H 1956 *Phys. Rev.* **101** 254 1966 *Phys. Lett.* **21** 78
[26] Tohsaki A Horiuchi H Schuck P and Röpke G 2001 *Phys. Rev. Lett.* **87** 192501
[27] Funaki Y Tohsaki A Horiuchi H Schuck P and Röpke G 2003 *Phys. Rev.* C **67** 051306(R)
[28] A very early attempt to consider for ^{12}C a three α-wavefunction of somewhat similar kind as in Ref. [26], but without the condensate aspect and no RGM solution, has been undertaken in: Hutzelmeyer H, Hackenbroich H H 1970 *Z. Phys.* **232** 356; see also discussion (p. 80) in *Clustering Phenomena in Nuclei* Wildermuth K Kramer P editors 1977 Vol. **1**; Wildermuth K Tang Y C 1977 *A Unified Theory of the Nucleus* Vieweg
[29] Yamada T and Schuck P 2005 *Eur. Phys. J.* **A 26** 185
[30] Tohsaki A 1994 *Phys. Rev.* **C 49** 1814
[31] Tohsaki A Horiuchi H Schuck P and Röpke G 2004 *Proc. of the 8th Int. Conf. on Clustering Aspects of Nuclear Structure and Dynamics*, Nara, Japan, 2003, ed. Ikeda K Tanihata I and Horiuchi H (*Nucl. Phys.* **A 738** 259
[32] Matsumura H and Suzuki Y 2004 *Nucl. Phys.* **A 739** 238
[33] The fact that this condensate state is actually an excited state of ^{12}C does not invalidate this picture, since, on nuclear scales, this state is long lived. Actually, also atomic Bose condensates in traps find themselves in meta-stable states, see e.g. [16].
[34] Funaki Y Horiuchi H Tohsaki A Schuck P and Röpke G 2005 *Eur. Phys. J.* **A 24** 321
[35] Itoh M *et al.* 2011 *Phys. Rev.* C **84** 054308
[36] Zimmerman W R *et al.* 2011 *Phys. Rev.* C **84** 027304
[37] Gai M for the UConn-Yale-Duke-Weizmann-PTB-UCL Collaboration 2011 J. Phys. G **267** 012046;for the UConn-Yale-Duke-Weizmann-PTB-UCL Collaboration 2011 Acta Phys. Pol. B **42** 775
[38] Kurakowa C and Kato K 2005*Phys. Rev.* C **71** 021301
[39] Funaki Y Yamada T Horiuchi H Röpke G Schuck P Tohsaki A 2008 *Phys. Rev. Lett.* **101** 082502
[40] Ali S Bodmer A R 1966 *Nucl. Phys.* **80** 99
[41] Pitaevskii L P 1961 *Zh. Eksp. Teor. Fiz.* **40** 646 [1961 *Sov. Phys. JETP* **13** 451]; Gross E P 1961 *Nuovo Cimento* **20** 454; 1963 *J. Math. Phys.* **4** 195
[42] Yamada T and Schuck P 2004 *Phys. Rev.* **C 69** 024309
[43] Kokalova Tz Itagaki N von Oertzen W and Wheldon C 2006 *Phys. Rev. Lett.* **96** 192502
[44] von Oertzen W et al. 2006 *Eur. Phys. J. A* **29** 133
[45] Ogloblin A A et al. 2005 *Proceedings of the International Nuclear Physics Conference, Peterhof, Russia, June 28-July 2*
[46] Brenner M W et al. 2000 *Proceedings of the International Conference "Clustering Phenomena in Nuclear Physics"* St. Petersburg, published in Physics of Atomic Nuclei (Yadernaya Fizika)
[47] Zhou Bo *et al.* 2013 *Phys. Rev. Lett.* **110** 262501 and arXiv:1312.1047
[48] Girod M Schuck P 2013 *Phys. Rev. Lett.* **111** 132503
[49] Harakeh M N 2012 private communication
[50] Akimune H *et al.* 2012 *Journal of Physics:Conference Series* Vol. **436** 012010
[51] Freer M et al. 2004 *Phys. Rev. C* **70** 064311
[52] Raduta Ad R *et al.* 2011 *Phys. Lett.* B **705** 65

Journal of Physics: Conference Series **529** (2014) 012014 doi:10.1088/1742-6596/529/1/012014

[53] Zarubin P I Artemenkov D A Orlova G I Cechak T et al. (eds.) 2006 *Nuclear Science and Safety in Europe* 189 (Springer-Verlag, Berlin)
[54] Nozières P Saint James D 1982 *J. Physique* **43** 1133
[55] Feldmeier H Neff T 2005 *Physik Journal* **4**, Nr. **1**, 29 www.physik-journal.de
[56] Rana T K *et al.* 2013 *Phys. Rev.* C **88** 021601
[57] Itoh M private communication
[58] Raduta A R *et al.* 2011 *Phys. Lett.* B **705** 65
[59] Epelbaum E Krebs H Lee D Meissner U G 2011 *Phys. Rev. Lett.* **106** 192501
[60] Wiringa B R 2012 Argonne Nat. Lab. report
[61] Yamada T Funaki Y Horiuchi H Röpke G Schuck P Tohsaki A 2011 arXiv:1103.3940, "Clusters in Nuclei" *Lecture Notes in Physics* Vol.2 Ed. by C. Beck (Springer Verlag, Berlin)

Journal of Physics: Conference Series **529** (2014) 012015 doi:10.1088/1742-6596/529/1/012015

Dissipation in quantum time dependent mean field

N Slama[1,2], P-M Dinh[1], P-G Reinhard[3], E Suraud[1,2,4]

1 Laboratoire de Physique Théorique (IRSAMC), CNRS, F-31062 Toulouse Cedex, France
2 Laboratoire de Physique Théorique (IRSAMC), Université de Toulouse, F-31062 Toulouse Cedex, France
3 Institut für Theoretische Physik II, Universität Erlangen-Nürnberg, Staudstr. 7, D-91058 Erlangen, Germany
4 Physics Department, University at Buffalo, The State University New York, Buffalo, NY 14260, USA

E-mail: suraud@irsamc.ups-tlse.fr

Abstract. We discuss an extension of Time dependent (TD) mean field theories such as TD Hartree Fock (TDHF) or TD Kohn-Sham (TDKS) in order to include dissipative effects as observed in many experimental situations. Applications to molecular problems are outlined. We present in particular a 1D model allowing to test a promising approach to include dissipative features in a quantum time dependent mean field. First results are presented in schematic cases.

1. Introduction

Mean field theory, and its time dependent (TD) extension, constitutes a sound starting basis for the description of numerous dynamical situations in a variety of many-fermion systems ranging from nuclei [1] to molecules and clusters [2]. Mean field theories in these systems have developed to a high level of sophistication, in particular in its stationary version, but dynamical versions thereof also attracted a lot of efforts especially for the description of low energy phenomena. In the nuclear context the keyword is Hartree Fock (HF) and its time dependent extension TDHF. Nuclear TDHF calculations have, to a large extent, been developed around effective density dependent functionals, typically Skyrme interactions, since several decades and have reached a high level of sophistication [3]. In clusters and molecules the underlying effective theory is Density Functional Theory (DFT) [4] and its Time Dependent extension TDDFT. The simplest mean field approximation thereof is known as Local Density Approximation (LDA and TDLDA), which again relies on a density dependent effective Hamiltonian [4]. Both nuclear TDHF and electronic TDLDA are thus using single particle density as a key input and thus bear strong formal resemblances, which actually reflect physical similarities in several observables [5]. One can mention as a typical example collective motion, in particular the Giant Dipole Resonance in nuclei [1] and its counterpart the Mie plasmon (or more generally speaking the optical response) in metal clusters [2].

Mean field theory constitutes a robust and sound basis for the description of many dynamical situations, but it fails to properly address dynamics beyond the linear regime, because it does not include dynamical correlations. Typical situations requiring such extensions are heavy-ion collisions in the Fermi energy range [6] or irradiation of clusters and molecules by intense laser fields [7, 8]. In spite of long standing efforts [9], the inclusion of dissipative effects via dynamical correlations in a quantum mean field theory remains to a large extent an open issue. Several semi-classical approximations were developed, especially in the nuclear case over the last decades [10, 5], and more recently in the case of clusters [11]. However these latter approaches are limited to cases where a (semi) classical approximation is justified, or, in other words, to situations in which quantum shell effects can be safely neglected, typically when excitation energies are large enough, which excludes numerous physically relevant situations. Although not fully satisfying from the formal point of view, and in spite of some failures, the use of such semi classical approximations is probably acceptable in the case of heavy ion collisions in the Fermi energy domain [6, 5]. The situation is quite different for most molecular situations where quantum shell effects may play a leading role. The interest in these systems has been recently renewed following the accumulation of experimental evidence via imagining techniques, in particular in the case of irradiation of clusters and molecules by intense laser beams [8]. There is thus a recent renewed interest in this question of including dynamical correlations in quantum mean field, especially in the TDDFT community [12].

In the following, we propose a stochastic extension of TDHF/TDLDA which should allow to envision on short term the inclusion of dissipative effects in a quantum mechanical framework, and applications to realistic test cases both in the nuclear and the molecular contexts. After a brief reminder of the original theory, we outline a simpler form allowing test cases in model systems and outline the next steps to be performed along that line.

2. Theoretical framework

The formal starting point is Stochastic Time-Dependent Hartree-Fock (STDHF) originally formulated in nuclear dynamics [13], but which can be used in any systems where mean-field theory provides a sound description of low energy phenomena, as outlined above. STDHF proposes a description of dynamical correlations in terms of a stochastic ensemble of TDHF (mean-field) states. Originally, STDHF was formulated in terms of density matrices on the basis of the quantum Liouville equation [13]. STDHF can be reduced to a quantum kinetic equation (even accounting for fluctuations) and thus contains all ingredients for describing the path to thermalization and the thermal fluctuations around, as discussed in [13, 14]. But it has never been turned into a practical scheme. The aim of this paper is to propose a reformulation of STDHF, reducing correlations to 2 particles-2 holes ($2ph$) excitations, which allows a practical implementation of STDHF at acceptable expense, as will be illustrated on one example.

The basic idea of STDHF is to build correlations perturbatively around TDHF/TDLDA trajectories. Starting from an initial time $t_0 = 0$ we expand the time dependent correlated state $|\psi(t)\rangle$ into n-particle-n-hole (nph) excitations around the underlying Slater state $|\Phi\rangle$:

$$|\psi(t)\rangle = |\Phi\rangle + \sum_{pp'hh'} c_{pp'hh'} \hat{a}_p^\dagger \hat{a}_{p'}^\dagger \hat{a}_{h'} \hat{a}_h |\Phi\rangle + ... \tag{1}$$

where all ingredients on r.h.s. ($|\Phi\rangle$, $\hat{a}_p^\dagger$, $\hat{a}_h$, c) depend explicitly on time t. The nph basis has the great advantage that all states in the expansion (1) are ortho-normalized. The first non vanishing term of the expansion is the $2ph$ excitation around mean field, as the $1ph$ excitation is already included in the TDHF propagation. Assuming that the $2ph$ excitations indeed correspond to the leading contribution we truncate the expansion to this first ($2ph$) term. There remains to compute the $2ph$ expansion coefficients $c_{pp'hh'}$ which account for the dynamical correlations beyond the mean field state built up during time evolution.

To that end, we consider not too large time intervals τ so that we can describe the built-up of correlations in time dependent perturbation theory. Following standard steps of derivation we finally obtain transition probabilities in terms of Fermi Golden rule as

$$|c_{pp'hh'}(\tau)|^2 \simeq \tau\delta(\varepsilon_{p'} + \varepsilon_p - \varepsilon_{h'} - \varepsilon_h) |\langle\Phi_{pp'hh'}|\hat{V}(\tau)|\Phi\rangle|^2 \tag{2}$$

where $|\Phi_{pp'hh'}\rangle$ denotes one particular $2ph$ excitation on top of $|\Phi\rangle$. The single particle (sp) energies ε_j are taken here as the expectation values $\varepsilon_j = \langle\varphi_j(t)|\hat{h}(t)|\varphi_j(t)\rangle$ where $\hat{h}(t)$ is the instantaneous mean field associated to state $|\Phi\rangle$.

Equation (2) can be read as the probability of transition from the state $|\Phi\rangle$ to the $2ph$ state $|\Phi_{pp'hh'}\rangle$ due to the residual interaction $\hat{V}$ not accounted for in the mean field. In fact, it is the probability accumulated during an interval τ of propagation. The δ function plays the role of a energy selection rule, where only $2ph$-transitions fulfilling (up to some small uncertainty) energy conservation are allowed, and give nonzero transition probabilities.

Practically speaking we propagate an ensemble of Slater states according to TDHF/TDLDA and occasionally (every τ) evaluate jump probabilities. At each time jump probabilities are evaluated we sample the thus computed jump probabilities and choose a new state of the system according to the sampling (namely a $2ph$ excited state) or keep the actual state, depending on the probability sampling. The process is repeated in time for each member of the ensemble and this produces *one* realization of STDHF (one event). The whole procedure is then repeated for each member of the ensemble thus producing an ensemble of STDHF events, the ensemble itself representing the correlated evolution of the system. The ensemble then provides the necessary content to compute relevant observables as is illustrated below.

3. Results

For simplicity, we work on a 1D dimensional system mocking up the scales of a typical molecular system (actually close to metallic). The associated mean field Hamiltonian in x representation is given by

$$\hat{h}_N = -\frac{\Delta}{2m} + V_{\text{conf}}(x) + \kappa\rho(x)^\alpha \tag{3}$$

where κ and α are parameters parametrizing the consistent component of the mean field (here we have chosen $\kappa = 3$ and $\alpha = 1$). The confining potential V_{conf} is chosen harmonic with frequency $\hbar w = 0.06Ry$ which delivers a rather dense spectrum with a large density of states thus a priori allowing numerous $2ph$ transitions even in a 1D system. We choose as residual interaction for two-body collisions a simple zero-range force

$$V_N(x,x') = \delta(x-x')V_0 \quad . \tag{4}$$

with $V_0 = 3Ry$ and reflecting boundary conditions. Starting from the ground state of the system we inject excitation energy E^* at initial time by applying some multi particle-hole excitation. The thus deposited excitation energy serves as seed for the branching of dynamical correlations which are evaluated each $\tau = 0.5fs$. We checked, by varying τ from half to double value, that the detailed choice of τ has minor effect. We propagate simultaneously 10 occupied wave functions (which constitute the actual Slater states) and 17 empty states which provide the reservoir for evaluating $2ph$ transitions. The examples presented below gather about 50 events.

As an illustrative example of how STDHF events do evolve in time we plot in Figure 1 the time evolution of single particle energies following an initial $3ph$ excitation delivering an excitation energy of a bit less than 2 Ry. The figure nicely shows how $2ph$ transitions do occur in the course of time. Of course, the figure is only illustrative to the extent that other events will deliver another sequence of $2ph$ transitions. But on the average the total number of $2ph$ transitions is a rather robust quantity (at least for such an excitation energy) and only varies little form one event to the next. The dense spectrum in that case allows a sizable number of such transitions and thus ensures an efficient dissipative scenario.

A key quantity to analyze dissipation and mixing are occupation numbers. This quantity can be evaluated from the ensemble of STDHF events via the one body density matrix $\hat{\rho}$ constructed

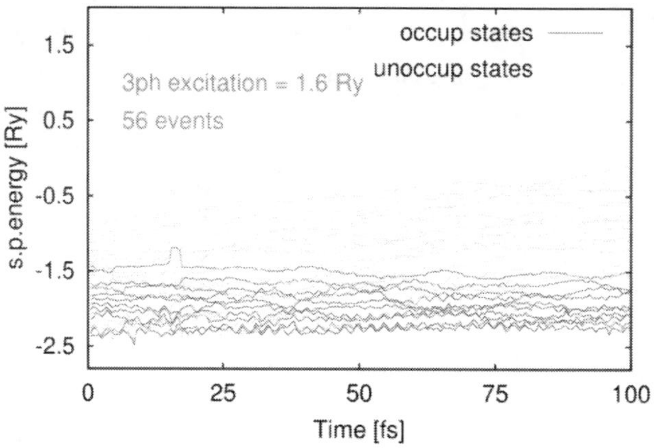

Figure 1: Time evolution of sp energies for one particular event. The excitation energy is 1.6 Ry in that case.

as the average of the one body density matrice of each event $\hat{\rho}_i$ as :

$$\hat{\rho} = \frac{1}{\mathcal{N}} \sum_{i=1}^{\mathcal{N}} \hat{\rho}_i \tag{5}$$

where $\mathcal{N}$ is the total number of events. The one body density matrix for a given event i is obtained by summing over the occupied states as

$$\hat{\rho}_i(x, x') = \sum_{\alpha=1}^{N} \varphi^*_\alpha(x')\varphi_\alpha(x) \tag{6}$$

where N represents the total number of occupied states. The density matrix (5), thus constructed, is then diagonalized and its eigenvalues provide the occupation numbers $n(\varepsilon)$ which can be plotted as a function of time to analyze the path to thermalization.

The time evolution of occupation numbers is illustrated in Figure 2 in the case of an initial $3ph$ excitation delivering $E^* = 1.6$ Ry. In the early times the figure nicely exhibits traces of the initial excitation with spiky structures. These structures are progressively smoothed as a consequence of the relaxation process. The asymptotic state finally exhibits a smooth profile resembling a Fermi distribution

$$n(\varepsilon) = \frac{1}{\exp(\frac{\varepsilon - \varepsilon_F}{k_B T}) + 1} \tag{7}$$

which signals a thermal equilibrium state. We have fitted the distribution finally reached in the STDHF propagation to the equilibrium profile (7), see figure 3. The agreement is satisfying in view of the small system under consideration. The fit allows to extract a temperature. In that it is $k_BT =\sim 0.17Ry$, which represents a significant fraction of the actual Fermi energy when measured from bottom of the potential well. We thus have clearly reached a thermal state. This statement can be corroborated by looking at the time evolution of the sp entropy (not shown here). It is found that the entropy increases steadily and levels of asymptotically with an exponential convergence rate.

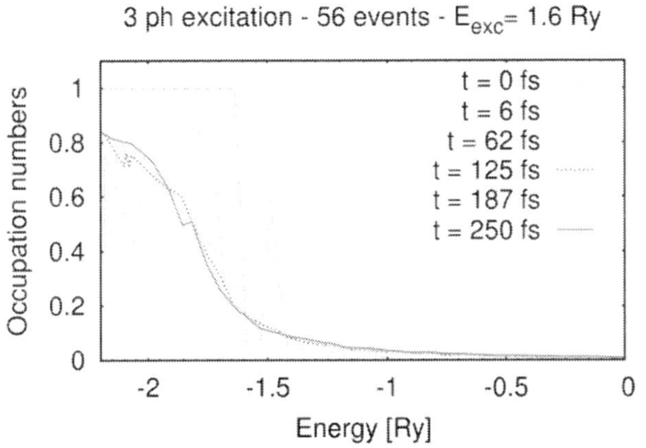

Figure 2: Snapshots of occupation numbers $n(\varepsilon)$ as a function of sp energy ε at selected times for 1.6 Ry excitation energy.

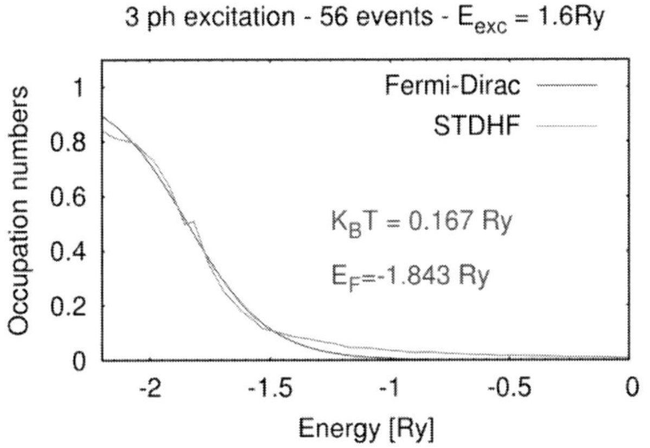

Figure 3: Fit of the final distribution of the example in figure 2, see red line, to a Fermi distribution (7), see blue line.

4. Conclusions and perspectives

We have introduced a stochastic approach to include dissipation (and fluctuations) on top of time dependent mean field theories. This approach treats correlations in time dependent perturbation theory on short time. Dynamical correlations are then sampled to provide an ensemble description of the dynamics, in terms of pure mean field states. We have illustrated this approach on a simple 1D model mocking up typical molecular situations. The first results obtained in this framework show that the adopted strategy is very promising. In spite of the 1D restriction the density of states is large enough to allow a significant number of $2ph$ transitions which are at the source of our dissipative mechanism. The system is excited at initial time by a random multi ph excitation which immediately deposits a (possibly large) excitation energy. The correlated one body density matrix obtained as the ensemble average of pure mean field density matrices displays the expected features of a system driven towards thermal equilibrium as expected in the cases explored in this study. This exploratory case raises a large number of questions. They concern for example analysis tools to be further developed in particular single particle entropy and associated relaxation times towards equilibrium, correlated two body density matrix..., to mention a few. It also concerns some technical issues on the preservation of conservation laws (total energy, centre of mass...) which are presently only preserved in average. They also concern formal developments to be performed in particular a reformulation of the theory in terms of mixed states to allow the treatment of low energy cases where the stochastic sampling reaches its limits in terms of statistics. These various aspects are presently being considered.

Acknowledgments

This work was supported by Institut Universitaire de France, Agence Nationale de la Recherche and Humboldt Gesellschaft.

References

[1] Ring P and Schuck P 1980 *The Nuclear Many-Body Problem* (Berlin: Springer)
[2] Reinhard P-G and Suraud E 2004 *Introduction to Cluster Dynamics* (Weinheim: Wiley-VCH Verlag)
[3] Bender M, Heenen P-H and Reinhard P-G 2003 *Rev. Mod. Phys.* **75** 121
[4] Dreizler R-M and Gross E-K-U 1990 *Density Functional Theory* (Berlin: Springer)
[5] Navarro J, Reinhard P-G and Suraud E 2006 *Eur. Phys. J. A* **30** 333
[6] Durand D, Suraud E and Tamain B 2000 *Nuclear Dynamics in the Nucleonic Regime* (London: Institute of Physics)
[7] Saalmann U, Siedschlag C and Rost J-M 2006 *J. Phys. B* **39** R39
[8] Fennel T, Meiwes-Broer K-H, Tiggesbäumker J, Reinhard P-G, Dinh P-M, and Suraud E 2010 *Rev. Mod. Phys.* **82** 1793
[9] Goeke K and Reinhard P-G 1982 *Proceedings of the International Symposium on Time-dependent Hartree–Fock and beyond, Bad Honnef, Germany, June 7–11, 1982*, Lecture Notes in Physics 171 (Berlin: Springer, Heidelberg and New York)
[10] Bertsch G-F and Das Gupta S 1988 *Phys. Rep.* **160** 190
[11] Domps A, Reinhard P-G and Suraud E 1998 *Phys. Rev. Lett.* **81** 5524
[12] Rajam A-K, Raczkowska I and Maitra N-T 2010 *Phys. Rev. Lett.* **105** 113002
[13] Reinhard P-G and Suraud E 1992 *Ann. Phys. (NY)* **216** 98
[14] Abe Y, Ayik S, Reinhard P-G and Suraud E 1996 *Phys. Rep.* **275** 49

Journal of Physics: Conference Series **529** (2014) 012016 doi:10.1088/1742-6596/529/1/012016

Strong-disorder magnetic quantum phase transitions: Status and new developments

T Vojta

Department of Physics, Missouri University of Science & Technology, Rolla, MO 65409, USA

E-mail: vojtat@mst.edu

Abstract. This article reviews the unconventional effects of random disorder on magnetic quantum phase transitions, focusing on a number of new experimental and theoretical developments during the last three years. On the theory side, we address smeared quantum phase transitions tuned by changing the chemical composition, for example in alloys of the type $A_{1-x}B_x$. We also discuss how the interplay of order parameter conservation and overdamped dynamics leads to enhanced quantum Griffiths singularities in disordered metallic ferromagnets. Finally, we discuss a semiclassical theory of transport properties in quantum Griffiths phases. Experimental examples include the ruthenates $Sr_{1-x}Ca_xRuO_3$ and $(Sr_{1-x}Ca_x)_3Ru_2O_7$ as well as $Ba(Fe_{1-x}Mn_x)_2As_2$.

1. Introduction

The question of how random disorder influences a zero-temperature quantum phase transition has been a subject of great interest during the last two decades. Pioneering theoretical work initially focused on model systems such as the random transverse-field Ising chain. Using a strong-disorder renormalization group, Fisher [1, 2] showed that the ferromagnetic quantum phase transition in this system is of exotic infinite-randomness type (for a review of this method, see, e.g., Ref. [3]). It is accompanied by strong off-critical singularities in various thermodynamic quantities, the so-called quantum riffiths singularities. They are caused by rare spatial regions that are locally in the "wrong phase." Analogous results were obtained using optimal fluctuation theory and numerical simulations [4, 5, 6]. Later, these findings were generalized to higher dimensions [7, 8, 9].

Castro Neto and Jones [10] pioneered the application of the riffiths singularity concept to itinerant (metallic) systems. The qualitative behavior of disordered metals close to a magnetic instability is determined by the order parameter symmetry together with the dissipative character of the order parameter dynamics [11]. For Ising symmetry, the quantum dynamics of a sufficiently large individual magnetic rare region stops completely because the magnetic fluctuations are overdamped [12]. This destroys the sharp global quantum phase transition by smearing [13, 14]. In contrast, for continuous (or Heisenberg) order parameter symmetry, individual rare regions continue to fluctuate to the lowest temperatures [15]; this leads to quantum riffiths singularities similar to those of the dissipationless transverse-field Ising chain.[1]

[1] This means, somewhat surprisingly, the dissipationless random transverse-field Ising model and the disordered metallic Heisenberg magnet are in the same universality class.

Note that long-range interactions between different rare regions can actually lead to a (cluster-glass) freezing transition at the lowest temperatures even in the continuous symmetry case [16]. These and many more results were summarized in Refs. [11, 17].

While significant progress was made since the mid-1990s in the theory of disordered quantum phase transitions, experimental verifications of the predicted exotic disorder effects were lacking for a long time. This changed only in the years 2007 to 2010, when quantum riffiths singularities were observed in the vicinity of ferromagnetic quantum phase transitions in the magnetic semiconductor $Fe_{1-x}Co_xS_2$ [18, 19, 20], the Kondo lattice system $CePd_{1-x}Rh_x$ [19, 21], and the transition metal alloy $Ni_{1-x}V_x$ [22, 23]. These experiments were reviewed in Ref. [17].

In the last three years, the field has made further progress, both in theory and in experiment. The purpose of the present paper is to review these new results and to relate them to the state of knowledge. The paper is organized as follows. Section 2 is devoted to new developments in the theory of disordered quantum phase transitions. In particular, we address smeared quantum phase transitions tuned by changing the chemical composition in systems of the type $A_{1-x}B_x$. We also consider the enhancement of ferromagnetic quantum riffiths singularities in disordered metallic systems as well as a semiclassical theory of transport properties in quantum riffiths phases. In Sec. 3, we review a number of exciting new experiments that show additional strong disorder effects. In particular, we discuss the smeared quantum phase transition in the ruthenate $Sr_{1-x}Ca_xRuO_3$, an unusual nearly ferromagnetic phase in $(Sr_{1-x}Ca_x)_3Ru_2O_7$, as well as a possible antiferromagnetic quantum riffiths phase in $Ba(Fe_{1-x}Mn_x)_2As_2$. We conclude in Sec. 4.

2. New developments: Theory

2.1. Composition-tuned smeared quantum phase transitions

Let us start by contrasting sharp and smeared transitions. Phase transitions are commonly defined as singularities of the free energy in parameter space. Such singularities can only occur in infinitely large systems; for continuous transitions they are associated with an infinite correlation length; and in the case of first-order transitions, they correspond to macroscopic phase coexistence.

What about phase transitions in the presence of randomness that locally favors one phase over the other Do they remain sharp in the above sense, or will the singularity be smeared over a range of the tuning parameter It turns out that phase transitions in random systems generically remain sharp. The reason is that finite-size regions normally cannot order independently of each other; the global phase transition thus occurs due to a collective effect of the entire sample. It was shown in Ref. [13], however, that this argument does not apply to certain magnetic quantum phase transitions in the presence of both disorder and dissipation. If a magnetic (Ising) cluster is coupled to an Ohmic dissipative bath, it can undergo the localization transition of the dissipative two-level system [24, 25]. This implies that sufficiently large clusters develop static magnetic order independently of the bulk because their quantum dynamics stops completely. As more and more clusters freeze, the global order parameter thus arises gradually, rather than as a collective effect of the entire system; and the correlation length remains finite. This defines a smeared phase transition. The prime experimental application of this concept are metallic quantum magnets in which the dissipation stems from the coupling of the magnetization to the fermionic degrees of freedom.

It is important to emphasize that the smearing of the quantum phase transition is a finite-length-scale phenomenon (there is a *finite* minimum size required for the magnetic clusters to freeze), in contrast to the usual critical phenomena that arise in the limit of infinite lengths. This also means that smeared transitions cannot be expected to display the same degree of universality as critical points. In particular, tuning the transition with different tuning parameters can lead to qualitatively different behavior.

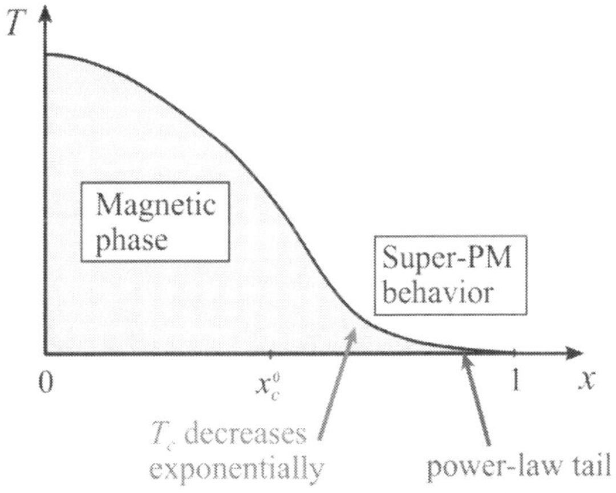

Figure 1. Schematic temperature-composition phase diagram of a binary alloy $A_{1-x}B_x$ displaying a smeared quantum phase transition. In the tail of the magnetic phase, which stretches to $x = 1$, the rare regions are aligned. Above T_c, they act as independent classical moments, resulting in super-paramagnetic (PM) behavior (after Ref. [26]).

The original theory of smeared (magnetic) quantum phase transitions [13] focused on tuning a given disordered sample through the transition by varying a parameter such as pressure. However, in almost all of the experimental systems in which unconventional disorder effects have been observed, the transition is actually tuned by changing the chemical composition. In Ref. [26], we therefore adapted the theory to alloys of the type $A_{1-x}B_x$ which are tuned through a quantum phase transition by changing x. A schematic of the resulting temperature-composition phase diagram is shown in Fig. 1. The ordered phase develops a pronounced tail that stretches all the way to $x = 1$. Two separate regimes can be distinguished. The major decrease of the critical temperature T_c near the mean-field critical concentration x_c^0 is of exponential type,

$$T_c \sim \exp\left[-C\frac{(x - x_c^0)^{2-d/\phi}}{x(1-x)}\right] \tag{1}$$

where C is a nonuniversal constant and ϕ is the finite-size shift exponent. In the far tail of the smeared transition, $x \to 1$, the critical temperature vanishes as a nonuniversal power law,

$$T_c \sim (1-x)^{L_{\rm min}^d} = (1-x)^{(D/x_c^0)^{d/\phi}} \tag{2}$$

where $L_{\rm min}$ is the minimum size a rare region must have to freeze, and D is another constant. The zero-temperature magnetization M behaves in a similar fashion. Close to x_c^0, M varies as

$$M \sim \exp\left[-C\frac{(x - x_c^0)^{2-d/\phi}}{x(1-x)}\right] \tag{3}$$

while it follows the power law

$$M \sim (1-x)^{L_{\rm min}^d} = (1-x)^{(D/x_c^0)^{d/\phi}} \tag{4}$$

for $x \to 1$. We thus find that M is nonzero in the entire composition range $0 \le x < 1$, illustrating the notion of a smeared quantum phase transition.

The fact that smeared phase transitions are nonuniversal also implies that spatial disorder correlations can have significant effects. This question was explored in Ref. [27]. It was found that even short-range correlations can qualitatively change the behavior of observable quantities at smeared transitions, in contrast to critical points, at which short-range correlations do not change the critical behavior.

2.2. Quantum Griffiths singularities in ferromagnetic metals

The theory of riffiths singularities started when riffiths [28] showed that the free energy of a disordered systems that does undergo a sharp phase transition is singular not only right at the transition point but in an entire parameter region around out. Today, this region is known as the riffiths region or riffiths phase, and the off-critical singularities are summarily called the riffiths singularities.

The theory of quantum riffiths phenomena in metallic systems initially focused on antiferromagnetic quantum phase transitions. This had two reasons First, there are many more materials featuring antiferromagnetic than ferromagnetic quantum phase transitions. Second, the theory for the antiferromagnetic case is considerably simpler (at least in three space dimensions). In the ferromagnetic case, extra complications arise from the order parameter conservation and from soft-mode induced long-range interaction between the order-parameter fluctuations [29, 30]. Ironically, all the recent experimental observations of quantum riffiths physics mentioned in Sec. 1 were in ferromagnetic conductors ($Fe_{1-x}Co_xS_2$, $CePd_{1-x}Rh_x$, $Ni_{1-x}V_x$) for which the theory was not well developed.

To remedy this problem, a theory of quantum riffiths singularities associated with itinerant ferromagnetic quantum phase transitions was developed in Ref. [31]. The crucial technical difference between the ferromagnetic and antiferromagnetic cases lies in the dynamic part of the two-point vertex of the order-parameter (L W) field theory [32]. In the antiferromagnetic case, it takes an Ohmic form $\sim |\omega_n|$ and becomes momentum-independent in the limit of small momenta and frequencies. In a disordered ferromagnet, the damping term instead has the structure $\sim |\omega_n|/\mathbf{q}^2$, reflecting the conserved nature of the total magnetization. As a result, the tunneling rate of a locally ordered ferromagnetic cluster decays as $\exp(-cL^5)$ with its linear size L, compared with a $\exp(-cL^3)$ decay in the antiferromagnetic case (c is a constant).

The form of the ferromagnetic quantum riffiths singularities can now be worked out by adapting the usual optimal fluctuation arguments (see, e.g., Ref. [11]) to the case at hand. The resulting low-energy density of states of the rare regions takes the form

$$\rho(\epsilon) \sim \frac{1}{\epsilon} \exp[\{-\tilde{\lambda} \log(\epsilon_0/\epsilon)\}^{3/5}] \, . \tag{5}$$

Here, ϵ_0 is a microscopic energy scale, and the nonuniversal parameter $\tilde{\lambda}$ takes the place of the usual riffiths exponent. In contrast to itinerant antiferromagnets (and models such as the transverse-field Ising chain), this quantum riffiths singularity in the density of states is not a pure power law. However, the $\exp[|\ln \epsilon|^{3/5}]$ structure implies that it is "almost a power law." Correspondingly, the quantum riffiths singularities in observables also differ from pure power laws. The low-temperature susceptibility, for example, is expected to behave as

$$\chi(T) \sim \frac{1}{T} \exp[\{-\tilde{\lambda} \log(T_0/T)\}^{3/5}] \, . \tag{6}$$

Compared to the power-law form $T^{\lambda-1}$, eq. (6) provides an additional upturn at low temperatures. This may be responsible for the slight upward deviations from power-law behavior observed at the lowest temperatures in the susceptibility data of $Ni_{1-x}V_x$ [22, 23]. We emphasize that eqs. (5) and (6) only hold in the asymptotic regime, when the relevant rare regions are so large that the order-parameter conservation significantly hampers their dynamics.

2.3. Semiclassical transport theory

Most theoretical studies of disordered quantum phase transitions and the accompanying quantum riffiths singularities have focused on thermodynamics while transport properties have received much less attention. This is particularly true for metallic systems whose transport properties such as the electrical resistivity are easily accessible in experiments but hard to calculate theoretically. In Ref. [33], we therefore developed a theory of the electrical resistivity in an antiferromagnetic quantum riffiths phase of a disordered metal. It is based on a semiclassical approach in which the conduction electrons are scattered by the spin fluctuations of the rare regions. Other transport quantities including the thermal resistivity, the thermopower and the Peltier coefficient were calculated in the same way.

The calculation starts from a two-band model of s and d electrons, $H = H_s + H_d + H_{s-d}$ [34]. Only the s electrons contribute to the electric transport; they are scattered by the spin fluctuations of the d electrons which are assumed to be in an antiferromagnetic quantum riffiths phase. This is described by means of a Boltzmann equation in which the dynamic susceptibility of the d electrons enters the collision term for the s electrons. The linearized Boltzmann equation is then solved using iman's variational principle [35]. The resulting rare region contribution to the electrical resistivity takes the form

$$\Delta\rho \sim T^{\lambda} . \tag{7}$$

Thus, the temperature-dependence of the resistivity follows a non-universal power law controlled by the same riffiths exponent λ that also governs the thermodynamics. (The reason is that the main contribution to the resistivity stems from scattering off the magnetic fluctuations whose temperature dependence is characterized by λ.) Analogous calculations for the thermal resistivity W, the thermopower S, and the Peltier coefficient Π yield

$$\Delta W \sim T^{\lambda-1}, \quad \Delta S \sim T^{\lambda+1}, \quad \Delta\Pi \sim T^{\lambda+2} . \tag{8}$$

All of these results apply to antiferromagnets in three space dimensions for which quasiparticles are (marginally) well-defined. Moreover, they hold in the case of strong randomness when the temperature-independent impurity part ρ_0 of the resistivity is large compared to the rare region contribution (see Ref. [36]).

The unusual temperature dependencies of thermodynamic and transport properties also affect various "universal ratios" that are commonly used to characterize Fermi liquid behavior [37]. The Wilson ratio $R_W = \chi T/C$ (which relates the susceptibility χ and the specific heat C) and the rüneisen parameter $\Gamma = \beta/C$ (the ratio of the thermal expansion coefficient β and C) both diverge logarithmically with T,

$$R_W = \chi T/C \sim [\log(1/T)]^2 , \qquad \Gamma = \beta/C \sim \log(1/T) . \tag{9}$$

While R_W and Γ only involve thermodynamics, further ratios can be calculated from the transport quantities. For example, the rare regions give a subleading contribution to the Lorenz number $L = \rho/WT$,

$$\Delta L \sim T^{\lambda} . \tag{10}$$

Moreover, the ratio $q = S/C$ between the thermopower and the specific heat as well as Kadowaki-Woods ratio $R_{KW} = (\rho - \rho_{\mathrm{imp}})/C^2$ both show nontrivial power laws in T,

$$q = S/C \sim T^{1-\lambda} , \qquad R_{KW} = (\rho - \rho_{\mathrm{imp}})/C^2 \sim T^{-\lambda} , \tag{11}$$

in contrast to the temperature-independent Fermi-liquid behavior.

The above results apply to quantum phase transitions in antiferromagnetic metals, the corresponding theory for the more complicated ferromagnetic quantum phase transition remains a task for the future.

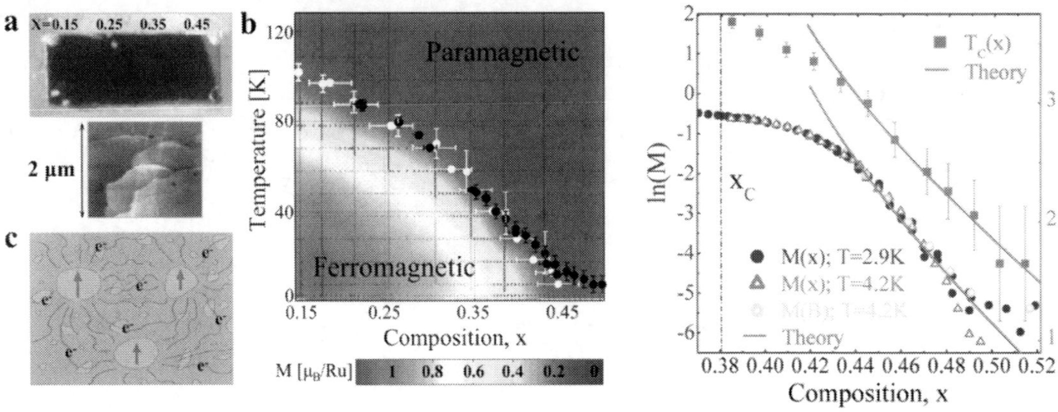

Figure 2. Left (a) Photographic image of the composition-spread $Sr_{1-x}Ca_xRuO_3$ epitaxial film. (b) Contour plot of the magnetization M over the composition-temperature phase diagram. The phase boundary, $T_c(x)$, derived from the susceptibility and magnetization data is shown by the black and white symbols, respectively. (c) Schematic of the magnetism in the tail of the smeared transition. The Sr-rich rare regions form locally ordered clusters whose dynamics freezes, giving rise to an inhomogeneous long-range ferromagnetic order. Right Magnetization and transition temperature as functions of the composition in the tail region. The symbols represent the experimental data while solid lines are fits to the theory of Sec. 2.1 (after Ref. [45]).

3. New developments: Experiment

3.1. Smeared quantum phase transition in $Sr_{1-x}Ca_xRuO_3$

The perovskite-type ruthenate $SrRuO_3$ is a ferromagnetic metal having a Curie temperature T_c^0 of about 165 K. In contrast, $CaRuO_3$ does not develop long-range magnetic order even at the lowest temperatures. By randomly substituting the smaller calcium atoms for the strontium atoms, one can thus tune the material through a ferromagnetic quantum phase transition. This system was extensively studied in the past, but different experiments [38, 39, 40, 41, 42, 43, 44] showed large variations of the phase boundary and especially in the critical calcium concentration x_c beyond which long-range magnetic order vanishes.

To study the quantum phase transition in $Sr_{1-x}Ca_xRuO_3$ with high precision, the group of Kezsmarki [45] used a high-quality composition-spread film whose calcium concentration changes from 13% to 53% from one end to the other (see Fig. 2a). The magnetic properties were measured using a magneto-optical Kerr microscope which gave a resolution $\delta x \approx 0.001$ in composition and about $6 \times 10^{-3}\mu_B$ per Ru atom in magnetization.

An overview over the results is given in Fig. 2b. The data show that the critical temperature $T_c(x)$ does not feature a singular drop at any calcium concentration. Instead, it develops a pronounced tail towards the calcium-rich side of the phase diagram. A similar tail also appears in the low-temperature magnetization as a function of temperature. These quantities thus provide strong evidence for the quantum phase transition to be smeared. The right panel of Fig. 2 shows a direct comparison of the magnetization and critical temperature with the theory of smeared quantum phase transitions discussed in Sec. 2.1. The magnetization data follow the prediction (3) over 1.5 orders of magnitude in M, giving $x_c^0 = 0.38$. The critical temperature can be fitted to eq. (1) using the same x_c^0.

Moreover, the smeared transition scenario provides a natural explanation for the strong variations in the magnetization and phase boundary between the different older experiments. As mentioned at the end of Sec. 2.1 and discussed in detail in Ref. [27], smeared phase transitions are particularly sensitive towards spatial disorder correlations. Model calculations have shown that a disorder correlation length of less than two lattice constants is sufficient to move the (seeming) critical concentration x_c from 0.5 to 1.

3.2. Nearly ferromagnetic $(Sr_{1-x}Ca_x)_3Ru_2O_7$

The double-layer ruthenate $(Sr_{1-x}Ca_x)_3Ru_2O_7$ has a rich phase diagram [46, 47]. For calcium concentration $x < 0.08$, it is in an itinerant metamagnetic state. In the region $0.08 < x < 0.4$, the system displays very strong ferromagnetic fluctuations but does not develop long-range order at low temperatures. Instead, it freezes into a cluster glass state. For $x > 0.4$, the material is an antiferromagnet consisting of ferromagnetic bilayers stacked antiferromagnetically along the c-axis. The Neel temperature increases from about 30 K at $x = 0.4$ to 56 K at $x = 1$.

Qu et al. [48] performed a detailed analysis of the magnetization M and specific heat C of $(Sr_{1-x}Ca_x)_3Ru_2O_7$ in the composition range $0.08 < x < 0.4$. Above the cluster glass freezing temperature (which ranges from less than 1 K at $x = 0.08$ to about 3 K at $x = 0.4$), both quantities exhibit anomalous power-law singularities. However, while the exponents of $M(H)$ and $C(T)$ agree, as expected in a quantum riffiths phase, the exponent characterizing the temperature dependence $M(T)$ in a small applied field differs significantly. This suggests that the quantum riffiths scenario may be insufficient to describe to observed behavior.

The quantum riffiths scenario assumes that different rare regions are far enough apart to be effectively independent of each other. The authors of Ref. [48] proposed that the observed deviations from this scenario may be due to interactions between the rare regions that exist in itinerant systems [16]. Specifically, the very strong strong ferromagnetic fluctuations in the composition range $0.08 < x < 0.4$ lead to a ferromagnetic coupling between the rare regions. The magnetization therefore increases faster upon cooling than the magnetization of isolated rare regions. Additional theoretical work is needed to provide a quantitative understanding of this phenomenon.

3.3. Possible antiferromagnetic Griffiths phase in $Ba(Fe_{1-x}Mn_x)_2As_2$

As discussed above, a number of explicit observations of rare region (riffiths) physics in itinerant ferromagnets have been reported in the last five years. In contrast, analogous observations in itinerant antiferromagnets have been missing (or, at least, clear-cut verifications of the expected power-law behaviors are not available). Very recently, Inosov et al. [49] reported a possible realization of an antiferromagnetic quantum riffiths phase in manganese-doped iron arsenide $Ba(Fe_{1-x}Mn_x)_2As_2$.

The antiferromagnetic spin-density wave order of the parent compound $BaFe_2As_2$ is suppressed with increasing manganese concentration x. Above a critical concentration x_c of about 10%, the character of the magnetic transition changes The long-range ordered phase, is preempted by a broad region of phase coexistence between antiferromagnetism, cluster glass and spin glass order. The authors of Ref. [49] interpret the coexistence of magnetically ordered and paramagnetic clusters as a manifestation of a riffiths phase and/or the smearing of the antiferromagnetic quantum critical point due to the random distribution of the manganese atoms.

Further research is required to firmly establish the rare region (riffiths) scenario in this system. In particular, various thermodynamic and transport quantities need to be studied and compared in the putative riffiths region above x_c. Are they controlled by nonuniversal power-laws in T and other control variables Are their exponents related These questions remain a task for the future.

4. Conclusions

To summarize, we have reviewed a number of theoretical and experimental developments regarding the unconventional disorder effects at magnetic quantum phase transitions. The focus has been on novel results that appeared after the review article [17]; this means roughly during the last three years.

The theory of smeared quantum phase transitions in itinerant magnets has been extended to the case of transitions tuned by changing the chemical composition in alloys of type $A_{1-x}B_x$. Recent high-precision measurements demonstrated that the ferromagnetic quantum phase transition in $Sr_{1-x}Ca_xRuO_3$ is a good example of such a smeared transition. Advances in theory also showed that quantum riffiths effects in ferromagnetic metals are enhanced because the conservation of the ferromagnetic order parameter hampers the relaxation of locally ordered clusters. The resulting quantum riffiths singularities are more singular than pure power laws. This may explain the low-temperature upturns observed in the susceptibility-temperature curves of $Ni_{1-x}V_x$. Moreover, a semiclassical theory of various transport quantities in antiferromagnetic quantum riffiths phases yielded nonuniversal power-law temperature dependencies controlled by the same riffiths exponent that governs the thermodynamics.

On the experimental side, several interesting results have been obtained in addition to the already mentioned smeared ferromagnetic quantum phase transition in $Sr_{1-x}Ca_xRuO_3$. The double-layer ruthenate $(Sr_{1-x}Ca_x)_3Ru_2O_7$ features a strongly fluctuating nearly ferromagnetic phase for $0.08 < x < 0.4$ in which magnetization and susceptibility are governed by nonuniversal power laws. However, the behavior of the exponents does not fully line up with the predictions of the quantum riffiths scenario, possibly because of strong interactions between the rare regions. Very recently, a possible antiferromagnetic riffiths phase was reported in $Ba(Fe_{1-x}Mn_x)_2As_2$. This system shows a region of coexistence between antiferromagnetic and paramagnetic clusters above a manganese concentration of about 10%. Further work will be necessary to determine whether this is indeed a manifestation of rare region effects.

In addition to these examples, magnetic cluster effects were also observed in d_5 e_4 and related intermetallic compounds [50, 51, 52]. They occur above the sizable critical temperature of the antiferromagnetic phase transition. The riffiths singularities should thus be of classical rather than quantum type. Such classical riffiths singularities generically require spatial disorder correlations to make them strong enough to be visible in experiment [53, 11]. The precise nature of these correlations and their effect on observables requires further experimental and theoretical work. The situation is thus somewhat similar to the riffiths effects reported in various manganites, see the discussion in Ref. [11].

To conclude, the last three years have seen considerable experimental and theoretical progress in the field of strongly disordered quantum phase transitions. However, many problems remain at best partially solved. Moreover, several new questions have emerged that will be attacked in the future.

Acknowledgments

This work has been supported in part by the National Science Foundation under rant No. DMR-1205803.

References

[1] Fisher D S 1992 *Phys. Rev. Lett.* **69** 534
[2] Fisher D S 1995 *Phys. Rev.* B **51** 6411
[3] Igloi F and Monthus C 2005 *Phys. Rep.* **412** 277
[4] Thill M and Huse D A 1995 *Physica* A **214** 321
[5] Rieger H and Young A P 1996 *Phys. Rev.* B **54** 3328
[6] Young A P and Rieger H 1996 *Phys. Rev.* B **53** 8486
[7] Senthil T and Sachdev S 1996 *Phys. Rev. Lett.* **77** 5292

[8] Pich C, Young A P, Rieger H and Kawashima N 1998 *Phys. Rev. Lett.* **81** 5916
[9] Motrunich O, Mau S C, Huse D A and Fisher D S 2000 *Phys. Rev.* B **61** 1160
[10] Castro Neto A H and Jones B A 2000 *Phys. Rev.* B **62** 14975
[11] Vojta T 2006 *J. Phys.* A **39** R143
[12] Millis A J, Morr D K and Schmalian J 2001 *Phys. Rev. Lett.* **87** 167202
[13] Vojta T 2003 *Phys. Rev. Lett.* **90** 107202
[14] Hoyos J A and Vojta T 2008 *Phys. Rev. Lett.* **100** 240601
[15] Vojta T and Schmalian J 2005 *Phys. Rev.* B **72** 045438
[16] Dobrosavljevic V and Miranda E 2005 *Phys. Rev. Lett.* **94** 187203
[17] Vojta T 2010 *J. Low Temp. Phys.* **161** 299
[18] Guo S, Young D P, Macaluso R T, Browne D A, Henderson N L, Chan J Y, Henry L and DiTusa J F 2007 *Phys. Rev. Lett.* **100** 017209
[19] Guo S, Young D P, Macaluso R T, Browne D A, Henderson N L, Chan J Y, Henry L L and DiTusa J F 2010 *Phys. Rev.* B **81** 144423
[20] Guo S, Young D P, Macaluso R T, Browne D A, Henderson N L, Chan J Y, Henry L L and DiTusa J F 2010 *Phys. Rev.* B **81** 144424
[21] Westerkamp T, Deppe M, Küchler R, Brando M, Geibel C, Gegenwart P, Pikul A P and Steglich F 2009 *Phys. Rev. Lett.* **102** 206404
[22] Ubaid-Kassis S, Vojta T and Schroeder A 2010 *Phys. Rev. Lett.* **104** 066402
[23] Schroeder A, Ubaid-Kassis S and Vojta T 2011 *J. Phys. Condes. Matter* **23** 094205
[24] Leggett A J, Chakravarty S, Dorsey A T, Fisher M P A, Garg A and Zwerger W 1987 *Rev. Mod. Phys.* **59** 1
[25] Weiss U 1993 *Quantum disspative systems* (Singapore: World Scientific)
[26] Hrahsheh F, Nozadze D and Vojta T 2011 *Phys. Rev.* B **83**(22) 224402
[27] Svoboda C, Nozadze D, Hrahsheh F and Vojta T 2012 *EPL (Europhysics Letters)* **97** 20007
[28] Griffiths R B 1969 *Phys. Rev. Lett.* **23** 17
[29] Kirkpatrick T R and Belitz D 1996 *Phys. Rev.* B **53** 14364
[30] Belitz D, Kirkpatrick T R and Vojta T 2005 *Rev. Mod. Phys.* **77** 579
[31] Nozadze D and Vojta T 2012 *Phys. Rev.* B **85**(17) 174202
[32] Hertz J 1976 *Phys. Rev.* B **14** 1165
[33] Nozadze D and Vojta T 2011 *EPL (Europhysics Letters)* **95** 57010
[34] Ueda K and Moriya T 1975 *J. Phys. Soc. Jpn.* **39** 605
[35] Ziman J 1960 *Electrons and Phonons* (Oxford: Clarendon)
[36] Rosch A 1999 *Phys. Rev. Lett.* **82**(21) 4280–4283
[37] Nozadze D and Vojta T 2012 *Journal of Physics: Conference Series* **391** 012162
[38] Cao G, McCall S, Shepard M, Crow J E and Guertin R P 1997 *Phys. Rev.* B **56** 321–329
[39] Kiyama T, Yoshimura K, Kosuge K, Mitamura H and Goto T 1999 *J. Phys. Soc. Japan* **68** 3372–3376
[40] Khalifah P, Ohkubo I, Christen H M and Mandrus D G 2004 *Phys. Rev.* B **70** 134426
[41] Okamoto J, Okane T, Saitoh Y, Terai K, Fujimori S I, Muramatsu Y, Yoshii K, Mamiya K, Koide T, Fujimori A, Fang Z, Takeda Y and Takano M 2007 *Phys. Rev.* B **76**(18) 184441
[42] Hosaka N, Yamada H, Shimada Y, Fujioka J, Bordács S, Kézsmárki I, Kawasaki M and Tokura Y 2008 *Appl. Phys. Express* **1** 113001
[43] Schneider M, Moshnyaga V and Gegenwart P 2010 *Phys. Status Solidi* B **247** 577
[44] Wissinger M, Fuchs D, Dieterle L, Leiste H, Schneider R, Gerthsen D and v Löhneysen H 2011 *Phys. Rev.* B **83** 144430
[45] Demkó L, Bordács S, Vojta T, Nozadze D, Hrahsheh F, Svoboda C, Dóra B, Yamada H, Kawasaki M, Tokura Y and Kézsmárki I 2012 *Phys. Rev. Lett.* **108**(18) 185701
[46] Cao G, McCall S C, Crow J E and Guertin R P 1997 *Phys. Rev.* B **56**(9) 5387–5394
[47] Qu Z, Spinu L, Yuan H, Dobrosavljević V, Bao W, Lynn J W, Nicklas M, Peng J, Liu T, Fobes D, Flesch E and Mao Z Q 2008 *Phys. Rev.* B **78**(18) 180407
[48] Qu Z, Peng J, Liu T, Fobes D, Dobrosavljević V, Spinu L and Mao Z Q 2012 *Phys. Rev.* B **86**(1) 014434
[49] Inosov D S, Friemel G, Park J T, Walters A C, Texier Y, Laplace Y, Bobroff J, Hinkov V, Sun D L, Liu Y, Khasanov R, Sedlak K, Bourges P, Sidis Y, Ivanov A, Lin C T, Keller T and Keimer B 2013 *Phys. Rev.* B **87**(22) 224425
[50] Pereira A M, Morellon L, Magen C, Ventura J, Algarabel P A, Ibarra M R, Sousa J B and Araújo J P 2010 *Phys. Rev.* B **82**(17) 172406
[51] Pérez N, Casanova F, Bartolomé F, García L M, Labarta A and Batlle X 2011 *Phys. Rev.* B **83**(18) 184411
[52] Hadimani R L, Melnikhov Y and Jiles D C 2013 *IEEE Transactions on Magnetics* **49** 820
[53] Imry Y 1977 *Phys. Rev.* B **15** 4448

Electrical conductivity of strongly degenerate plasma with the account of electron-electron scattering

V S Karakhtanov[1]

[1] Institute of Problems of Chemical Physics, Russian Academy of Sciences, 142432 Chernogolovka, Russia

E-mail: karakhtanovvs@gmail.com

Abstract. The influence of electron-electron scattering on the strongly degenerate plasma conductivity is investigated with a linear response theory. In the present work the temperature dependence of the electron-electron scattering term of the electrical conductivity and further modification of the Ziman formula are discussed.

1. Introduction

It is known that electron-electron scattering makes a significant contribution to the conductivity of low-density non-degenerate plasma [1] and is completely absent at zero temperature. A temperature influence on the conductivity is usually taken into account by means of a well-known finite-temperature modification of the Ziman formula [2, 3] for electron-ion scattering. At the same time electron-electron scattering can contribute an income comparable with that due to the distribution function deviation from the Fermi step. The temperature behaviour of the electron-electron scattering influence on transport properties is interesting both for the description of experiments on laser heating of plasma [4] and the construction of interpolation formulas for the conductivity [5, 6, 7].

2. Basic approximations

A neutral two-component plasma consisting of free singly charged particles with charges e_i and e_e ($e_i = -e_e = e$) at temperature T and density $n = n_\mathrm{e} = n_\mathrm{i}$, interacting via Coulombic forces is considered. Indices i and e denote ions and electrons, respectively. The value of the mass ratio $\gamma = m_\mathrm{e}/m_\mathrm{i}$ tends to zero (the adiabatic limit). The dimensionless degeneracy parameter $\Theta = (2m_\mathrm{e}k_\mathrm{B}T/\hbar^2)(3\pi^2 n_\mathrm{e})^{-2/3}$ and the electron-ion coupling constant $\Gamma = (e^2/4\pi\epsilon_0 k_\mathrm{B}T)(4\pi n_\mathrm{e}/3)^{1/3}$ are introduced. For the strong degenerate plasma $\Theta \ll 1$. Within the linear response theory in the formulation of Zubarev [8], the transport properties are expressed via force-force correlation functions. The procedure of their derivation is well-known (see [9, 10, 11, 12, 13, 14]). As a result, the static electrical conductivity is related to the Onsager transport coefficients $L_{ik} = L_{ki}$ according to

$$\sigma = e^2 L_{11} \tag{1}$$

where

$$L_{ik} = -\frac{(-h)^{i+k-2}}{\Omega \det(d)} \begin{vmatrix} 0 & \frac{k-1}{\beta h}N_1 - N_0 \\ \frac{i-1}{\beta h}\overline{N_1} - \overline{N_0} & d \end{vmatrix}, \tag{2}$$

$$N_n = \begin{pmatrix} N_{n0} & N_{n1} & \dots & N_{nl} \end{pmatrix}, \tag{3}$$

$$\overline{N_n} = \begin{pmatrix} N_{n0} \\ N_{n1} \\ \vdots \\ N_{nl} \end{pmatrix}, d = \begin{pmatrix} d_{00} & d_{01} & \dots & d_{0l} \\ d_{10} & d_{11} & \dots & d_{1l} \\ \vdots & \vdots & \ddots & \vdots \\ d_{l0} & d_{l1} & \dots & d_{ll} \end{pmatrix}. \tag{4}$$

In (2)-(4) Ω denotes the system volume, N_{mn}, d_{mn} are correlation functions for the thermodynamic equilibrium, h is enthalpy per one electron, N_e - the number of electrons and $\beta = (k_B T)^{-1}$. The dimension of the matrix d coincides with the number of moments in the corresponding relevant statistical operator (see [8]). In the adiabatic limit we can omit the ion flux [14] and obtain for the matrix elements in Eq.(4)

$$d_{mn} = d_{mn}^{ei} + d_{mn}^{ee}, \tag{5}$$

$$N_{mn} = N_e \frac{\Gamma(m+n+5/2)}{\Gamma(5/2)} \frac{I_{m+n+1/2}(\beta\mu_e^{id})}{I_{1/2}(\beta\mu_e^{id})}, \tag{6}$$

with $I_\nu(y) = \frac{1}{\Gamma(\nu+1)} \int\limits_0^\infty \frac{x^\nu dx}{e^{x-y}+1}$ - the Fermi integrals, μ_e^{id}- the ideal part of the electronic chemical potential.

The correlation functions d_{mn} are evaluated using thermodynamic Green's functions. The first Born approximation for the screened Coulomb interaction $V(q) = e^2(q^2\Omega\epsilon_0)^{-1}$ [15, 13] with omitting the exchange term in electron-electron scattering leads to the Lenard-Balescu-type [16, 17, 18] collision integrals, obtained in [9]:

$$\begin{aligned} d_{mn}^{ei} &= \pi\beta\hbar \sum_{\vec{k}\vec{p}\vec{q}} \int\limits_{-\infty}^{\infty} d\omega \left| \frac{V(q)}{\epsilon(q,\omega)} \right|^2 f_k^e(1 - f_{|\vec{k}+\vec{q}|}^e) f_p^i(1 - f_{|\vec{k}-\vec{p}|}^i) \times \\ & \delta(\omega - E_{|\vec{k}+\vec{q}|}^e + E_k^e)\delta(\omega - E_p^i + E_{|\vec{p}-\vec{q}|}^i) K_n(\vec{k},\vec{q}) K_m(\vec{k},\vec{q}), \end{aligned} \tag{7}$$

$$\begin{aligned} d_{mn}^{ee} &= \frac{\pi\beta\hbar}{2} \sum_{\vec{k}\vec{p}\vec{q}} \int\limits_{-\infty}^{\infty} d\omega \left| \frac{V(q)}{\epsilon(q,\omega)} \right|^2 f_k^e(1 - f_{|\vec{k}+\vec{q}|}^e) f_p^e(1 - f_{|\vec{k}-\vec{p}|}^e) \times \\ & \delta(\omega - E_{|\vec{k}+\vec{q}|}^e + E_k^e)\delta(\omega - E_p^e + E_{|\vec{p}-\vec{q}|}^e)(K_n(\vec{k},\vec{q}) + K_n(\vec{p},-\vec{q}))(K_m(\vec{k},\vec{q}) + K_n(\vec{p},-\vec{q})), \end{aligned} \tag{8}$$

where $E_k^c = \hbar^2 k^2/(2m_c), K_n(\vec{k},\vec{q}) = k_z(\beta E_k^e)^n - (k_z + q_z)(\beta E_{|\vec{k}+\vec{q}|}^e)^n$, $\epsilon(q,\omega)$ - the dielectric function, $f_k^c = (\exp((E_k^c - \mu_c^{id})\beta) + 1)^{-1}$ - the Fermi distribution function.

In the adiabatic limit

$$d_{mn}^{ei} = \frac{4m_e^2}{3\pi^2\beta^2\hbar^3} \int\limits_0^\infty dx x^{n+m+2} f_k^e(1 - f_k^e) Q_{ei}(x) \tag{9}$$

$$Q_{ei}(x) = \frac{\beta^2\Omega^2}{16\pi x^2} \int\limits_0^{2k} \left| \frac{V(q)}{\epsilon_e(q,0)} \right|^2 S_{ii}(q) q^3 dq \tag{10}$$

$$\epsilon_e(q,0) = 1 + \Omega V(q)(1 - G_e(q))\chi_e^{(0)}(q,0) \tag{11}$$

with $x = \frac{\beta\hbar^2k^2}{2m_e}$, $Q_{ei}(x)$ - the transport cross-section for electron-ion scattering in the first Born approximation, $\epsilon_e(q,0)$ - the effective static electronic dielectric function, $S_{ii}(q)$ - the ion-ion structure factor, $\chi_e^{(0)}(q,\omega)$ the free-electron polarizability [19, 20], $G_e(q)$ - the static electronic local field correction. For d_{00}^{ei} this result was presented in [9, 21, 22].

3. The limit of the strong degeneration

Expressions (9)-(10) for the electron-ion correlation functions are valid in any degeneration degree. Electron-electron correlation functions can be reduced to the two-dimension integrals in the Boltzmann limit [23]. For the strong degeneracy conductivity calculations electron-electron terms are usually omitted due to the restrictions of the Fermi statistics, and as a result the values of dc conductivity are equal to each other at any values of l in (3), (4), if the temperature in the Fermi distribution functions is taken equal to zero. With increasing temperature, the frequently used method for the calculation of the conductivity consists in substituting the non-zero temperature into distribution functions. In this case, at first, the simple proportionality between electron-ion correlation functions disappears, that leads to the dependence of the result on the dimension of determinants, even in the absence of electron-electron scattering. And, at second, the electron-electron correlation functions begin to grow.

To trace the influence of both factors, consider the simplest approximation that includes them. We restrict ourselves to $l = 1$ in (3), (4). For the preliminary estimation of integrals we take for the Fermi distribution function $f(z) = 1$ for $z < 0$ and $f(z) = \exp(-z)$ for $z > 0$, where $z = (E_k^c - \mu_c^{id})\beta$. Introducing the dimensionless variables $Q = \hbar q(\beta/m_e)^{1/2}, \nu = (\beta m_e)^{1/2}\omega/q$ we obtain that d_{11}^{ee} is proportional to the sum of the following integrals:

$$d_{11kl} = \int_0^{2Q_F} \frac{dQ}{Q^3} \left[\int_0^{Q_F - \frac{Q}{2}} \frac{I_{1k} I_{1l} \mathrm{e}^{-Q\nu} d\nu}{|\epsilon(Q,\nu)|^2} + \int_{Q_F - \frac{Q}{2}}^{Q_F + \frac{Q}{2}} \frac{I_{2k} I_{2l} \mathrm{e}^{-Q\nu} d\nu}{|\epsilon(Q,\nu)|^2} + \int_{Q_F + \frac{Q}{2}}^{\infty} \frac{I_{3k} I_{3l} \mathrm{e}^{-Q\nu} d\nu}{|\epsilon(Q,\nu)|^2} \right] \tag{12}$$

$$+ \int_{2Q_F}^{\infty} \frac{dQ}{Q^3} \left[\int_0^{\frac{Q}{2} - Q_F} \frac{I_{3k} I_{3l} \mathrm{e}^{-Q\nu} d\nu}{|\epsilon(Q,\nu)|^2} + \int_{\frac{Q}{2} - Q_F}^{Q_F + \frac{Q}{2}} \frac{I_{2k} I_{2l} \mathrm{e}^{-Q\nu} d\nu}{|\epsilon(Q,\nu)|^2} + \int_{Q_F + \frac{Q}{2}}^{\infty} \frac{I_{3k} I_{3l} \mathrm{e}^{-Q\nu} d\nu}{|\epsilon(Q,\nu)|^2} \right] \tag{13}$$

where

$$I_{1l} = \left(\int_0^{\frac{Q_F^2}{2} - S_+} \exp(x + S_+ - \frac{Q_F^2}{2}) x^l dx + \int_{\frac{Q_F^2}{2} - S_+}^{\frac{Q_F^2}{2} - S_-} x^l dx + \int_{\frac{Q_F^2}{2} - S_-}^{\infty} \exp(-x - S_- + \frac{Q_F^2}{2}) x^l dx \right) \tag{14}$$

$$I_{2l} = \left(\int_0^{\frac{Q_F^2}{2} - S_-} x^l dx + \int_{\frac{Q_F^2}{2} - S_-}^{\infty} \exp(-x - S_- + \frac{Q_F^2}{2}) x^l dx \right) \tag{15}$$

$$I_{3l} = \int_0^{\infty} \exp(-x - S_- + \frac{Q_F^2}{2}) x^l dx \tag{16}$$

In Eqs. (12)-(16) $Q_F = \hbar k_F(\beta/m_e)^{1/2} = \left(\frac{2}{\Theta}\right)^{1/2} \gg 1$, $S_\pm = \frac{1}{2}\left(\nu \pm \frac{Q}{2}\right)^2$.

The main contribution to d_{11kl} at $\Theta \ll 1$ comes from the first integral in (12):

$$d_{11kl} = \int_0^{2Q_F} \frac{dQ}{Q^3} \int_0^{Q_F - \frac{Q}{2}} \frac{I_{1k} I_{1l} \mathrm{e}^{-Q\nu} d\nu}{|\epsilon(Q,\nu)|^2} \tag{17}$$

Due to the multiplier $e^{-Q\nu}$ and the frequency dependence of the dielectric function in the form $f(\frac{\nu}{Q_F})$, we can neglect the frequency dependence of $\epsilon(q,\omega)$ in the overwhelming part of the domain of integration.

Bearing in mind that the non-vanishing part of $(K_1(\vec{k},\vec{q}) + K_1(\vec{p},-\vec{q}))^2$ due to its symmetry properties with respect to the permutation of the variables of integration x and y is proportional to $Q^2\left((y-x)^2 + 4Q^2\nu^2\right)$, and restoring dimensional multipliers, we have:

$$d_{11}^{ee} = 37.3\Theta^{7/2}k_F d \int_0^{2k_F} \frac{dq}{q^2\epsilon_e^2(q,0)}, \quad (18)$$

where $d = \frac{8}{3}\frac{m_e^{1/2}e^4N^2\beta^{3/2}}{\Omega(4\pi\epsilon_0)^2}$

For a more accurate determination of the numerical multiplier when integrating with the real Fermi distribution function, it is convenient to present the latest in the form of an expansion in Chebyshev polynomials [24]. Limited to three polynomials in the decomposition, we obtain:

$$d_{11}^{ee} = 17.3\Theta^{7/2}k_F d \int_0^{2k_F} \frac{dq}{q^2\epsilon_e^2(q,0)}, \quad (19)$$

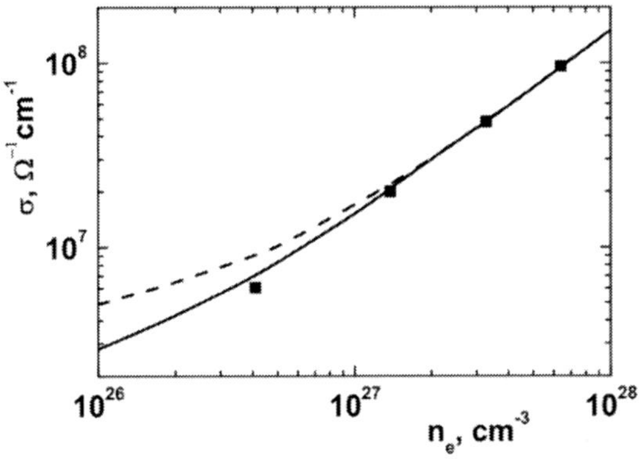

Figure 1. Electrical conductivity of degenerate plasma dependent on electron density in the two-moment representation. $T = 10^7K$. Solid line - with e-e scattering, dashed line - without e-e scattering, dots - data of [27] in the Ziman approximation.

4. Results and discussion

In the Fig. 1 the results of calculations of the electrical conductivity of the fully ionized plasma in the described approximation with and without electron-electron scattering are presented. For the local field correction $G_e(q)$ the simplest approximation of the VS type [25] is chosen. As it turned out, the results for the electrical conductivity do not depend on the choice of $G_e(q)$. For $S_{ii}(q)$ in Eq. (10) the HNC approximation is used with the numerical scheme described in [26]. The study of the dependence of the results from the model for the structure factor (see, e.g., [9, 27]) is not the subject of this work. The HNC method is quite popular and there are numerous data obtained in the Ziman approximation.

The present calculations are made for temperatures and densities corresponding to final states of systems obtained from the laser compression [27].

The boundary $\Theta = 1$ corresponds to $n_e = 10^{26}cm^{-3}$, for $n_e = 10^{27}cm^{-3}$ Θ is near 0.2. Note that the simple Ziman approximation [2, 3] ($l = 0$ in the Eqs.(3), (4)) is more close to the result for $l = 1$ with the electron-electron scattering than without it (within a few percent). The results are similar for $T = 10^6 K$ and $T = 10^8 K$. Thus, at least for the considered case of plasma with singly charged ions, the Ziman approximation for finite temperatures amazingly well describes the electrical conductivity of the degenerate plasma in a wide range of degeneration parameter values. However, with the expansion of the basis of the relevant variables in the theory, we must simultaneously take into account electron-electron scattering which also has a separate application in the study of other plasma properties (e.g., thermal conductivity, see [4]).

Acknowledgments

R. Redmer focused the author's attention to the problem. Useful discussions were held with G. Röpke and H. Reinholz during the MBT-17 conference. All of them the author expresses his gratitude. This work was supported by the Program No.2 "Matter Under High Energy Density" of Russian Academy of Sciences.

References

[1] Spitzer L and Härm R 1953 *Phys. Rev.* **89** 977
[2] Ziman J M 1961 *Philos. Mag.* **6** 1013
[3] Faber T E 1974 *Introductions to the Theory of Liquid Metals* (Cambridge: Cambridge University Press)
[4] Inogamov N A, Petrov Y V 2010 *J. Exp Theor. Phys.* **110** 446
[5] Esser A and Röpke G 1998 *Phys. Rev.* E **58** 2446
[6] Röpke G 2003 *Contrib. Plasma Phys.* **43** 33
[7] Ichimaru S and Tanaka S 1985 *Phys. Rev.* A **32** 1790
[8] Zubarev D N 1974 *Nonequilibrium Statistical Mechanics* (New York: Plenum)
[9] Meister C-V and Röpke G 1982 *Ann. Phys.* A **494** 133
[10] Röpke G 1983 *Physica* A **121** 92
[11] Röpke G 1988 *Phys. Rev.* A **38** 3001
[12] Höhne F E, Redmer R, Röpke G and Wegener H 1985 *Physica* A **128** 643
[13] Redmer R, Röpke G, Morales F and Kilimann K 1990 *Phys. Fluids* B **2** 390
[14] Reinholz H, Redmer R and Tamme D 1989 *Contrib. Plasma Phys.* **29** 395
[15] Gould H A and DeWitt E H 1967 *Phys. Rev.* **155** 68
[16] Balescu R 1960 *Phys. Fluids* **3** 52
[17] Lenard A 1960 *Ann. Phys.(NY)* **10** 390
[18] Gurnsey R L 1964 *Phys. Fluids* **7** 1600
[19] Khanna F C and Glyde H R 1976 *Can. J. Phys.* A **54** 648
[20] Gouedard C and Deutch C 1978 *J. Math. Phys.* A **19** 32
[21] Boerker D B 1981 *Phys. Rev.* A **23** 1969
[22] Boerker D B, Rogers F J and DeWitt E H 1982 *Phys. Rev.* A **25** 1623
[23] Karakhtanov V S, Redmer R, Reinholz H and Röpke G 2013 *Contrib. Plasma Phys.* **53** 639
[24] Luke Y L 1975 *Mathematical Functions and their Approximations* (New York: Academic Press)
[25] Vashishta P and Singwi K S 1972 *Phys. Rev.* B **6** 875
[26] Springer J F, Pokrant M A and Stevens F A 1973 *J. Chem. Phys.* **58** 4863
[27] Minoo H, Deutsch C and Hansen J P 1976 *Phys. Rev.* A **14** 840

The influence of boundary conditions on the excitation of instabilities in magnetohydrodynamic systems

C-V Meister, B R Lee and D H H Hoffmann

Technische Universität Darmstadt, Institut für Kernphysik, Schlossgartenstr. 9, 64289 - Darmstadt, and Graduate School of Excellence Energy Science and Engineering, Jovanka-Bontschits-Str. 2, 64287 - Darmstadt, Germany

E-mail: c.v.meister@skmail.ikp.physik.tu-darmstadt.de

Abstract. The recent stage of the magnetohydrodynamic energy principle applied to laboratory and space plasmas is briefly reviewed. In detail, the energy principle is presented for an internally homogeneous pinch in a perfectly conducting wall. The plasma is separated from the wall by a vacuum. The principle is applied to ITER-type and lightning systems. Thereat, a system of mathematical equations of motion for fluid elements is derived using a cylindrical coordinate system. But the obtained equations may be also applied to plasmas with disturbances of non-cylindrical symmetry. From the equations of motion, an analytical relation for the radial displacements of the fluid elements is presented, which describes magnetohydrodynamic waves as e.g. sausage and kink ones. The numerical results here presented are, as a first step, only performed for plasma disturbances with cylindrical symmetry and outer azimuthal magnetic fields directed parallely to the conducting wall. Thus, the dispersion relations for sausage instabilities in ITER-type and lightning plasmas are solved. It is shown for which values of the inner and external magnetic fields of the systems instabilities occur. In case of lightnings, the radial displacements in the plasma are estimated.

1. Introduction

In nuclear fusion systems, plasma instabilities getting unstable for different reasons interrupt the fusion process and make the application of nuclear fusion for an energy gain in reactor systems up to now unachievable. In the Earth's magnetosphere, plasma instabilities modify the physical parameters and vary the transport of solar energy and radio signals from vertical sounding stations. Thus, it is essential to study plasma instabilities in both nuclear fusion and magnetospheric systems. Using the energy principle of magnetohydrodynamics (MHD), this all can be done without knowing the exact values of growth rates of unstable waves. According to [1], the growth rates of MHD instabilities in inertial fusion plasmas are, anyway, on the order or smaller than 5 percent of the possible observation times. And if even shorter observation times will be possible in future, the application of the MHD energy principle will often be much easier as a kinetic calculation of the growth rate.

Applying the MHD energy principle means to analyse, if the potential energy of the plasma system may decrease for any of its allowable small displacements. If this is the case, the system is unstable. Thus, applying the MHD energy principle, one has to solve variational problems for the potential energy.

The MHD energy principle is only applicable if the force on the particles is self-adjoint in the plasma. Because of this self-adjointness, a decrease of the potential energy is equivalent to a negative square of a frequency of a wave excited in the system. Thus, the analysis of the self-adjointness of the force is an important problem. Recent applications of the energy principle to systems with different plasma boundaries, especially to space systems, show that the self-adjointness reduces the number of the allowable small displacements which may be taken into account.

The present work considers instabilities in plasmas which are described within the frame of ideal MHD, that means the system has an infinite electrical conductivity. Hence, in section 2 the energy principle of ideal MHD is briefly introduced. Section 3 reviewes recently used extended energy principles which are applied for plasmas not bounded by a vacuum and/or perfectly conducting wall, and where the magnetic field at the plasma boundary is not directed parallelly to the boundary. In section 4, the MHD energy principle is applied to an internally homogeneous pinch. There an analytical equation is derived to describe axial fluid displacements in systems with non-cylindrical disturbances. Numerical applications are performed for systems with cylindrically symmetric disturbances in sections 5 and 6 for ITER-type and lightning plasmas, respectively. Some conclusions are made in section 7.

2. Energy principle of ideal MHD

The present work considers instabilities in plasmas which may be described within the frame of ideal magnetohydrodynamics, the basic equations of which are the continuity equation

$$\frac{d\rho}{dt} = -\rho\nabla\vec{v}, \tag{1}$$

the momentum balance

$$\rho\frac{d\vec{v}}{dt} = \vec{j} \times \vec{B} - \nabla p + \rho\vec{g}, \tag{2}$$

the equation of state

$$\frac{d}{dt}\left[\frac{p}{\rho^{\gamma}}\right] = 0, \tag{3}$$

and the Maxwell equations without displacement current

$$\frac{\partial\vec{B}}{\partial t} = -\nabla \times \vec{E}, \tag{4}$$

$$\nabla \times \vec{B} = \mu_o\vec{j}. \tag{5}$$

From the Ohm's law

$$\vec{j} = \sigma(\vec{E} + \vec{v} \times \vec{B}) \tag{6}$$

follows in the approximation of ideal MHD

$$\frac{\partial\vec{B}}{\partial t} = \nabla \times (\vec{v} \times \vec{B}). \tag{7}$$

The 14 scalar expressions (1-3, 4-5, 7) form a full system of equations of ideal magnetohydrodynamics to determine the 14 variables plasma pressure p, mass density ρ, fluid velocity $\vec{v}$, vectores of magnetic induction $\vec{B}$ and electric field $\vec{E}$, and electrical current $\vec{j}$. Besides, the Maxwell equation

$$\nabla\vec{B} = 0 \tag{8}$$

is valid.

Expressing the momentum balance of the linearized MHD system of equations (2) by the Lagrange displacement $\vec{\xi}$, using the relation between the Euler velocity $\vec{u}(\vec{r},t)$ and the Lagrange velocity of a fluid element (which was situated at $\vec{r}_o$ at time t_o)

$$\frac{\partial\vec{\xi}(\vec{r},t)}{\partial t} = \vec{u}(\vec{r},t) \approx \vec{u}(\vec{r}_o,t) + (\vec{\xi}\cdot\nabla)\vec{u} + ... \approx \vec{u}(\vec{r}_o,t), \tag{9}$$

the momentum balance may be written in the form

$$\rho_o \frac{\partial^2\vec{\xi}}{\partial t^2} = \vec{F}(\vec{\xi}). \tag{10}$$

$$\vec{F}(\vec{\xi}) = \nabla(\gamma p_o \nabla\cdot\vec{\xi} + (\vec{\xi}\cdot\nabla)p_o)$$

$$+\frac{1}{\mu_o}\Big((\nabla\times\nabla\times(\vec{\xi}\times\vec{B}_o))\times\vec{B}_o + (\nabla\times\vec{B}_o)\times(\nabla\times(\vec{\xi}\times\vec{B}_o))\Big) - \vec{g}((\vec{\xi}\cdot\nabla)\rho_o + \rho_o\nabla\cdot\vec{\xi})$$

gives the force on a fluid element in the plasma ($\vec{g}$ - gravity acceleration).
Further it is assumed that the separation of variables

$$\vec{\xi}(\vec{r}_o,t) = \vec{\xi}_k(\vec{r}_o)\tau_k(t) \tag{11}$$

is possible and the function $\tau_k(t)$ describes a plain wave

$$\tau_k \sim \exp[i(\omega_k t + \varphi_k)]. \tag{12}$$

Then, ω_k is the solution of the equation

$$-\rho_o\omega_k^2\vec{\xi}_k = \vec{F}(\vec{\xi}_k). \tag{13}$$

For the potential energy density W, $\vec{F} = -\nabla W$, it follows

$$W = -\int_0^{\vec{\xi}} \vec{F}(\vec{\eta})\, d\vec{\eta} = -\frac{\vec{\xi}}{2}\vec{F}. \tag{14}$$

Thus, for the stability of the MHD system it is necessary and sufficient that the potential energy caused by the plasma displacement $\vec{\xi}$ has no negative values.
Taking the MHD stability condition for ideal plasmas into account,

$$\nabla\, p_o = \frac{1}{\mu_o}(\nabla\times\vec{B}_o)\times\vec{B}_o, \tag{15}$$

I.B. Bernstein, J.P. Freidberg [2-4], and Spatschek [5] found, that eq. (14) may be used for the determination of instabilities in plasmas, surrounded by a vacuum, which are located in a fixed conducting wall, provided that in addition to the volume contribution to the potential energy

$$W_F = \frac{1}{2}\int\limits_{V_p}\left[\frac{Q^2}{\mu_o} + \frac{1}{\mu_o}(\nabla\times\vec{B}_o)(\vec{\xi}\times\vec{Q}) + (\nabla\cdot\vec{\xi})(\vec{\xi}\cdot\nabla)p_o + \gamma p_o(\nabla\cdot\vec{\xi})^2\right] d\vec{r}, \tag{16}$$

a vacuum contribution

$$W_V = \frac{1}{2\mu_o}\int\limits_{V_V}(\delta\vec{B}_V)^2\, d\vec{r}, \tag{17}$$

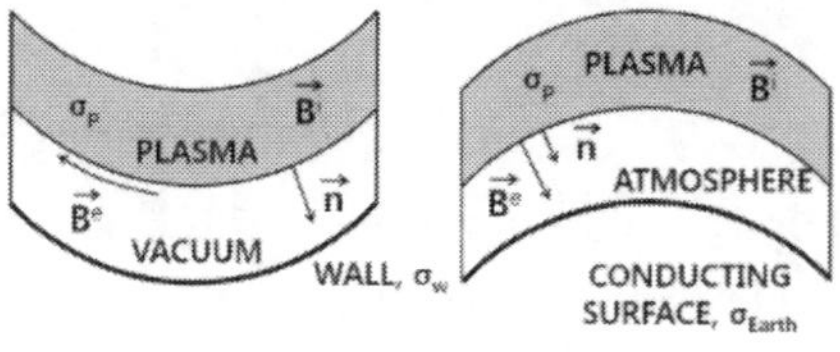

Figure 1. Comparison of the boundary condition of a laboratory plasma at a conducting wall with a vacuum barrier sheet (left side) with the boundary condition of the magnetosphere at the Earth's surface with an ionospheric barrier sheet. σ_P - electrical conductivity of the plasma, $\vec{B}^i$ - magnetic field of the plasma, $\vec{B}^e$ - magnetic field at the plasma-vacuum boundary and in the Earth's atmosphere, respectively. An analogous figure, the work [6] contains.

and a surface contribution

$$W_S = -\frac{1}{2}\int_S (\vec{\xi}\cdot\vec{n})^2\vec{n}\cdot\nabla\left(p_o + \frac{B_o^2}{2\mu_o} - \frac{B_V^2}{2\mu_o}\right)\, d\vec{S} \tag{18}$$

are taken into account.

$$W = W_F + W_S + W_V, \qquad \vec{Q} = \nabla\times(\vec{\xi}\times\vec{B}_o). \tag{19}$$

3. Boundary conditions

Indeed, eqs. (16-19) may be applied for laboratory fusion systems, where the magnetic field is parallel to the walls. But for magnetospheric systems, with non-parallel magnetic field lines, additional conditions follow for the allowable displacements $\vec{\xi}$. Besides, in case of the Earth's magnetosphere, the plasma is surrounded by a vacuum at lower altitudes, and by an ionospheric or magnetospheric boundary at larger altitudes (see Fig. 1). Such boundary conditions were first studied by Miura [6], who introduced two approximations. First he took only geomagnetic field lines into account, which are perpendicular to the ionospheric boundary, and second, he neglected the neutral atmosphere. In doing so he found that when the vectors $\vec{\xi}$ and $\vec{\eta}$ satisfy the boundary conditions

$$\eta_{\parallel} = 0 \quad \text{or} \quad \nabla\cdot\vec{\xi} = 0 \tag{20}$$

and

$$\eta_{\perp} = 0 \quad \text{or} \quad (\vec{B}\cdot\nabla)\xi_{\perp} = 0 \tag{21}$$

at the unperturbed ionospheric boundaries, the force operator of the plasma of the Earth's atmosphere is self-adjoint. But, taking the neutral atmosphere into account, $\nabla\cdot\vec{\xi}\neq 0$ is valid, and the second condition of eq. (20) is not satisfied.

4. Application to an internally homogeneous pinch

Now a homogeneous plasma is considered. It is assumed that the plasma is located in a cylinder in the equilibrium state. In the plasma, a uniform axial magnetic field $B_z\vec{n}_z$ exists. Further, it is supposed that the plasma is surrounded by a vacuum region, and in the vacuum only an azimuthal magnetic field is located. The whole system consisting of the cylindrical plasma core and the surrounding vacuum is contained in a cylindrical conducting shell. That means, the plasma does not contact the shell at all, there is a vacuum boundary between the plasma and the shell.

In the homogeneous plasma of the core of the pinch, $\nabla \times \vec{B}_z = 0$ and $\nabla p_o \approx 0$. The approximation $\nabla \times \vec{B}_z = 0$ was first introduced by Sturrock [7] and Spatschek [5], and it means that no current exists in the inner part of the plasma (i.e. along the z-axis). Here, the same approximation is used to compare partial results with that of the works [5, 7]. Indeed, in the case of the θ-pinch with an axial magnetic field in the plasma core, the currents are almost negligible along the symmetry axis (i.e. at $r = 0$ in cylindrical coordinates (r, φ, z)), and the pressure gradients vanish (see Figs. 11.12 and 11.13, p. 266, in [7]).
Under such conditions, one finds for the minimum potential energy of the plasma in the cylinder [8]

$$\text{Min } W_F \approx \frac{1}{2}\int \left(\gamma p_o(\nabla\cdot\vec{\xi}) - \frac{(\vec{B}_z\cdot\vec{Q})}{\mu_o}\right)(\vec{\xi}\cdot d\vec{F}). \tag{22}$$

The equation of motion (10) in the plasma reduces to

$$-\omega^2\rho_o\xi_z = \gamma p_o\frac{\partial}{\partial z}(\nabla\cdot\vec{\xi}), \tag{23}$$

$$-\omega^2\rho_o\xi_r = \gamma p_o\frac{\partial}{\partial r}\nabla\vec{\xi} + \frac{B_z^2}{\mu_o}\Big[\frac{\partial}{\partial r}\frac{1}{r}\frac{\partial\xi_\varphi}{\partial\varphi} + \frac{\partial^2\xi_r}{\partial z^2} + \frac{\partial}{\partial r}\frac{1}{r}\frac{\partial}{\partial r}(r\xi_r)\Big], \tag{24}$$

$$-\omega^2\rho_o\xi_\varphi = \frac{\gamma p_o}{r}\frac{\partial}{\partial\varphi}\nabla\vec{\xi} + \frac{B_z}{\mu_o}\Big[\frac{1}{r^2}\frac{\partial^2}{\partial\varphi\partial r}(r\xi_r) + \frac{\partial^2\xi_\varphi}{\partial z^2} + \frac{1}{r^2}\frac{\partial^2}{\partial\varphi^2}(\xi_\varphi)\Big]. \tag{25}$$

Further, to find the dispersion relation of the waves in the plasma, one may use the ansatz

$$\vec{\xi}(\vec{r}) = \vec{\xi}_o(r)\exp(im\varphi + ik_z z) \tag{26}$$

and simplify the spatial derivatives with respect to φ and z in eqs. (23-25)

$$(v_s^2k_z^2 - \omega^2)\xi_z = ik_z\frac{v_s^2}{r}\frac{\partial}{\partial r}(r\xi_r) - k_z m v_s^2\frac{\xi_\varphi}{r}, \tag{27}$$

$$(v_A^2k_z^2 - \omega^2)\xi_r = (v_A^2 + v_s^2)\frac{\partial}{\partial r}\left[\frac{1}{r}\frac{\partial}{\partial r}(r\xi_r)\right] + im(v_A^2 + v_s^2)\frac{\partial}{\partial r}\left(\frac{\xi_\varphi}{r}\right) + ik_z v_s^2\frac{\partial\xi_z}{\partial r}, \tag{28}$$

$$(v_A^2k_z^2 - \omega^2)\xi_\varphi = (v_A^2 + v_s^2)\frac{im}{r^2}\frac{\partial}{\partial r}(r\xi_r) + \frac{imv_s^2 - m^2v_A^2}{r}\frac{\xi_\varphi}{r} + \frac{imk_z v_s^2}{r}\xi_z, \tag{29}$$

$$v_A^2 = \frac{B_z^2}{\mu_o\rho_o}, \quad v_s^2 = \frac{\gamma p_o}{\rho_o}. \tag{30}$$

With

$$X = \frac{1}{r}\frac{\partial}{\partial r}(r\xi_r), \quad Y = \frac{\xi_\varphi}{r}, \tag{31}$$

one finds from eqs. (27, 29)

$$X = \frac{(v_s^2k_z^2 - \omega^2)}{ik_z v_s^2}\xi_z + \frac{m}{i}Y, \quad Y = \frac{r(v_A^2k_z^2 - \omega^2)}{(m^2 + im)v_s^2}\xi_\varphi - \frac{[ik_z^2v_s^4 + (v_s^2k_z^2 - \omega^2)(v_A^2 + v_s^2)]}{k_z(m+i)v_s^4}\xi_z \tag{32}$$

Substituting X and Y into eq. (28), and acting on (28) with the operator $(1/r)(\partial/\partial r)r$, one obtains

$$(v_A^2k_z^2 - \omega^2)X = \frac{v_A^2v_s^2k_z^2 - \omega^2(v_A^2 + v_s^2)}{ikv_s^2}\frac{1}{r}\frac{\partial}{\partial r}\left(r\frac{\partial\xi_z}{\partial r}\right), \tag{33}$$

$$X = \frac{-iv_s^2\omega^2 + iv_s^4k_z^2 + m[v_s^2v_A^2k_z^2 - ik_z^2v_s^4 - \omega^2(v_A^2 + 2v_s^2)]}{k_z(im-1)v_s^4} + \frac{v_A^2k_z^2 - \omega^2}{(im-1)v_s^2}r\xi_\varphi. \tag{34}$$

Considering only equilibrium plasma systems with cylindrical symmetry ($\partial/\partial\varphi = 0$) and neglecting eq. (25), which then describes pure Alfvén waves in the plasma ($\omega^2 = k^2 v_A^2$ at any ξ_φ), one finds from (23, 24) at locally constant magnetic fields B_z (wave number $k = k_z$)

$$\frac{\partial^2 \xi_z}{\partial r^2} + \frac{1}{r}\frac{\partial \xi_z}{\partial r} - \alpha^2 \xi_z = 0, \quad \alpha^2 = \frac{(v_s^2 k^2 - \omega^2)(v_A^2 k^2 - \omega^2)}{v_s^2 v_A^2 k^2 - (v_s^2 + v_A^2)\omega^2}, \quad v_A^2 = \frac{B_z^2}{\mu_o \rho_o}, \quad v_s^2 = \frac{\gamma p_o}{\rho_o}, \tag{35}$$

which was already derived by Sturrock [9] studying the sausage instability. Thus, one has

$$\xi_z(r) = \frac{\text{const}}{\alpha^2} I_o(\alpha r), \tag{36}$$

where I_o is the modified Bessel function of first kind of order zero.
The equations (23, 24, 36) result then into a solution for $\xi_r(z)$,

$$\xi_r = \frac{v_s^2(v_A^2 k^2 - \omega^2) - v_A^2\omega^2}{ikv_s^2(k^2 v_A^2 - \omega^2)}\frac{\partial \xi_z}{\partial r} = -\frac{i\ \text{const}}{\alpha k}\frac{[v_s^2(v_A^2 k^2 - \omega^2) - v_A^2\omega^2]}{v_s^2(k^2 v_A^2 - \omega^2)}\frac{\partial I_o(\alpha\ r)}{\partial(\alpha\ r)}. \tag{37}$$

The analysis of α^2 described by eq. (35) gives the following maximum and minimum values:

$\text{Max}(\alpha^2) = \omega^2(v_s - v_A)^2/(v_A^2 v_s^2)$ at $\omega = \text{const}$, $k^2 = \omega^2(v_s^2 + v_A^2 - v_s v_A)/(v_s^2 v_A^2)$,

$\text{Max}(\alpha^2) = k^2(v_s^2 - v_A^2)^2/(v_s^2 + v_A^2)^2$ at $k = \text{const}$, $\omega^2 = 2v_s^2 v_A^2 k^2/(v_s^2 + v_A^2)$,

$\text{Min}(\alpha^2) = \omega^2(v_s + v_A)^2/(v_A^2 v_s^2)$ at $\omega = \text{const}$, $k^2 = \omega^2(v_s^2 + v_A^2 + v_s v_A)/(v_s^2 v_A^2)$,

$\text{Min}(\alpha^2) = k^2$ at $k = \text{const}$, $\omega^2 = 0$.

Thus, at constant k or ω the α^2-values of the minima are larger than the values of the maxima. At $\omega^2 = v_s^2 v_A^2 k^2/(v_s^2 + v_A^2)$, α^2 diverges (see the application for ITER, Fig. 2).

The perturbation of the vector potential $\delta\vec{A}$ between the plasma surface and the outer shell, that means within the vacuum region, satisfies the vacuum wave equation. The perturbation of the electric field within the vacuum equals $\delta\vec{E} = -\partial(\delta\vec{A})/\partial t$, and the magnetic field perturbation is $\delta\vec{B} = \nabla \times (\delta\vec{A})$.

Further, combining the equation of motion (27-29)) with the boundary condition for the electromagnetic field at the plasma-vacuum interface (see [9]: eq. (15.3.3), the index i describes the plasma region and the index e designates the vacuum region, $B^{(i)} = B_z$, $B^{(e)} = B_\varphi$)

$$\vec{n} \times \left[\delta\vec{E}^{(i)} + \vec{v}^{(i)} \times \vec{B}^{(i)}\right] = \vec{n} \times \left[\delta\vec{E}^{(e)} + \vec{v}^{(e)} \times \vec{B}^{(e)}\right], \tag{38}$$

one finds the dispersion relation for the existing magnetohydrodynamic waves $\omega(k)$ under the influence of a plasma boundary (see [9]: $\partial/\partial\varphi = 0$ and $B_z = \text{const}$, eq. (15.5.14) without k^2 in the numerator of the second term on the right side)

$$\omega^2 = \frac{B_z^{i2} k^2}{\mu_o \rho_o} - \frac{B_\varphi^{e2}}{\mu_o \rho_o R^2}\frac{\alpha R \partial I_o(\alpha R)/\partial(\alpha R)}{I_o(\alpha R)}. \tag{39}$$

R is the radius of the plasma-vacuum boundary. According to the dispersion relation (39), there would be no plasma instability, but a normal Alfvén wave, in case that the outer azimuthal magnetic field would vanish. That means, the plasma-vacuum interface determines the stability of the plasma. The plasma may become unstable at sufficiently large values of $\kappa = B_\varphi^{e2}/B_z^{i2}$.

Figure 2. Function $\alpha^2(\omega^2)$ for an ITER-type plasma.

5. ITER

Since 2005 in Caderache, France, the International Thermonuclear Experimental Reactor ITER is under construction. It is a project of Europe, Japan, the former Soviet Union and the United States, joint in 2003 by China and South Korea. ITER is a Tokamak-type reactor with a total radius of 10.7 m and a height of 30 m.

ITER is designed to deliver a fusion power of 500 MW. It contains 18 superconducting toroidal and 6 poloidal field coils. The strength of its toroidal magnetic field will be about 5.3 T (here locally described by B_z). It can reach up to 11.8 T. A device volume of 840 m^3 will contain a plasma of 0.5 g at mean temperature of $2 \cdot 10^8$ K. Thus, in case of ITER, one has to consider a plasma with a mass density of about $\rho = 6 \cdot 10^{-4}$ gm^{-3}. The Alfvén and sound velocities of the plasma amount to $v_A = 6 \cdot 10^6$ m/s and $v_s = 1.2 \cdot 10^6$ m/s, respectively.

Studying the possibility of magnetohydrodynamic instabilities of an ITER-type plasma, one has first to determine the maximum and minimum values of the α-function (eq. (35)), as well as the location of its divergency. In Fig. 2, α^2 is presented for the ITER plasma as function of ω^2. It is to be seen, that the α^2-value at small ω^2 is also small, so that the argument of the Bessel function $R\alpha$ is smaller than one, and one may approximate $I_o(\alpha R)$ by

$$I_o(\alpha R) \approx 1 + \frac{\alpha^2 R^2}{4}, \quad \frac{\partial I_o(\alpha R)}{\partial(\alpha R)} = \frac{\alpha R}{2}. \tag{40}$$

Substituting eq. (40) for the Bessel function into eq. (39), one has

$$\omega^2 - k^2 v_A^2 = -\kappa v_A^2 \frac{\alpha^2}{(2 + \alpha^2 R^2/2)}. \tag{41}$$

Neglecting the $\alpha^2 R^2$ dependence on the right side of eq. (41) and expressing α^2 by eq. (35), two solutions of eq. (41) follow for ω^2,

$$\omega_1^2 = k^2 v_A^2 \tag{42}$$

and

$$\omega_2^2 = \frac{v_s^2 v_A^2 k^2 (2 - \kappa)}{2v_s^2 + v_A^2 (2 - \kappa)}. \tag{43}$$

Journal of Physics: Conference Series **529** (2014) 012018 doi:10.1088/1742-6596/529/1/012018

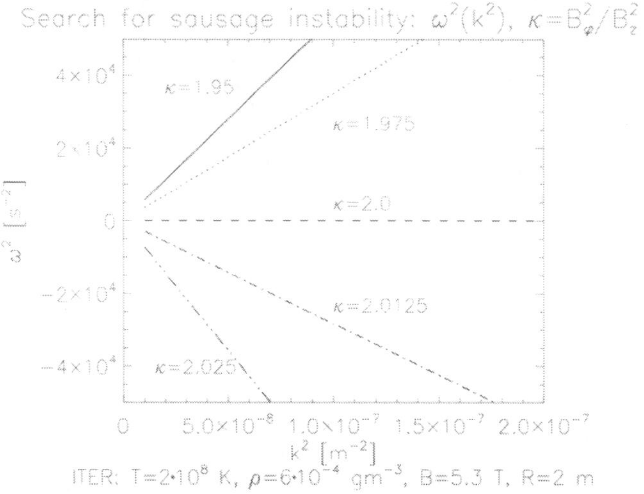

Figure 3. Dispersion relation of waves in an ITER-type plasma at different ratios of azimuthal to axial magnetic field B_φ/B_z. Solution for $\alpha R < 1$. At $\omega^2 < 0$, the sausage instability is excited.

The relation (42) describes a stable Alfvén wave. Equation (43) is the dispersion relation of an additional wave which is unstable in the case of $2 \leq \kappa < 2 + 2v_s^2/v_A^2$. At $\kappa = 2 + 2v_s^2/v_A^2$, the applied approximation eq. (40) is not valid anymore, ω_2^2 - and thus also α^2 - diverge. Further, using the approximation of the Bessel function eq. (40), it follows from the dispersion relation eq. (39) at $\omega^2 = 0$ (i.e. $\alpha^2 = k^2$) that $k^2R^2 = \kappa - 2$. Thus, one obtains for $\alpha R < 1$ that $\kappa < 3$. Consequently, studying plasmas with $\kappa \geq 3$, the formulae (40) cannot be used at frequencies $\omega^2 \approx 0$.

Results for ω_2^2 are presented in Fig. 3 for the case of the ITER-type plasma. It is to be seen that the strength of the instability grows monotonically with increasing κ at small wave numbers k. Having found an instability for a given κ and a special wave number, the system is obtained to be unstable for the κ at any wave number. But the wave numbers studied in Fig. 3 are not of real importance for ITER. They are too small. Nevertheless, the obtained results are very helpful as limiting cases for numerical analyses, when the nonlinear dynamics of ITER-type plasma instabilities at smaller wavelengths is studied.

Considering large $\alpha R > 1$, the Bessel function may be approximated by Hankels expansion

$$I_o(\alpha R) \approx \frac{\exp(\alpha R)}{\sqrt{2\pi\alpha R}} \left(1 + \frac{1}{8\alpha R}\right). \tag{44}$$

Therefrom follows

$$\frac{\partial I_o(\alpha R)/\partial(\alpha R)}{I_o(\alpha R)} \approx \frac{8\alpha R - 3}{8\alpha R + 1} \approx 1 \tag{45}$$

and one finds

$$\omega^2 = \frac{2v_s^2 v_A^2 k^2 + v_A^4 k^2 \pm \sqrt{\beta^2 - 4\beta v_s^4 k^2}}{2(v_s^2 + v_A^2)}, \quad \beta = \frac{\kappa^2 v_A^4}{R^2}. \tag{46}$$

As ω^2 is real, one has $k^2 < \beta/(4v_s^4)$. or $k^2 < 46\kappa^2$ under ITER-like conditions. From eq. (46) follows for $\omega^2 < 0$ (sign minus), that the relation

$$k^2 < \frac{2\kappa^2}{R^2(2+\chi)^2}\left[-1 + \sqrt{1 + \chi^2 + \chi^3 + 0.25\chi^4}\right], \quad \chi = \frac{v_A^2}{v_s^2}. \tag{47}$$

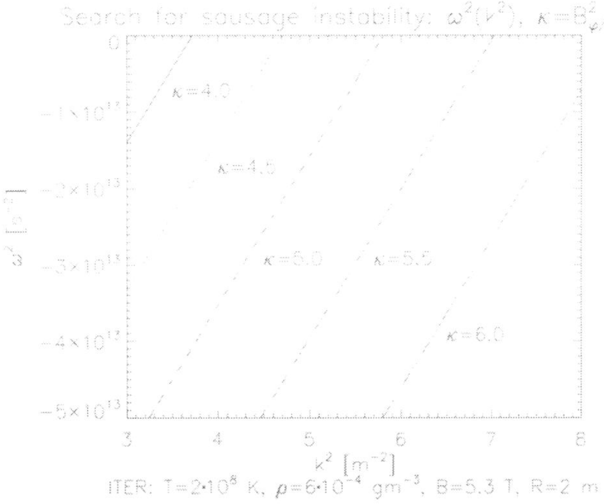

Figure 4. Dispersion relation of waves in an ITER-type plasma at different ratios of azimuthal to axial magnetic field B_φ/B_z. Solution for $\alpha R > 1$. At $\omega^2 < 0$, the sausage instability is excited.

has to be satisfied. In case of the ITER plasma with $v_A > v_s$, eq. (47) may be approximated by

$$k^2 < \frac{\kappa^2}{R^2} = \frac{\beta}{v_A^4}. \tag{48}$$

Thus instabilities in ITER-type plasmas at $\alpha R > 1$ should occur at wavelengths $\lambda > 2\pi R/\kappa = 4\pi/\kappa$ m. For $\kappa = 2$ (as studied before for $\alpha R < 1$) unstable waves have much smaller wavelengths, but they are larger than 6 m. And if κ increases, the wavelength will further decrease. In Figure 4, results of the numerical calculation of ω^2 as function of the square of the wave number k^2 are presented for different values of the parameter κ. To describe smaller wavelengths than that considered in Fig. 3 for $\alpha R < 1$, much larger k^2-values are chosen. They correspond to wavelengths of about 3 m. Results are only shown for $\omega^2 < 0$, as then also $\alpha^2 > 0$. Indeed for the k^2-interval studied, $\alpha R = 1$ is satisfied at $\omega^2 \approx 10^{12} - 10^{15}$ s^{-2}. For stable plasmas with $\omega^2 > 0$, values of α^2 below zero occur. From Figure 4 follows, that, in contradiction to the case $\alpha R < 1$, under the condition of $\alpha R > 1$ instabilities may be obtained for all considered κ-parameters, but they do not occur at any wave number k. The smaller the wave number the stronger the instability. The transition from the stable to the unstable ITER-type plasmas at large αR has to be described using Bessel functions of complex arguments. This will be done in a future work.

6. Lightnings

In nature, pinches occur very often in connection with the appearance of lightnings. Here, temperatures up to some 30000 K are found, and electrical currents are of the order of 40 kA [10]. The lifetime of lightnigs amounts to 1 ms. The discharge radii with values of a few centimeters are rather small. This results into magnetic fields of about 0.4 T and a magnetic pressure of 0.6 atm (60795 Pa). Thus, at MHD pressure balance, the pinch pressure is about 1.6 atm (162120 Pa) and the pinch Alfvén speed has a value of 4000 m/s [10].

Concerning the plasma model considered in the present work, the $\vec{B}_z$ in the inner plasma core, i.e. the hot lightning plasma, equals 0.4 nT. Of course, the hot lightning plasma is not surrounded by a vacuum, but by a plasma with much lower ionization intensity, which is here approximated by a vacuum. The magnetic field of the surrounding plasma is the geomagnetic one with an intensity of $5 \cdot 10^{-5}$ T. In the Earth's lower atmosphere, always an electrical field exists, which is directed vertically to the Earth's surface. It causes field-aligned electrical currents which may generate azimuthal magnetic fields. It is known that thunderclouds mainly help to conserve the vertical electrical field of the Earth's atmosphere. But lightnings may be indeed directed to the Earth or in the opposite direction, so they may increase or decrease the normal electric field locally. In the Earth's atmosphere also horizontal electric fields are observed. But these fields are not taken into account in the present model. Besides, the gravitational acceleration is neglected.

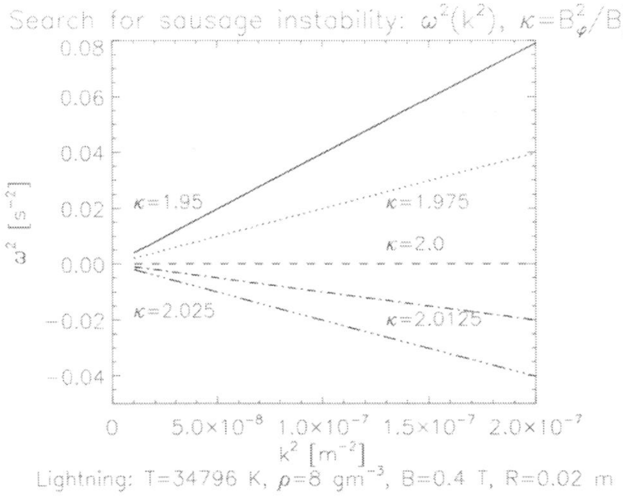

Figure 5. Dispersion relation of waves in an lightning plasma at different ratios of azimuthal to axial magnetic field B_φ/B_z. At $\omega^2 < 0$, the sausage instability is excited.

For the above mentioned plasma parameters in a lightning, the dispersion equation (43) is solved and the results are presented in Fig. 5 for the same κ-parameters and wave numbers k as have been chosen for the ITER study at $\alpha R < 1$ in section 5. It is to be seen, that again under the condition $\kappa > 2$, the plasma becomes unstable. But the square of the frequencies of the waves ω^2 is about 6 orders of magnitude smaller than in the case of the ITER plasma. Besides, ω^2 increases more slowly with the wave number than in the ITER case. The parameter αR of the Bessel function is smaller than unity in the considered plasma region. Thus, the radial displacement of the plasma fluid in the lightning was found using eq. (37). The results are shown in Fig. 6. One may conclude that the amplitudes of the radial plasma displacements grow with increasing wave number k. In the studied plasma parameter region they have a maximum value of 10 m.

7. Conclusions

The magnetohydrodynamic energy principle is presented for an internally homogeneous pinch and applied to ITER-type and lightning systems. In doing so, a system of mathematical equations of motion for fluid elements (eqs. (23-25)) is derived. The expressions are written in a cylindrical coordinate system, but they may be also applied to plasmas with disturbances of non-cylindrical symmetry. From the equations of motion, analytical formulae to be solved to

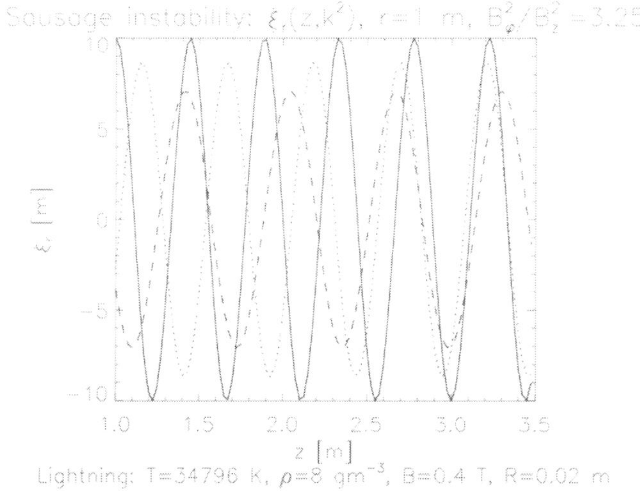

Figure 6. Radial displacement in a lightning as function of the coordinate in direction of the local axial magnetic field. The ratio of azimuthal to axial magnetic field equals $\kappa = 1.8$, that means a sausage instability is excited. $\omega^2 = -2.8 \cdot 10^9$ s^{-2}. —— $k^2 = 200$ m^{-1}, ······ $k^2 = 150$ m^{-1}, - - - - $k^2 = 100$ m^{-1}.

obtain the radial fluid displacements and the wave dispersion are presented, which describe the excitation of magnetohydrodynamic waves, e.g. of sausage and kink ones.

To simplify the first numerical solutions presented in the present work, the obtained dispersion relation is only solved for plasma disturbances with cylindrical symmetry and homogeneous axial magnetic fields in the inner plasma core. Using the model of a pinch plasma separated from the outer conducting wall by a vacuum, values of the ratio of the inner axial to the external azimuthal magnetic fields are found, at which a sausage instability will occur. Numerical solutions of the dispersion equation are presented for stable and unstable systems within the limit of small values of αR, where α is a function of the wave frequency and wave number, and R describes the radial dimension of the plasma system. In the case of ITER-type plasmas, small αR correspond to rather large wavelengths in unstable systems. Thus, to describe more realistic technical problems, the calculations were further extended to the region of $\alpha R > 1$. Under such conditions, in dependence on the wave number, unstable waves with wavelengths of the order of 2-3 m occur. But therefore the external azimuthal magnetic field at the outer wall has to be two times stronger than the inner axial magnetic field. Thus the performed numerical analysis shows the importance of the outer boundary conditions of a plasma for the excitation of instabilities in the system. Besides, the results may help to chose such magnetic field ratios that sausage instabilities will not appear in ITER-type plasmas.

Further, the model of the pinch plasma separated from the outer conducting wall by a vacuum is applied to lightnigs in the Earth's atmosphere. In this case, the mathematical approximations made for $\alpha R < 1$ are more appropriate, and it was possible to calculate the radial displacements in the unstable plasma. But in cases of planetary atmospheres, the lightning plasma with the larger degree of ionization is not bounded to a vacuum, but to other material with a lower degree of ionization. Besides, at lightning altitudes particle collisions have to be taken into account in the plasma model, which was not done here. Therefore, the magnetospheric energy principle has to be further developed for weakly-collisional, partially ionized systems in future. Some steps to determine instabilities in the Earth's atmosphere taking also weak collisions between charged and neutral particles into account were already performed by the authors and presented in [11] and [12] (misprints occur in eqs. (35-37) of [12], Fig. 2 of [12] is correct). There the MHD energy principle was not used.

Last not least, consequences of an extended magnetospheric energy principle derived by A. Miura [6], that means of a principle which one recently tries to apply for much larger scales than the lightning ones in the Earth's magnetosphere, are briefly reviewed in the present work.

8. Acknowledgments

B.R. Lee kindly acknowledges financial support of the present work by the Deutsche Forschungsgemeinschaft in the framework of the Excellence Initiative Darmstadt Graduate School of Energy Science and Engineering (GSC1070).

9. References

[1] Simakov A N 2008 *Formal theory of MHD stability: Energy principle* CMPD & CMSO Winter School http://home.physics.ucla.edu/calendar/conferences/cmpd/talks/simakov.pdf (downloaded 01.01.14)

[2] Freidberg J P 1982 Ideal magnetohydrodynamic theory of magnetic fusion systems *Rev. Modern Phys.* **54** pp 801-854

[3] Freidberg J P 1978 *Ideal magnetohydrodynamics* (New York: Springer)

[4] Bernstein I B, Frieman M D, Kruskal D and Kulsrud R M 1958 An energy principle for hydromagnetic stability problems *Proc. R. Soc. London* **A244** pp 17-40

[5] Spatschek K-H 1990 *Theoretische Plasmaphysik. Eine Einführung* (Stuttgart: B G Teubner)

[6] Miura A 2007 A magnetospheric energy principle for hydromagnetic stability problem *J. Geophys. Res.* **112** A06234 doi:10.1029/2006JA011992

[7] Freidberg J P 2007 *Plasma physics and fusion energy* (Cambridge: Cambridge University Press)

[8] Meister C-V. and Hoffmann D H H 2013 *Application of the MHD energy principle to pinch plasmas* 33dr International Workshop on Physics of High Energy Density in Matter, 13-18.01.13, Waldemar-Petersen-Haus, Hirschegg, Austria

[9] Sturrock P A 2004 *Plasma physics. An introduction to the theory of astrophysical, geophysical and laboratory plasmas* (Cambridge-New York-Melbourne: Cambridge University Press)

[10] Manheimer W M, Lampe M and Boris J P 1973 Effect of a surrounding gas on magnetohydrodynamic instabilities in Z pinch *Phys. Fluids* **16** pp 1126-1134

[11] Meister C-V, Erhard H and Hoffmann D H H 2013 Dielectric model for electromagnetic waves generated in the E-layer of the Earth's ionosphere by infrasound *Contr. Plasma Phys.* **53** pp 406-413 doi:10.102/ctpp.201200104

[12] Meister C-V, Mayer B, Dziendziel P, Fülbert F, Hoffmann D H H and Liperovsky V A 2011 On the acoustic model of lithosphere-atmosphere-ionosphere coupling before earthquakes, *Nat. Hazards Earth Syst. Sciences* **11** pp 1011-1017 doi: 10.5194/nhess-11-1011-2011

Magnetic-field-induced quantum criticality in a planar ferromagnet with single-ion anisotropy

M T Mercaldo[1], **I Rabuffo**[1], **L De Cesare and A Caramico D'Auria**[2]

[1] Dipartimento di Fisica "E. R. Caianiello", Università di Salerno, I-84084 Fisciano (Salerno), Italy

[2] Dipartimento di Fisica, Università di Napoli Federico II, I-80125 Napoli, Italy

E-mail: mtm@physics.unisa.it

Abstract. We analyze the effects induced by single-ion anisotropy on quantum criticality in a d-dimensional spin-3/2 planar ferromagnet. To tackle this problem we employ the two-time Green's function method, using the Tyablikov decoupling for exchange interactions and the Anderson-Callen decoupling for single-ion anisotropy. In our analysis the role of non-thermal control parameter which drives the quantum phase transition is played by a longitudinal external magnetic field. We find that the single-ion anisotropy has substantial effects on the structure of the phase diagram close to the quantum critical point.

The study of quantum magnetic systems is a very active research subject in condensed matter physics [1, 2]. In particular, field-induced quantum phase transitions (QPTs) and anisotropy effects have been the focus of a great amount of theoretical and experimental investigations [3, 4]. Moreover, quantum critical points (QCPs) [3] are currently considered responsible for the unusual low-temperature properties observed in many low-dimensional magnetic compounds. Furthermore, there are also clear evidences that magnetic anisotropies of different nature play a fundamental role and lead to new interesting phenomena [5, 6]. Indeed, anisotropic Heisenberg models have attracted increasing attention as a basic tool to understand relevant properties of several magnetic materials. In particular, Heisenberg models with easy-plane or easy-axis anisotropies [7] have attracted a lot of interest, especially for low-dimensional ferromagnets (FM) and antiferromagnets (AFM). In this context, in constructing spin models for several anisotropic magnetic samples, one has also to face the presence of additional crystal anisotropic fields as single-ion anisotropy (SIA) [8].

Intensive studies [9, 10] have been focused on the XXZ model with exchange anisotropies and in the presence of an external longitudinal field h. In this paper we consider also the inclusion of SIA [8, 11], hence we are going to study the following Hamiltonian (in convenient units)

$$\mathcal{H} = -\frac{1}{2}\sum_{i,j=1}^{N}\left[J_{ij}(S_i^x S_j^x + S_i^y S_j^y + \Delta S_i^z S_j^z)\right] - h\sum_{i=1}^{N} S_i^z - D\sum_{i=1}^{N}(S_i^z)^2. \quad (1)$$

Here $S_i^\alpha(\alpha = x, y, z)$ denote the components of the vector spin operator $\mathbf{S}_i$ at site i of a $d-$dimensional hypercubic lattice with N sites, satisfying the commutation relations $[S_i^\alpha, S_j^\beta] = i\varepsilon_{\alpha\beta\gamma}\delta_{ij}S_i^\gamma$ ($\varepsilon_{\alpha\beta\gamma}$ is the usual Levi-Civita tensor), J_{ij} (with $J_{ii} = 0$) is the planar exchange coupling between the spins at sites i and j, Δ gives the strength of the exchange anisotropy,

h is the external longitudinal field and D is the SIA paramter. Positive or negative values of J_{ij} correspond to FM or AFM cases. Our attention is devoted to FM models and one has a uniaxial ferromagnet if $\Delta > 1$ (easy-axis exchange anisotropy), the isotropic Heisenberg model when $\Delta = 1$, and the planar ferromagnet (PFM) if $\Delta < 1$ (easy-plane exchange anisotropy). As concerning the SIA parameter D, we have that $D > 0$ gives rise to easy-axis SIA and $D < 0$ to easy-plane one. This kind of anisotropy has to be necessarily included in several situations to obtain a more accurate correspondence between theoretical finding and experimental data and occur only in system with spin $S \geq 1$. In the present paper we are going to consider the peculiar case $S = 3/2$, also because there are certain ferromagnetic compounds, containing Fe^{3+}, which assumes the unusual spin S=3/2 state and where crystal field effects are important [12].

In absence of SIA, the PFM exhibits a magnetic-field-induced QPT [4, 9, 13] with characteristics similar to those of the XY model in a transverse field [14]. The model was first introduced and studied, at a mean-field approximation (MFA) level, by Matsubara and Matsuda [15] as a pseudospin model for a gas of hard core bosons with attractive interaction to explain superfluidity in ^{4}He. On the experimental front, several magnetic compounds as spin dimer $Cs_3Cr_2Br_9$ [16], the cuprate $Sr_{14}Cu_{24}O_{41}$ [17], the crystal CaV_2O_5 with layers of coupled two-leg ladders [18], and the laddered $SrCu_2O_5$ [19], exhibit field-induced QPTs which appear to be qualitatively well described by the model (1) with easy-plane exchange anisotropy.

However, due to the complex crystalline structure of magnetic materials and, in several cases, the quantitatively unclear validity of the conventional paradigm of quantum criticality [3], it seems reasonable to explore the effects of additional anisotropic crystal fields on the ideal quantum criticality in the PFM.

We use the two-time Green function (GF) method by adopting the Tyablikov decoupling (TD) [20] for the exchange anisotropy and the Anderson-Callen decoupling (ACD) [21] for the single-ion term in the equations of motion chain. At this level of approximation one must keep in mind that the TD was shown to achieve nearly exact predictions close to the field-induced QCP of the PFM in absence of the SIA [10, 9], while the ACD is widely used for the SIA term, although this approximation may be not valid for large values of the parameter D [11, 22].

We follow the Callen approach for the isotropic Heisenberg model [23], which has been used also when anisotropies are presents [9, 24]. We introduce the retarded two-time GF for commutator (in units $\hbar = 1$)

$$G_{ij}(t-t') = \langle\langle S_i^+(t-t'); e^{aS_j^z} S_j^- \rangle\rangle = -i\theta(t-t')\langle [S_i^+(t-t'), e^{aS_j^z} S_j^-] \rangle \tag{2}$$

where $A(t) = e^{i\mathcal{H}t} A e^{-i\mathcal{H}t}$, $\theta(x)$ is the step function, $\langle \cdots \rangle$ denotes the equilibrium average with Hamiltonian (1) and $\beta = 1/T$ is the inverse temperature. The auxiliary Callen parameter a has to be set to zero at the end of calculations for extracting the physics of the model.

The equation of motion for the time Fourier transform $G_{ij}(\omega)$ is

$$\begin{aligned}(\omega - h)\langle\langle S_i^+ | e^{aS_j^z} S_j^- \rangle\rangle_\omega = \psi(a)\delta_{ij} - \sum_h \langle\langle J_{hi}(S_i^z S_h^+ \\ -\Delta S_h^z S_i^+) | e^{aS_j^z} S_j^- \rangle\rangle_\omega + D\langle\langle S_i^z S_i^+ + S_i^+ S_i^z | e^{aS_j^z} S_j^- \rangle\rangle_\omega,\end{aligned} \tag{3}$$

where

$$\psi(a) = \langle [S_i^+, e^{aS_i^z} S_i^-] \rangle, \tag{4}$$

is a site independent quantity to be determined as a function of the Callen parameter a. In particular, $\psi(0) = \langle [S_i^+, S_i^-] \rangle = 2m$, where $m = \langle S_i^z \rangle$ is the longitudinal magnetization per spin.

In order to get a closed equation of motion for $G_{ij}(\omega)$, the higher order Green's functions on the right hand side of eq. (3) has to be properly decoupled. We adopt here the TD [20], for the exchange interaction term

$$\langle\langle S_h^z S_k^+ | e^{aS_j^z} S_j^- \rangle\rangle_\omega \simeq \langle S_h^z \rangle \langle\langle S_k^+ | e^{aS_j^z} S_j^- \rangle\rangle_\omega \tag{5}$$

and the ACD [21] to decouple the single-ion term,

$$\langle\langle S_i^z S_i^+ + S_i^+ S_i^z | e^{aS_j^z} S_j^- \rangle\rangle_\omega \simeq 2C_S \langle S_i^z \rangle \langle\langle S_i^+ | e^{aS_j^z} S_j^- \rangle\rangle_\omega, \tag{6}$$

with $C_S = 1 - \frac{1}{2S^2}\left[S(S+1) - \langle (S_i^z)^2 \rangle\right]$. The term $\langle (S_i^z)^2 \rangle$ in the coefficient C_S can be expressed in terms of $\langle S_i^z \rangle$ and $\langle S_i^- S_i^+ \rangle$ using the identity $(S_i^z)^2 = S(S+1) - S_i^z - S_i^- S_i^+$. With $S = 3/2$, as in the case under study, it is $C_{3/2} = 1 - 2/9[\langle S_i^z \rangle + \langle S_i^- S_i^+ \rangle]$.

Using the above decoupling approximation, the equation of motion eq. (3) reduces to

$$\left[\omega - h - m(\Delta J(0) + 2C_{3/2}D)\right] G_{ij}(\omega) = \psi(a)\delta_{ij} - m\sum_l J_{li}^{\perp} G_{lj}(\omega) \tag{7}$$

where $J(0)$ is the zero-wave-vector Fourier component of J_{ij}.

Using Fourier transforms, with the wave-vectors $\mathbf{k}$ ranging within the first Brillouin zone (1BZ), the equation of motion is readily diagonalized providing the formal solution

$$G_{\mathbf{k}}(\omega) = \frac{\psi(a)}{\omega - \omega_{\mathbf{k}}}, \tag{8}$$

with $\omega \to \omega + i\varepsilon$ ($\varepsilon \to 0^+$) for retarded GF where

$$\omega_{\mathbf{k}} = \omega_0 + m[J(0) - J(\mathbf{k})] \tag{9}$$

is the energy spectrum of the undamped spin-excitations and

$$\omega_0 = h - m[J(0)(1-\Delta) - 2C_{3/2}D] \tag{10}$$

is the corresponding energy gap. The quantity $C_{3/2}$ can be consequently written, via the spectral theorem [20], as $C_{3/2} = 1 - \frac{2m}{9}(1 + 2\Phi)$, where

$$\Phi = \frac{1}{N}\sum_{\mathbf{k}} \frac{1}{e^{\beta\omega_{\mathbf{k}}} - 1} \overset{N\to\infty}{=} \int_{1BZ} \frac{d^d k}{(2\pi)^d} \frac{1}{e^{\beta\omega_{\mathbf{k}}} - 1}\,. \tag{11}$$

The key point in our scheme is to determine the expression of $\psi(a)$ as a function of the Callen parameter a and hence of $m = \psi(0)/2$.

Following ref.[24] we find for the magnetization

$$m = \frac{3}{2} - \Phi + \frac{4}{(1+\Phi^{-1})^4 - 1} \equiv M(\Phi)\,. \tag{12}$$

Eqs. (9)-(12) constitute a set of self-consistent equations to determine m, $G_{\mathbf{k}}(\omega)$ (at $a = 0$), and other relevant thermodynamic quantities as functions of T and h, for fixed values of the SIA parameter D.

To study the static and dynamic properties, we need the transverse GF

$$G_{\perp}(\mathbf{k},\omega) = \langle\langle S_i^+ | S_j^- \rangle\rangle_{\mathbf{k},\omega} \equiv G_{\mathbf{k}}(\omega)|_{a=0}\,, \tag{13}$$

which, at our level of approximation, is given by

$$G_{\perp}(\mathbf{k},\omega) = \frac{2m}{\omega - \omega_{\mathbf{k}}}. \tag{14}$$

This leads to the dynamical transverse susceptibility

$$\chi_\perp(\mathbf{k},\omega) = -G_\perp(\mathbf{k},\omega) = \frac{2m}{\omega_{\mathbf{k}} - \omega}\,, \tag{15}$$

and, as a particular case, the static one

$$\chi_\perp(T,h) = \chi_\perp(\mathbf{k}=0,\omega=0) = \frac{2m}{\omega_0}\,. \tag{16}$$

The stability condition on susceptibility ($\chi_\perp > 0$) implies that $\omega_0 \geq 0$. Thus the equation $\omega_0 = 0$ ($\chi_\perp \to \infty$) defines the stability boundary and hence the possible critical points.

Hereafter we will consider only the case of easy-plane exchange anisotropy ($\Delta < 0$), in order to explore the effects of the SIA on the quantum criticality in a planar ferromagnet (related to the field-driven in-plane ordering described by the quantities ω_0 and $\chi_\perp$ as function of T and h [9, 10, 24].

In absence of SIA ($D = 0$) the systems exhibit a QPT at the value of the field $h_{0c} = \frac{3}{2}J(0)(1-\Delta)$. In the presence of the SIA, the QPT does not survive for all the value of the parameter D. Let us rewrite the expression for ω_0

$$\omega_0 = h - M(\Phi)\left[J(0)(1-\Delta) - 2D\left(1 - \frac{2M(\Phi)}{9}(1+2\Phi)\right)\right], \tag{17}$$

which is a self-consistent expression, reminding the definition of Φ (see (11)).

As first step we study the problem at zero temperature. Since as $T \to 0$ we have that $\Phi \to 0$, then from eq.(12), we get $M(0) = m(T=0) = 3/2$. Hence, for the gap equation at $T = 0$ we have

$$\omega_0 \;=\; h - \frac{3}{2}\left[J(0)(1-\Delta) - \frac{4}{3}D\right] = h - h_{0c} + 2D. \tag{18}$$

Thus, for $h > h_{0c} - 2D$, we have for the static transverse susceptibility

$$\chi_\perp = \frac{3}{h - h_{0c} + 2D} \equiv \frac{3}{h - h_c}. \tag{19}$$

This expression clearly shows that a QCP exist for any dimensionality of the systems and the critical magnetic field at $T = 0$ is $h_c = h_{0c} - 2D$, which depends on the strength of the SIA parameter D. Moreover we obtain for susceptibility a mean-field like behavior, with critical exponent $\gamma = 1$, hence at zero temperature the SIA produce just a shift of the QCP. However if $D = h_{0c}/2$ we have a zero critical field, i.e. a suppression of the QPT. Thus, quantum criticality is realized only if $D < \bar{D} = \frac{3}{4}J(0)(1-\Delta)$.

We now analyze the system at finite temperature. We limit ourselves to the case of short-range exchange interaction and d-dimensional hypercubic lattice, with interaction $J(\mathbf{k}) = J(0) - Jk^2$, with $J(0) = 2dJ$. From now on all quantities will be written in unity of J. The basic quantity to solve our problem is the function Φ (11). On the phase boundary, being $\omega_0 = 0$, we have

$$\Phi_c = \Phi|_{\omega_0=0} = \int_{1BZ} \frac{d^dk}{(2\pi^d)} \frac{1}{e^{\frac{3}{2}\beta M(\Phi_c)k^2} - 1}\,. \tag{20}$$

It is evident that,while a QCP exists for any d, a finite temperature critical line ending in the QCP, may take place only for dimensionalities which assure the convergency of the integral (20). Indeed, we see that only for $d > 2$ we have a finite temperature critical line, ending in the QCP

$(T = 0, h = h_c)$, consistently with the Mermin-Wagner theorem [26]. In such a case we can write a critical line equation (using eqs. (17) and (20))

$$h - M(\Phi_c)\left[\frac{2}{3}h_{0c} - 2D\left(1 - \frac{2M(\Phi_c)}{9}(1 + 2\Phi_c)\right)\right] = 0. \tag{21}$$

This self-consistent equation has been solved numerically for $d = 3$, for three different value of the exchange anisotropy parameter Δ, and changing the SIA parameter D. The results are shown in fig. 1.

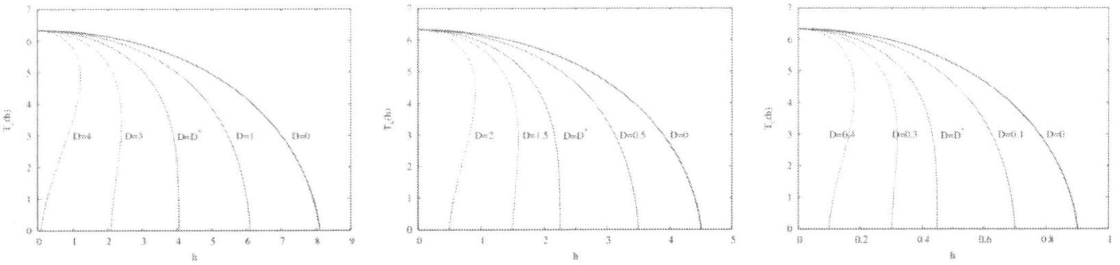

Figure 1. Critical lines for different values of single-ion anisotropy parameter D in the (h, T)-plane as obtained by solving numerically the general equation (21) at $d = 3$ for three different values of the parameter Δ: namely, from left to right, $\Delta = 0.1$, $\Delta = 0.5$ and $\Delta = 0.9$. The borderline value D^* of the single-ion anisotropy parameter D is defined in the text.

By inspection of the figure it is evident the effect played by the SIA on the phase diagram of the ideal PFM ($D = 0$). The typical behavior of the known critical line ending at the QCP of PFM with $D = 0$ breaks down when D exceeds the value D^*, which numerically for $d = 3$ is found to be $D^* = \frac{9}{4}(1 - \Delta)$. For smaller values of D the shape of the critical line remains essentially similar to that of the PFM in absence of the SIA, hence we expect that also the low-temperature behavior around the critical line should be preserved.

In the following we will consider only $D < D^*$, in order to study the effects of SIA on conventional quantum criticality of PFM.

Analytical expressions can be derived in asymptotical regimes. Hence we consider the two following cases: (i) $T \to 0$ and (ii) $h \to 0$. In the first case, i.e. close to the QCP, we have [24]

$$h_c(T) \simeq h_c - b_d(h_c - 2D)T^{d/2} \tag{22}$$

where b_d is a coefficient which depends on the spin S and on the dimensionality of the system. In the present case, with $S = 3/2$ $b_d = \zeta(d/2)/[2^{(d-2)/2}3^{(d+2)/2}\pi^{d/2}]$, in particular for $d = 3$ we have $b_3 = \zeta(1.5)/(9\sqrt{6}\pi^{3/2}) \simeq 0.0213$. By inspection of equation (22), we see that there exists a value of D above which the coefficient of $T^{d/2}$ changes sign, i.e. for $D > h_c$, and hence the low-temperature part of the critical line shows the reentrant behavior. This provides the analytical expression for the borderline value D^* of the SIA parameter D, which is $D^* = h_{0c}/4 = 3J(0)(1 - \Delta)/8$, that coincides with the numerical result for $d = 3$.

For case (ii), i.e. in the limit $h \to 0$, we have [24]

$$T_c(h) \simeq T_c(0)\left[1 - \left(12 + \frac{5d^2}{d^2 - 4}\right)\frac{3}{125}\frac{h^2}{h_c + \frac{2}{3}D}\right], \tag{23}$$

where

$$T_c(0) = \frac{5}{16\pi}\frac{d}{d-2}\left(\frac{d}{2}\Gamma(d/2)\right)^{-2/d}. \tag{24}$$

The critical temperature at zero field $T_c(0)$ does not depend on the value of the SIA parameter D.

We proceed now by studying the low-temperature properties and crossovers of our anisotropic XXZ ferromagnet in the easy-plane disordered phase close to the QCP where the energy gap ω_0 is small. In this low-T regime, being Φ very small, we can use a simplified expression for magnetization $M(\Phi) \simeq \frac{3}{2} - \Phi$. Furthermore, considering the limit $T \to 0$ in the integral (11) which defines Φ and with $\omega_{\mathbf{k}} = \omega_0 + mk^2$, we have

$$\Phi \simeq \frac{K_d}{2}\left(\frac{T}{m}\right)^{d/2}\int_0^\infty dx \frac{x^{(d-2)/2}}{e^{\beta\omega_0+x}-1}. \tag{25}$$

The integral which appear in the above equation is the Bose-Einstein function $F_\nu(y) = \int_0^\infty dx \frac{x^{\nu-1}}{e^{x+y}-1}$ for which the asymptotic expansions in y for different values of ν are known [9]. This provides for the magnetization

$$m \simeq \frac{3}{2} - \frac{K_d}{2}\left(\frac{2}{3}T\right)^{d/2} F_{d/2}(\omega_0/T). \tag{26}$$

We can then write a self-consistent equations for the energy gap ω_0, to leading order in T

$$\omega_0 \simeq h - h_c + \frac{K_d}{3}(h_c - 2D)\left(\frac{2}{3}T\right)^{d/2} F_{d/2}(\omega_0/T)\,. \tag{27}$$

For any d, with $h > h_c$ in the quantum regime we find for the energy gap (with $F_{d/2}(\omega_0/T) \simeq \Gamma(d/2)e^{-\omega_0/T}$)

$$\omega_0(T,h) \simeq (h-h_c) + \Gamma(d/2)C_{d,D}T^{d/2}e^{-(h-h_c)/T}\,, \tag{28}$$

which differs from the MF result $\omega_0 \simeq (h-h_c)$ for an exponentially small correction in $(h-h_c)/T \gg 1$. In eq.(28) it is $C_{d,D} = \frac{K_d}{3}(2/3)^{d/2}(h_c - 2D)$.

We now determine the asymptotic solution of (27) in the classical ($\omega_0/T \ll 1$) and quantum ($\omega_0/T \gg 1$) regimes, and related crossovers, for $d=2$ and $d>2$.

• $d = 2$

For this value of dimensionality we do not have a critical line, but only a QCP for $D < \bar{D}$. Close to the QCP, the energy gap equation is obtained setting $d = 2$ and $F_1(\omega_0/T) = \ln(1-e^{-\omega_0/T})^{-1}$ in eq. (27). This equation, which is valid for arbitrary ω_0/T, cannot be solved in general but one can obtain explicit asymptotic expressions in relevant regimes.

We start with the classical regime. Simple considerations show that no physical solution occurs for $h \geq h_c$ while, for $h < h_c$ and $(h_c - h) \gg T$, one easily gets

$$\omega_0(T,h) \simeq T\exp\left[-\frac{h_c-h}{C_{2,D}T}\right]. \tag{29}$$

This result implies that the transverse susceptibility (and hence the correlation length $\xi_\perp \propto \chi_\perp^{1/2}$) diverges exponentially as $T \to 0$ at fixed $h < h_c$.

Within the region of the (h,T)-plane delimited by the two crossover lines $T^- = h_c - h$ and $T^+ = h - h_c$, we can obtain a consistent asymptotic solution for $|h - h_c| \ll T$ only assuming ω_0/T finite. Let us first rewrite eq.(27) for $d = 2$

$$\frac{\omega_0}{T} = \frac{h-h_c}{T} + \frac{h_c - 2D}{9\pi}\ln\left(\frac{1}{1-e^{-\omega_0/T}}\right). \tag{30}$$

We are interested in the behavior on and around the so-called quantum critical trajectory (QCT) $h = h_c$, where we can rewrite eq. (30) in the form $\omega_0 = bT$ where b is determined by the equation $b = C_{2,D} \ln\left(\frac{1}{1-e^{-b}}\right)$. Then by expanding eq. (30) in $(h - h_c)/T$ with $\omega_0/T = b$ as zero order solution, to leading order in $h - h_c$, the energy gap reads

$$\omega_0(T, h) \simeq bT + [1 + C_{2,D}(e^b - 1)^{-1}]^{-1}(h - h_c) \,. \tag{31}$$

The phase diagram for $d = 2$ is shown in fig. 2. It is worth noting that the previous scenario breaks down as $D \to D^*$ (i.e. $C_{d,D} \to 0$), as expected.

• $d > 2$

For such dimensionalities a critical line exists ending in the field-induced QCP, and it is convenient to rewrite eq. (27) in terms of the distance $g(T) = h - h_c(T) \geq 0$ from the critical line in the disordered phase. This is simply performed by combining eqs. (22) and (27) to write

$$\omega_0 = g(T) + C_d(D)T^{d/2}G_{d/2}(\omega_0/T) \,, \tag{32}$$

where $G_{d/2}(y) = F_{d/2}(y) - F_{d/2}(0)$. The full low-temperature physics for $d > 2$ follows from this equation but we limit ourselves to $2 < d < 4$ since for $d \geq 4$ MF-like findings can be trivially obtained with logarithmic corrections at $d = 4$, consistently with the theory of critical phenomena. We first explore the classical regime $\omega_0/T \ll 1$, where $G_{d/2}(\omega_0/T) \simeq -\pi|\sin(\pi d/2)|^{-1}(\omega_0/T)^{d/2-1}$. This lead us to the equation for the ω_0

$$\omega_0 \simeq g(T) - \frac{\pi C_{3,D}}{|\sin(\pi d/2)|} T\omega_0^{d/2-1} \tag{33}$$

whose asymptotic solution are

$$\omega_0(T, h) \simeq \begin{cases} g(T) & \text{far from the critical line} \\ \left(\frac{|\sin(\pi d/2)|}{\pi C_{3,D}}\right)^{\frac{2}{d-2}} \left(\frac{g(T)}{T}\right)^{\frac{2}{d-2}} & \text{close to the critical line.} \end{cases} \tag{34}$$

For transverse susceptibility, in the region close to the critical line, we then have

$$\chi_\perp(T, h) \simeq 3\frac{h}{h_c(T)} \left(\frac{\pi C_{3,D}}{|\sin(\pi d/2)|}\right)^{\frac{2}{d-2}} T^{\frac{2}{d-2}} (h - h_c(T))^{-\frac{2}{d-2}} \,, \tag{35}$$

hence the susceptibility diverges with a power-law in $h - h_c(T)$, with the spherical critical exponent $\gamma = 2/(d - 2)$, independently of the anisotropy parameter. Far from the critical line we have instead the mean-field critical exponent $\gamma = 1$. The crossover line which separates this two asymptotic behavior is

$$h_G(T) \simeq h_c(T) + \left[\frac{\pi C_{d,D}}{|\sin(\pi d/2)|}\right]^{\frac{2}{4-d}} T^{\frac{2}{4-d}}. \tag{36}$$

Let us finally consider the behavior along the QCT, where $h = h_c$. Here we have

$$\omega_0 = b_d(h_c - 2D)T^{d/2} - \frac{\pi C_{3,D}}{|\sin(\pi d/2)|} T\omega_0^{d/2-1} \tag{37}$$

and hence, at low temperature, we have for the susceptibility $\chi_\perp \sim T^{-d/2}$. Within a narrow region around $h = h_c$ the energy gap and all the related thermodynamic quantities behave essentially as along the QCT, except for negligible corrections in $(h - h_c)/T$.

Journal of Physics: Conference Series **529** (2014) 012019 doi:10.1088/1742-6596/529/1/012019

Increasing $h - h_c > 0$ a crossover to the quantum regime $\omega_0/T \gg 1$ occurs which is characterized by the solution (28), valid for any d. In fig. 2 the phase diagram with crossover line for $d = 3$ is shown and compared with that for $d = 2$.

Previous results for $2 < d < 4$ close to the field-induced QCP in the presence of SIA with $D < D^*$ is similar to those found for conventional paradigmatic quantum models [3] by means of the RG approach. However, if $D \to D^*$ the SIA tends to destroy the conventional quantum critical scenario.

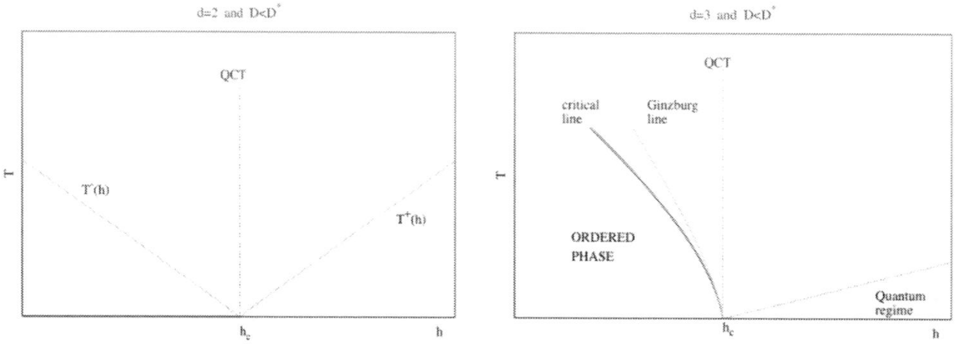

Figure 2. Phase diagram with crossover lines for dimensionalities $d = 2$ (left) and $d = 3$ (right). In both cases we consider small value of single-ion anisotropy parameter D, namely $D < D^*$. The phase diagram looks qualitatively similar to the one obtained in absence of single-ion anisotropy [9].

In conclusion, we used the two-time GFs method to explore the effects of the SIA on the magnetic-field-induced quantum criticality of a d-dimensional spin-3/2 PFM with short-range exchange couplings. The full phase diagram in the (h, T)-plane and the main physical properties at fixed single-ion parameter D have been obtained at the level of the TD and the ACD for higher order exchange interactions and SIA GFs, respectively. We stress that in our study the applied magnetic field h is assumed as the non-thermal control parameter driving quantum criticality, in contrast with recent literature where this role is played by the parameter D at fixed h (frequently assumed equal to zero). The emerging feature is that the phase boundary of the PFM is significantly influenced by the SIA, especially close to the QCP, but the quantum criticality of PFM for different dimensionalities remains essentially unchanged for values of the SIA parameter smaller than the threshold value $D^* = 3J(0)(1-\Delta)/8$. Indeed we obtain, within our approximation, mean-field-like critical exponents at the QPT and critical exponents belonging to the spherical model universality class at finite temperature. These are the same results obtained in absence of SIA [9, 10]. Hence, the inclusion of SIA does not affect the universality class of the model. For $D^* < D < \bar{D}$, where $\bar{D} = h_{0c}/2$ is the value of D at which the QCP is suppressed, our self-consistent equations are substantially modified. As a consequence, for $d > 2$, a reentrant low-temperature structure of the phase diagram appears (see fig. 1) and a modified quantum critical scenario is expected. The detailed analysis of the reentrant feature in the the phase diagram will be the subject of a future study. One may argue that the reentrant behavior may be due to the approximation used. Unfortunately, to our knowledge, no landmarks are available from literature about the SIA effects on quantum criticality controlled by an external magnetic field. Hence, at the present time, one cannot draw definitive statements about the role played by large easy-axis anisotropy in the vicinity of a magnetic-field-induced QCP. On the other hand, there is no clear reasons to consider "a priori" the ACD inadequate to provide physical findings around a QCP. Moreover, reentrant features in the phase diagram usually arise when there are competing interactions. Furthermore, in the light of our numerical and analytical results, we speculate that the existence of D^* and of the easy-axis anisotropy regime for values of D in the limited range $D^* < D < \bar{D}$ are physically plausible. Indeed, the feature that the zero-temperature critical field h_c vanishes as $D \to \bar{D}$ suggests that increasing of the easy-axis anisotropy parameter tends, as expected, to suppress quantum critical fluctuations until, for $D \geq \bar{D}$, no field-induced QPT takes place. In this picture, D^* represents a borderline value of D between an anisotropy regime for $0 < D < D^*$, where the conventional quantum critical scenario is preserved, and that for $D^* < D < \bar{D}$, where reentrant phenomena appear in the low-temperature phase diagram close to the QCP and a non-conventional magnetic-field-induced quantum critical scenario is expected.

[1] Schollwöck U, Richter J, Farnell D J J, and Bishop R F 2004 *Quantum magnetism* Lecture Notes in Physics **645** (Springer, Berlin)
[2] Fröbich P and Kunz P J 2006 *Phys. Rep.* **432** 223
[3] Sachdev S 1999 *Quantum Phase Transitions* (Cambridge University Press, Cambridge); Löhneysen H V, Rosch A, Vojta M and Wölfle P 2007 *Rev. Mod. Phys.* **79** 1015 ; Mercaldo M T, De Cesare L, Rabuffo I and Caramico D'Auria A 2007 *Phys. Rev. B* **75** 014105
[4] Caramico D'Auria A, De Cesare L, Mercaldo M T and Rabuffo I 2005 *Physica A* **351** 294; Rabuffo I, Mercaldo M T, De Cesare L and Caramico D'Auria A 2006 *Phys. Lett. A* **356** 174; Mercaldo M T, Caramico D'Auria A, De Cesare L and Rabuffo I 2008 *Phys. Rev B* **77** 184424
[5] Gambardella P *et al* 2002 *Nature* **416** 301
[6] Mukherjee C J *et al* 2004 *J. Magn. Magn. Mater.* **272-276** 920
[7] Cuccoli A, Roscilde T, Vaia R and Verrucchi P 2003 *Phys. Rev. Lett.* **90** 167205
[8] Vogt B and Kettermann S 2009 *Ann. Phys.* (Berlin) **18** 759
[9] De Cesare L, Mercaldo M T, Rabuffo I, Campana L S, and Esposito U 2010 *Phys. Rev. B* **82** 024409
[10] De Cesare L, Mercaldo M T, Campana L S and Esposito U 2009 *Physica A* **388** 1446
[11] Devlin J F 1971 *Phys. Rev. B* **1** 136
[12] De Fotis G C, Laccheo M L, Aruga Katori H 2002 *Phys. Rev. B* **65** 094403
[13] Syromyatnikov A V 2007 *Phys. Rev. B* **75** 134421
[14] Gerber P B 1978 *J. Phys. C* **11** 5005
[15] Matsubara T and Matsuda H 1956 *Progr. Theor. Phys.* **16** 569; *ibidem* 1957 **17** 19
[16] Grenier B *et al* 2004 *Phys. Rev. Lett.* **92** 177202
[17] Regnault L P *et al* 1999 *Phys. Rev. B* **59** 1055
[18] Uhrig G S, Schmidt K P and Grüninger M 2004 *Phys. Rev. Lett.* **93** 267003
[19] de Graaf C, de P. R. Moreira I, Illas F, and Martin R L 1999 *Phys. Rev. B* **60** 3457
[20] Tyablikov S V 1967 *Methods in the Quantum Theory of Magnetism* (Plenum Press, New York)
[21] Anderson F B and Callen H 1964 *Phys. Rev.* **136** A1068 (1964)
[22] Fröbich P and Kunz P J 2006 *Phys. Rep.* **432** 223
[23] Callen H B 1963 *Phys. Rev.* **130** 890
[24] Mercaldo M T, Rabuffo I, De Cesare L, and Caramico D'Auria A 2013 *Eur. Phys. J. B* **86** 340
[25] Nolting W and Ramakanth A 2009 *Quantum Theory of Magnetism* (Springer-Verlag, Berlin Heidelberg)
[26] Mermin N D and Wagner H 1966 *Phys. Rev. Lett.* **17** 1133

Longitudinal Excitations in Bipartite and Hexagonal Antiferromagnetic Spin Lattices

Y Xian[1] **and M Merdan**[1,2]

[1]School of Physics and Astronomy, The University of Manchester, Manchester M13 9PL, UK
[2]Department of Physics, College of Science, University of Babylon, Babylon, Iraq

Abstract. Based on our recently proposed magnon-density-waves using the microscopic many-body approach, we investigate the longitudinal excitations in quantum antiferromagnets by including the second order corrections in the large-s expansion. The longitudinal excitation spectra for a general spin quantum number using the antiferromagnetic Heisenberg Hamiltonian are obtained for various spin lattice models. For bipartite lattice models, we find that the numerical results for the energy gaps for the longitudinal modes at $q \to 0$ and the magnetic ordering wavevector $\mathbf{Q}$ are reduced by about 40-50 % after including the second order corrections. Thus, our estimate of the energy gaps for the quasi-one-dimensional (quasi-1D) antiferromagnetic compound $KCuF_3$ is in better agreement with the experimental result. For the quasi-1D antiferromagnets on hexagonal lattices, the full excitation spectra of both the transverse modes (i.e., magnons) and the longitudinal modes are obtained as functions of the nearest-neighbor coupling and the anisotropy constants. We find two longitudinal modes due to the non-collinear nature of the triangular antiferromagnetic order, similar to that of the phenomenological field theory approach by Affleck. We compare our results for the longitudinal energy gaps at the magnetic wavevectors with the experimental results for several antiferromagnetic compounds with both integer and non-integer spin quantum numbers, and also find good agreement after the higher-order contributions are included in our calculations.

1. introduction

The dynamics of the two-dimensional (2D) and three-dimensional (3D) quantum antiferromagnetic systems with long-ranged order at low temperature can be considered as that of a dilute gas of weakly interacting spin-wave quasiparticles (magnons) with its density given by the quantum correction to the classical Néel order [1, 2, 3]. These magons are transverse modes with spin $S = \pm 1$. The longitudinal fluctuations with spin $S = 0$ present in these systems consist of the multi-magnon continuum [4]. The question concerning long-lived, well-defined longitudinal modes in quantum antiferromagnetic systems with long-ranged order remains open. This is our main focus in this paper.

In case of the quantum antiferromagnetic systems without long-ranged order, the triplet excitation states (two transverse and one longitudinal) of the spin-1 Heisenberg chain with non-zero gap, first predicted by Haldane [5], are well known. This theoretical prediction of an energy gap separating the singlet ground state from the triplet excitation states has been confirmed experimentally in the quasi-1D antiferromagnetic compounds such as $CsNiCl_3$ and $RbNiCl_3$ of the spin-1 [6]. Some subsequent experimental investigations [7, 6, 8, 9, 10] and theoretical calculations [11, 12, 13, 14, 3] also support Haldane's conjecture. In these experiments, the temperature is high enough so the quasi-1D systems have no long-ranged magnetic order and the dynamics can be described by the 1D models. At lower temperatures, these quasi-1D antiferromagnets behave as 3D systems with long-ranged magnetic orders. The key question is whether or not the longitudinal modes survive in present of the long-ranged order and, if the answer is yes, how we describe them in general terms. In this regard, the observation of an energy gap for the quasi-1D compound $CsNiCl_3$ at low temperature in 1986 generated much theoretical interest [6]. This energy gap was initially explained by a uniaxial single-ion anisotropy but now it is widely accepted that the gapped excited state belongs to longitudinal excitation modes, first proposed by Affleck [in the quasi-1D hexagonal antiferromagnetic compounds of the ABX_3-type with both spin quantum number $s = 1$ $CsNiCl_3$ and $RbNiCl_3$ [15, 16]. A field theory approach focusing on the spin frustrations of the hexagonal antiferromagnetic systems has also been proposed [17]. Clearly, such longitudinal modes are beyond the usual spin-wave theory (SWT) which only predicts the transverse spin-wave excitations (magnons). There have been several theoretical investigations in these longitudinal modes, all using the field theory approach, such as the sine-Gordon theory for the spin-1/2 systems in bipartite quasi-1D antiferromagnetic systems treating the inter-chain couplings as perturbation [18, 19]. The theoretical estimate for the energy gap of the longitudinal mode is in good agreement with the experimental results for the compound $KCuF_3$ [20]. More recently, a longitudinal mode was also observed in the dimerized antiferromagnetic compound $TlCuCl_3$ under pressure with a long-range Néel order [21] and tetrahedral spin system $Cu_2Te_2O_5Br_2$ using Raman scattering [22] . To our knowledge, no observation of longitudinal modes in 2D or quasi-2D antiferromagnets has been reported yet.

Our theoretical investigation of the longitudinal modes in quantum antiferromagnets is based on a microscopic theory for a generic spin-s Hamiltonian system [23]. We identify the longitudinal excitation states in a quantum antiferromagnet with a Néel-like order as the collective modes of the magnon-density waves, which represent the fluctuations in the magnitude of the long-range order and are supported by the interactions between magnons. The basic idea in our analysis is similar to Feynmann's theory on the low-lying excited states of the helium-4 superfluid [24], thereby employing the magnon-density operator S^z for the antiferromagnets in place of the particle density operator for the helium-4 superfluid. Hence, the longitudinal excitation states in antiferromagnets are constructed by the S^z spin operators, contrast to the transverse spin operators $S^{\pm}$ of the magnon states in Anderson's SWT [1]. Our preliminary calculation for the two dimensional triangular model [25] have been extended to the quasi-1D Hexagonal structures of $CsNiCl_3$ and $RbNiCl_3$, where we find that our numerical results for the energy gap values at the magnetic wavevector are in good agreement with experimental results after inclusion of the high-order contributions in the large-s expansion [26]. In this article, we extend similar high-order calculations to the bipartite antiferromagnetic systems where the long-ranged order is collinear [27].

We organize this article as follows. Sec. II-V cover various bipartite systems with new results from the high-order calculations. In particular, Sec. V focuses on the tetragonal quasi-1D compound $KCuF_3$ with spin-1/2 and we find that our estimate of the minimum energy gap for the longitudinal mode after inclusion of the high-order contributions is in good agreement with the experimental result. For completeness, in Sec. VI we include our earlier analysis [26] for the quasi-1D hexagonal systems where there are a number of experimental results for comparison. In Sec. VII we conclude this article by a summary and a discussion of a possible longitudinal mode in a 2D square lattice model, relevant to the parent compound La_2CuO_4 of the high-T_c superconductors.

2. Longitudinal Excitation in bipartite quantum antiferromagnets

The longitudinal excitations in a quantum antiferromagnetic system with a Néel-like long range order correspond to the fluctuations in the order parameter. Since the quantum correction in the order parameter is given by the magnon density [1], we identify the longitudinal modes as the

Journal of Physics: Conference Series **529** (2014) 012020 doi:10.1088/1742-6596/529/1/012020

magnon-density waves (MDW), supported by the interactions between the magnons. It is clear that the states of MDW may not be well defined in the high-order dimensional systems where the magnon density is very dilute and the long-range order is near the classical value with little quantum correction. However, the magnon density may be high enough in the low dimensional systems to support the longitudinal waves. In terms of microscopic many-body language, the MDW states are constructed by applying the magnon density operator S^z on the ground state in a form as $S^z|\Psi_g\rangle$, similar to Feynmann's theory of the phonon-roton excitation state of the helium superfluid, where the density operator is the usual particle density operator [24, 28].

We consider a general spin-s XXZ Heisenberg model on a bipartite lattice with Hamiltonian given by

$$H = J\sum_{\langle i,j\rangle}\left[\frac{1}{2}(S_i^+S_j^- + S_i^-S_j^+) + AS_i^zS_j^z\right], \tag{1}$$

where the notation $\langle i,j\rangle$ indicates the nearest-neighbor couplings only and A is the anisotropy parameter $A\,(\geq 1)$. The usual isotropic Heisenberg Hamiltonian is given by $A = 1$. The classical ground state of Eq. (1) is given by the Néel state with two alternating sublattices, one with all spin-up and the other with all spin-down. We leave the discussion of the hexagonal systems to Sec. VI. We perform spin rotation on the spin-up sublattice by 180° so that all spins align in the same down direction. This is equivalent to the transformation

$$S_i^\mp \to -S_i^\pm, \quad S_i^z \to -S_i^z, \tag{2}$$

for all sites of the spin-up sublattice. The Hamiltonian (1) after this transformations is given as

$$H = -\frac{1}{4}J\sum_{l\varrho}\left[(S_l^+S_{l+\varrho}^+ + S_l^-S_{l+\varrho}^-) + 2AS_l^zS_{l+\varrho}^z\right], \tag{3}$$

where l runs through all N sites, ϱ is the nearest neighbor index vector with coordination number $z = 2, 4$ and 6 for linear chain, square lattice and cubic lattice respectively. The quantum ground-state of the Hamiltonian H is different from the the classical Néel state. This difference can be quantify by a correction in the sublattice magnetization M as,

$$M = \frac{1}{N}\sum_l \langle S_l^z\rangle_g = s - \rho, \tag{4}$$

where $\langle S_l^z\rangle_g$ indicated the ground-state expectation, s is the classical value, and ρ is the quantum correction with the physical meaning of the magnon density. Therefore, the operator S^z corresponds to the magnon-density operator, contrast to the spin-flip operators $S^\pm$ which creates or destroys magnons. In this article, we employ Anderson's SWT for our approximation of the ground state $|\Psi_g\rangle$. Anderson's SWT can be most simply formulated by expressing S^z and $S^\pm$ for all sites in terms of boson operators $a^\dagger$ and a, as

$$S^z = -s + a^\dagger a, \quad S^+ = \sqrt{2s}fa, \quad \sqrt{2s}a^\dagger f, \tag{5}$$

with $f = \sqrt{1 - a^\dagger a/2s}$. For example, the linear SWT by setting $f = 1$ produces the values of $\rho = 0.078$ per lattice site for the spin-1/2 isotropic Heisenberg model on a simple cubic lattice, and $\rho = 0.198$ per lattice site for the same model but on a square lattice. For the same model on 1D, however, SWT fails to produce the exact result of $\rho = 1/2$, which represents the maximum, saturated value of the magnon density. We consider the longitudinal modes of the quantum antiferromagnetic systems in terms of the fluctuations in these magnon densities as described below.

Following Feynman as mentioned earlier, the longitudinal excitation state is approximated by applying the magnon density fluctuation operator X_q to the ground state $|\Psi_g\rangle$ as

$$|\Psi_e\rangle = X_q|\Psi_g\rangle, \tag{6}$$

where X_q is given by the Fourier transformation of S^z operators,

$$X_q = \frac{1}{\sqrt{N}}\sum_l e^{i\mathbf{q}\cdot\mathbf{r}_l} S_l^z, \quad q > 0, \tag{7}$$

with index l running over all lattice sites. The condition $q > 0$ in Eq. (7) ensures the orthogonality to the ground state. The energy spectrum for the trial excitation state of Eq. (6) can be written as

$$E(q) = \frac{\langle\Psi_g|\tilde{X}_q[H, X_q]|\Psi_g\rangle}{\langle\Psi_e|\Psi_e\rangle},$$

where $\tilde{X}_q$ is the Hermitian of X_q and where we have used the ground state equation, $H|\Psi_g\rangle = E_g|\Psi_g\rangle$. We notice that operator S_l^z in X_q of Eq. (7) is a Hermitian operator, hence $\tilde{X}_q = X_{-q}$. By considering the similar excitation state $X_{-q}|\Psi_g\rangle$ with the energy spectrum $E(-q) = E(q)$, it is straightforward to derive [29],

$$E(q) = \frac{N(q)}{S(q)}, \tag{8}$$

where $N(q)$ is given by the ground-state expectation value of a double commutator as

$$N(q) = \frac{1}{2}\langle[X_{-q}, [H, X_q]]\rangle_g, \tag{9}$$

and the state normalization integral $S(q)$ is the structure factor of the lattice model

$$S(q) = \langle\Psi_e|\Psi_e\rangle = \frac{1}{N}\sum_{l,l'} e^{i\mathbf{q}.(\mathbf{r}_l - \mathbf{r}_{l'})}\langle S_l^z S_{l'}^z\rangle_g. \tag{10}$$

In the following sections, we apply the SWT for the approximation of the ground state $|\Psi_g\rangle$ to evaluate these expectation values.

3. Magnon-density waves in simple lattices

Using the Hamiltonian (1) and the usual spin commutation relations, it is straightforward to derive the following result for $N(q)$

$$N(q) = zsJ(1 + \gamma_q)\,\tilde{g}_\varrho, \tag{11}$$

where γ_q is defined as usual as

$$\gamma_q = \frac{1}{z}\sum_\varrho e^{i\mathbf{q}\cdot r_\varrho}. \tag{12}$$

In Eq. (11) $\tilde{g}_\varrho$ is the transverse spin correlation functions defined as

$$\tilde{g}_\varrho = \langle S_l^+ S_{l+\varrho}^+\rangle = \Delta_\varrho - \frac{2\rho\,\Delta_\varrho + \mu_\varrho\,\delta}{2s}, \tag{13}$$

where

$$\rho = \langle a_l^\dagger a_l \rangle = \frac{1}{N}\sum_q \rho_q, \quad \mu_\varrho = \langle a_l^\dagger a_{l+\varrho} \rangle = \frac{1}{N}\sum_q e^{i\mathbf{q}\cdot\varrho}\rho_q,$$
$$\Delta_\varrho = \langle a_l a_{l+\varrho} \rangle = \frac{1}{N}\sum_q e^{i\mathbf{q}\cdot\varrho}\Delta_q, \quad \delta = \langle a_l a_l \rangle = \frac{1}{N}\sum_q \Delta_q, \tag{14}$$

and where

$$\Delta_q = \frac{1}{2}\frac{\gamma_q/A}{\sqrt{1-\gamma^2/A^2}}, \quad \rho_q = \frac{1}{2}\left(\frac{1}{\sqrt{1-\gamma^2/A^2}} - 1\right). \tag{15}$$

In evaluating the correlation function $\tilde{g}_\varrho$ of Eq. (13), we keep up to second order in the large-s expansions. The double commutator in general behaves as near the magnetic wavevector: $N(q+\mathbf{Q}) \propto q^2, \;\; q \to 0$, similar to that of the helium superfluid [24, 28].

The structure factor is thus given by

$$S(q) = \rho + \frac{1}{N}\sum_{q'} \rho_{q'}\rho_{q+q'} + \frac{1}{N}\sum_{q'} \Delta_{q'}\Delta_{q+q'}. \tag{16}$$

We notice that the integral in the structure factor involving the function $\gamma_{q'}\gamma_{q-q'}$ indicate the couplings between magnons, with the summation over q is given by the integration

$$\sum_q = \frac{1}{(2\pi)^D}\int_{-\pi}^{\pi} d^D q, \tag{17}$$

where D is the dimensionality of the system. The longitudinal excitation $E(q)$ of Eq. (8) is hence obtained by evaluating the the double commutator and the structure factor of Eqs. (11)-(16). We can compare the longitudinal spectrum $E(q)$ with the transverse spin-wave spectrum given by

$$\mathcal{E}(q) = szJA\sqrt{1-\gamma_q^2/A^2}. \tag{18}$$

3.1. Results for linear chain (1D) Model

The SWT breaks down for the isotropic 1D case as most integrals involving the ground-state expectation diverge. For example, $\rho \to \infty$ as $A \to 1$ in the linear SWT. Furthermore, we notice that the spin-wave spectrum of Eq. (18) is doublet while the exact result by Bethe ansatz [30] is triplet for the spin-1/2 model. Nevertheless, the value of Eq. (18), $J\sin q$, is not far off the exact triplet spectrum derived as [31]

$$\mathcal{E}^{\text{exact}}(q) = \frac{\pi}{2}J\sin q. \tag{19}$$

For the longitudinal mode, we have examined the behaviors of in $N(q)$ and $S(q)$ in the isotropic limit ($A \to 1$) in our earlier paper [27] and found that both have the same divergence as $\tilde{g}_1 \propto -\frac{1}{2\pi}\ln(A-1)$, and $S(q) \to -\frac{1}{2\pi}\ln(A-1)/\sqrt{A-1}$, as $q \to 0$. Therefore, the longitudinal excitation sectrum in the isotropic limit, $E(q) \to J\sin q$, coincides precisely with spin-wave spectrum of Eq. (18). Thus, together they form the triplet excitation states, in agreement with the exact result of Bethe ansatz [31]. For $A > 1$, the longitudinal excitation spectrum $E(q)$ is higher than those of the doublet spin-wave spectrum as shown in Fig. 1, where we plot $E(q)$ for $A = 1.1$. The gaps for the longitudinal mode $E(q)$ are about $1.16szJ$ and $1.64szJ$ at $q \to 0$

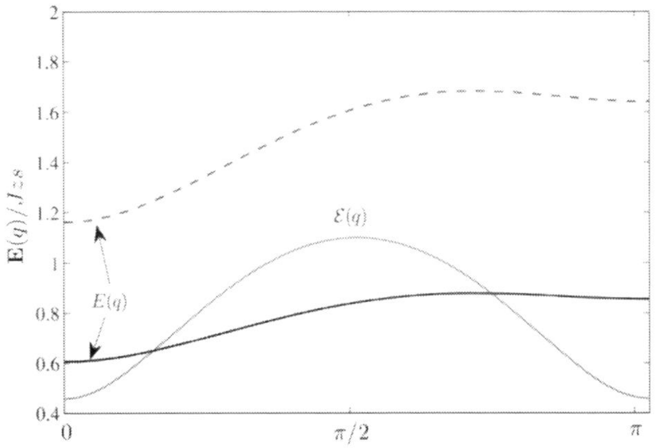

Figure 1. The longitudinal excitation spectrum of Eq. (8) for the linear chain (1D), together with the spin-wave spectrum of Eqs. (18) both with anisotropy $A = 1.1$. The longitudinal spectra calculated from the first-order and second-order approximations are indicated by the dash and solid lines respectively.

and $q = \pi$ respectively in the first order approximation of Eq. (13) [27]. After including the second-order terms, the gap values are $0.61szJ$ and $0.86szJ$ at $q \to 0$ and $q = \pi$ respectively, comparing with $0.46szJ$ of the spin-wave spectrum at the both points. We notice that the longitudinal spectrum is reduced by about half after including the second order correction, and is even lower than that of the spin-wave spectrum around $q = \pi/2$.

3.2. Results for the square (2D) and cubic (3D) lattices

For the square lattice model at the isotropic point $A = 1$, we obtain, for all ϱ, $\tilde{g}_\varrho \approx 0.28$ and 0.17 for the first-order and the second-order approximations respectively. In both cases, the longitudinal excitation spectrum $E(q)$ becomes gapless due to the divergence of the structure factor in logarithmic manner, i.e. $S(q) \to -\ln q$ at $q \to 0$ and the magnetic ordering wavevector $\mathbf{Q} = (\pi, \pi)$ as discussed in our previous paper [?, 23, 27]. We also confirmed this logarithmic behaviour of $E(q)$ in the triangular lattice model at $q \to 0$ and the triangular magnetic ordering wave vector $\pm[\mathbf{Q} = (4\pi/3, 0)]$ [25]. We have referred this behavior in the spectrum as "quasi-gapped" since any tiny anisotropy or finite-size effect will produce a significant gap. For example, we consider a tiny anisotropy with a value $A - 1 = 1.5 \times 10^{-4}$, which in fact is a typical value for the high-T_c compound La_2CuO_4 [32], the gap values at $q \to 0$ and the magnetic ordering wavevector $\mathbf{Q}$ increase to $E(q) \sim 0.44szJ$ and $E(\mathbf{Q}) \sim 0.76szJ$ in the first order approximation, and $E(q) \sim 0.27szJ$ and $E(\mathbf{Q}) \sim 0.47szJ$ in the second-order approximation, as shown in Fig. 2. We notice that these gap values at both points are much larger than the corresponding spin-wave gap of $0.02szJ$. At the two particular momenta $(\pi/2, \pi/2)$ and $(\pi, 0)$, where $\gamma_q = 0$ and the spin-wave spectrum of Eq. (18) gives the same value of szJ, the longitudinal spectrum $E(q)$ has slightly different values, $1.36szJ$ and $1.40szJ$ in the first-order approximation, and $0.84szJ$ and $0.87szJ$ in the second-order approximation.

For the simple cubic lattice model (3D), the numerical results for $\tilde{g}_\varrho$ of Eq. (13) are about 0.13 and 0.11 in the first-order and second-order approximations respectively. The gap values at $q \to 0$ and the magnetic ordering wavevector $\mathbf{Q}$ are $E(q) \sim 0.99szJ$ and $E(\mathbf{Q}) \sim 1.40szJ$ in the first-order approximation, and $E(q) \sim 0.84szJ$ and $E(\mathbf{Q}) \sim 1.19szJ$ in the second-order approximation.

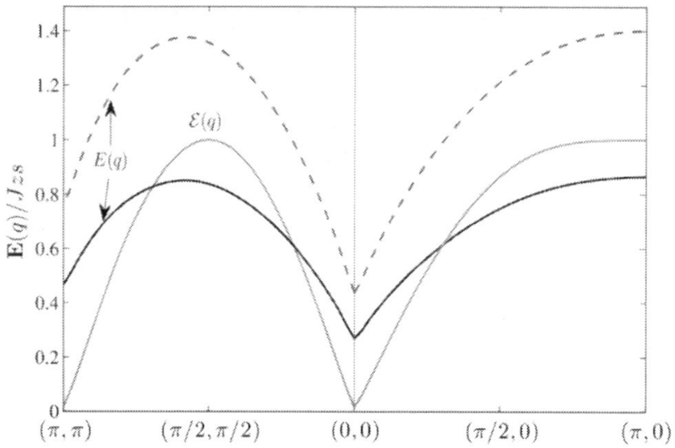

Figure 2. (Color online) The longitudinal excitation spectrum of Eq. (8) for square lattice (2D) , together with the spin-wave spectrum of Eqs. (18) both with anisotropy $A = 1.00015$. The longitudinal spectra calculated from the first-order and second-order approximations are indicated by the dash and solid lines respectively.

4. Tetragonal quasi-1D and quasi-2D antiferromagnets

The Hamiltonian for tetragonal quasi-1D and quasi-2D antiferromagnets is given by

$$H = \frac{1}{2}J\big(\sum_{l,\varrho}^{\text{chain}} \mathbf{S}_l \cdot \mathbf{S}_{l+\varrho} + \xi \sum_{l,\varrho'}^{\text{plane}} \mathbf{S}_l \cdot \mathbf{S}_{l+\varrho'}\big), \tag{20}$$

where l runs over all the lattice sites, and ϱ and ϱ' are the nearest-neighbor vectors along the chain and on the basal plane of the tetragonal structure respectively, and $\xi = J_\perp/J$ is the ratio between the coupling constants on the basal plane $J_\perp$ and along the chain J. Both these coupling constants are positive for antiferromagnetic systems. The quasi-1D and quasi-2D models correspond to the cases of $\xi \ll 1$ and $\xi \gg 1$ respectively, whereas the simple cubic 3D model is given by $\xi = 1$. The Hamiltonian operator of Eq. (20) can be expressed in terms of the rotated coordinates of Eq. (2). The spin-wave spectrums and all the previous formula for the longitudinal mode with with anisotropy A remain the same after the following replacements:

$$\begin{aligned} z \to z' &= 2(1+2\xi)A, \\ \gamma_q \to \gamma'_q &= \frac{2}{z'}\left[\cos q_z + \xi(\cos q_x + \cos q_y)\right], \end{aligned} \tag{21}$$

where $z' = 2(1+\xi)$. In Fig. 3 we present the results for quasi-1D model at $A = 1$ together with the spin-wave spectrum of Eq. (18) for $\xi = 0.05$ as an example. The gaps for $E(q)$ at $q \to 0$ and $\mathbf{Q} = (\pi, \pi, \pi)$ are $0.78\,Jsz'$ and $1.20\,Jsz'$ in the first-order approximation and $0.35Jsz'$ and $0.54\,Jsz'$ in the second-order approximation.

We also presented the result for the quasi-2D system with $A = 1$ and $\xi = 10^3$ in Fig. 4. The gaps for $E(q)$ at $q \to 0$ and $\mathbf{Q} = (\pi, \pi, \pi)$ are $0.47\,Jsz'$ and $0.80Jsz'$ respectively in the first-order approximation and $0.29Jsz'$ and $0.50\,Jsz'$ respectively in the second-order approximation. The longitudinal excitation spectrum at two particular momenta $(\pi/2, \pi/2, 0)$ and $(\pi, 0, 0)$, has slightly different values of $1.36Jsz'$ and $1.40Jsz'$ respectively in the first-order approximation, and $0.85Jsz'$ and $0.88Jsz'$ respectively in the second-order approximation. The spin-wave spectra of Eq. (18) are the same at these momentum points with the value of Jsz'.

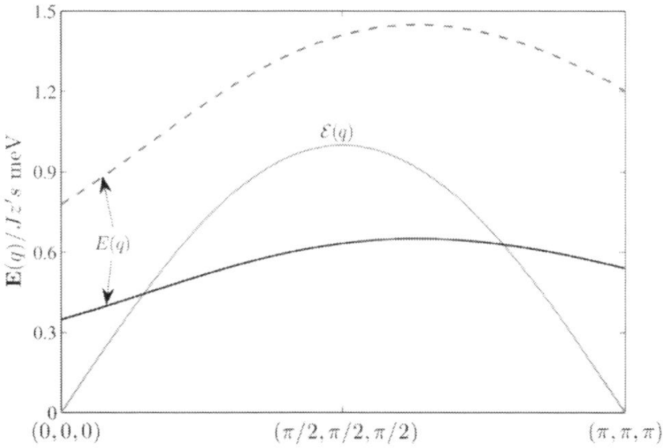

Figure 3. (Color online) The longitudinal excitation spectra of Eq. (8) for quasi-1D antiferromagnet with $\xi = 0.05$, together with the spin-wave spectrum of Eqs. (18), both at $A = 1$. The longitudinal spectra calculated from the first-order and second-order approximations are indicated by the dash and solid lines respectively.

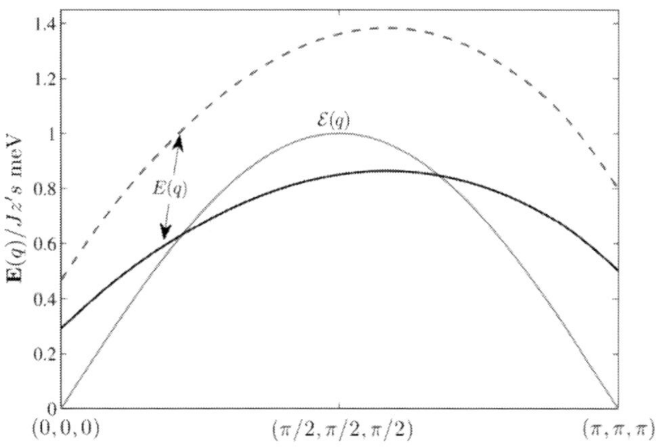

Figure 4. (Color online) The longitudinal excitation spectra of Eq. (8) for quasi-2D antiferromagnet with $\xi = 10^3$, together with the spin-wave spectrum of Eqs. (18), both at $A = 1$. The longitudinal spectra calculated from the first-order and second-order approximations are indicated by the dash and solid lines respectively.

5. Quasi-1D antiferromagnetic $KCuF_3$ structure

The $KCuF_3$ compound of the $s = 1/2$ chains is crystallized in the tetragonal ($Pnma$) structure with a lattice parameter $a = 4.126$ Å and $c = 3.914$ Å at temperature of 10K. The magnetic moment of this compound is carried by the Cu^{2+} ion. Some experimental studies on this material for the magnetic excitation spectrum using inelastic neutron scattering confirm the spinon picture [33, 20, 34, 35]. The magnetic interaction is represented by the antiferromagnetic strong couplings along the chain with coupling constant J and ferromagnetic weak couplings on the basal plane with couplings constant $J_\perp$. The Heisenberg Hamiltonian in terms of the lowering and raising operators with the rotated coordinates of the chains is given by

$$H = -\frac{1}{4}J\Big[\sum_{l,\varrho}^{\text{chain}} (S_l^+ S_{l+\varrho}^+ + S_l^- S_{l+\varrho}^- + 2S_l^z S_{l+\varrho}^z) - \xi \sum_{l,\varrho'}^{\text{plane}} (S_l^+ S_{l+\varrho'}^- + S_l^- S_{l+\varrho'}^+ + 2S_l^z S_{l+\varrho'}^z)\Big], \tag{22}$$

again where $\xi = J_\perp/J$. The spin-wave spectrum $\mathcal{E}_q$ in the linear SWT can then be obtained as

$$\mathcal{E}_q = 2Js\sqrt{\Gamma_q^2 - \cos^2 q_z}, \tag{23}$$

where

$$\Gamma_q = 1 + 2\xi(1 - \gamma_q^{2\text{D}}), \tag{24}$$

and

$$\gamma_q^{2\text{D}} = \frac{1}{2}(\cos q_x + \cos q_y). \tag{25}$$

The spin wave spectrum for the ferromagnetic square lattice model is recovered by setting $J = 0$ and that of the linear antiferromagnetic chain is recovered by setting $J_\perp = 0$. For the longitudinal excitation, the double commutator is derived as

$$N(q) = -2sJ(1 + \cos q_z)\tilde{g}_\varrho + 4sJ_\perp(1 - \gamma_q^{2D})\tilde{g}'_{\varrho'}, \tag{26}$$

where the correlation functions $\tilde{g}_\varrho$ is as defined before in Eqs. (13) and (14) with Δ_q and ρ_q given by

$$\Delta_q = -\frac{1}{2}\frac{\cos q_z}{\sqrt{\Gamma_q^2 - \cos^2 q_z}} \;\; ; \;\; \rho_q = \frac{1}{2}\Big(\frac{\Gamma}{\sqrt{\Gamma_q^2 - \cos^2 q_z}} - 1\Big). \tag{27}$$

The new correlation function $\tilde{g}'_\varrho$ is defined as

$$\tilde{g}'_\varrho = \langle S_l^+ S_{l+\varrho}^- \rangle = \mu_\varrho - \frac{2\rho\,\mu_\varrho + \Delta_\varrho \delta}{2s}, \tag{28}$$

with the parameters calculated by Eqs. (14) and (27).

The structure factor can be calculated from Eq. (16) using Eq. (27) for Δ_q and ρ_q. The longitudinal excitation spectrum can thus be calculated by Eq. (8). We present the numerical results for the longitudinal spectrum in Fig. 5 for both the first-order and the second-order approximations, using the experimental value of the couplings constants $J \approx 34$ meV and $J_\perp \approx 1.6$ meV [20]. We notice the difference in the first-order approximation between our current results and those reported earlier in [27] where a wrong assumption was made in one of the integrals. Now the overall spectrum in the first-order approximation is higher than those reported earlier. In particular, the minimum energy gap at $q \to 0$ for the first-order approximation is 28.9 meV and is reduced to 12.7 meV after including the second-order correction of Eq. (13) and (28). This second value of 12.7 meV is close to the experimental value of about 15 meV. The field theory by Essler *et al* produces a gap value of 17.4 meV [19]. At the antiferromagnetic wavevector, the gaps our first-order and second-order approximations are 44.7 and 19.7 meV respectively.

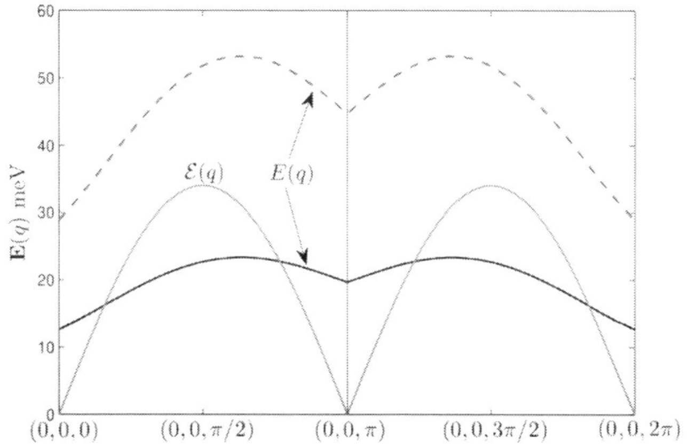

Figure 5. (Color online) The longitudinal spectra of Eq. (8) for $KCuF_3$ antiferromagnetic components with $\xi \approx 0.047$, together with the spin-wave spectrum of Eqs. (23). The longitudinal spectra calculated from the first-order and second-order approximations are indicated by the dash and solid lines respectively.

6. Hexagonal quasi-1D ABX_3-type antiferromagnetic systems

The quasi-1D materials such as $CsNiCl_3$ crystallize in the hexagonal ABX_3 structure with space group $P6_3/mmc$, where A is an alkaline-metal cation, B is a cation of the $3d$ group, and X is a halogen anion. The magnetic ions B constructs the hexagonal lattice in the ab plane with adjacent spins forming angles of $\theta = 2\pi/3$, and antiparallel adjacent spins along the chain, thus forming three-sublattice structure, in contrast to the bipartite systems discussed earlier. The lattice constants of $CsNiCl_3$, for example, are $a = 7.14$ Å and $c = 5.90$ Å, and the magnetic moments are carried by Ni^{2+}. The superexchange interaction between B (Ni^{2+}) ions is modeled by an N-spin Heisenberg Hamiltonian with a strong intrachain interaction J and weak interchain interaction J' such as

$$H = 2J \sum_{\langle i,j \rangle}^{\text{chain}} \mathbf{S}_i \cdot \mathbf{S}_j + 2J' \sum_{\langle i,j \rangle}^{\text{plane}} \mathbf{S}_i \cdot \mathbf{S}_j + D \sum_i (S_i^z)^2, \tag{29}$$

where we have added an Ising-like single-ion anisotropy term with constant $D(< 0)$. Most of the intrachain couplings in ABX_3 compounds are antiferromagnetic such as in $CsNiCl_3$ or $RbNiCl_3$ with easy single-site anisotropy, or $CsMnBr_3$ and $RbMnBr_3$ with hard anisotropy [36, 37]. These intrachain couplings can also be ferromagnetic (i.e., $J < 0$) as in $CsNiF_3$ [38, 39] or $CsCuCl_3$ [40]. We consider only the antiferromagnetic couplings here. Therefore, the classical ground state of each linear chain along the c axis (also denoted as y-axis) is a Néel state with alternating spin-up and spin-down alignments.

As before, we employ the the spin-wave approximation for the ground state in our analysis for the longitudinal modes. We therefore first consider a spin-wave theory for the Hamiltonian (29) based on the one-boson approach by performing two spin rotations. Firstly, we rotate the local axes of all up-spins by 180° so that all spins along each chain align in the same down direction. This spin rotation is applied by using the transformation of Eq. (2) to the first terms in Eq. (29), leaving the last two terms unchanged. The second rotation is on the hexagonal lattice of the ab plane (or xz-plane) on the second and third terms of Eq. (29). Following Singh and Huse [41] and Miyake [42], for every triangle of the hexagonal lattices, we rotate the local axes of two spins along the classical direction in the xz-plane to align with that of the third spin [43, 44]. This is equivalent to the rotation of the i-sites of Eq. (29) by the following transformation

$$\begin{aligned} S_i^x &\to S_i^x \cos(\theta_i) + S_i^z \sin(\theta_i), \\ S_i^y &\to S_i^y, \\ S_i^z &\to S_i^z \cos(\theta_i) - S_i^x \sin(\theta_i), \end{aligned} \tag{30}$$

where $\theta_i \equiv \mathbf{Q_z} \cdot \mathbf{r}_i$ and $\mathbf{Q_z} = (4\pi/3, 0, q_z)$ with $\mathbf{Q}_z$ at $q_z = \pi$ defined as the magnetic-ordering wavevector of the quasi-1D hexagonal systems. The Hamiltonian (29) after these two transformations is given as

$$\begin{aligned} H = &-\frac{1}{2}J\sum_{l,\varrho}^{\text{chain}} [S_l^+ S_{l+\varrho}^+ + S_l^- S_{l+\varrho}^- + 2S_l^z S_{l+\varrho}^z] - \frac{1}{2}J'\sum_{l,\varrho'}^{\text{plane}} [S_l^z S_{l+\varrho'}^z + \frac{3}{4}(S_l^+ S_{l+\varrho'}^+ + S_l^- S_{l+\varrho'}^-) \\ &- \frac{1}{4}(S_l^+ S_{l+\varrho'}^- + S_l^- S_{l+\varrho'}^+) - 2\sin(\theta_l - \theta_{l+\varrho'})(S_l^z S_{l+\varrho'}^x - S_l^x S_{l+\varrho'}^z)] + \tilde{\mathcal{H}}^D, \end{aligned} \tag{31}$$

where l runs through all sites, ϱ and ϱ' are the nearest neighbor index vectors with coordination numbers $z = 2$ along the chain and $z' = 6$ on the hexagonal basal planes respectively, and $\tilde{\mathcal{H}}^D$ is the rotated anisotropy term. In order to perform the second rotation of Eq. (30) involving rotations of the axes of the two spins to align with the axis of the third spin on the triangular planes, we rewrite the anisotropy term of the Hamiltonian (29) in the following equivalent, suitable form

$$\sum_i (S_i^z)^2 = \frac{1}{z'}\sum_{l,\varrho'}\left[\frac{1}{3}(S_l^z)^2 + \frac{2}{3}(S_{l+\varrho'}^z)^2\right]. \tag{32}$$

The transformation of Eq. (30) to the second term in Eq. (31) gives

$$\begin{aligned} \tilde{\mathcal{H}}^D = &\frac{1}{z'}\sum_{l,\varrho'}[\frac{1}{3}D(S_l^z)^2 + \frac{2}{3}D[(S_{l+\varrho'}^z)^2\cos^2\theta_{l+\varrho'} \\ &+ (S_{l+\varrho'}^x)^2\sin^2\theta_{l+\varrho'} - \cos\theta_{l+\varrho'}\sin\theta_{l+\varrho'}(S_{l+\varrho'}^z S_{l+\varrho'}^x \\ &+ S_{l+\varrho'}^x S_{l+\varrho'}^z)]]. \end{aligned} \tag{33}$$

After application of the usual boson transformation for the spin operators as given by Eqs. (5), the Hamiltonian of Eq. (31) can be expressed in terms of polynomials of the boson operators. After Fourier transformations of the boson operators with the Fourier component operators a_q and $a_q^\dagger$, we have, to the order of $(2s)$ in the large-s expansion

$$H \approx H_0 + H_2, \tag{34}$$

where H_0 is the classical energy

$$H_0 = -2JNs^2 - 3J'Ns^2 + \frac{1}{3}DNs^2(1 + 2\cos^2\theta + \frac{1}{s}\sin^2\theta) \tag{35}$$

with $\theta = 2\pi/3$ and H_2 is given by the quadratic terms in the boson operators as

$$H_2 = s\sum_q [A_q a_q^\dagger a_{-q} - \frac{1}{2}B_q(a_q^\dagger a_{-q}^\dagger + a_q a_{-q})] \tag{36}$$

with constants A_q and B_q defined by

$$\begin{aligned} A_q &= 4J + 6J'(1 + \frac{1}{2}\gamma_q) - \frac{2}{3}D(1 + 2\cos^2\theta - \sin^2\theta), \\ B_q &= 4J\cos q_z + 9J'\gamma_q - \frac{2}{3}D\sin^2\theta, \end{aligned} \tag{37}$$

and γ_q defined by

$$\gamma_q = \frac{1}{z'}\sum_{\varrho'} e^{i\mathbf{q}\cdot\mathbf{r}_{\varrho'}} = \frac{1}{3}\Big(\cos q_x + 2\cos\frac{q_x}{2}\cos\frac{\sqrt{3}}{2}q_y\Big). \tag{38}$$

The quadratic Hamiltonian H_2 of Eq. (36) is diagonalized by the usual Bogoliubov transformation and can be written in terms of the new boson operators α_q and $\alpha_q^\dagger$ as

$$H_2 = \Delta H_0 + \sum_q \mathcal{E}_q\left(\alpha_q^\dagger\alpha_q + \frac{1}{2}\right), \tag{39}$$

where ΔH_0 is the quantum correction to the classical ground state energy of Eq. (35)

$$\Delta H_0 = -2JNs - 3J'Ns + \frac{1}{3}DNs(1 + 2\cos^2\theta - \sin^2\theta), \tag{40}$$

and $\mathcal{E}_q$ is the spin-wave excitation spectra

$$\mathcal{E}_q = s\sqrt{A_q^2 - B_q^2}. \tag{41}$$

The spin-wave energy spectra with different polarizations are obtained by folding of the wavevectors. In Fig. 6, several branches along the symmetry direction of $(0, 0, \eta + 1), (\eta, \eta, 1)$, and $(1/3, 1/3, 1+\eta)$ are shown, where η is the reduced wave vector component in the reciprocal lattice unit (r.l.u) with $q_z = (2\pi l/c)\cdot(c/2) = \pi l$, and $\gamma = 1/3[\cos 2\pi h + \cos 2\pi k + \cos 2\pi(h + k)]$. Using Eq. (38) the moving in the paramagnetic Brillouin zone can be written as for $q_x = 4\pi\eta$ and $q_z = \pi + \pi\eta$, and the corresponding symmetry directions to those in reciprocal lattice unit are $(0, 0, \pi + \pi\eta), (4\pi\eta, 0, \pi)$ and $(4\pi/3, 0, \pi + \pi\eta)$ respectively. The three transverse spin-wave branches are obtained from Eq. (41) as follows. The y-mode has the polarization along the y-axis of the hexagonal lattice where the quantum fluctuation is at q; the other two modes are found in the xz-plane by translating the wavevector by a magnetic wavevector as $q \to (q \pm Q)$ and are denoted as $zx_\pm$ respectively.

As can be seen from Fig. 6, at the magnetic wavevector $\mathbf{Q}$, the y-mode is gapless for zero anisotropy ($D = 0$). However, as mentioned earlier, an energy gap about $0.41(2J)$ has been observed by the neutron scattering experiments for $CsNiCl_3$ [6]. This energy gap can be reproduced in the y-mode excitation by introducing an anisotropy with $D = -0.0285$ using our approximation of Eq. (33), also plotted in Fig. 6. If we use the simple form of Ref. [45] corresponding to setting $\theta = 0$ in Eq. (33), the required anisotropy is reduced by a little more than half with the value $D = -0.0141$. Both of these values are now considered too large for $CsNiCl_3$ which has negligible anisotropy. The conclusion is that the observed gaps are not of the transverse spin-wave spectra, but belong to the longitudinal modes, as first proposed by Affleck [15, 16].

Using the Hamiltonian of Eqs. (31), it is straightforward to derive the following double commutator with zero anisotropy (i.e. $D = 0$) as

$$\begin{aligned} N(q) = 2sJ\sum_{\varrho}(1 + \cos q_z)\tilde{g}_\varrho + \frac{1}{2}J's\sum_{\varrho'}\Big[&3(1 + \gamma_q)\tilde{g}_{\varrho'} \\ &- (1 - \gamma_q)\tilde{g}'_{\varrho'}\Big], \end{aligned} \tag{42}$$

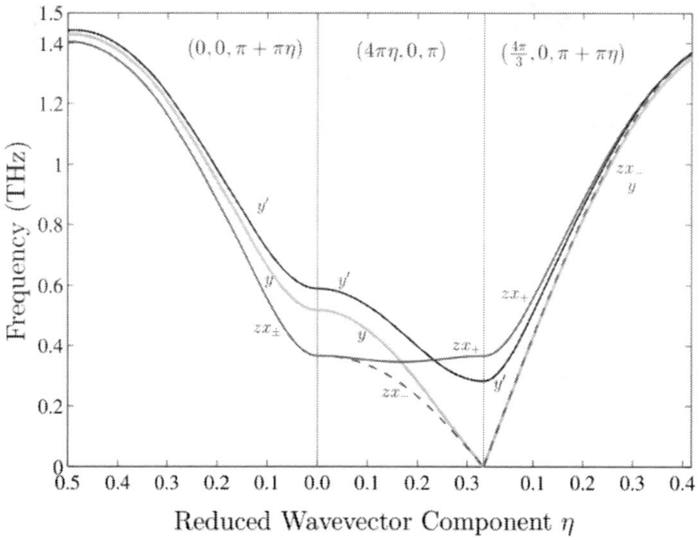

Figure 6. The three spin-wave excitation spectra (in colors) for $CsNiCl_3$ with $J = 0.345$, $J' = 0.0054$ and $D = 0$ THz, along the symmetry direction $(0, 0, \pi + \pi\eta)$, $(4\pi\eta, 0, \pi)$ and $(\frac{4\pi}{3}, 0, \pi + \pi\eta)$. Also included is the gapped y-mode (black, denoted as y') with $D = -0.0285$ using the anisotropy term of Eq. (33). The solid and dash with the blue color on the lines indicate the zx_+-mode and zx_--mode respectively.

where γ_q is as defined in Eq. (38) and the transverse correlation functions $\tilde{g}_\varrho$ and $\tilde{g}'_\varrho$ are defined in Eqs. (13) and (28) respectively, all independent of index l due to the lattice translational symmetry. Also, the contribution from the three-boson operators with $\sin(\theta_l - \theta_{l+\varrho})$ (the so-called cubic term) is zero. We notice that this cubic term has been included in perturbation theory for the correction in spin-wave spectrum [42, 43]. In the evaluation of the correlation functions $\tilde{g}_\varrho$ and $\tilde{g}'_\varrho$, we use the definition of Eqs. (13), (14) and (28) with the following expression

$$\Delta_q = \frac{1}{2}\frac{B_q}{\sqrt{A_q^2 - B_q^2}}, \quad \rho_q = \frac{1}{2}\left(\frac{A_q}{\sqrt{A_q^2 - B_q^2}} - 1\right), \tag{43}$$

with A_q and B_q as given before by Eqs. (37). The structure factor within the linear spin-wave approximation is independent of s, and is given by Eq. (16) with the results of Eq. (43).

We first discuss the general behaviors of the longitudinal spectrum of Eq. (8) as a function of the ratio of the two nearest-neighbor coupling constants, $\xi = J'/J$. In the limit $\xi \to 0$, the Hamiltonian of (29) becomes the pure 1D systems; the longitudinal spectrum is gapless and identical to the doublet spin-wave spectra thus forming a triplet excitation state as discussed in Sec. III(A). This demonstrates the limitation by the spin-wave ground-state employed, particularly when applied to the integer-spin Heisenberg chain where the Haldane gap is expected as discussed in Sec. I. In the other limit, $\xi \to \infty$, the Hamiltonian is a pure triangular antiferromagnet with the quasi-gapped longitudinal modes as discussed in details in our previous paper [25] where we keep only the first order term in Eqs. (13) and (28) in the large s-expansion, similar to the case of the square lattice model.

For the quasi-1D materials with intermediate values of ξ, we expect that the spin-wave ground state is a reasonable approximation. We obtain nonzero energy gaps for the longitudinal excitation spectra of Eq. (41). As discussed before, following Affleck [15, 16], two longitudinal

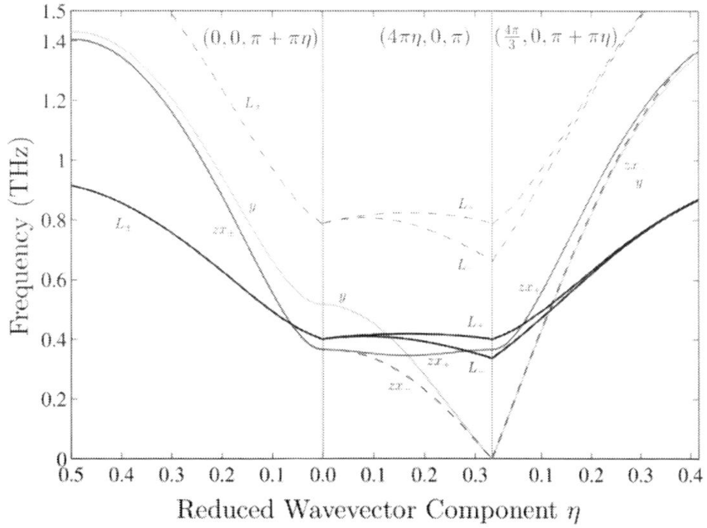

Figure 7. The longitudinal modes $L_\pm$ as derived from Eq. (8) together with the spin-wave y- and $zx_\pm$ modes as derived from Eq. (15) for $CsNiCl_3$ along the symmetry direction $(0, 0, \pi + \pi\eta)$, $(4\pi\eta, 0, \pi)$ and $(\frac{4\pi}{3}, 0, \pi + \pi\eta)$. The longitudinal modes $L_\pm$ calculated from the first-order and second-order approximations are indicated by the dash and solid lines respectively.

modes for the quasi-1D hexagonal antiferromagnets can be obtained by folding of the wavevector. We denote one as L_- with the spectrum $E(q - Q)$ and the other as L_+ with the spectrum $E(q + Q)$. We plot these two longitudinal spectra in the first and second order approximations together with the three spin-wave spectra of Eq. (41) in Fig. 7 near the magnetic wavevector Q for the compound $CsNiCl_3$. Our numerical result for the energy gap of the lower longitudinal mode L_- at Q is $0.96(2J)$ in the first order approximation in Eqs. (13) and (28). After including the second order terms the energy gap value is now $(0.49)2J$, in agreement with the experimental results of $0.41(2J)$. We also notice that the upper mode L_+ is higher than the L_- mode by about $(0.092)2J$ at Q.

For the compound $RbNiCl_3$ also with $s = 1$, using the exchange parameters $J = 0.485$ and $J' = 0.0143$ THz with a larger ratio $\xi = J'/J = 0.0295$ [46], we obtain similar longitudinal modes as those of $CsNiCl_3$. The numerical result for the energy gap of the L_- mode is 1.16 THz in the first order approximation and 0.69 THz after including the second order contributions at the magnetic wavevector. This later result is in better agreement with the experimental result of about 0.51 THz. We like to point out that there is some difficulty in fitting of Affleck's model with the experimental results for $RbNiCl_3$ [16, 46].

Finally we turn to the longitudinal modes for the non-integer-spin quasi-1D hexagonal systems. The superexchange interactions in the hexagonal compound $CsMnI_3$ can be described by the Hamiltonian of (29) with spin quantum number $s = 5/2$ and the nearest-neighbor coupling constants $J = 0.198$ and $J' = 0.001$ THz and negligible anisotropy [47]. This system is very close to the pure 1D system with a very small ratio $\xi = J'/J \approx 0.005$. The linear spin-wave theory may be a poor approximation for such a system. Nevertheless, with a similar analysis as before based on the spin-wave ground state, we obtain the L_- mode energy gap value of 0.64 THz at the magnetic wavevector Q in the first order approximation, and of 0.47 THz after including the second order contributions. This later value is still much larger than the experimental value of about 0.1 THz by Harrison *et al* [47], which was used to fit a modified spin-wave theory by Plumer and Cailé [17]. Clearly, for such systems as $CsMnI_3$, we need a better ground state than that of the spin-wave theory in our analysis.

7. Conclusion

In this paper we have extended our high-order calculations for the longitudinal modes in the hexagonal quantum antiferromagnetic systems [26] to a number of bipartite systems, including the the quasi-1D compound $KCuF_3$ where good agreement in the minimum energy gap is found between the experimental result and our estimate after inclusion of the high-order contributions.

We notice that all the longitudinal modes in antiferromagnetic systems with long-ranged order have so far been observed only on a few quasi-1D systems near the critical points. This is not surprising since in these systems the magnon density is high and the longitudinal modes is well-defined and long-lived. We also notice that there is no report of longitudinal modes in 2D or quasi-2D quantum antiferromangetc systems. In particular, as we have discussed in Sec. III(B), the longitudinal energy spectrum after inclusion of the high-order contributions, as shown in Fig. 2, is comparable to the spin-wave spectrum for the 2D square model with a tiny anisotropy same in value to the parent compound La_2CuO_4 of the high-T_c superconductor. It will therefore be interesting to examine possible longitudinal modes in this compound near the transition to the superconducting phase when doping.

References

[1] Andrson P W 1952 Phys. Rev. **86** 694
[2] Oguchi T 1960 Phys. Rev. **117** 117
[3] Takahashi M 1989 Phys. Rev. Lett. **62** 2313
[4] Huberman T *et al.* 2005 Phys. Rev. B **72** 014413
[5] Haldane F D M 1983 Phys. Rev. Lett. **50** 1153
[6] Buyers W J L *et al.* 1986 Phys. Rev. Lett. **56** 371
[7] Regnault L P, Zaliznyak I, Renard J P and Vettier C 1994 Phys. Rev. B **50** 9174
[8] Ma S, Broholm C, Reich D H, Sternlieb B J and Erwin R W 1992 Phys. Rev. Lett. **69** 3571
[9] Steiner M, Kakurai K, Kjems J K, Petitgrand D and Pynn R 1987 J. Appl. Phys **61** 3953
[10] Kenzelmann M *et al.* 2001 Phys. Rev. Lett. **87** 17201
[11] Sørensen E S and Affleck I 1994 Phys. Rev. B **49** 13235
[12] Golinelli O, Jolicoeur T and Lacaze R 1992 Phys. Rev. B **46** 10854
[13] Yamamoto S 1995 Phys. Rev. Lett. **75** 3348
[14] Deisz J, Jarrell M and Cox D L 1993 Phys. Rev. B **48** 10227
[15] Affleck I 1989 Phys. Rev. Lett. **62** 474
[16] Affleck I and Wellman G F 1992 Phys. Rev. B **46** 8934
[17] Plumer M L and Caillé A 1992 Phys. Rev. Lett. **68** 1042
[18] Schulz H J 1996 Phys. Rev. Lett. **77** 2790
[19] Essler F H L, Tsvelik A M and Delfino G 1997 Phys. Rev. B **56** 11001
[20] Lake B, D.Tennant D A and Nagler S E 2005 Phys. Rev. B **71** 134412
[21] Rüegg C *et al.* 2008 Phys. Rev. Lett. **100** 205701
[22] Gros C *et al.* 2003 Phys. Rev. B **67** 174405
[23] Xian Y 2006 Phys. Rev. B **74** 212401
[24] Feynman R P 1954 Phys. Rev. **94** 262
[25] Merdan M and Xian Y 2012 J. of Low Temp. Phys. **171** 797
[26] Merdan M and Xian Y 2013 Phys. Rev. B **87** 174434
[27] Xian Y 2011 Journal of Physics: Condensed Matter **23** 346003
[28] Feynman R P and Cohen M 1956 Phys. Rev. **102** 1189
[29] Xian Y 2007 Journal of Physics: Condensed Matter **19** 216221
[30] Bethe H 1931 Zeitschrift für Physik **71** 205
[31] des Cloizeaux J and Pearson J J 1962 Physical Review **128** 2131
[32] Keimer B *et al.* 1992 Phys. Rev. B **46** 14034
[33] Nagler S E, Tennant D A, Cowley R A, Perring T G and Satija S K 1991 Phys. Rev. B **44** 12361
[34] Tennant D A, Perring T G, Cowley R A and Nagler S E 1993 Phys. Rev. Lett. **70** 4003
[35] Tennant D A, Cowley R A, Nagler S E and Tsvelik A M 1995 Phys. Rev. B **52** 13368

[36] Santini P, Fath G, Domanski Z and Erdos P 1997 Phys. Rev. B **56** 5373
[37] Santini P, Domanski Z, Dong J and Erdos P 1996 Phys. Rev. B **54** 6327
[38] Delica T, de Jonge W J M, Kopinga K, Leschke H and Mikeska H J 1991 Phys. Rev. B **44** 11773
[39] Baehr M *et al.* 1996 Phys. Rev. B **54** 12932
[40] Rastelli E and Tassi A 1994 Phys. Rev. B **49** 9679
[41] Singh R R P and Huse D A 1992 Phys. Rev. Lett. **68** 1766
[42] Miyake S J 1992 Journal of the Physics Society Japan **61** 983
[43] Chubukov A V, Sachdev S and Senthil T 1994 Journal of Physics: Condensed Matter **6** 8891
[44] Chernyshev A and Zhitomirsky M 2009 Physical Review B **79** 144416
[45] Feile R *et al.*1984 Solid State Communications **50** 435
[46] Tun Z, Buyers W J L, Harrison A and Rayne J A 1991 Phys. Rev. B **43** 13331
[47] Harrison A, Collins M E, Abu-Dayyeh J, and Stager C V 1991 Phys. Rev. B **43** 679

17th International Conference on Recent Progress in Many-Body Theories (MBT17) IOP Publishing
Journal of Physics: Conference Series **529** (2014) 012021 doi:10.1088/1742-6596/529/1/012021

Formulation of effective interaction in terms of the non-perturbative and analytic $\widehat{Q}$ box

R Okamoto[1,+], **K Suzuki**[1,†], **H Kumagai**[2,−], **M Matsuzaki**[3,*]

[1] Senior Academy, Kyushu Institute of Technology, Kitakyushu 804-8550, Japan
[2] Faculty of Information Engineering, Fukuoka Institute of Technology, Fukuoka 811-0295, Japan
[3] Department of Physics, Fukuoka University of Education, Munakata, Fukuoka 811-4192, Japan

E-mail: [+]okamoto.ryoji.munakata@gmail.com, [†]suz93@mocha.ocn.ne.jp, [−]kumagai@fit.ac.jp, [*]matsuza@fukuoka-edu.ac.jp

Abstract. One of the useful and practical methods for solving quantum-mechanical many-body systems is to recast the full problem into a form of the effective interaction acting within a model space of tractable size. Many of the effective-interaction theories in nuclear physics have been formulated by use of the so called $\widehat{Q}$ box introduced by Kuo et al. It has been one of the central problems how to calculate the $\widehat{Q}$ box accurately and efficiently. Introducing new basis states, the Hamiltonian is transformed to a block-tridiagonal form in terms of submatrices with small dimension. With the transformed Hamiltonian we derive a recursion method of calculating the $\widehat{Q}$ box non-perturbatively and analytically. The $\widehat{Q}$ box given in this study corresponds to a non-perturbative solution for the energy-dependent Hamiltonian interaction which is often referred to as the Bloch-Horowitz or the Feshbach form. The present approach has a possibility of resolving many of the theoretical and practical difficulties encountered in the calculation of effective-interaction and/or Hamiltonian.

1. Introduction

In nuclear many-body physics various methods have been proposed, on the basis of the shell model, to solve the Schrödinger equations for nuclear many-body systems starting with realistic nucleon-nucleon(NN) interactions. These methods, which are called the *ab initio* calculations, include the Green's function Monte Carlo method, the no-core shell model, the effective interaction for hyperspherical harmonics method, the coupled cluster method and the unitary-model-operator approach. Much effort has been made also to diagonalize a matrix of a many-body shell-model Hamiltonian in a huge dimensional Hilbert space on the basis of, or alternatively to, the Lanczos method.

The effective interaction theory has been one of the *ab initio* method which has aimed at solving the nuclear properties starting from realistic nucleon-nucleon forces. Most of the effective-interaction theories given to date have been formulated in terms of the $\widehat{Q}$ box introduced by Kuo and his collaborators. Originally the $\widehat{Q}$ box has been defined as the sum of linked and unfolded diagrams. In the algebraic or non-diagrammatical approach the $\widehat{Q}$ box is equivalent to the energy-dependent effective Hamiltonian given by Bloch and Horowitz and Feshbach which has been studied extensively on the Brillouin-Wigner perturbation theory.

It has been established that the effective interaction can be expressed as a series expansion in terms of the $\widehat{Q}$ box and its energy derivatives. The series can be summed up by using either the Krenciglowa-Kuo (KK) or the Lee-Suzuki(LS) methods. It has been known that, in general, two methods have different convergence properties: Many of the numerical calculations have shown that the KK method yields the eigenvalues for the eigenstates which have the largest overlaps with the chosen model space. However, it has been pointed out that the rigorous convergence condition for the KK method has not yet been clarified. On the other hand the LS method reproduces the eigenvalues which lie closest to the chosen unperturbed energy. Both of the two approaches reproduce only certain of the eigenvalues of the original Hamiltonian. This restriction is not, in general, desirable.

Recently a new vertex function, called the $\widehat{Z}$ box, has been introduced to yield the eigenvalues of a given Hamiltonian H regardless of the properties of the eigenstates [1]. The $\widehat{Z}$ box itself has also been defined as a function of the $\widehat{Q}$ box. At present the most important remaining task would be to establish a method of how to calculate the $\widehat{Q}$ box rigorously and efficiently. The perturbative calculation method for the $\widehat{Q}$ box has been established and applied widely. In the derivation of the nuclear effective interaction, the convergence of the order-by-order calculation was confirmed in many of the numerical studies. However, a basic problem of the convergence of its perturbation expansion has not been made clear theoretically for general cases. Main concern of the present study is to propose a non-perturbative method for obtaining the $\widehat{Q}$ box for any of the starting NN interactions. Detailed discussions have been given in Ref.[2].

2. Effective-interaction theory by means of similarity transformation

2.1. Basic equation for effective interaction

Let us begin with a Hamiltonian H defined in a Hilbert space. We divide the space into a model space (P space) and its complementary space (Q space). When all the eigenvalues of an operator H_{eff} given in the P space coincide with those of H, we call H_{eff} an effective Hamiltonian.

There are various ways of constructing H_{eff}. We adopt the following standard one. First we introduce an operator ω that maps states in the P space and those in the Q space to each other, with the properties,

$$\omega = Q\omega P, \tag{2.1}$$

$$\omega^n = 0 \qquad (n \geq 2). \tag{2.2}$$

Suppose that the operator ω satisfies the following equation;

$$QHP + QHQ\omega - \omega PHP - \omega PHQ\omega = 0. \tag{2.3}$$

This equation for ω was first derived by Okubo. Once a solution ω to Eq.(2.3) is given, H_{eff} is written as

$$H_{\text{eff}} = PHP + PHQ\omega. \tag{2.4}$$

It has been proved that the effective Hamiltonian H_{eff} can be derived as in Eq.(2.4) with the solution ω to Eq.(2.3)

2.2. Krenciglowa-Kuo's solution

Since Eq.(2.3) is a nonlinear matrix equation for ω, it is difficult to find a general solution. The following formal solution, however, has been known and is enough for applications. We rewrite Eq.(2.3) as

$$QHP + QHQ\omega - \omega H_{\text{eff}} = 0, \tag{2.5}$$

using Eq.(2.4). Here the eigenvalue equation for $H_{\rm eff}$ is given by

$$H_{\rm eff}|\phi_k\rangle = E_k|\phi_k\rangle. \tag{2.6}$$

If the operator ω is a solution to Eq.(2.3), we can verify that the eigenstates $\{|\phi_k\rangle\}$ belong to the P space and each eigenvalue E_k coincides with one of those of H.

Using Eq.(2.5), ω is given by

$$\omega = \sum_{k=1}^{d} \frac{1}{E_k - QHQ} QHP|\phi_k\rangle\langle\widetilde{\phi}_k|, \tag{2.7}$$

and from Eq.(2.4) $H_{\rm eff}$ becomes

$$H_{\rm eff} = PHP + \sum_{k=1}^{d} PHQ\frac{1}{E_k - QHQ} QHP|\phi_k\rangle\langle\widetilde{\phi}_k|, \tag{2.8}$$

where $\langle\tilde{\phi}_k|$ is the biorthogonal states of $|\phi_k\rangle$. Here we introduce an operator in the P space called the $\widehat{Q}$ box

$$\widehat{Q}(E) = PHP + PHQ\frac{1}{E - QHQ}QHP, \tag{2.9}$$

where E is an energy variable. The $\widehat{Q}$ box thus defined is equivalent to the energy-dependent effective Hamiltonian referred to as the Bloch-Horowitz and /or the Feshbach forms. In terms of $\widehat{Q}(E)$, $H_{\rm eff}$ is expressed as

$$H_{\rm eff} = \sum_{k=1}^{d} \widehat{Q}(E_k)|\phi_k\rangle\langle\widetilde{\phi}_k|, \tag{2.10}$$

from which the following self-consistent equation can be derived

$$\widehat{Q}(E_k)|\phi_k\rangle = E_k|\phi_k\rangle. \tag{2.11}$$

The $H_{\rm eff}$ in Eq.(2.10) is just a formal solution in the sense that unknown E_k, $|\phi_k\rangle$, and $\langle\widetilde{\phi}_k|$ appear on the right-hand side, but the following method of solving is available: In order that the solutions to Eq.(2.6) coincide with those given by Eq.(2.10), they selfconsistently satisfy the iterative equation

$$\widehat{Q}(E_k^{(n)})|\phi_k^{(n+1)}\rangle = E_k^{(n+1)}|\phi_k^{(n+1)}\rangle, \tag{2.12}$$

where $E_k^{(n+1)}$ and $|\phi_k^{(n+1)}\rangle$ are the $(n+1)$-th order eigenvalue and eigenstate of the $\widehat{Q}$ box, respectively, given by the n-th order eigenvalue $E_k^{(n)}$. This iterative scheme has been called the Kreciglowa-Kuo method.

From Eq.(2.9), we see that $\widehat{Q}(E)$ has poles at energies $\{\varepsilon_q\}$, where ε_q is one of the eigenvalues of QHQ,

$$QHQ|q\rangle = \varepsilon_q|q\rangle. \tag{2.13}$$

These singularities of the $\widehat{Q}$ box lead to some difficulties in numerical calculations. These arguments suggest that some further improvements are desired for the $\widehat{Q}$-box method although it has been applied widely to practical problems.

2.3. The $\widehat{Z}$ box method

In order to resolve the pole problem we have introduced a vertex function $\widehat{Z}$, called the $\widehat{Z}$ box, as[1]

$$\widehat{Z}(E) = \frac{1}{1-\widehat{Q}_1(E)}\left[\widehat{Q}(E) - E\widehat{Q}_1(E)\right], \tag{2.14}$$

where E is an energy variable and

$$\widehat{Q}_1(E) = \frac{d\widehat{Q}(E)}{dE}. \tag{2.15}$$

The $\widehat{Z}$ box has the following properties:

(i) The $\widehat{Z}$ box satisfies the selfconsistent equation

$$\widehat{Z}(E_k)|\phi_k\rangle = E_k|\phi_k\rangle, \tag{2.16}$$

where $\{E_k\}$ are the eigenvalues of H and $\{|\phi_k\rangle\}$ the model-space eigenstates which are the same as in Eqs.(2.10) and (2.11).

(ii) The $\widehat{Z}(\varepsilon_q)$ also satisfies the self-consistent equation for the eigenvalues $\{\varepsilon_q\}$ of QHQ as

$$\widehat{Z}(\varepsilon_q)|\mu_q\rangle = \varepsilon_q|\mu_q\rangle. \tag{2.17}$$

(iii) Both of $\{E_k\}$ and $\{\varepsilon_q\}$ satisfy the selfconsistent equations as in Eqs.(2.16) and (2.17), respectively, but the derivative of the $\widehat{Z}$ box takes different values as

$$\frac{d\widehat{Z}(E)}{dE} = \begin{cases} 0 & \text{for } E = E_k, \\ 2 & \text{for } E = \varepsilon_q. \end{cases} \tag{2.18}$$

Using these properties we can divide the solutions of the selfconsistent equations into two parts, namely, the true eigenvalues $\{E_k\}$ of H and the pole energies $\{\varepsilon_q\}$.

(iv) The $\widehat{Z}(E)$ is finite and differentiable at any energy variable E, even at pole positions $\{\varepsilon_q\}$ of the $\widehat{Q}$ box. This may be clear from Eqs.(2.17) and (2.18).

3. Calculation of the $\widehat{Q}$ box by means of recurrence relations

3.1. Block tridiagonalization of Hamiltonian

Before discussing the calculation procedure of the $\widehat{Q}$ box, we transform the Hamiltonian H into a tractable form by changing basis vectors. First we introduce

$$Y_P = PHQ \cdot QHP. \tag{3.1}$$

The Y_P is an operator in the P space, which is Hermitian and positive semi-definite, that is, $y_k^{(1)} \geq 0$ in the eigenvalue equation

$$Y_P|p_k\rangle = y_k^{(1)}|p_k\rangle. \tag{3.2}$$

Suppose that d_1 eigenvalues are nonzero among $\{y_k^{(1)}\}$. In terms of the eigenvectors $\{|p_k\rangle, k = 1, 2, \cdots, d_1\}$ with nonzero eigenvalues, we define normalized vectors $\{|q_k^{(1)}\rangle\}$ in the Q space as

$$|q_k^{(1)}\rangle = \frac{1}{\sqrt{y_k^{(1)}}} QHP|p_k\rangle, \quad (k = 1, 2, \cdots, d_1). \tag{3.3}$$

They are orthogonal to each other and span the d_1-dimensional subspace Q_1 in the Q space. Then the projection operator onto the Q_1 space becomes

$$Q_1 = \sum_{k=1}^{d_1} |q_k^{(1)}\rangle\langle q_k^{(1)}|. \tag{3.4}$$

The complement of the Q_1 space in the Q space is given by

$$\overline{Q}_1 = Q - Q_1. \tag{3.5}$$

Equation (3.3) indicates that

$$QHP = \sum_{k=1}^{d_1} \sqrt{y_k^{(1)}}|q_k^{(1)}\rangle\langle p_k|, \tag{3.6}$$

then we have

$$QHP = Q_1HP \tag{3.7}$$

which leads to

$$\overline{Q}_1HP = 0. \tag{3.8}$$

Next, a similar manipulation with replacing P and Q with Q_1 and $\overline{Q}_1$, respectively, leads to another orthogonal system and the subspace, namely the Q_2 space. Repeating the same procedures leads to the following: Decompose the Q space as

$$Q = Q_1 + Q_2 + \cdots + Q_n + \cdots. \tag{3.9}$$

Basis vectors of the subspace Q_m, namely, $\{|q_k^{(m)}\rangle, k = 1, 2, \cdots, d_m\}$, define the projection operator

$$Q_m = \sum_{k=1}^{d_m} |q_k^{(m)}\rangle\langle q_k^{(m)}|. \tag{3.10}$$

The basis vectors $\{|q_k^{(m)}\rangle\}$ are given as follows: Introduce $Y_{Q_{m-1}}$ as

$$Y_{Q_{m-1}} = Q_{m-1}H\overline{Q}_{m-1} \cdot \overline{Q}_{m-1}HQ_{m-1} \tag{3.11}$$

with

$$\overline{Q}_{m-1} = Q - (Q_1 + Q_2 + \cdots + Q_{m-1}). \tag{3.12}$$

Its eigenvalue equation is

$$Y_{Q_{m-1}}|q_k'^{(m-1)}\rangle = y_k^{(m)}|q_k'^{(m-1)}\rangle. \tag{3.13}$$

In general new orthogonal bases

$$|q_k^{(m)}\rangle = \frac{1}{\sqrt{y_k^{(m)}}}\overline{Q}_{m-1}HQ_{m-1}|q_k'^{(m-1)}\rangle \tag{3.14}$$

are derived from the eigenvectors $\{|q_k'^{(m-1)}\rangle\}$ with nonzero eigenvalues $\{y_k^{(m)}\}$. They span the subspace Q_m. With the projection operators Q_m and Q_{m-1} we obtain, from Eq.(3.14), an expression written as

$$Q_m H Q_{m-1} = \sum_{k=1}^{d_m} \sqrt{y_k^{(m)}} |q_k^{(m)}\rangle\langle q_k'^{(m-1)}|. \tag{3.15}$$

We conclude from the above discussion that

$$PHQ_m = Q_m HP = 0 \quad (m \geq 2), \tag{3.16}$$

$$Q_m H Q_{m+k} = Q_{m+k} H Q_m = 0 \quad (k \geq 2) \tag{3.17}$$

hold for the subspaces $\{P, Q_1, Q_2, \cdots, Q_m, \cdots\}$. This means that the given Hamiltonian H is transformed to a block-tridiagonal matrix

$$H = \begin{pmatrix} PHP & PHQ_1 & 0 & 0 & \cdots \\ Q_1HP & Q_1HQ_1 & Q_1HQ_2 & 0 & \cdots \\ 0 & Q_2HQ_1 & Q_2HQ_2 & Q_2HQ_3 & \cdots \\ 0 & 0 & Q_3HQ_2 & Q_3HQ_3 & \cdots \\ \vdots & \vdots & \vdots & \vdots & \vdots \end{pmatrix}, \tag{3.18}$$

where each block matrix is at most d-dimensional.

3.2. Calculation of the $\widehat{Q}$ box

We introduce an operator $\chi(E)$ defined as

$$\chi(E) = \frac{1}{e(E)} QHP \tag{3.19}$$

with

$$e(E) = Q(E - H)Q. \tag{3.20}$$

The Q space is decomposed as in Eq.(3.9), and also $\chi(E)$ is as

$$\chi(E) = \chi_1(E) + \chi_2(E) + \cdots + \chi_n(E) + \cdots, \tag{3.21}$$

where

$$\chi_n(E) = Q_n\, \chi(E) P. \tag{3.22}$$

The operators $\{\chi_n(E)\}$ obey the recursion relation written as[2]

$$Q_1\, e(E)\{\chi_1(E) + \chi_2(E)\} = Q_1 HP, \tag{3.23}$$

$$Q_n\, e(E)\{\chi_{n-1}(E) + \chi_n(E) + \chi_{n+1}(E)\} = 0, \ (n \geq 2). \tag{3.24}$$

Note that, from Eq.(3.7), the $\widehat{Q}$ box in Eq.(2.9) can be written as

$$\widehat{Q}(E) = PHP + PHQ_1\chi_1(E). \tag{3.25}$$

Then the calculation procedure of the $\widehat{Q}$ box is reduced to solving the recursive equations and finding the solution for $\chi_1(E)$.

It has been proved that the solution for $\chi_1(E)$ is given by

$$\chi_1(E) = \frac{1}{\widetilde{e}_1(E)} Q_1 HP, \tag{3.26}$$

where the energy denominator is determined from the following descending recurrence relations;

$$\widetilde{e}_{n-1}(E) = e_{n-1}(E) - H_{n-1,n} \frac{1}{\widetilde{e}_n(E)} H_{n,n-1}. \tag{3.27}$$

with

$$e_{n-1}(E) = Q_{n-1}(E - H)Q_{n-1}, \tag{3.28}$$
$$H_{ij} = Q_i H Q_j. \tag{3.29}$$

Suppose that, for a sufficiently large number N, the condition

$$\|\frac{1}{e_N(E)} H_{N,N-1}\| \ll 1 \tag{3.30}$$

is satisfied, where the symbol $\|X\|$ means the norm of a matrix X. If we start with $n = N$ in Eq.(3.28), we have a sequence $\{\widetilde{e}_{N-1}(E), \widetilde{e}_{N-2}(E), \cdots\}$ and finally we have $\widetilde{e}_1(E)$. Then we may write the $\widehat{Q}$ box as

$$\widehat{Q}(E) = PHP + PHQ_1 \frac{1}{\widetilde{e}_1(E)} Q_1 HP. \tag{3.31}$$

The above expression indicates that a certain energy denominator $\widetilde{e}_1(E)$ exists such that the $\widehat{Q}(E)$ can be represented by a sum of only second-order perturbation terms[2].

4. Model calculation

We solve the eigenvalue problem for the Hamiltonian H in the framework of the $\widehat{Z}$-box method[1, 2]. The calculation procedure is as follows: We define d functions $\{F_k(E), k = 1, 2, \cdots, d\}$ through the eigenvalue equation for $\widehat{Z}(E)$ as

$$\widehat{Z}(E)|\zeta_k\rangle = F_k(E)|\zeta_k\rangle, \quad k = 1, 2, \cdots, d. \tag{4.32}$$

We further introduce functions $\{F'_k(E), k = 1, 2, \cdots, d\}$ as

$$F'_k(E) = \left\langle \zeta_k \right| \frac{d\widehat{Z}}{dE} \left| \zeta_k \right\rangle. \tag{4.33}$$

Using the properties in Eqs.(2.16)- (2.18), we can prove that the true eigenvalues $\{E_k\}$ of H can be obtained by solving the equation

$$g_k(E) = \left\{ \frac{F_k(E) - E}{F_0} \right\}^2 + \{F'_k(E)\}^2 = 0, \tag{4.34}$$

where F_0 is a parameter chosen suitably. We solve Eq.(4.34) by means of the parabolic-interpolation method given in Ref.[2].

Journal of Physics: Conference Series **529** (2014) 012021 doi:10.1088/1742-6596/529/1/012021

In order to obtain some assessments of the present approach we study a model problem. We start with a model Hamiltonian H of which matrix elements are given by

$$\langle i|H|j\rangle = (\alpha i + \beta i^2)\delta_{ij} + \gamma x_{ij} \tag{4.35}$$

with

$$x_{ij} = 2\left\{\sqrt{\sqrt{2}(i+j)} - \left[\sqrt{\sqrt{2}(i+j)}\right]\right\} - 1, \tag{4.36}$$

where $[X]$ is Gauss's symbol which means the integer part of a real number X. A set of $\{x_{ij}\}$ are recognized as pseudo random numbers satisfying

$$-1 \leq x_{ij} \leq 1. \tag{4.37}$$

The α, β and γ are the parameters chosen suitably. The total dimension of H is taken to be $N_h = 100$. As for the P space we choose a two-dimensional space ($d = 2$) spanned by the two states which have the lowest and second lowest diagonal energies of H. We suppose that the subspaces $\{Q_k,\ k = 1, 2, \cdots, N_q\}$ are all d-dimensional and the number of the subspace $\{Q_k\}$ is given by $N_q = (N_h - 2)/2 = 49$.

In Table 1 we show the results for the lowest two eigenvalues of H calculated by the parabolic-interpolation method[2]. The convergence is markedly fast. With three times of the changes of the interval $[a, b]$, convergence is reached with accuracy better than 10 decimal places.

Table 1. Correct digits of the lowest two eigenvalues of H calculated by the parabolic-interpolation method. Initial intervals are taken to be $[a, b] = [0.0, 1.0]$ and $[2.5, 3.5]$ for E_1 and E_2, respectively. The numbers of repeats means the number of changes of the interval [a, b]. All the values of E_1 and E_2 are dimensionless.

E_i	No. of repeats	Calculated value
E_1	1	0.365
	2	0.365550
	3	0.365550151994574
E_2	1	2.999
	2	2.9994240
	3	2.99942408730107

It would be interesting to examine the dependence of the calculated eigenvalues of H on the number of the subspaces $\{Q_k\}$ taken into consideration. Let $K_{\max}$ be the maximum number of the subspaces $\{Q_k\}$. The dependence on $K_{\max}$ is shown in Figs.1. It is clear that, as $K_{\max}$ approaches to N_q=49, the eigenvalues converge to the exact values. These results suggest a possibility of introducing a new way of truncation in the calculation of effective interaction, instead of making it according to the magnitude of energies of intermediate states.

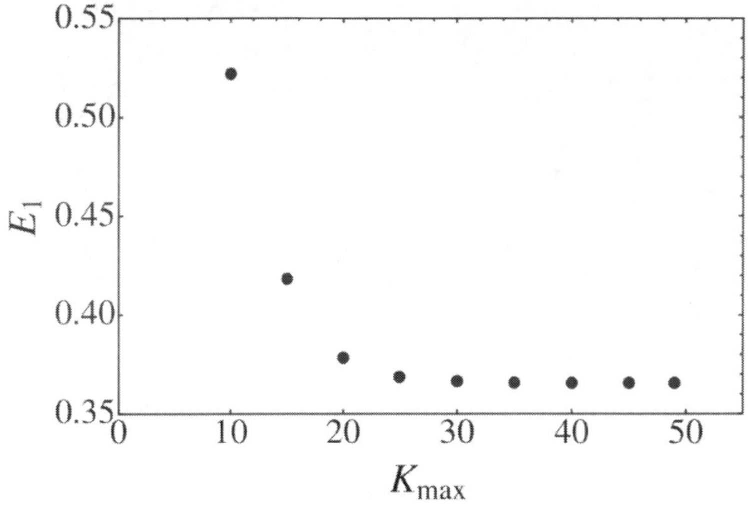

Figure 1. Convergence of E_1 as a function of K_{max}. The K_{max} denotes the block dimension which means the number of the subspaces $\{Q_k, k{=}1{,}2{,}\cdots{,}K_{max}\}$ taken into calculation. In this model calculation K_{max} is in the range $1 \leq K_{max} \leq 49$. The exact value of E_1 is $0.36555\cdots$ as given in Table 1. The value of E_1 is dimensionless.

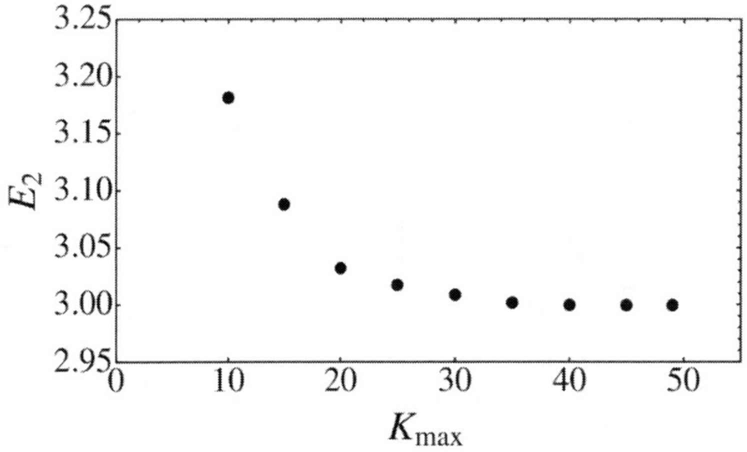

Figure 2. Convergence of E_2 as a function of K_{max}. The exact value of E_2 is $2.9994\cdots$. Other notations are the same as in Fig.1.

5. The relation of the present method to the block Lanczos method

The calculation of the $\widehat{Q}$ box can be carried out without matrix inversion of QHQ which is usually a huge-dimensional matrix. All the procedures for obtaining the $\widehat{Q}$ box are reduced to calculations of small-dimensional submatrices in the block-tridiagonalized Hamiltonian.

Regarding the block tridiagonalization of the Hamiltonian, the present approach has a common aspect to the so called block Lanczos method based on the theory of the Krylov subspaces. For a given model space P and a Hamiltonian H, the subspaces leading to a block-tridiagonal form of H are determined uniquely. Therefore, the subspaces given in the present study are the same as those of Krylov. However, the choice of basis states of each subspace is ambiguous. For determining the basis states we employ a different calculation procedure from the usual one in the block Lanczos method, as shown in Eq.(3.14). We show that the basis states thus introduced are suitable for the purpose of calculating not only the $\widehat{Q}$ box but also the eigenstates of H[1].

It is well known that roundoff errors makes the Lanczos method somewhat difficult to use in practice[3],[4]. The central problem is a loss of orthogonality among the Lanczos vectors that are produced by iteration. There are several ways, say selective or full reorthonormalization to cope with this problem[3],[4]. However, there is no need of additional orthonormalization of the relevant basis in the present formulation, as long as we concern the effective Hamiltonian and its eigenvalues in the model space.

6. Summary and outlook

The main purpose of the present study has been to derive a new method of calculating the $\widehat{Q}$ box as accurately as possible even if the original Hamiltonian H is given in a huge-dimensional space.

The present formulation[2] consists of two steps: First one is to transform a given Hamiltonian H to a block-tridiagonal form by dividing the complementary space Q of the P space into subspaces $\{Q_k,\ k = 1, 2, \cdots\}$ with tractable dimensions. If the subspaces are chosen suitably the Hamiltonian is transformed to a block-tridiagonal form. With the Hamiltonian thus transformed, the next step is to derive coupled equations for determining the $\widehat{Q}$ box. We have shown that these coupled equations can be solved by means of a recursion method.

Given the $\widehat{Q}$ box, we have applied the $\widehat{Z}$-box method for solving the eigenvalue problem of a Hamiltonian H[2]. In order to assess the present method we have made a test calculation by introducing a 100×100 model Hamiltonian. We have confirmed that the present method reproduces successfully the eigenvalues of the original Hamiltonian.

There would be two applicabilities of the present approach: One is to solve the eigenvalue problem for a Hamiltonian given in a huge dimensional space. Once the $\widehat{Q}$ box is given, the eigenvalues and the corresponding eigenstates can be calculated according to the procedures in the present framework. The other is to apply the present formalism to the derivation of the effective interaction to be used in the shell-model calculations.

The present study is based essentially on the algebraic approach to the effective Hamiltonian. When we want to calculate the effective interaction acting among valence nucleons, it is necessary to represent all the terms contained in the $\widehat{Q}$ box in terms of linked diagrams. A general and rigorous relation is not made clear between the present approach and the linked-and-folded-diagram theory. We note that algebraic approaches can not always exclude unlinked diagrams in general. Therefore, this formal relation is an interesting problem to be clarified.

References

[1] Suzuki K, Okamoto R, Kumagai H and Fujii S 2011 *Phys. Rev.* C **83** 024304
[2] Suzuki K, Kumagai H, Matsuzaki M and Okamoto R 2013 *Phys. Rev.* C **87** 024001
[3] Golub G H and Van Loan C F 1996 *Matrix Computations* (The Johns Hopkins University Press)
[4] Arbenz P 2012 *Numerical Methods for Solving Large Scale Eigenvalue Problems* chap.9 (http://people.inf.ethz.ch/arbenz/ewp/lnotes.html) and references therein

Journal of Physics: Conference Series **529** (2014) 012022 doi:10.1088/1742-6596/529/1/012022

Many–body Bose systems and the hard-sphere model: dynamic properties from the weak to the strong interaction regime

R Rota [1], F Tramonto[2], D E Galli[2] and S Giorgini[1]

[1] Dipartimento di Fisica and INO-CNR BEC Center, Università degli Studi di Trento, via sommarive 14, 38123 Trento, Italy

[2] Dipartimento di Fisica, Università degli Studi di Milano, via Celoria 16, 20134 Milano, Italy

E-mail: rota@science.unitn.it

Abstract. We obtain *ab-initio* estimations of the dynamic structure factor, $S(\mathbf{q}, \omega)$, of Bose gases at zero temperature. More precisely, we use the Genetic Inversion via Falsification of Theories (GIFT) algorithm to perform analytic continuations of imaginary time correlation functions computed via an exact Path Integral projector method. Using the hard-sphere potential to model the two-body interactions between the atoms, we compute $S(\mathbf{q}, \omega)$ changing the gas parameter from the dilute regime ($na^3 = 10^{-4}$) up to the density corresponding to superfluid ^{4}He at equilibrium ($na^3 = 0.2138$). With increasing density, we observe the emergence of a broad multiphonon contribution accompanying the quasiparticle peak and a crossover of the dispersion of elementary excitations from a Bogoliubov-like spectrum to a phonon-maxon-roton curve. Apart from the low wave vector region, for $na^3 = 0.2138$ the energy-momentum dispersion relation and the static density response function, $\chi(\mathbf{q})$, turns out to be in good agreement with the superfluid ^{4}He experimental data at equilibrium density.

1. Introduction

The study of the collective modes and the dynamic properties of ultracold gases has always represented a very important issue in quantum many-body theories. In particular, in the case of Bose systems, this study is fundamental to understand the phenomenon of superfluidity and, starting from the pioneering studies of Landau and Bogoliubov in the 1940s, it has been subject of a huge number of works, both theoretical and experimental [1].

A physical insight on the dynamic behavior of quantum many-body systems can be given estimating the dynamic structure factor, $S(\mathbf{q}, \omega)$. This quantity, indeed, contains a wealth of information about the nature and the energy spectrum of the excitations coupled to density fluctuations and, for a system made up of N particles described by the Hamiltonian $\hat{H}$, at zero temperature it can be written as:

$$S(\mathbf{q}, \omega) = \frac{1}{2\pi N} \int_{-\infty}^{+\infty} dt\, e^{i\omega t} \frac{\langle \Psi_0 | \rho_{-\mathbf{q}}(t) \rho_{\mathbf{q}}(0) | \Psi_0 \rangle}{\langle \Psi_0 | \Psi_0 \rangle} = \frac{1}{N} \sum_{n \geq 0} \delta\left(\omega - \frac{E_n - E_0}{\hbar}\right) \frac{|\langle \Psi_n | \rho_{\mathbf{q}} | \Psi_0 \rangle|^2}{\langle \Psi_0 | \Psi_0 \rangle}, \quad (1)$$

where $\rho_{\mathbf{q}}(t) = e^{i\hat{H}t/\hbar} \rho_{\mathbf{q}} e^{-i\hat{H}t/\hbar}$ is the time evolution of the density fluctuation operator $\rho_{\mathbf{q}} = \sum_{i=1}^{N} e^{-i\mathbf{q}\cdot\mathbf{r}_i}$ and $|\Psi_0\rangle$ is the ground state of the many-body system. Equivalently, introducing a complete set of eigenstates $|\Psi_n\rangle$, $S(\mathbf{q},\omega)$ can be expressed in terms of all the excited states which have non-zero overlap with the density perturbation, $\rho_{\mathbf{q}}|\Psi_0\rangle$, as in the right term of Eq. 1.

For dilute gases at low temperature, we can safely assume that the s-wave scattering length, a, is much smaller than the mean inter-particle distance, $n^{-1/3}$ (being n the density of the gas), and the ground state of the many-body system can be described in terms of a macroscopic wave function [2]. In this regime, mean-field theories are expected to provide an accurate description of the system and the elementary excitations can be expressed in terms of a single quasiparticle. Therefore, the dynamic structure factor reduces to a single peak, whose dispersion follows the Bogoliubov spectrum

$$\varepsilon_B(\mathbf{q}) = \frac{\hbar^2}{2m\xi^2}\sqrt{(q\xi)^4 + 2(q\xi)^2}\ , \tag{2}$$

in which $\xi = 1/\sqrt{8\pi na}$ is the healing length. This result, predicted for the first time in 1947, has been confirmed experimentally in 2002 by Steinhauer *et al.*, who have been able to measure the excitation spectrum of a dilute condensate of ^{87}Rb atoms by means of Bragg spectroscopy [3].

As the density of the system increases, the interactions among the particles become more and more important and the picture provided by mean-field theories is no longer valid. Indeed, experimental measurements [4, 5, 6, 7, 8, 9] of the dynamic structure factor in liquid ^{4}He, that is a well-known benchmark of a dense Bose fluid, show a different behavior from the one predicted for dilute systems. First of all, the $S(\mathbf{q},\omega)$ in liquid ^{4}He presents, in addition to a sharp "quasiparticle" peak, a broad contribution at higher frequencies, usually called multiphonon branch, that indicates the possibility to induce incoherent excitations in the many-body system. Furthermore, the dispersion of the main peak of $S(\mathbf{q},\omega)$ differs from the Bogoliubov spectrum, Eq. 2, and presents a phonon-maxon-roton behavior, which is linear at low q and displays a relative minimum for a non-zero wave vector.

To perform a reliable study of strongly interacting Bose gases, it is fundamental to develop many-body theories able to include all the relevant correlations among the particles. For instance, *ab-initio* numerical techniques based on Quantum Monte Carlo (QMC) simulations have been widely used to calculate "exactly" the equilibrium properties of dense Bose fluids, such as superfluid ^{4}He, both at zero [10] and finite temperature [11]. Nevertheless, the study of the dynamic properties of quantum many-body systems by means of QMC methods has to face the problem of analytical continuation from purely imaginary to real time of the correlation function entering the Fourier transform in Eq. 1, which strongly limits any possibility of a precise determination of $S(\mathbf{q},\omega)$. Several methods have been developed in order to extract information about the real-time dynamics of quantum systems from the limited and noisy data achievable in QMC simulations. The most popular one is the Maximum Entropy (ME) method, a stochastic technique based on the bayesian inference [12]. ME has been used in the study of several physical systems, but it provides only qualitative results when applied in the analysis of quantum fluids [13, 14]. Recently, a powerful algorithm called Genetic Inversion via Falsification of Theories (GIFT) has been proposed; this algorithm can provide very accurate results for liquid and solid ^{4}He, being able to recover both the sharp quasiparticle peak and the multiphonon branch, each with the correct relative spectral weight [15].

In this work, we want to study, by means of QMC methods, the dynamic properties of Bose gases at zero temperature, focusing on the crossover from weak to strong interaction regime. To model the interactions among the particles, we use the two-body hard-sphere (HS) potential

$$V(r) = \left\{ \begin{array}{ll} \infty & (r \le a) \\ 0 & (r > a) \end{array} \right. , \tag{3}$$

where the range of the potential a coincides with the s-wave scattering length. This model has been widely used for the study of many-body systems with short-range repulsive interactions, not only in the dilute regime, where the details of the interatomic potential are irrelevant, but also in the dense regime [16, 17, 18, 19]. The HS model, indeed, serves as a reference for those systems in which the leading part of the two-body potential is the repulsive hard-core at short distances and it has been used to characterize semi-quantitatively the static properties of superfluid ^{4}He [17, 18]. Moreover, the HS model provides one with a well defined system, where quantum correlations can be investigated from the weak to the strong interaction regime by varying a single parameter, i.e. the reduced density in units of the HS range.

The structure of the paper is as follows. In Section 2 we introduce the numerical methods used in our study, namely the Path Integral Ground State (PIGS) method, applied to the calculation of the imaginary-time correlation functions, and the GIFT algorithm, for the estimation of the dynamic structure factors from the PIGS data. In Section 3 we show our results for the dynamic structure factor, for the dispersion law of the elementary excitations and for the static density response function. Finally, in Section 4 we draw our conclusions.

2. Numerical Methods

2.1. Path Integral Ground State method

In this work, we use the Path Integral Ground State (PIGS) method [10] to evaluate ground-state expectation values of static and imaginary time dynamic properties of the many-body system. PIGS is a QMC algorithm in which the many-body ground state wave function $\Psi_0(\mathrm{R})$ is expressed as an evolution in imaginary time of a trial wave function $\Psi_T(\mathrm{R})$, where $\mathrm{R} = \{\mathbf{r}_i\}_{i=1}^N$ represents the set of the positions of the N particles. Indeed, if the trial state is not orthogonal to the ground state, the imaginary time evolution operator $e^{-\tau\hat{H}}$ acts on Ψ_T, allowing the exponential decay of any overlap with the excited states with respect to the overlap with Ψ_0. Thus, when the imaginary time τ is large enough, it projects $\Psi_T(\mathrm{R})$ onto the ground-state wave function, according to the formula

$$\frac{\Psi_0(\mathrm{R})}{\sqrt{\langle\Psi_0 \mid \Psi_0\rangle}} = \lim_{\tau\to\infty} \mathcal{N}_\tau \int d\mathrm{R}'\, G(\mathrm{R},\mathrm{R}';\tau)\,\Psi_T(\mathrm{R}') \,, \tag{4}$$

where $\mathcal{N}_\tau$ is a τ-dependent normalization factor and $G(\mathrm{R},\mathrm{R}';\tau) = \langle \mathrm{R}|e^{-\tau\hat{H}}|\mathrm{R}'\rangle$ is the imaginary time propagator.

In general, $G(\mathrm{R},\mathrm{R}';\tau)$ is not known, but it is possible to develop analytical expressions which approximate it accurately in the limit of short imaginary times [11, 20, 21, 22]. For this reason, it is convenient to rewrite the propagator in Eq. 4 as a convolution of M propagators $G(\mathrm{R}_l,\mathrm{R}_{l+1};\delta\tau)$ on a short time $\delta\tau = \tau/M$. Making use of a certain scheme to approximate $G(\mathrm{R}_l,\mathrm{R}_{l+1};\delta\tau)$, we are able to obtain an accurate analytical form for the ground state wave function

$$\frac{\Psi_0(\mathrm{R_M})}{\sqrt{\langle\Psi_0 \mid \Psi_0\rangle}} \simeq \mathcal{N}_\tau \int d\mathrm{R}_1\ldots d\mathrm{R}_{M-1} \prod_{l=0}^{M-1} G(\mathrm{R}_l,\mathrm{R}_{l+1};\delta\tau)\,\Psi_T(\mathrm{R}_0) \,, \tag{5}$$

and finally we are able to calcualte a generic expectation value $\langle\Psi_0\,|\hat{O}|\Psi_0\rangle$ as a statistical average over many random sets of $2M+1$ configurations, $\{\mathrm{R}_l\}_{l=0}^{2M}$, obtained by sampling the probability density

$$p(\mathrm{R}_0,\mathrm{R}_1,\ldots,\mathrm{R}_{2M}) = \mathcal{N}_\tau^2\,\Psi_T(\mathrm{R}_0) \prod_{l=0}^{2M-1} G(\mathrm{R}_l,\mathrm{R}_{l+1};\delta\tau)\Psi_T(\mathrm{R}_{2M}) \,, \tag{6}$$

with a Metropolis-like algorithm. In order to sample correctly and efficiently the Bose symmetry of the ground-state wave function, we adopt a canonical worm algorithm [20, 23]. For Bose systems, the PIGS method can be considered "exact": indeed, controlling the parameters M and $\delta\tau$, it is possible to systematically improve the accuracy of Eq. 5 and to reach a regime in which the approximations affect the numerical results to an extent which is below the statistical error.

Even though PIGS results do not depend on the approximation for the imaginary time propagator and on the trial wave function used [20, 24], a wise choice of these two ingredients can notably increase the efficiency of the algorithm. In our work, we use the pair-action approximation for the propagator at short imaginary times [11, 25, 26]:

$$G(\mathrm{R},\mathrm{R}',\delta\tau) \simeq G_0(\mathrm{R},\mathrm{R}',\delta\tau)\prod_{i<j}\frac{g_{rel}(\mathbf{r}_{ij},\mathbf{r}'_{ij},\delta\tau)}{g^0_{rel}(\mathbf{r}_{ij},\mathbf{r}'_{ij},\delta\tau)} \,. \tag{7}$$

Here, $G_0(\mathrm{R},\mathrm{R}',\delta\tau)$ is the N-particle free propagator, $g_{rel}(\mathbf{r}_{ij},\mathbf{r}'_{ij},\delta\tau)$ is the two-body propagator of the interacting system, which depends on the relative coordinates $\mathbf{r}_{ij}=\mathbf{r}_i-\mathbf{r}_j$ and $\mathbf{r}'_{ij}=\mathbf{r}'_i-\mathbf{r}'_j$, and $g^0_{rel}(\mathbf{r}_{ij},\mathbf{r}'_{ij},\delta\tau)$ is the free propagator for two particles. In the study of the HS system, it is convenient to approximate the two-body interacting propagator with its high energy expansion, proposed for the first time by Cao and Berne [22]:

$$\frac{g_{rel}(\mathbf{r},\mathbf{r}',\delta\tau)}{g^0_{rel}(\mathbf{r},\mathbf{r}',\delta\tau)} = 1-\frac{a(r+r')-a^2}{rr'}e^{-m[rr'+a^2-a(r+r')](1+\cos\theta)/(2\hbar^2\delta\tau)} \,, \tag{8}$$

where θ is the angle between $\mathbf{r}$ and $\mathbf{r}'$, $r=|\mathbf{r}|$ and $r'=|\mathbf{r}'|$.

As trial wave function, we choose the translationally invariant Jastrow wave function

$$\Psi_T(\mathrm{R})=\prod_{i<j}f(|\mathbf{r}_i-\mathbf{r}_j|)\,, \quad \text{with } f(r)=\begin{cases}0 & (r\le a)\\ \frac{\sin[k(r-a)]}{r} & (r>a)\end{cases} \,. \tag{9}$$

The function f is the solution of the two-body s-wave scattering problem with a HS potential and the wave vector k is chosen such as the derivative $f'(r)$ vanishes at $r=L/2$, where L is the size of the cubic simulation box, in order to fulfill the periodic boundary conditions [19].

2.2. *GIFT algorithm*

With the PIGS method described above, we can obtain the imaginary time intermediate scattering function, $F(\mathbf{q},\tau)$, i.e. the imaginary time correlation function of the density fluctuation operator, $\hat{\rho}_\mathbf{q}=\sum_{i=1}^{N}e^{-i\mathbf{q}\cdot\hat{\mathbf{r}}_i}$:

$$F(\mathbf{q},\tau)=\frac{1}{N}\frac{\langle\Psi_0|e^{\tau\hat{H}}\hat{\rho}_{-\mathbf{q}}e^{-\tau\hat{H}}\hat{\rho}_\mathbf{q}|\Psi_0\rangle}{\langle\Psi_0\mid\Psi_0\rangle} \,. \tag{10}$$

This quantity is related to the dynamic structure factor, $S(\mathbf{q},\omega)$, via the Laplace transform

$$F(\mathbf{q},\tau)=\int_0^\infty d\omega\, e^{-\omega\tau}\,S(\mathbf{q},\omega) \,. \tag{11}$$

Therefore, if we want to estimate $S(\mathbf{q},\omega)$ from the PIGS data for $F(\mathbf{q},\tau)$, it is necessary to invert this integral equation.

However, the inversion of Eq. 11 is a well-known ill-posed problem: indeed, given the smoothing nature of the kernel $\mathcal{K}(\omega,\tau)=e^{-\omega\tau}$, several spectral functions $S(\mathbf{q},\omega)$, even with very different features, can give similar results for $F(\mathbf{q},\tau)$. Furthermore, we have to notice that, in QMC simulations, $F(\mathbf{q},\tau)$ is accessible only in correspondence of a finite number of imaginary time values affected by statistical uncertainties. As a consequence, there exists an

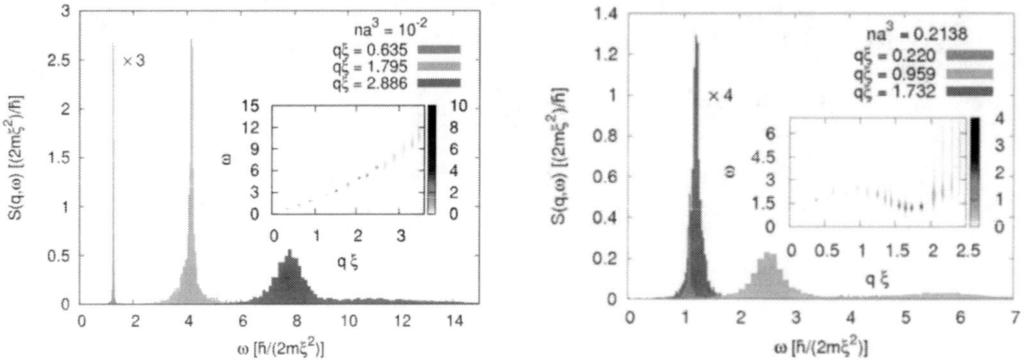

Figure 1. GIFT results for the dynamic structure factor $S(\mathbf{q},\omega)$ at $na^3 = 10^{-2}$ (left panel) and at $na^3 = 0.2138$ (right panel), for three different values of the wave vector q. The peak at $q\xi = 0.635$ in the left panel and the peak at $q\xi = 1.732$ in the right panel have been rescaled by a factor 0.33 and 0.25 respectively. The inset figures show color maps of $S(\mathbf{q},\omega)$ as a function of q.

infinite number of spectral functions which are compatible with a given numerical data set for the imaginary time correlation function.

To tackle this problem, we have used the Genetic Inversion via Falsification of Theories (GIFT) method [15]. The basic idea of this algorithm is that, instead of finding a unique solution for the inverse problem, it explores a wide space of models by means of a genetic algorithm in order to search and collect a large set of spectral functions compatible with a given QMC estimation for $F(\mathbf{q},\tau)$.

The final estimation of $S(\mathbf{q},\omega)$ is obtained averaging all the spectral functions collected: this procedure washes out the unphysical spurious features, which arise from statistical fluctuations, and reinforce the features shared by most of the collected spectral functions, which can be attributed to the real dynamic structure factor.

The GIFT technique has been successfully applied to a number of quantum many-body systems: liquid and solid ^{4}He [15, 27], bosons with soft-core repulsive potentials [28], two-dimensional ^{4}He [29] and two-dimensional normal liquid ^{3}He [30]. In these applications, GIFT has allowed to accurately determine the energy-momentum dispersion relation, $E(\mathbf{q})$, to discern the multi-excitations and to obtain the static density response function, $\chi(\mathbf{q})$ [31, 32, 33]. More details of the GIFT method can be found in Ref. [15].

3. Results

In order to calculate the imaginary time correlation function $F(\mathbf{q},\tau)$ in a HS Bose gas at a given density, we performed PIGS simulations of $N = 400$ particles interacting with the pair potential in Eq. 3 and confined inside a cubic box with periodic boundary conditions. Typical results of the dynamic structure factor obtained with the GIFT algorithm are shown in Fig. 1.

In a regime where the interactions among the particles are moderately weak (gas parameter $na^3 = 10^{-2}$, left panel of Fig. 1), $S(\mathbf{q},\omega)$ presents, at low q, a single narrow peak with a linear dispersion in q: this indicates clearly the excitation of phonons in the quantum system. As the wave vector increases, the excitation peak broadens, suggesting the emergence of physical processes involving the damping of these phonons. For large values of the wave vector ($q\xi \gtrsim 2.886$), $S(\mathbf{q},\omega)$ shows, in addition to the main peak, also a broad tail at high frequencies, indicating the emergence of a multiphonon branch in the spectrum of the excitations.

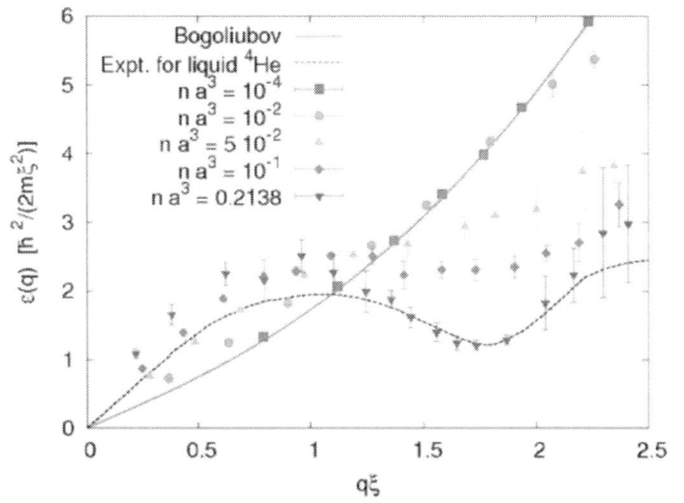

Figure 2. Dispersion of the central position of the main peak in $S(\mathbf{q}, \omega)$ for different values of the gas parameter.The orange solid line is the Bogoliubov prediction for the excitation spectrum (Eq. 2) and the black dashed line is the experimental dispersion of the elementary excitations of superfluid ^{4}He at SVP [34]. Where not shown, statistical uncertainties are below the symbol size.

When the interaction strength increases, the behavior of the dynamic structure factor changes notably. In the right panel of Fig. 1, we show $S(\mathbf{q}, \omega)$ for the HS gas at the density $na^3 = 0.2138$: this system is particularly interesting since it has been used as a reference system for the simulation of superfluid ^{4}He at saturated vapor pressure (SVP) in a previous work which considers the hard-wall potential as the leading part of the He-He interaction and the attractive tail as a weak perturbation [18]. For small values of the wave vector q, we can see that $S(\mathbf{q}, \omega)$ is exhausted by a single sharp peak, whose frequency follows the phonon dispersion $\omega(\mathbf{q}) = cq$ (being c the speed of sound), similarly to the behavior of the dilute gas at low q. However, the broadening of the main peak and the appearance of the multiphonon contribution become relevant already at values of the wave vector which are smaller compared to the case of the dilute gas ($q\xi \gtrsim 0.5$). Furthermore, we notice that the dynamic structure factor shows again a sharp peak for $1.5 \lesssim q\xi \lesssim 2$, indicating the presence of well defined quasiparticles even in this range of wave vectors. The dispersion of these elementary excitations is not monotonous with q, but it displays a relative minimum, showing that the sharp peak in $S(\mathbf{q}, \omega)$ can be associated to the excitation of a roton.

The emergence of the roton, as the density of the gas increases, shows up if we plot the dispersion $\varepsilon(\mathbf{q}) = \hbar\omega(\mathbf{q})$ of the main peak of the dynamic structure factor as a function of the wave vector q (Fig. 2). At the lowest gas parameter that we have studied, $na^3 = 10^{-4}$, the dispersion law $\varepsilon(\mathbf{q})$ follows very well the Bogoliubov prediction (Eq. 2). At higher densities, we clearly notice some deviations from this behavior, both for small and large values of q. In the phonon region, we notice that the excitation energy is higher than the one expected from $\varepsilon_B(\mathbf{q})$, indicating that the speed of sound c increases as the density of the gas increases. For wave vectors $1.3 \lesssim q\xi \lesssim 2.5$, instead, we notice that the energy of the excitation decreases for increasing gas parameter. We can also notice that, for $na^3 \simeq 5 \times 10^{-2}$, the spectrum of elementary excitations shows a change of curvature, which develops into a roton minimum when the gas parameter $na^3 \gtrsim 10^{-1}$.

It is interesting to compare the dispersion law of elementary excitations for the HS gas at $na^3 = 0.2138$ (i.e. the gas parameter corresponding to superfluid ^{4}He at equilibrium density) to the one obtained experimentally from inelastic neutron scattering in superfluid ^{4}He [34], which is also shown in Fig. 2. For small wave vectors, our results for the energy of the excitation is larger than the experimental data. This indicates that, as one should expect, the attractive tail of the He-He potential plays a relevant role in determining the speed of sound in the system.

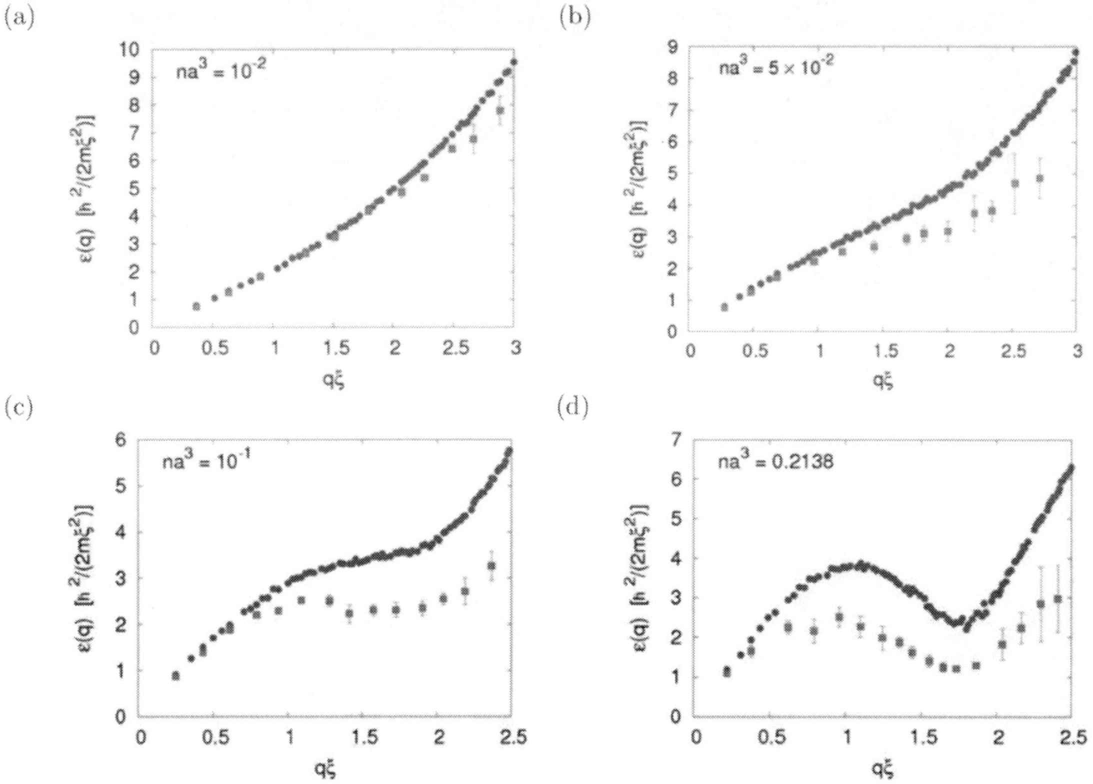

Figure 3. Spectrum of the elementary excitations $\varepsilon(\mathbf{q})$ of the HS-gas for different values of the gas paramenter: $na^3 = 10^{-2}$ (panel a), $na^3 = 5 \times 10^{-2}$ (panel b), $na^3 = 10^{-1}$ (panel c), $na^3 = 0.2138$ (panel d). GIFT results (red squares) are compared with the estimation of $\varepsilon_F(\mathbf{q})$ provided within the Feynman approximation, Eq. 12 (blue circles). Where not shown, statistical uncertainties are below the symbol size.

On the contrary, the dispersion law of elementary excitations obtained with our QMC approach is in good agreement with the experimental measurements in the roton region and for higher momenta: this suggests that the HS model is able to reproduce the density fluctuation spectrum of superfluid ^{4}He for wave vectors reciprocal to the mean interparticle and shorter distances [35].

A qualitative description of the appearance of the roton in the excitation spectrum for increasing na^3 can be obtained via the Feynman's approximation [36]

$$\varepsilon_F(\mathbf{q}) = \hbar\omega_F(\mathbf{q}) = \frac{\hbar^2 q^2}{2mS(\mathbf{q})} , \tag{12}$$

in which $S(\mathbf{q})$ is the static structure factor. This relation is easily obtained from the f-sum rule

$$\int_0^\infty d\omega\, \omega S(\mathbf{q},\omega) = \frac{\hbar q^2}{2m} \tag{13}$$

with the assumption that the dynamic structure factor can be written as a single delta-peak $S(\mathbf{q},\omega) = S(\mathbf{q})\delta\left(\omega - \omega_F(\mathbf{q})\right)$ and indicates that the emergence of a minimum in $\varepsilon(\mathbf{q})$ can be associated to the emergence of a maximum in $S(\mathbf{q})$ and thus to the formation of the microscopic local structures typical of dense fluids [35]. In Ref. [37], the excitation spectrum of the HS gas

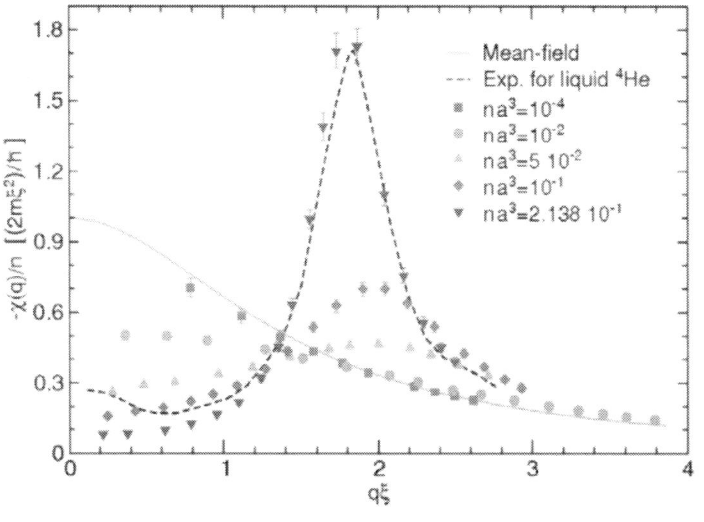

Figure 4. Static density response function $\chi(\mathbf{q})$ for the HS gas at different values of the gas parameter. The orange solid line is the mean-field result for $\chi(\mathbf{q})$ and the black solid line is the experimental static density response function of superfluid ^{4}He at SVP [40]. Where not shown, statistical uncertainties are below the symbol size.

has been obtained from PIGS calculation of $S(\mathbf{q})$, making use of the Feynman's approximation, and it has been shown that, even within this approximated scheme, a roton minimum appears at high density. In order to test this approximation from a quantitative point of view, we compare in Fig. 3 the accurate result for $\varepsilon(\mathbf{q})$, obtained from the GIFT estimation of the dynamic structure factor, with the approximated result $\varepsilon_F(\mathbf{q})$ obtained from the PIGS results for the static structure factor, according to Eq. 12. We can see that the Feynman's approach is able to describe accurately the dynamics of the HS gas only for small values of the wave vector q. At large q, instead, we notice a discprepancy between the Feynman's prediction $\varepsilon_F(\mathbf{q})$ and the GIFT results, which becomes larger as the gas parameter increases. From this comparison, we can conclude that the assumption of describing the dynamic structure factor in terms of a single peak is accurate only in the phonon region, while the secondary multiphonon branch gives a relevant contribution at large wave vectors, especially for strongly interacting systems.

In our study, we also calculate the static density response function $\chi(\mathbf{q})$. This quantity describes the linear response in density to a weak perturbation which is spatially modulated with a wave vector $\mathbf{q}$. In the limit $q \to 0$, $\chi(\mathbf{q})$ converges to the isothermal compressibility [38]. From the knowledge of the dynamic structure factor, the static density response function can be obtained using the following formula:

$$\chi(\mathbf{q}) = -2n \int_0^\infty d\omega \, \frac{S(\mathbf{q},\omega)}{\omega} \, . \tag{14}$$

In figure 4 we present the static density response function calculated from the reconstructed dynamic structure factors of the HS system for different gas parameters. In the weakly interacting regime ($na^3 = 10^{-4}$), $\chi(\mathbf{q})$, for the available wave vectors, is monotonically decreasing, consistently with the behavior of the dilute gas. Comparing our numerical results with the curve for $\chi(\mathbf{q})$ obtained within the mean-field approximation based on the Bogoliubov dispersion (Eq. 2), that is

$$\chi(\mathbf{q}) = \frac{2m\xi^2}{\hbar} \frac{2}{(q\xi)^2 + 2} \, , \tag{15}$$

we notice an excellent agreement at large q, while at small q we find that the QMC estimates are systematically slightly below the mean-field prediction. This discrepancy may be due to finite-size effects, which are not easy to evaluate since the set of the wave vectors q achievable in our QMC simulations depends on the geometry of the box and thus a calculation of $\chi(\mathbf{q})$ at the same q but with different sizes of the box is not straightforward. However, the statistical uncertainties of our results do not allow to exclude a statistical origin of such discrepancy.

By increasing the gas parameter, we observe a decreasing of the value of $\chi(\mathbf{q})$ at small q, which indicates a decreasing of the compressibility of the gas, while at intermediate values of q, we can see the emergence of a peak, which becomes clear for $na^3 = 10^{-1}$, i.e. the same gas parameter at which the roton minimum appears in $\varepsilon(\mathbf{q})$. The behavior displayed by the static density response function confirms that in a strongly interacting many–body Bose system, characterized by the presence of a hard-core interaction, the preferred modulation of the system, revealed by a peak in $\chi(\mathbf{q})$, corresponds to a wave vector in the vicinity of the roton minimum. Such peak is a precursor of a Bragg–like peak which appears during crystallization [39].

As we did for the dispersion relation of the elementary excitations, it is interesting to compare our numerical result for the static response function of the HS gas at $na^3 = 0.2138$ with the experimental measurements for superfluid ^{4}He at SVP [40]. As we can see in figure 4, there is a significative difference for $q\xi \lesssim 1.4$, which reflects a discrepancy in the compressibility, and thus in the speed of sound (see Fig. 2), between the HS gas and superfluid ^{4}He. On the contrary, we can notice that, for momenta in the roton and post-roton regions, the agreement between the two curves is good. This result confirms the hypothesis that superfluid ^{4}He behaves like a HS system for wave vectors corresponding to the inverse of the mean interatomic distance and above [35].

4. Conclusions

In this work, we have estimated the dynamic structure factor and the spectrum of elementary excitations for Bose HS gases at zero temperature, ranging from the dilute to the dense regime, by means of *ab-initio* calculations based on QMC methodologies. The numerical approach followed, which makes use of the GIFT algorithm to perform the analytical continuation of the "exact" PIGS results for the imaginary time correlation functions, is one of the most powerful methods presently available and allow us to get very accurate results for the spectral functions, even for strongly interacting systems, where it is difficult to describe properly all the relevant correlations arising among the quantum particles. We have been able to see, in the dispersion of the elementary excitations, a crossover from the Bogoliubov spectrum, predicted in mean-field theories and accurate for weakly interacting gases, to the phonon-maxon-roton spectrum, experimentally seen in dense Bose liquids, such as superfluid ^{4}He. We also show that, for large values of the wave vector of the density fluctuation, the dynamic structure factor presents a broad multiphonon contribution at high frequencies, which becomes relevant as the density of the system increases, making quantitatively inaccurate the Feynman's approximation for the spectrum of the elementary excitations. Remarkably, for $na^3 = 0.2138$, the gas parameter corresponding to superfluid ^{4}He at equilibirum density, the hard-sphere model turns out to describe accurately the energy–momentum dispersion relation and the static density response of superfluid ^{4}He in the roton region; this suggests that, for wavelengths comparable to interatomic distances, the low–energy dynamic properties are dominated by the hard-core repulsive part of the interaction potential.

Acknowledgments

This work has been supported by Regione Lombardia and CINECA Consortium through a LISA Initiative (Laboratory for Interdisciplinary Advanced Simulation) 2012 grant [http://www.hpc.cineca.it/services/lisa], by ERC through the QGBE grant and by Provincia Autonoma di Trento. Partial support by the Italian MIUR under Contract Cofin-2009 "Quantum gases beyond equilibrium" is also acknowledged. We thank the Aurora-Science project (funded by PAT and INFN) for allocating part of the computing resources for this work and for technical support.

References

[1] Glyde H R 1994 *Excitations in Liquid and Solid Helium* (Oxford: Clarendon Press)
[2] Dalfovo F, Giogrini S, Pitaevskii L P and Stringari S 1999 *Rev. Mod. Phys.* **71** 463
[3] Steinhauer J, Ozeri R, Katz N and Davidson N 2002 *Phys. Rev. Lett.* **88** 120407
[4] Svensson E C, Martel P, Sears V F and Woods A D B 1976 *Can J. Phys* **54** 2178
[5] Talbot E F, Glyde H R, Stirling W G and Svensson E C 1988 *Phys. Rev. B* **38** 11229
[6] Stirling W G and Glyde H R 1990 *Phys. Rev. B* **41** 4224
[7] Andersen K H, Stirling W G, Scherm R, Stunault A, Fak B, Godfrin H and Dianoux A J 1994 *J. Phys: Condensed Matter* **6** 821
[8] Andersen K H and Stirling W G 1994 *J. Phys: Condensed Matter* **6** 5805
[9] Gibbs M R, Andersen K H, Stirling W G and Schober H 1999 *J. Phys: Condensed Matter* **11** 603
[10] Sarsa A, Schmidt K E and Magro W R 2000 *J. Chep. Phys* **113** 1366
[11] Ceperley D M 1995 *Rev. Mod. Phys.* **67** 279
[12] Jarrell M and Gubernatis J E 1996 *Phys. Rep.* **269** 133
[13] Boninsegni M and Ceperley D M 1996 *J. Low Temp. Phys* **104** 339
[14] Baroni S and Moroni S 1999 *Phys. Rev. Lett.* **82** 4745
[15] Vitali E, Rossi M, Reatto L and Galli D E 2010 *Phys. Rev. B* **82** 174510
[16] Huang K and Yang C N 1957 *Phys. Rev.* **105** 767
[17] Hansen J P, Levesque D and Shiff D 1971 *Phys. Rev. A* **3** 776
[18] Kalos M H, Levesque D and Verlet L 1957 *Phys. Rev. A* **9** 2178
[19] Giorgini S, Boronat J and Casulleras J 1999 *Phys. Rev. A* **60** 5129
[20] Rossi M, Nava M, Reatto L and Galli D E 2009 *J. Chem. Phys.* **131** 154108
[21] Zillich R E, Mayrhofer J M and Chin S A 2010 *J. Chem. Phys.* **132** 044103
[22] Cao J and Berne B J 1992 *J. Chem. Phys.* **97** 2382
[23] Boninsegni M, Prokof'ev N V and Svistunov B V 2006 *Phys. Rev. Lett.* **96** 070601
[24] Rota R, Casulleras J, Mazzamti F and Boronat J 2010 *Phys. Rev. E* **81** 016707
[25] Pollock E L and Ceperley D M 1984 *Phys. Rev. B* **30** 2555
[26] Pollock E L and Ceperley D M 1987 *Phys. Rev. B* **36** 8343
[27] Rossi M, Vitali E, Reatto L and Galli D E 2012 *Phys. Rev. B* **85** 014525
[28] Saccani S, Moroni S and Boninsegni M 2012 *Phys. Rev. Lett.* **108** 175301
[29] Arrigoni F, Vitali E, Galli D E and Reatto L 2013 *Low Temp. Phys.* **39** 793 [*Fiz. Niz. Temp.* **39**, 1021]
[30] Nava M, Galli D E, Moroni S and Vitali E 2013 *Phys. Rev. B* **87** 144506
[31] Minoguchi T and Galli D E 2011, *J. low temp phys* **162**, 160
[32] Minoguchi T, Galli D E, Rossi M, and Yoshimori A 2012, *J. Phys.: Conf. Ser.* **400**, 012050
[33] Minoguchi T, Nava M, Tramonto F and Galli D E 2013, *J. Low. Temp. Phys.* **171**, 259
[34] Donnelly R J, Donnelly J A and Hills R N 1981 *J. Low Temp. Phys.* **44** 471
[35] Rota R, Tramonto F, Galli D E and Giorgini S 2013 *Phys. Rev. B* **88**, 214505
[36] Feynman R P 1954 *Phys. Rev.* **94** 262
[37] Rossi M and Salasnich L 2013 *Phys. Rev. A* **88** 053617
[38] See *e.g.* Pitaevskii L and Stringari S 2003 *Bose-Einstein Condensation* (Oxford: Clarendon Press).
[39] Vitali E, Galli D E, and Reatto L 2008, in *Advanced in Quantum Many-body Theory* **11**, 251
[40] Cowley R A and Woods A D B 1971, *Cam. J. Phys.* **49**, 187

Magnetic susceptibility of frustrated spin-s J_1-J_2 quantum Heisenberg magnets: High-temperature expansion and exact diagonalization data

J Richter[1], A Lohmann[1], H-J Schmidt[2], D C Johnston[3]

[1]Institut für Theoretische Physik, Otto-von-Guericke-Universität Magdeburg, PF 4120, D - 39016 Magdeburg, Germany

[2]Universität Osnabrück, Fachbereich Physik, Barbarastr. 7, D - 49069 Osnabrück, Germany

[3] Ames Laboratory and Department of Physics and Astronomy, Iowa State University, Ames, Iowa 50011

E-mail: `johannes.richter@physik.uni-magdeburg.de`

Abstract. Motivated by recent experiments on low-dimensional frustrated quantum magnets with competing nearest-neighbor exchange coupling J_1 and next nearest-neighbor exchange coupling J_2 we investigate the magnetic susceptibility of two-dimensional J_1-J_2 Heisenberg models with arbitrary spin quantum number s. We use exact diagonalization and high-temperature expansion up to order 10 to analyze the influence of the frustration strength J_2/J_1 and the spin quantum number s on the position and the height of the maximum of the susceptibility. The derived theoretical data can be used to get information on the ratio J_2/J_1 by comparing with susceptibility measurements on corresponding magnetic compounds.

1. Introduction

The investigation of frustrated magnetic systems is currently a field of active theoretical and experimental research [1, 2]. Systems with competing nearest-neighbor (NN) exchange coupling J_1 and next nearest-neighbor (NNN) exchange coupling J_2 can serve as model systems to study the interplay of quantum effects, thermal fluctuations and frustration. The quantum J_1-J_2 Heisenberg models on the square-lattice exhibit several ground-state phases including non-classical non-magnetic ground states, see, e.g., [3]. The corresponding Hamiltonian reads

$$H = J_1 \sum_{\langle i,j \rangle} \mathbf{S}_i \cdot \mathbf{S}_j + J_2 \sum_{[i,j]} \mathbf{S}_i \cdot \mathbf{S}_j, \qquad (1)$$

where $(\mathbf{S}_i)^2 = s(s+1)$, and $\langle i,j \rangle$ denotes NN and $[i,j]$ denotes NNN bonds. For antiferromagnetic NNN bonds, $J_2 > 0$, the spin system is frustrated irrespective of the sign of J_1. Due to frustration the theoretical treatment of this model is challenging.

The numerous theoretical studies of the ground state phase diagram so far did not lead to a consensus on the nature of the quantum ground state and on the nature of the quantum phase transitions present in the model, see, e.g., [4] and references therein. Interestingly there are also various compounds well described by square-lattice J_1-J_2 Heisenberg models, such as oxovanadates [5] and iron pnictides [6]. In experiments, typically temperature-dependent quantities are reported. Hence reliable (and flexible) tools are desirable to calculate thermodynamic quantities such as the uniform magnetic susceptibility χ. In this paper we present two methods, namely the full exact diagonalization, see, e.g., [7], and the high-temperature expansion [8, 9, 10] to calculate the temperature dependence of the magnetic susceptibility for the square-lattice J_1-J_2 spin-s Heisenberg model with both ferromagnetic (FM) and antiferromagnetic (AFM) NN coupling J_1 and AFM NNN bonds J_2 for arbitrary spin quantum number s. In particular, we analyze the position and the height of the maximum in the susceptibility in dependence on J_1, J_2 and s.

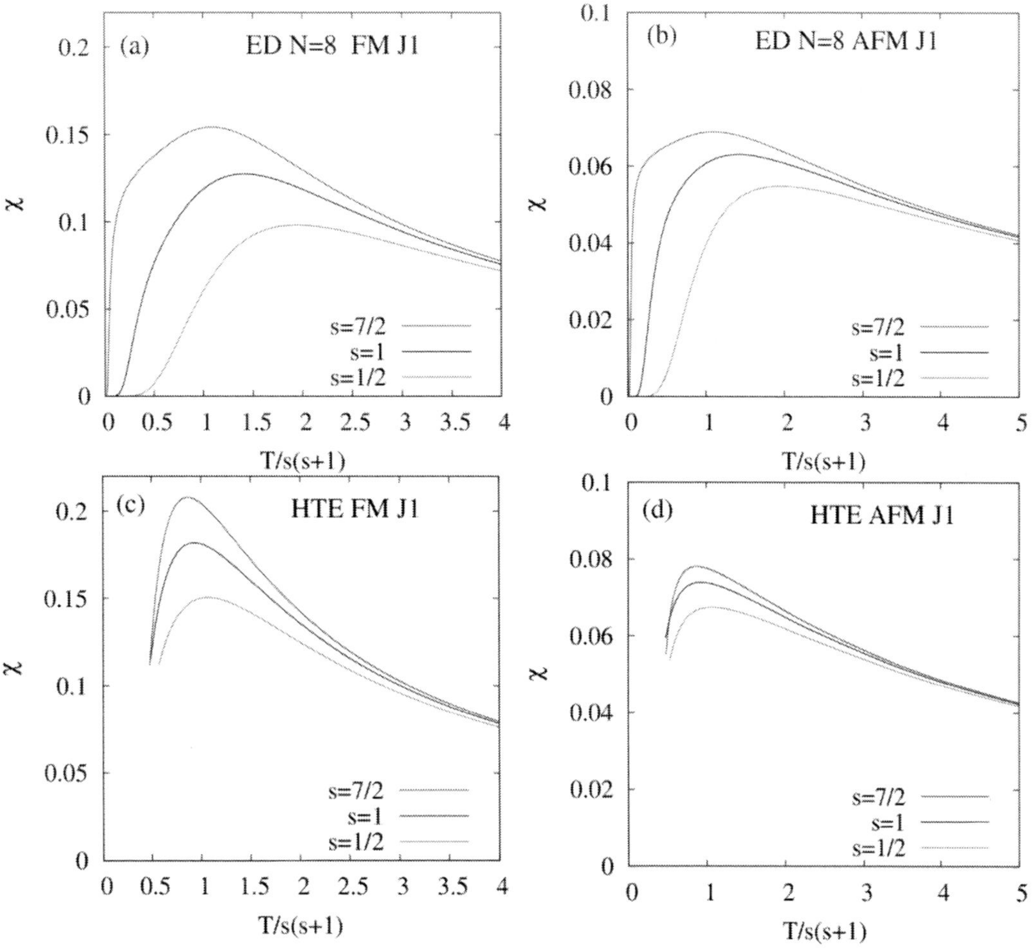

Figure 1. Uniform susceptibility χ as a function of renormalized temperature $T/s(s+1)$ for NNN exchange $J_2 = 1$ and three values of the spin quantum number $s = 1/2, 1$, and $7/2$. (a) Numerical exact data for a finite square lattice of $N = 8$ sites and antiferromagnetic $J_1 = 1$. (b) Numerical exact data for a finite square lattice of $N = 8$ sites and ferromagnetic $J_1 = -1$. (c) [6,4] Padé approximant of the 10th order HTE series for an infinite square lattice and ferromagnetic $J_1 = -1$. (d) [6,4] Padé approximant of the 10th order HTE series for an infinite square lattice and antiferromagnetic $J_1 = 1$. The order of labeling in each legend is the same as the order of the plots (top to bottom).

2. Methods

The full exact diagonaliazion (ED) yields numerical exact results at arbitrary temperature T, but it is typically limited to about $N = 22$ sites for $s = 1/2$ models. For larger spin quantum numbers s the system size N accessible for ED shrinks significantly. Hence, ED is used preferably for $s = 1/2$ and $s = 1$. In the present study we exploit the special symmetry properties of the finite square-lattice of $N = 8$ sites and perform full ED for the J_1-J_2 model for $s = 1/2, 1, .., 9/2$, thus allowing to study the role of the spin quantum number. Since the ED approach suffers from the finite-size effect, the ED calculations do not yield quantitatively correct results for the thermodynamic limit. Nevertheless, they will give insight into the qualitative behavior of the susceptibility. The high-temperature expansion (HTE) for the J_1-J_2 model up to 10th order was presented in [9], however, restricted to $s = 1/2$. This restriction can be overcome by using our general HTE scheme for Heisenberg models with arbitrary exchange patterns and arbitrary spin quantum number s up to order 8 [10]. The scheme is encoded in a simple C++-program and can be downloaded [11] and freely used by interested researchers.

Very recently the present authors have extended this general HTE scheme up to 10th order [12]. Here we use this 10th order HTE as an alternative method to the ED. We use here three different subsequent Padé approximants, namely Padé [4,6], [5,5], and [6,4], see e.g. [8, 10]. Such a Padé approximant extends the region of validity of the HTE series down to lower temperatures. Since the HTE approach is designed for infinite systems the HTE data for the susceptibility maximum, in principle, can be quantitatively correct, if the maximum is not located at too low temperatures. Indeed, it was found [10] that for the unfrustrated ($J_2 = 0$) square-lattice spin-1/2 Heisenberg antiferromagnet the Padé [4,4] approximant of the 8th order HTE series yields correct data for the susceptibility maximum located at $T \approx 0.94 J_1$. However, it may happen that a certain Padé approximant does not work for some particular values of J_1, J_2, and s, since Padé approximants may exhibit unphysical poles for temperatures in the region of interest. Hence we show in the next section only those Padé data not influenced by poles.

3. Results

First we present the temperature dependence of the susceptibility χ in Fig. 1 for a particular value of J_2 and both FM and AFM J_1. In this paper the symbol χ means $\chi |J_1|/N g^2 \mu_B^2$, where N is the number of spins and μ_B is the Bohr magneton. The temperature is measured in terms of $|J_1|$, i.e. the symbol T means $T/|J_1|$. The qualitative behavior of $\chi(T)$ shown in Figs. 1(a-d) is similar, there is the broad maximum in $\chi(T)$ that is typical for a two-dimensional antiferromagnet (note that for $J_2/|J_1| = 1$ the system is in the AFM ground state irrespective of the sign of J_1). The various $\chi(T)$ curves give an impression on the finite-size effects, the effect of the sign of the NN exchange J_1, and the influence of spin quantum number s. The height, χ_{max}, and the position, T_{max}, of the maximum in the $\chi(T)$ curve are interesting features for the comparison with experimental data, in particular to get information on the ratio $J_2/|J_1|$ from susceptibility measurements, see e.g. [14]. Therefore we will discuss χ_{max} and the T_{max} now in more detail.

We present our data for the susceptibility maximum for both FM and AFM NN exchange J_1 in Figs. 2 (ED data) and 3 (HTE and ED data). For FM J_1, χ_{max} (T_{max}) becomes larger (smaller) upon lowering J_2. Finally, when approaching the critical value J_2^c, where the transition to the ferromagnetic ground state takes place, χ_{max} diverges and T_{max} goes to zero. The critical point for $s = 1/2$ is $J_2^c = 0.333\,|J_1|$ for $N = 8$ (but it is $J_2^c \approx 0.4|J_1|$ for $N \to \infty$ [13]). It increases with growing s and becomes $J_2^c = 0.5|J_1|$ for $s \to \infty$. The data for $N = 8$ and $N \to \infty$ are in qualitative agreement. Although the finite-size effects are obviously large, the general features of χ_{max} and T_{max} as functions of J_2 and s are quite similar. Naturally the HTE fails when approaching J_2^c, since in this limit low temperatures become relevant. Note that the HTE data for FM J_1 and $s = 1/2$ are also in qualitative agreement with recently reported data calculated

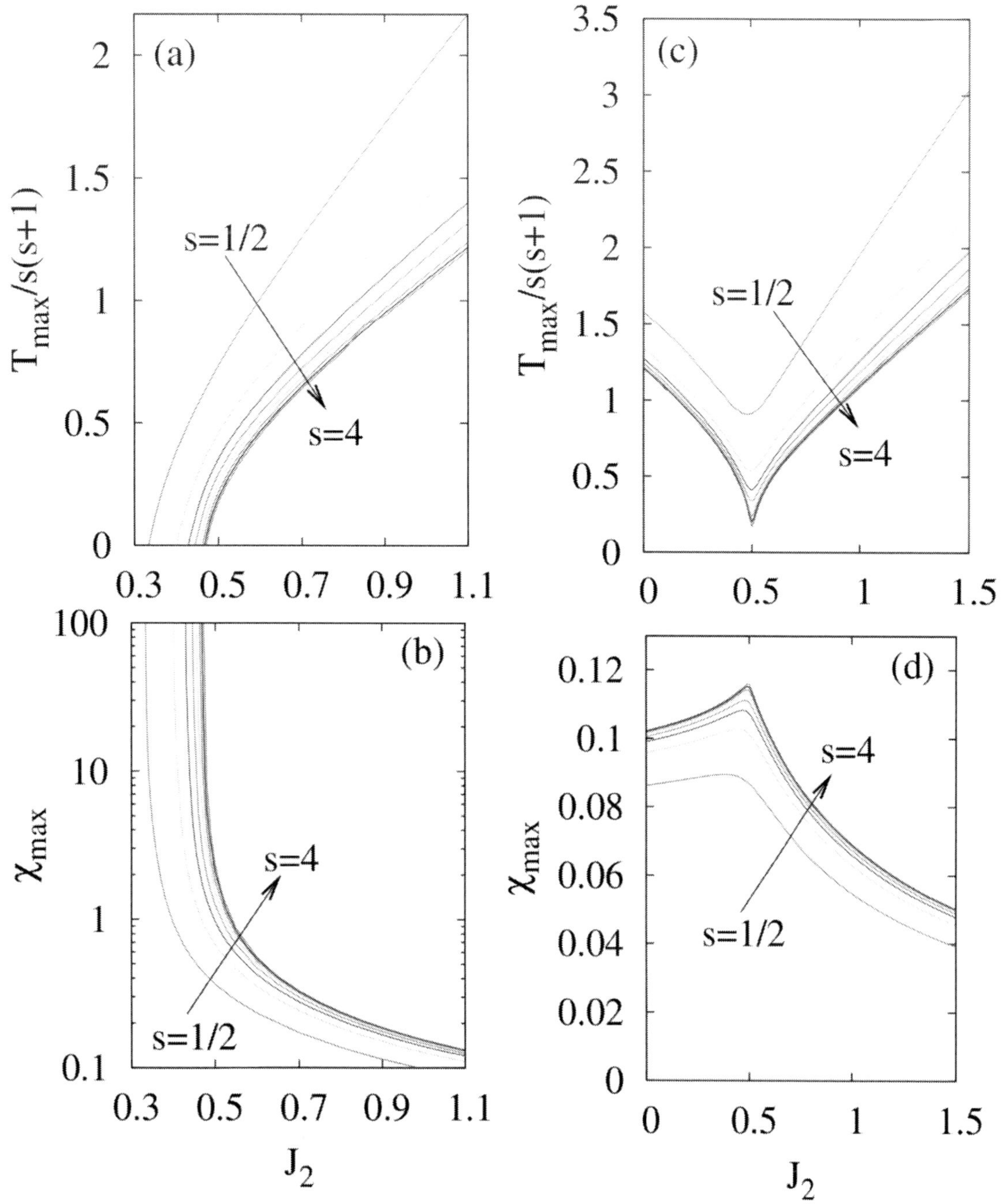

Figure 2. Position T_{max} (a and c) and height χ_{max} (b and d) of $\chi(T)$ for the finite $N = 8$ square-lattice J_1-J_2 model (left panels FM $J_1 = -1$, right panels AFM $J_1 = +1$).

by second-order Green's function approach [14]. We discuss now the case of AFM J_1 (right panels in Figs. 2 and 3). For large J_2 the behavior of χ_{max} and T_{max} is very similar to that for

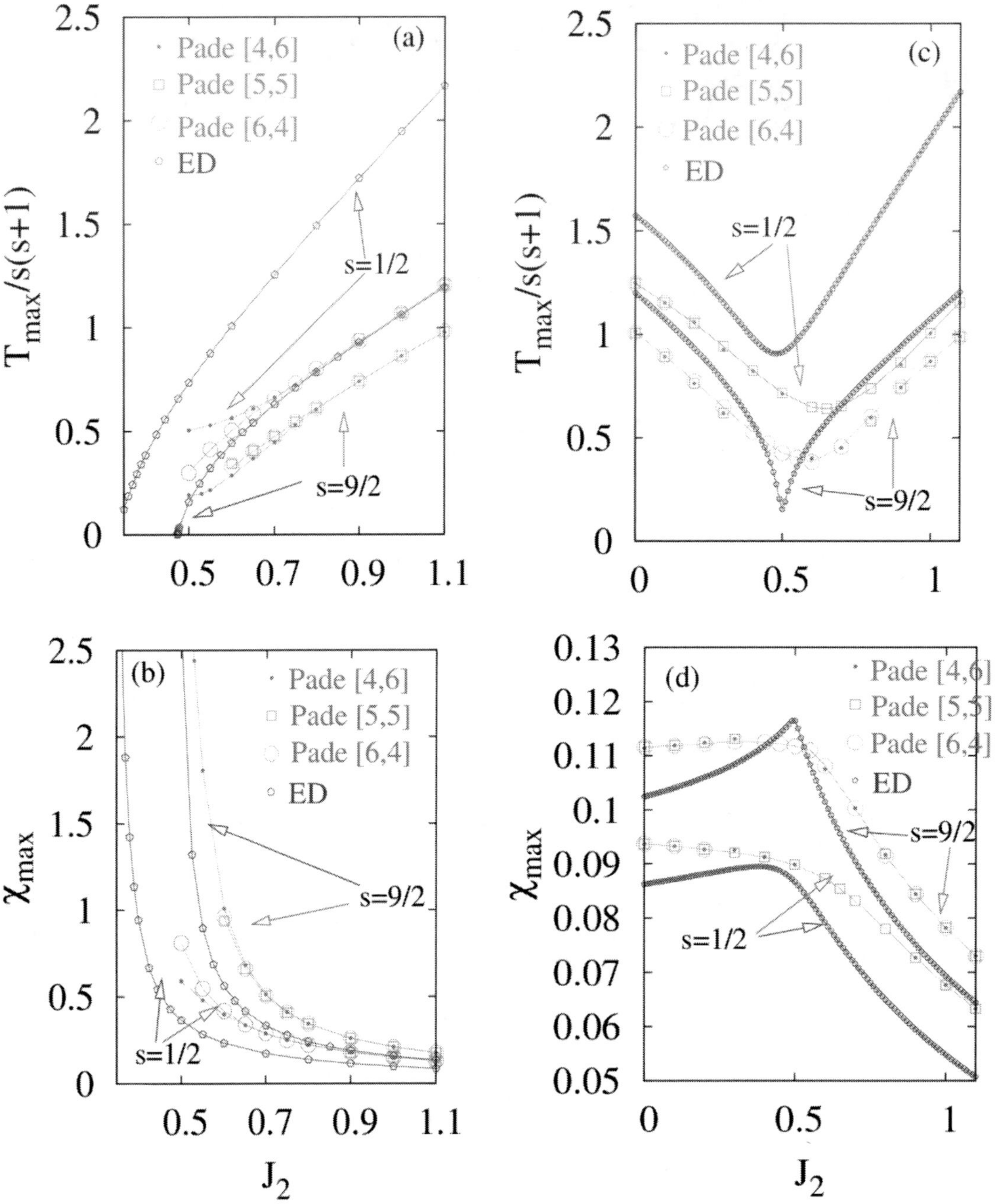

Figure 3. Position T_{max} (a and c) and height χ_{max} (b and d) of $\chi(T)$ for an infinite square-lattice J_1-J_2 model obtained by 10th order HTE (left panels FM $J_1 = -1$, right panels AFM $J_1 = +1$). For comparison we show the ED data for $N = 8$.

FM J_1, i. e. the sign of J_1 becomes irrelevant, cf. Ref. [13]. On the other hand, for smaller values of J_2 naturally both cases behave completely different, since J_1 dominates the physics. We find

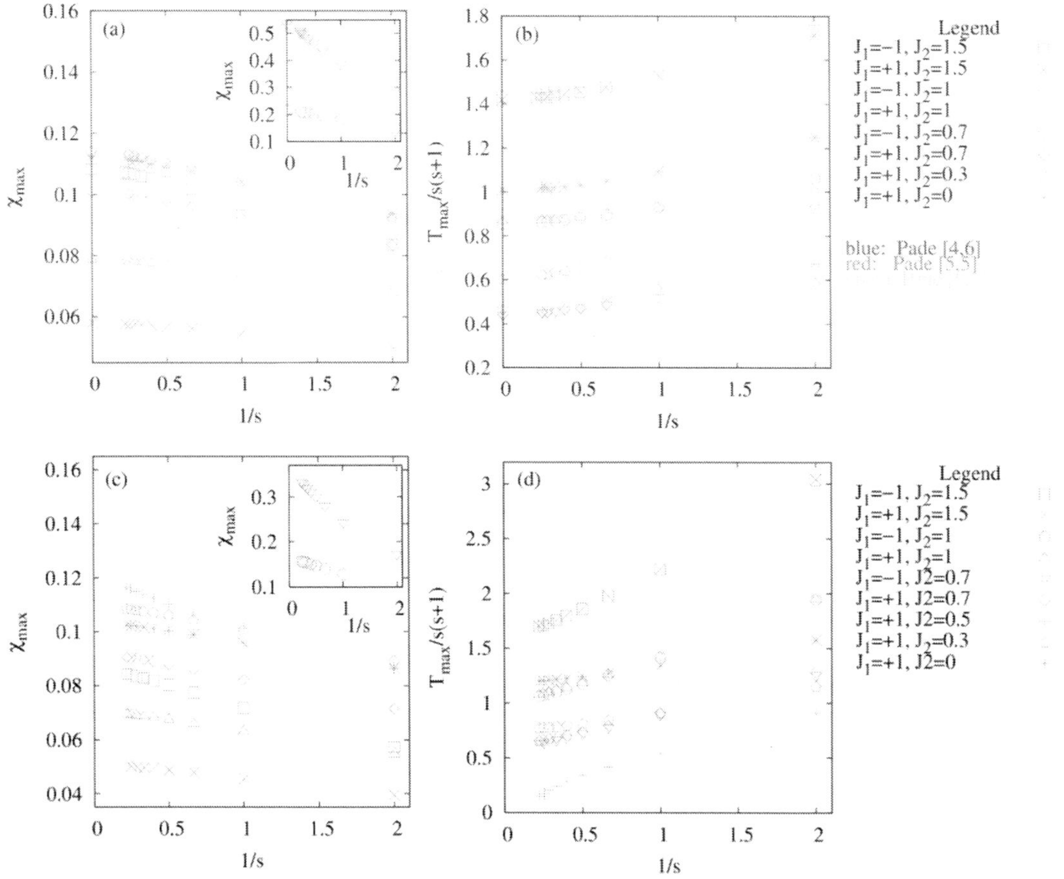

Figure 4. Height χ_{max} (a and c) and position T_{max} (b and d) of $\chi(T)$ for the square-lattice J_1-J_2 model for various sets of parameters J_1 and J_2 as a function of the inverse spin quantum number s obtained from Padé approximants of 10th order HTE series of an infinite system (a and b) and from ED for $N = 8$ (c and d).

a well pronounced minimum in T_{max} in the region of strongest frustration around $J_2 = 0.5$. For the finite system χ_{max} exhibits a maximum in this region, whereas for the infinite system χ_{max} is almost constant in the region $0 \leq J_2 \leq 0.5$.

To take a closer look on the role of the spin quantum number s we present in Fig. 4 the quantities χ_{max} and T_{max} as a function of $1/s$ for particular values of J_2. Obviously, there is monotonous increase (decrease) of χ_{max} ($T_{max}/s(s+1)$) with growing s. For FM $J_1 = -1$ the increase of χ_{max} is particular strong for $J_2 = 0.7$ (see the insets in panels a and c), since for large s this value of J_2 becomes quite close to the transition point to the FM ground state. From Figs. 4(a-d) it is also seen that the position T_{max} of the maximum for $J_2 \gtrsim 0.7|J_1|$ is almost independent of the sign of J_1, whereas the height χ_{max} strongly depends on the sign of the NN coupling. Let us finally mention the special s-dependence of the maximum in $\chi(T)$ for $J_1 = 1$ and $J_2 = 0.5$, where the classical ground state exhibits a large non-trivial degeneracy. The position $T_{max}/s(s+1)$ of the maximum shifts to zero in the limit $s \to \infty$, whereas the height remains finite. This behavior is quite similar to that found for the pyrochlore AFM [10, 15, 16], where the classical ground state is also highly degenerate.

4. Summary

Using high-temperature expansion and full exact diagonalization we have calculated the uniform susceptibility χ of the spin-s J_1-J_2 square-lattice Heisenberg magnet in a wide parameter regime of FM and AFM J_1 and frustrating AFM J_2. Especially, we have studied the height and the position of the maximum in the $\chi(T)$ curve as functions of J_2/J_1 and the spin quantum number s. These data can be used to get information on the ratio $J_2/|J_1|$ from susceptibility measurements, e.g. on oxovanadates which are well described by the square-lattice J_1-J_2 model.

Acknowledgments

The work at Ames Laboratory was supported by the U.S. Department of Energy under Contract No. DE-AC02-07CH11358.

References

[1] *Quantum Magnetism* 2004 ed Schollwöck U, Richter J, Farnell D J J, and Bishop R F, Lecture Notes in Physics **645** (Berlin: Springer)
[2] *Frustrated Spin Systems* 2005 ed Diep H T (Singapore: World Scientific)
[3] Schmalfuß D, Darradi R, Richter J, Schulenburg J, and Ihle D 2006 Phys. Rev. Lett. **97** 157201; Jiang H-C, Yao H, and Balents L 2012 Phys. Rev. B **86** 024424
[4] Sirker J, Weihong Z, Sushkov O P, and Oitmaa J 2006 Phys. Rev. B **73** 184420; Darradi R, Derzhko O, Zinke R, Schulenburg J, Krüger S E, and Richter J 2008 Phys. Rev. B **78** 214415; Richter J and Schulenburg J 2010 Eur. Phys. J. B **73** 117; Wang L, Poilblanc D, Gu Z-C, Wen X-G , and Verstraete F 2013 Phys. Rev. Lett. **111** 037202; Hu W J, Becca F, Parola A, and Sorella S 2013 Phys. Rev. B **88** 060402; Ren Y Z, Tong N H, Xie X C 2014 J. Phys. Cond. Matter **26** 115601
[5] Nath R, Tsirlin A A, Rosner H, and Geibel C 2008 Phys. Rev. B **78** 064422; Bossoni L, Carretta P, Nath R, Moscardini M, Baenitz M, and Geibel C 2011 Phys. Rev B **83** 014412; Roy B, Furukawa Y, Nath R, and Johnston D C 2011 J. Phys. Conf. Series **320** 012048
[6] Si Q and Abrahams E 2008 Phys. Rev. Lett. **101** 076401; Johnston D C 2010 Adv. Phys. **59** 803
[7] Härtel M, Richter J, Ihle D, and Drechsler S-L 2008 Phys. Rev. B **78** 174412
[8] Oitmaa J, Hamer C J, and Zheng W H 2006 *Series Expansion Methods* (Cambridge: Cambridge University Press)
[9] Rosner H, Singh R R P, Zheng W H, Oitmaa J, and Pickett W E 2003 Phys. Rev. B **67** 014416
[10] Schmidt H J, A. Lohmann A, and Richter J 2011 Phys. Rev. B **84** 104443
[11] see http://www.uni-magdeburg.de/jschulen/HTE/
[12] Lohmann H, Schmidt H J, and Richter J 2014 Phys. Rev. B **89** 014415
[13] Shannon N, Momoi T, and Sindzingre P 2006 Phys. Rev. Lett. **96** 027213; Richter J, Darradi R, Schulenburg J, Farnell D J J, and Rosner H 2010 Phys. Rev. B **81** 174429
[14] Härtel M, Richter J, Götze O, Ihle D, and Drechsler S-L 2013 Phys. Rev. B **87** 054412
[15] Moessner R and Berlinsky A J 1999 Phys. Rev. Lett. **83** 3293
[16] Garcia-Adeva A J and Huber D L 2001 Phys. Rev. B **63** 140404

Journal of Physics: Conference Series **529** (2014) 012024 doi:10.1088/1742-6596/529/1/012024

Three-body losses of repulsively interacting three-component fermionic atoms in optical lattices

S Suga[1] **and K Inaba**[2,3]

[1]Graduate School of Engineering, University of Hyogo, Himeji 671-2280, Japan
[2] NTT Basic Research Laboratories, NTT Corporation, Atsugi 243-0198, Japan
[3] JST, CREST, Sanbancho, Chiyoda-ku, Tokyo 102-0075, Japan

E-mail: suga@eng.u-hyogo.ac.jp

Abstract. We investigate the effects of a repulsive three-body interaction on the Mott transition of the repulsively interacting three-component fermionic atoms in optical lattices by means of the self-energy functional approach. We find that the three-body repulsion hardly affects the qualitative features of the Mott transition, because the three-body repulsion does not compete with the two-body repulsions. When the three-body repulsion is extremely strong, the triple occupancy vanishes in the Fermi liquid state. This situation is equivalent to that caused by strong three-body losses. Our results imply that three-body losses have little influence on the Mott transitions in the repulsively interacting three-component fermionic atoms in optical lattices.

1. Introduction

High controllability of cold atoms in optical lattices paves a way for studying novel quantum many-body systems that are difficult to realize in condensed matter systems. Actually, degenerate fermionic atoms with three-component internal degrees of freedom were actively investigated and their interesting features have been revealed in recent experiments [1, 2]. For three-component (color) fermionic atoms in optical lattices, theoretical studies based on unperturbative methods have pointed out various interesting features. For attractively interacting systems, a phase transition appears between the color superfluid and the trionic state, namely, the singlet bound state of three-color atoms [3, 4, 5]. For repulsively interacting systems, a characteristic Mott phase called the paired Mott insulator (PMI) appears at non-integer half filling [6, 7, 9], and also a superfluid phase appears in the vicinity of half filling [8, 9]. However, most of these previous studies have not dealt with three-body losses, which intrinsically exist in the realistic cold atom systems. Some recent theoretical studies considering thee-body losses for the attractively interacting system have suggested that strong three-body losses induce a phase separation instead of a trion formation [10, 11].

In this paper, we discuss how three-body losses affect the PMI transition in repulsively interacting three-component fermionic atoms in optical lattices. For this purpose, we investigate the three-component fermionic Hubbard-type model including a repulsive three-body interaction term. We find that the extremely strong three-body repulsion projects out the triple occupancy effectively. This situation is equivalent to that caused by very large three-body losses [10, 11].

2. Model and method

According to the conventional model for cold atoms in optical lattices [12], we include only the nearest-neighbor tunneling and set the on-site interaction between different colors. The model Hamiltonian reads as

$$\mathcal{H} = -t \sum_{\langle i,j \rangle} \sum_{\alpha=1}^{3} \hat{a}_{i\alpha}^{\dagger} \hat{a}_{j\alpha} - \sum_{i} \sum_{\alpha=1}^{3} \mu_{\alpha} \hat{n}_{i\alpha} + \frac{1}{2} \sum_{i} \sum_{\alpha \neq \beta} U_{\alpha\beta} \hat{n}_{i\alpha} \hat{n}_{i\beta} + V \sum_{i} \hat{n}_{i1} \hat{n}_{i2} \hat{n}_{i3}, \tag{1}$$

where the subscript $\langle i,j \rangle$ is the summation over the nearest-neighbor sites, and $\hat{a}_{i\alpha}^{\dagger}(\hat{a}_{i\alpha})$ and $\hat{n}_{i\alpha}$ are creation (annihilation) and number operators of a fermion with color $\alpha(=1,2,3)$ at the ith site, respectively. The first three terms describe the three-component Hubbard Hamiltonian as usual. The last term denotes the three-body interaction with $V > 0$. The chemical potentials μ_{α} are adjusted to choose half filling with a balanced population of each color atoms ($\langle \hat{n}_{i\alpha} \rangle = 1/2$), leading to $\mu_{\alpha} = (U_{\alpha\beta} + U_{\gamma\alpha})/2$. Thus, 3/2 atoms exist at each site on the average at half filling. For simplicity, we set $U_{12} \equiv U(>0)$ and $U_{23} = U_{31} \equiv U'(>0)$, and introduce parameters $v \equiv V/U'$ and $R \equiv U/U'$. We focus on the Mott transition by neglecting the symmetry breaking phases and the effects of the confinement potential.

To take account of the local correlation effects precisely, we employ an unperturbative method, a self-energy functional approach (SFA) [13, 14]. This method has been successfully applied to investigate Mott transitions at zero and finite temperatures in strongly correlated electron systems [13, 14, 15]. Furthermore, the SFA allows us to deal with superfluid transitions in the repulsively [8] and attractively [4] interacting three-component fermionic atoms in optical lattices. The SFA is based on the Luttinger-Ward variational principle $\partial\Omega/\partial\Sigma = 0$, where Ω is the grand potential of the system and Σ is the *physical* self-energy. We introduce a reference system $\mathcal{H}_{\rm ref}$ that has the same interaction terms as those of the original Hamiltonian (1), so that the grand potential Ω can be written as follows:

$$\Omega = \Omega_{\rm ref} + {\rm Tr}\ln\left[-(\omega + \mu - \mathbf{t} - \mathbf{\Sigma}_{\rm ref})^{-1}\right] - {\rm Tr}\ln\left[-(\omega + \mu - \mathbf{t}_{\rm ref} - \mathbf{\Sigma}_{\rm ref})^{-1}\right], \tag{2}$$

where $\Omega_{\rm ref}$ and $\mathbf{\Sigma}_{\rm ref}$ are the grand potential and the self-energy of the reference system $\mathcal{H}_{\rm ref}$, and $\mathbf{t}_{\rm ref}$ and $\mathbf{t}$ are parameter matrices of the one-body terms of the reference Hamilitonial and the original Hamiltonian, respectively. By choosing the parameter matrix $\mathbf{t}_{\rm ref}$ to satisfy the variational condition $\partial\Omega/\partial\mathbf{t}_{\rm ref} = (\partial\Omega/\partial\mathbf{\Sigma}_{\rm ref})(\partial\mathbf{\Sigma}_{\rm ref}/\partial\mathbf{t}_{\rm ref}) = 0$, we obtain a reference self-energy $\mathbf{\Sigma}_{\rm ref}$ that properly reproduces the physical properties of the original system.

We use the following local reference Hamiltonian: $\mathcal{H}_{\rm ref} = \sum_i \mathcal{H}_{\rm ref}^{(i)}$,

$$\begin{aligned} \mathcal{H}_{\rm ref}^{(i)} &= \sum_{\alpha=1}^{3} \left(\epsilon_{c\alpha} \hat{c}_{i\alpha}^{\dagger} \hat{c}_{i\alpha} + \epsilon_{a\alpha} \hat{a}_{\alpha}^{\dagger} \hat{a}_{\alpha} \right) + \sum_{\alpha=1}^{3} \left(\Gamma_{\alpha} \hat{c}_{i\alpha}^{\dagger} \hat{a}_{\alpha} + H.c. \right) \\ &\quad + \frac{1}{2} \sum_{\alpha \neq \beta = 1}^{3} U_{\alpha\beta} \hat{n}_{i\alpha} \hat{n}_{i\beta} + V \sum_{i} \hat{n}_{i1} \hat{n}_{i2} \hat{n}_{i3}, \end{aligned} \tag{3}$$

where $\hat{a}_{\alpha}^{\dagger}(\hat{a}_{\alpha})$ is the creation (annihilation) operator of a noninteracting fermion with color α connecting to the ith site in the original lattice. Here, the variational parameters are $\epsilon_{c\alpha}$, $\epsilon_{a\alpha}$, and Γ_{α} ($\alpha = 1,2,3$). In the present local approximation, the hybridization (Γ_{α}) between the noninteracting fermions and the original lattice fermions effectively captures the itineracy of each color atoms. By searching for the optimized parameters $\epsilon_{c\alpha}$, $\epsilon_{a\alpha}$, and Γ_{α}, we can thus discuss characteristic Mott transitions in the three-component repulsive fermionic atoms with the three-body repulsion in optical lattices.

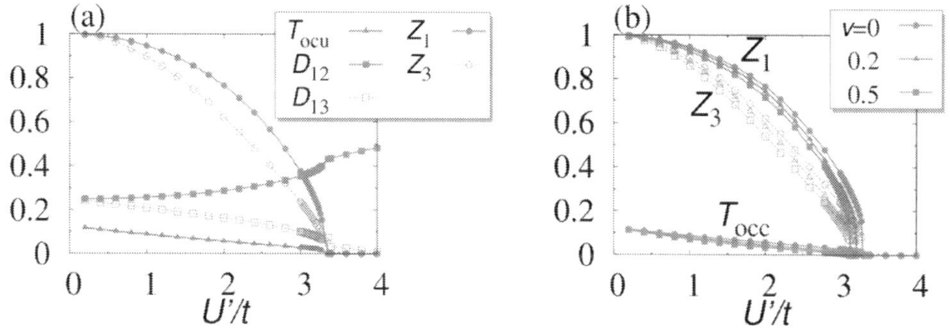

Figure 1. (Color online) (a) Quasiparticle weight Z_α, double occupation $D_{\alpha\beta}$, and triple occupation $T_{\rm occ}$ for $v = 0$ at half filling. (b) Quasiparticle weight Z_α and triple occupation $T_{\rm occ}$ for $v = 0, 0.2$, and 0.5 at half filling, where filled and open symbols for Z_α represent $Z_1(= Z_2)$ and Z_3, respectively. These calculations are performed for $R = 0.1$ at zero temperature. Lines are guide to eyes.

In the following, for noninteracting atoms, we use a semi-circular density of states, $\rho_0(x) = \sqrt{4t^2 - x^2}/(2\pi t^2)$, which is independent of color. We calculate the quasiparticle weight $Z_\alpha = [1 - \partial\Sigma_{{\rm ref},\alpha}(\omega)/\partial\omega]|_{\omega=0}^{-1}$, using the reference self-energy $\Sigma_{\rm ref}$. We also calculate the double occupation $D_{\alpha\beta} = \partial\Omega/\partial U_{\alpha\beta} = \langle\hat{n}_{i\alpha}\hat{n}_{i\beta}\rangle$ and the triple occupation $T_{\rm occ} = \partial\Omega/\partial V = \langle\hat{n}_{i1}\hat{n}_{i2}\hat{n}_{i3}\rangle$, where $D_{\alpha\beta}$ and $T_{\rm occ}$ are independent of the site index i. Note that $D_{\alpha\beta} = 1/4$ and $T_{\rm occ} = 1/8$ for noninteracting atoms. Because of $U_{12} \neq U_{23} = U_{31}$, $Z_1 = Z_2$ and $D_{13} = D_{23}$ are satisfied. The hopping integral t is used in units of energy.

3. Results

We first analyze the system without the three-body interaction ($V = 0$) for $R = 0.1$ at zero temperature. Figure 1(a) exhibits that Z_1, Z_3, and D_{13} become zero at $U'_c/t \sim 3.4$, while D_{12} approaches the maximum $D_{12} = 1/2$ in $U' > U'_c$. This is a manifestation of the quantum phase transition to the PMI, where pairs of color-1 and 2 atoms are formed to avoid the two stronger repulsions. This Mott transition at non-integer half filling occurs as a result of the pair formation. Here, the total number of the effective particles at each site becomes an integer as $n_{\rm pair} + n_3 = 1$, where $n_{\rm pair}$ is the average number of the paired particles, and n_3 is that of the color-3 atoms [6]. We also find that the triple occupation $T_{\rm occ}$ decreases with increasing U'/t and becomes zero at the PMI transition point.

Next, we discuss the effects of a weak three-body interaction for $v = 0.2$ and 0.5. Figure 1(b) shows that, as v is increased, Z_1 and Z_3 decrease and the critical value of the PMI transition shifts to smaller U'/t. The triple occupation $T_{\rm occ}$ also reduces with increasing v. Importantly, no qualitative difference is observed between the $v = 0$ and $v \neq 0$ cases. This is because the repulsive three-body and two-body interactions do not compete with each other.

The decrease in the triple occupation allows us to deal with the repulsive three-body interaction by using the mean-field approximation. The mean-field three-body interaction reduces to $\sim V\left(\langle\hat{n}_{i1}\rangle\hat{n}_{i2}\hat{n}_{i3} + \langle\hat{n}_{i2}\rangle\hat{n}_{i1}\hat{n}_{i3} + \langle\hat{n}_{i3}\rangle\hat{n}_{i1}\hat{n}_{i2}\right)$. This suggests that the three-body repulsion can be renormalized into the two-body repulsions U and U'. Using the condition $\langle\hat{n}_{i\alpha}\rangle = 1/2$, the renormalized two-body interactions are evaluated as $\tilde{U} = (R + v/2)U'$ and $\tilde{U}' = (1 + v/2)U'$. Thus, the slight shift of the PMI transition point with increasing v shown in Fig. 1(b) can be qualitatively explained by this mean-field approximation. To discuss the validity of this mean-field approximation, we further perform numerical calculations on the basis

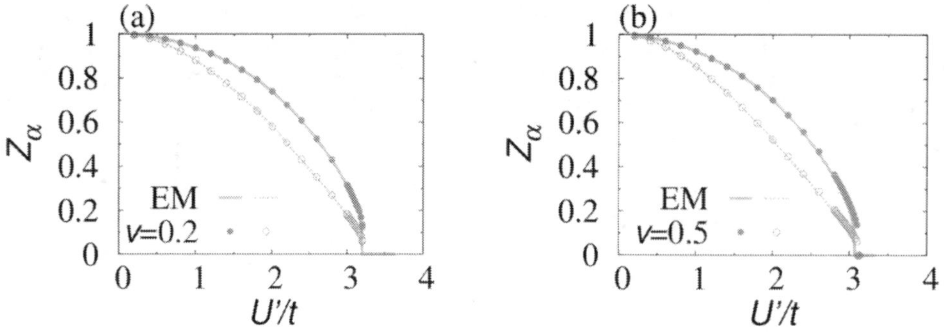

Figure 2. (Color online) Quasiparticle weight Z_α derived from the effective model (EM) in Eq. (2) and from the original model in Eq. (1) for (a) $v = 0.2$ and (b) $v = 0.5$ at half filling. Filled and open symbols represent $Z_1(= Z_2)$ and Z_3, respectively, obtained from the original model, while solid and broken lines represent those obtained from the effective model. Calculations are performed for $R = 0.1$ at zero temperature.

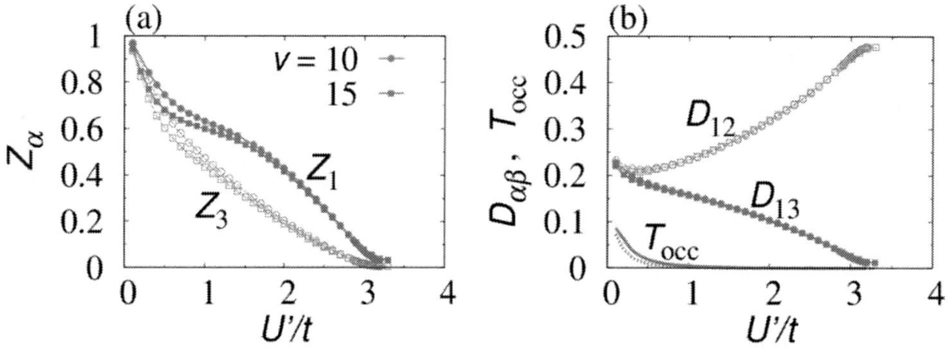

Figure 3. (Color online) (a) Quasiparticle weight Z_α, and (b) double occupation $D_{\alpha\beta}$ and triple occupation $T_{\rm occ}$ for $v = 10$ and $v = 15$ at half filling. Calculations are performed for $R = 0.1$ at zero temperature. Lines are guide to eyes.

of the effective three-component Hubbard model with the renormalized parameters $\tilde{U}$ and $\tilde{U}'$. The effective model reads as

$$\mathcal{H}_{\rm eff} = -t\sum_{\langle i,j\rangle}\sum_{\alpha=1}^{3}\hat{a}^\dagger_{i\alpha}\hat{a}_{j\alpha} - \sum_{i}\sum_{\alpha=1}^{3}\mu_\alpha\hat{n}_{i\alpha} + \sum_{i}\left(\tilde{U}\hat{n}_{i1}\hat{n}_{i2} + \tilde{U}'\hat{n}_{i2}\hat{n}_{i3} + \tilde{U}'\hat{n}_{i1}\hat{n}_{i3}\right). \qquad (4)$$

We compare Z_1 and Z_3 obtained from this effective model with those obtained from the original model. Figure 2 shows that the effective model well reproduces the feature of the original one quantitatively for both $v = 0.2$ and $v = 0.5$, indicating that the mean-field approximation well captures the effects of the weak three-body repulsion.

We further investigate the effects of an extremely strong three-body repulsion. We show the quasiparticle weights Z_1 and Z_3 in Fig. 3(a), and the double and triple occupations in Fig. 3(b) for $v = 10$ and $v = 15$. We find that the triple occupation decreases quickly and vanishes at $U'/t \sim 1.5$, where the system is in the Fermi liquid. This behavior of $T_{\rm occ}$ is in contrast with that in the weak v shown in Fig. 1(b), where the $T_{\rm occ}$ vanishes at the PMT transition point.

For $U'/t > 1.5$, the quasiparticle weights and double occupations are almost independent of v. These findings for extremely strong three-body repulsions mean that the properties close to the PMI transition point can be described by the constrained Hamiltonian $\hat{\mathcal{P}}^\dagger \mathcal{H} \hat{\mathcal{P}}$, where $\hat{\mathcal{P}}$ is a projection operator to exclude the triple occupancy at any site [10, 11]. Significantly, the PMI transition still occurs under the extremely strong three-body repulsion, which can be confirmed from the behavior that Z_1 and Z_2 go to zero, and D_{12} and D_{13} approach 1/2 and 0, respectively, towards the transition point $U'_c/t \sim 3.2$. This result also originates form the absence of the competition between the repulsive three-body and two-body repulsions.

4. Discussion and conclusion

We have shown that the repulsive three-body interaction hardly influences the qualitative features of the PMI transition, because the three-body repulsion does not compete with the two-body repulsions. For the extremely strong three-body repulsion, we find that the triple occupancy is projected out close to the PMI transition point. This situation is equivalent to the system with very large three-body losses [10, 11]. Therefore, we conclude that large three-body losses have little influence on the PMI transition in the repulsively interacting three-component fermionic atoms in optical lattices. This conclusion is in contrast with that obtained in the attractively interacting three-component fermionic atoms, where the trionic state becomes unstable due to three-body losses, and the phase separation occurs towards paired and unpaired atoms [10, 11]. Its driving force is the competition between the repulsive three-body interaction and the attractive two-body interactions.

We finally comment on the quantum degenerate three-component fermionic ^{6}Li gases, which have been intensively investigated [1]. In this system, the small three-body loss rate was observed in the repulsively interacting region under 550G-600G magnetic field, while the very large loss rate was observed in the attractive region around 500G [1]. In the repulsive region, in addition to the small loss coefficient, the two-body repulsive interactions itself decrease the triple occupation $T_{\rm occ}$ as shown in Fig. 1(a), which implies the suppression of three-body losses. Note that the decrease in the number of color-α atoms per site caused by three-body losses is described by $dn_\alpha/dt \propto -\gamma T_{\rm occ}$, where γ is a (positive) loss coefficient that includes the effects of the bare three-body loss coefficient K_3 [1] and the overlap integration $\int d\mathbf{r}|W(\mathbf{r})|^6$ of the Wannier function ($W(\mathbf{r})$) of atoms in the lattice. We have shown that $T_{\rm occ}$ is negligibly small close to the PMI transition point. We thus consider that the ^{6}Li atom system in optical lattices is a possible candidate for observing the PMI.

Acknowledgments

This work was supported by Grants-in-Aid for Scientific Research (C) (No. 23540467) and (B) (No. 25287104) from the Japan Society for the Promotion of Science.

References

[1] Ottenstein T B, Lompe T, Kohnen M, Wenz A N and Jochim S 2008 *Phys. Rev. Lett.* **101** 203202
[2] Huckans J H, Williams J R, Hazlett E L, Stites R W and O'Hara K M 2009 *Phys. Rev. Lett.* **102** 165302
[3] Rapp A, Zarand G, Honerkamp C and Hofstetter W 2007 *Phys. Rev. Lett.* **98** 160405
[4] Inaba K and Suga S 2009 *Phys. Rev.* A **80** 041602(R)
[5] Inaba K and Suga S 2011 *Mod. Phys. Lett.* B **25** 987
[6] Inaba K, Miyatake S and Suga S 2010 *Phys. Rev.* A **82** 051602(R)
[7] Suga S and Inaba K 2011 *J. Phys.: Conference Series* **273** 012016
[8] Inaba K and Suga S 2012 *Phys. Rev. Lett.* **108** 255301
[9] Inaba K and Suga S 2013 *Mod. Phys. Lett.* B **27** 1330008
[10] Privitera A, Titvinidze I, Chang S-Y, Diehl S, Daley A-J and Hofstetter W 2011 *Phys. Rev.* A **84** 021601
[11] Titvinidze I, Privitera A, Chang S-Y, Diehl S, Baranov M A, Daley A and Hofstetter W 2011 *New J. Phys.* **13** 035013
[12] Jaksch D, Bruder C, Cirac J I, Gardiner C W and Zoller P 1998 *Phys. Rev. Lett.* **81** 3108

[13] Potthoff M 2003 *Eur. Phys. J.* B. **32** 429
[14] Potthoff M 2003 *Eur. Phys. J.* B. **36** 335
[15] Koga A, Inaba K and Kawakami N 2005 *Prog. Theor. Phys. Suppl.* **160** 253

Journal of Physics: Conference Series **529** (2014) 012025 doi:10.1088/1742-6596/529/1/012025

Explicit Energy Functional for Infinite Nuclear Matter with the Tensor Force

M Takano[1,2], K Kato[2] and M Yamada[1]
[1]Research Institute for Science and Engineering, Waseda University
[2]Department of Physics and Applied Physics, Waseda University
3-4-1 Okubo, Shinjuku-ku, Tokyo 169-8555 Japan.
E-mail: `takanom@waseda.jp`

Abstract. We have applied the variational method using explicit energy functionals (EEFs) to energy calculations of infinite nuclear matter. In EEFs, the energy per nucleon is explicitly expressed with spin-isospin-dependent two-body distribution functions, which are regarded as variational functions, and fully minimized energies are conveniently calculated with the EEF. A remarkable feature of this approach is that EEFs guarantee non-negativeness of structure functions. In this study, we extend the EEF variational method so as to consider state-independent three-body forces for neutron matter at finite temperatures following the procedure proposed by Schmidt and Pandharipande. For neutron matter, the free energies obtained with the Argonne v4' two-body potential and the repulsive part of the Urbana IX (UIX) three-body potential are quite reasonable. Furthermore, we improve the EEF of nuclear matter using the two-body central and tensor forces by considering the main three-body cluster terms and guaranteeing non-negativeness of tensor structure functions. In addition, healing distances are introduced for two-body distribution functions so that Mayer's condition is satisfied. The obtained energies per neutron of neutron matter with the Argonne v6' two-body potential and the repulsive part of the UIX potential are in good agreement with those obtained by auxiliary field diffusion Monte Carlo calculations.

1. Introduction

The variational method is one of the most powerful techniques to calculate energies per particle of infinite fermion systems in which fermions interact through strong short-range forces. In the Fermi hypernetted chain (FHNC) method [1], which is a typical variational method, the expectation value of the Hamiltonian with the Jastrow wave function is cluster expanded, and parts of the cluster terms are resummed infinitely. Then, the expectation value of the Hamiltonian is minimized with respect to the correlation functions included in the Jastrow wave function. This method has been applied to liquid ^{3}He [2, 3] and nuclear matter [4]. In recent years, quantum Monte Carlo methods have also been applied to these systems [5, 6, 7, 8]. In the case of liquid ^{3}He, the intermolecular force between ^{3}He atoms is expressed as state-independent two-body central force. In the case of nuclear matter, on the contrary, the two-body nuclear force is strongly state dependent, and the three-body force is also necessary for a realistic nuclear equation of state (EOS). Therefore, in the FHNC calculations for nuclear matter, the correlation functions should be state-dependent operators, and this requirement necessitates a laborious resummation of the cluster terms. In fact, the correlation functions are usually parameterized, and the total energy per nucleon is minimized with respect to these parameters reflecting that the full minimization is difficult to perform. Another problem of the FHNC method is that it is difficult to treat asymmetric nuclear matter with arbitrary proton fractions.

In response to these difficulties associated with the FHNC method, we have been studying a new type of variational method using explicit energy functionals (EEFs). In this method, we first propose an EEF in which the energy per particle of an infinite fermion system is explicitly expressed with variational functions. Once a reliable EEF is acquired, the Euler-Lagrange equations of the variational functions are derived, and, by solving them, the fully minimized energy is obtained.

We first applied this variational method to liquid ^{3}He and obtained reasonable results [9]. We also applied the EEF to hypothetical neutron matter [9] and symmetric nuclear matter [10], where, in both cases, two nucleons interact only through static two-body central forces. By applying this method to more realistic nuclear matter, we next treated the tensor, spin-orbit, and quadratic-angular-momentum parts of the two-body nuclear force [11]. In this case, since the extension of the EEF was insufficient with respect to the kinetic energy expression, the minimized energy was too low, and the obtained two-body distribution functions possessed unrealistically long tails. In Ref. [11], we also proposed an effective method so as to avoid these unphysical solutions. Furthermore, we extended this variational method to arbitrarily spin-polarized liquid ^{3}He [12] and (spin-unpolarized) asymmetric nuclear matter with arbitrary proton fractions [13]. However, it remains to refine the EEF with respect to the tensor, spin-orbit, and quadratic-angular-momentum parts of the two-body force. Moreover, treatment of the three-body nuclear force is imperative to obtain a realistic nuclear EOS. Furthermore, it is of interest to extend the theory to nuclear matter at finite temperatures, as has been performed with other variational methods [14, 15, 16, 17].

In this paper, we first report recent extensions of the EEF variational method, such as treatment of the three-body force and extension to nuclear matter at finite temperatures. We also report an improvement of the EEF using the two-body central and tensor forces by considering the proper three-body cluster terms and the non-negativeness of tensor structure functions.

2. Explicit Energy Functional of Neutron Matter at Finite Temperatures

In this section, we review the EEF of neutron matter in which neutrons interact through spin-dependent two-body central forces, and extend the EEF so as to treat three-body forces. Furthermore, we extend this variational method to neutron matter at finite temperatures following the method proposed by Schmidt and Pandharipande [18].

The EEF of neutron matter at zero temperature proposed in Ref. [9] represents the energy per neutron as a functional of spin-dependent radial distribution functions $F_s(r)$ defined by

$$F_s(r) = \Omega^2 \sum_{\text{spin}} \int \Psi^\dagger(x_1, x_2, ..., x_N) P_{s12} \Psi(x_1, x_2, ..., x_N) d\mathbf{r}_3 d\mathbf{r}_4 ... d\mathbf{r}_N, \tag{1}$$

where Ω is the volume of the system, and the summation is conducted over spin coordinates of all the neutrons. The function Ψ in Eq. (1) is the wave function of neutron matter, and x_i represents the space and spin coordinates of the i-th neutron. Here N is the total number of neutrons, and P_{sij} is the spin-projection operator projecting the (i, j) neutron pair state onto the spin triplet ($s = 1$) or spin singlet ($s = 0$) state. We also introduce the following structure functions

$$S_{c1}(k) = \frac{1}{N} \left\langle \left| \sum_{i=1}^{N} \exp(i\mathbf{k} \cdot \mathbf{r}_i) \right|^2 \right\rangle = 1 + S_1(k) + S_0(k) \geq 0, \tag{2}$$

$$S_{c2}(k) = \frac{1}{3N} \left\langle \left| \sum_{i=1}^{N} \sigma_i \exp(i\mathbf{k} \cdot \mathbf{r}_i) \right|^2 \right\rangle = 1 + \frac{S_1(k)}{3} - S_0(k) \geq 0, \tag{3}$$

$$S_s(k) = 4\pi\rho \int_0^\infty [F_s(r) - F_s(\infty)]\, j_0(kr) r^2 dr \tag{4}$$

with ρ being the neutron number density.

Then, the energy per neutron E_2/N is expressed as follows

$$\frac{E_2}{N} = \frac{3}{5}E_{\rm F} + \frac{E_{\rm V}}{N} + \frac{E_{\rm TC}}{N}. \tag{5}$$

The first term on the right-hand side of Eq. (5) is the one-body kinetic energy term, and $E_{\rm F}$ is the Fermi energy. The second term $E_{\rm V}/N$ is the potential energy expressed as

$$\frac{E_{\rm V}}{N} = 2\pi\rho \sum_{s=0}^{1} \int_0^\infty F_s(r) V_{{\rm C}s}(r) r^2 dr, \tag{6}$$

where $V_{{\rm C}s}(r)$ is the spin-dependent two-body central potential with r being the distance between two neutrons. The term $E_{\rm V}/N$ represents the expectation value of the potential energy per neutron exactly. The term $E_{\rm TC}/N$ in Eq. (5) is the kinetic energy caused by the correlation among neutrons, and given as follows

$$\begin{aligned}\frac{E_{\rm TC}}{N} &= \frac{\pi\hbar^2\rho}{2m} \sum_{s=0}^{1} \int_0^\infty F_s(r) \left[\frac{1}{F_s(r)}\frac{dF_s(r)}{dr} - \frac{1}{F_{{\rm F}s}(r)}\frac{dF_{{\rm F}s}(r)}{dr}\right]^2 r^2 dr \\ &\quad - \frac{\hbar^2}{16\pi m\rho} \sum_{n=1}^{2} \int_0^\infty (2n-1) \frac{[S_{cn}(k)-1]\,[S_{cn}(k)-S_{c{\rm F}}(k)]^2}{S_{cn}(k)/S_{c{\rm F}}(k)} k^4 dk,\end{aligned} \tag{7}$$

where m is the neutron mass. The functions $F_{{\rm F}s}(r)$ and $S_{c{\rm F}}(k)$ in Eq. (7) are $F_s(r)$ and $S_{cn}(k)$ for the Fermi gas, respectively. In order to determine the expression of $E_{\rm TC}/N$, we first temporarily assume the Jastrow wave function

$$\Psi = {\rm Sym}\left[\prod_{i<j} f_{ij}\right] \Phi_{\rm F}, \tag{8}$$

where Sym[ ] is the symmetrizer with respect to the order of the factors in the products and f_{ij} is the correlation function for the (i, j) neutron pair, which is given by

$$f_{ij} = \sum_{s=0}^{1} f_s(r_{ij}) P_{sij}. \tag{9}$$

The function $\Phi_{\rm F}$ in Eq. (8) is the wave function of the Fermi gas. Then, we expand the expectation value of the Hamiltonian per neutron $\langle H\rangle/N = \langle\Psi|H|\Psi\rangle/N/\langle\Psi|\Psi\rangle$ into cluster terms and construct $E_{\rm TC}/N$ so as to include the two-body kinetic energy cluster terms and the main part of the three-body cluster (TBC) direct non-nodal kinetic energy terms that are the lowest order in $h_s(r) = f_s(r) - 1$. We note that the denominator in the integrand of the last term on the right-hand side of Eq. (7) plays an important role in guaranteeing the non-negativeness of $S_{cn}(k)$, as shown in Eqs. (2) and (3) [19].

Next, we extend this EEF so as to treat three-body forces. As the first step of this extension, we treat state-independent three-body potential $V_3(\mathbf{r}_1, \mathbf{r}_2, \mathbf{r}_3)$. To treat this potential, the corresponding three-body distribution function $F_3(\mathbf{r}_1, \mathbf{r}_2, \mathbf{r}_3)$ is necessary. Here we employ the following extended Kirkwood's assumption

$$F_3(\mathbf{r}_1, \mathbf{r}_2, \mathbf{r}_3) = F(r_{12})F(r_{23})F(r_{31}) \frac{F_{3\rm F}(\mathbf{r}_1, \mathbf{r}_2, \mathbf{r}_3)}{F_{\rm F}(r_{12})F_{\rm F}(r_{23})F_{\rm F}(r_{31})}, \tag{10}$$

where $F(r) = F_1(r) + F_0(r)$ and $F_{3F}(\mathbf{r}_1, \mathbf{r}_2, \mathbf{r}_3)$ is the three-body distribution function $F_3(\mathbf{r}_1, \mathbf{r}_2, \mathbf{r}_3)$ for the Fermi gas. Then, the total energy per neutron E/N is expressed as

$$\frac{E}{N} = \frac{E_2}{N} + \frac{\rho^2}{6} \int F(\mathbf{r}_1, \mathbf{r}_2, \mathbf{r}_3) V(\mathbf{r}_1, \mathbf{r}_2, \mathbf{r}_3) d\mathbf{r}_{12} d\mathbf{r}_{23}, \tag{11}$$

and is minimized with respect to $F_s(r)$, i.e., the Euler-Lagrange equations for $F_s(r)$ are solved.

Furthermore, we extend this variational method so as to treat neutron matter at finite temperatures following the method proposed by Schmidt and Pandharipande (SP) [18]. For this treatment, the free energy per neutron is expressed as $F/N = \langle H \rangle_{\mathrm{T}} /N - TS/N$ with$\langle H \rangle_{\mathrm{T}}$, T, and S being the internal energy, temperature, and entropy, respectively. According to SP, $\langle H \rangle_{\mathrm{T}} /N$ is constructed by replacing $F_{\mathrm{F}s}(r)$ and $S_{\mathrm{cF}}(k)$ in E/N of Eq. (11) at zero temperature by $F_{\mathrm{F}s}(T, r)$ and $S_{\mathrm{cF}}(T, k)$, respectively. Here $F_{\mathrm{F}s}(T, r)$ is expressed as

$$F_{\mathrm{F}s}(T, r) = \frac{(2s+1)}{4} \left\{ 1 + (-1)^s \left[l(T, r) \right]^2 \right\}, \tag{12}$$

with $l(T, r)$ being the Slater function at finite temperature

$$l(T, r) = \frac{1}{\pi^2 \rho} \int_0^\infty n(k, T) j_0(kr) k^2 dk. \tag{13}$$

In Eq. (13), $n(k, T)$ is the average occupation probability of single-neutron states expressed as

$$n(k, T) = \left\{ 1 + \exp\left[\frac{\epsilon(k) - \mu}{k_{\mathrm{B}} T} \right] \right\}^{-1}, \tag{14}$$

where $\epsilon(k) = \hbar^2 k^2 / (2m^*)$ is the single-neutron energy with m^* being the effective mass, and μ is determined by the normalization condition. $S_{\mathrm{cF}}(T, k)$ is constructed through Eqs. (2), (3), and (4) by replacing $F_s(r)$ in Eq. (4) by $F_{\mathrm{F}s}(T, r)$. Furthermore, S/N is expressed as in the case of non-interacting quasi-neutrons with the effective mass m^*. Then, F/N is minimized with respect to $F_s(r)$ and m^*.

Figure 1 shows F/N of neutron matter with the Argonne v4'(AV4') two-body potential [21] and the repulsive part of the Urbana I (UI) three-body potential [22]. The result without the three-body potential is also shown. Since the repulsive part of the UI potential is rather strong, F/N with the three-body force becomes significantly higher than that without the three-body force at high densities. At finite temperatures, F/N decreases with T The behavior of F/N is quite reasonable.

3. Improvement of the Explicit Energy Functional with the Tensor Force

In this section, we report on an improvement of the EEF with the tensor force. The previous EEF is reported in Ref. [11], where the tensor distribution function $F_{\mathrm{T}}(r)$ is introduced.

$$F_{\mathrm{T}}(r) = \Omega^2 \sum_{\mathrm{spin}} \int \Psi^\dagger(x_1, x_2, ..., x_N) S_{\mathrm{T}12} \Psi(x_1, x_2, ..., x_N) d\mathbf{r}_3 d\mathbf{r}_4 ... d\mathbf{r}_N. \tag{15}$$

Here $S_{\mathrm{T}ij}$ is the tensor operator operating on the (i, j) neutron pair. We also introduce the following auxiliary functions, i.e., the intrinsically central distribution functions $F_{\mathrm{C}s}(r)$ and the dressed tensor correlation function $g_{\mathrm{T}}(r)$ defined as the solutions to the following equations

$$\begin{aligned} F_s(r) &= F_{\mathrm{C}s}(r) + 8s \left[g_{\mathrm{T}}(r) \right]^2 F_{\mathrm{F}s}(r), \\ F_{\mathrm{T}}(r) &= 16 \sqrt{g_{\mathrm{T}}(r) F_{\mathrm{C}1}(r)} - 16 \left[g_{\mathrm{T}}(r) \right]^2 F_{\mathrm{F}1}(r). \end{aligned} \tag{16}$$

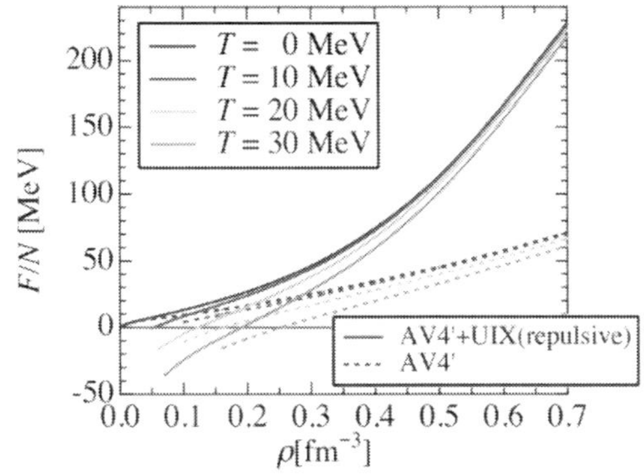

Figure 1. Free energy per neutron of neutron matter with the AV4' two-body potential and the repulsive part of the UI three-body potential as a function of the number density ρ.

We note that, without the tensor force, $F_{\mathrm{C}s}(r)$ reduce to $F_s(r)$, and $g_{\mathrm{T}}(r)$ vanishes.

In the proposed improvement of the EEF with the tensor force, we newly introduce the following tensor structure functions

$$S_{\mathrm{cT1}}(k) = \frac{1}{Nk^2}\left\langle\left|\sum_{i=1}^{N}(\sigma_i\cdot\mathbf{k})\exp(i\mathbf{k}\cdot\mathbf{r}_i)\right|^2\right\rangle = S_{\mathrm{c2}}(k)-\frac{S_{\mathrm{T}}(k)}{3}\geq 0, \tag{17}$$

$$S_{\mathrm{cT2}}(k) = \frac{1}{2Nk^2}\left\langle\left|\sum_{i=1}^{N}(\sigma_i\times\mathbf{k})\exp(i\mathbf{k}\cdot\mathbf{r}_i)\right|^2\right\rangle = S_{\mathrm{c2}}(k)+\frac{S_{\mathrm{T}}(k)}{6}\geq 0, \tag{18}$$

with

$$S_{\mathrm{T}}(k) = 4\pi\rho\int_0^\infty F_{\mathrm{T}}(r)j_2(kr)r^2dr. \tag{19}$$

By using these functions, we propose an improved EEF for neutron matter with the tensor force as in Eq. (5), where E_{V}/N and E_{TC}/N are expressed, respectively, as

$$\frac{E_{\mathrm{V}}}{N} = 2\pi\rho\int_0^\infty\left[\sum_{s=0}^{1}F_s(r)V_{\mathrm{C}s}(r)+F_{\mathrm{T}}(r)V_{\mathrm{T}}(r)\right]r^2dr, \tag{20}$$

and

$$\begin{aligned}\frac{E_{\mathrm{TC}}}{N} &= \frac{\pi\hbar^2\rho}{2m}\int_0^\infty\sum_{s=0}^{1}F_{\mathrm{C}s}(r)\left[\frac{1}{F_{\mathrm{C}s}(r)}\frac{dF_{\mathrm{C}s}(r)}{dr}-\frac{1}{F_{\mathrm{F}s}(r)}\frac{dF_{\mathrm{F}s}(r)}{dr}\right]^2r^2dr\\ &+\frac{16\pi\hbar^2\rho}{m}\int_0^\infty\left\{\left[\frac{dg_{\mathrm{T}}(r)}{dr}\right]^2+\frac{6}{r^2}\left[g_{\mathrm{T}}(r)\right]^2\right\}F_{\mathrm{F1}}(r)r^2dr+\frac{E_{\mathrm{nod}}}{N}\\ &-\frac{\hbar^2}{16\pi m\rho}\int_0^\infty\frac{[S_{\mathrm{c1}}(k)+2S_{\mathrm{cF}}(k)-3]\left[S_{\mathrm{c1}}(k)-S_{\mathrm{cF}}(k)\right]^2}{S_{\mathrm{c1}}(k)/S_{\mathrm{cF}}(k)}k^4dk\\ &-\frac{\hbar^2}{16\pi m\rho}\int_0^\infty\sum_{n=1}^{2}n\frac{[S_{\mathrm{cT}n}(k)+2S_{\mathrm{cF}}(k)-3]\left[S_{\mathrm{cT}n}(k)-S_{\mathrm{cF}}(k)\right]^2}{S_{\mathrm{cT}n}(k)/S_{\mathrm{cF}}(k)}k^4dk.\end{aligned} \tag{21}$$

In Eq. (20), $V_{\mathrm{C}s}(r)$ and $V_{\mathrm{T}}(r)$ are the two-body central and tensor potentials, respectively. The term E_{nod}/N in Eq. (21) is given by

$$\begin{aligned}\frac{E_{\mathrm{nod}}}{N} &= \frac{\hbar^2\rho^2}{2m}\int d\mathbf{r}_{12}\frac{3}{4}\left[\left[\frac{dG(r_{12})}{dr_{12}}\right]^2+\frac{1}{18}\left\{\left[\frac{dF_{\mathrm{T}}(r_{12})}{dr_{12}}\right]^2+\frac{6}{r_{12}^2}\left[F_{\mathrm{T}}(r_{12})\right]^2\right\}\right]\\ &\times\int d\mathbf{r}_{23}\left\{\left[G(r_{23})\right]^2+\frac{1}{18}\left[F_{\mathrm{T}}(r_{23})\right]^2\right\},\end{aligned} \tag{22}$$

with

$$G(r) = \frac{1}{3}F_1(r) - F_0(r) - \frac{1}{3}F_{F1}(r) + F_{F0}(r). \tag{23}$$

As in the case of the central forces alone, the expectation value of the potential energy is exactly expressed with Eq. (20). The functional form of E_{TC}/N representing the kinetic energy caused by the correlations among neutrons is the improved form, which is constructed by considering the following two points

i) When we temporarily assume the Jastrow wave function given in Eq. (8) with the following correlation function

$$f_{ij} = \sum_{s=0}^{1} f_s(r_{ij})P_{sij} + f_T(r_{ij})S_{Tij}, \tag{24}$$

E_{TC}/N then properly includes the two-body kinetic-energy cluster terms and the main part of the TBC kinetic-energy terms that are the lowest order in $h_s(r)$ and $f_T(r)$ as cluster terms of the expectation value of the Hamiltonian per neutron $\langle H \rangle /N$. Here we note that the EEF proposed in Ref. [11] is insufficient with respect to the inclusion of the TBC terms that consist of $f_T(r)$. Furthermore, in this refinement of the EEF, we correctly consider the TBC two-particle exchange terms as well as the direct terms that are the lowest order in the correlations. We have studied, in Ref. [23], the effect of the TBC two-particle exchange terms on the EEF in a more sophisticated way. However, in the present improvement, we simply include all the TBC two-particle exchange terms in E_{TC}/N. Other TBC terms included in this refinement are the so-called nodal diagrams, which have not been considered in the previous studies. Though it is well known that the nodal diagrams cancel out in the case of state-independent central forces, it is found that they play an important role in the present improvement. Therefore, we consider the TBC nodal diagrams that are the lowest order in the correlations as E_{nod}/N.

ii) We guarantee non-negativeness of the tensor structure function $S_{cTn}(k)$. It is obvious that the relation $S_{c2}(k) \geq 0$ is automatically guaranteed when $S_{cTn}(k) \geq 0$ $(n = 1, 2)$ is satisfied. Therefore, rather than employing $S_{c2}(k)$ as the denominator of the integrand of the last term of Eq. (7), the denominators $S_{cTn}(k)$ are adopted in the integrand of the last term of Eq. (21).

Using the new EEF shown above, we derive the Euler-Lagrange equations for $F_{Cs}(r)$ and $g_T(r)$. When we solve them, Mayer's condition $S_{c1}(0) = 0$ is considered. Since this condition is not explicitly imposed in the present variational method, the solutions obtained may violate the condition. In general, when Mayer's condition is violated, the distribution functions exhibit long tails. This corresponds to the problem of treating long-range correlations in neutron matter. As reported in Ref. [3], it is difficult to treat long-range correlations exactly in the FHNC method. Therefore, in FHNC calculations [4, 8], healing distances with respect to the correlation functions in the Jastrow wave function are introduced so as to avoid unphysical long-range correlation. In the present EEF variational method, we adopt a similar technique, i.e., we introduce healing distances r_{hCs} and r_{hT} for $F_s(r)$ and $F_T(r)$, respectively, so that $F_s(r)$ and $F_T(r)$ satisfy the following conditions $F_s(r_{hCs}) = F_{Fs}(r_{hCs})$, $dF_s(r)/dr|_{r=r_{hCs}} = dF_{Fs}(r)/dr|_{r=r_{hCs}}$ and $F_T(r_{hT}) = dF_T(r)/dr|_{r=r_{hT}} = 0$. Furthermore, we set $F_s(r) = F_{Fs}(r)$ at $r \geq r_{hCs}$ and $F_T(r) = 0$ at $r \geq r_{hT}$. In order to guarantee these conditions, following the FHNC method, we replace $V_{Cs}(r)$ and $V_T(r)$ in the Euler-Lagrange equations by $V_{Cs}(r) - \lambda_{Cs}$ and $V_T(r) - \lambda_T$, respectively, where the constants λ_{Cs} and λ_T represent corrections with respect to the long-range correlations. The values of λ_{Cs} and λ_T are chosen so that the above-mentioned continuity conditions of $F_s(r)$ at $r = r_{hCs}$ and $F_T(r)$ at $r = r_{hT}$ are satisfied. Finally, the values of the healing distances r_{hCs} and r_{hT} are determined so that the quantity $E_2/N + \lambda|S_{c1}(0)|^2$ is minimized. Here, we choose $\lambda = 1000$ MeV This choice is similar to the method adopted in the FHNC calculations [8, 20].

The obtained E_2/N of neutron matter with the Argonne v6' (AV6') potential [21] is shown in Fig. 2. E/N calculated with the AV6' and the repulsive part of the UI three-body potential is also shown in this figure. In this calculation, the contribution from the three-body force is

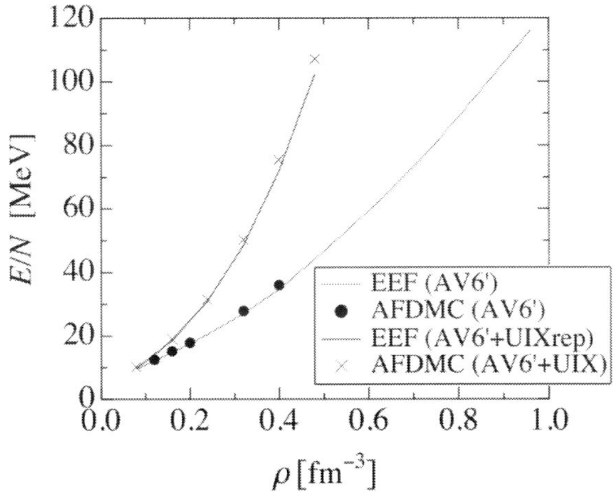

Figure 2. Energy per neutron of neutron matter with the AV6' two-body potential and the UI repulsive three-body potential as a function of the number density ρ.

considered as shown in Eq. (11). It is seen that, in both cases, the present results are in good agreement with those obtained by auxiliary field diffusion Monte Carlo (AFDMC) calculations [7, 8]. Here we note that, in the AFDMC calculations, the rest of the UI potential, i.e., the 2π-exchange part, is also considered. The agreement observed between our E/N calculations and that by AFDMC may imply that the contribution from the 2π-exchange part of the UI potential is negligibly small. It is also noted that we have improved the EEF for symmetric nuclear matter using the two-body central and tensor forces in a manner similar to that for neutron matter reported here, and that the improved E/N are also in good agreement with those obtained by the AFDMC method, as will be reported in detail elsewhere.

4. Concluding Remarks

In this study, we have extended the EEF variational method for simplified neutron matter to finite temperatures including the three-body force. Furthermore, we have improved the EEF using the two-body central and tensor forces. The extension and improvement were successful. In order to treat the full Argonne v18 two-body potential [24] and the UI three-body potential, the next improvement is to treat the spin-orbit forces and corresponding correlations. Then, we will proceed further to consider asymmetric nuclear matter toward systematic calculations of hot asymmetric nuclear EOS with the EEF variational method.

Acknowledgments

The numerical computations in this work were carried out on SR16000 at ITP in Kyoto University and on SR16000 at the University of Tokyo. This work is supported by JSPS (No. 25400275) and the rants-in-Aid for Scientific Research on Innovative Areas of ME T (Nos. 20105003 and 21105510).

References

[1] Clark J W 1979 *Prog. Part. Nucl. Phys.* **2** 89
[2] Viviani M, Buendia E, Fantoni S and Rosati S 1988 *Phys. Rev.* B **38** 4523
[3] Krotscheck E 2000 *J. Low Temp. Phys.* **119** 103
[4] Akmal A, Pandharipande V R and Ravenhall D G 1998 *Phys. Rev.* C **58** 1804
[5] Boronat J and Casulleras J 2001 *Int. J. Mod. Phys.* B **15** 1591
[6] Carlson J, Morales J Jr., Pandharipande V R and Ravenhall D G 2003 *Phys. Rev.* C **68** 025802
[7] Sarsa A, Fantoni S, Schmidt K E and Pederiva F 2003 *Phys. Rev.* C **68** 024308
[8] Lovato A, Benhar O, Fantoni S, Illarionov A Yu. and Schmidt K E 2011 *Phys. Rev.* C **83** 054003
[9] Takano M and Yamada M 1994 *Prog. Theor. Phys.* **91** 1149

Journal of Physics: Conference Series **529** (2014) 012025 doi:10.1088/1742-6596/529/1/012025

[10] Takano M 1995 *Prog. Theor. Phys.* **93** 745
[11] Takano M and Yamada M 1998 *Prog. Theor. Phys.* **100** 745
[12] Takano M, Endo T, Kimura R and Yamada M 2003 *Prog. Theor. Phys.* **109** 213
[13] Takano M and Yamada M 2006 *Prog. Theor. Phys.* **116** 545
[14] Kanzawa H, Oyamatsu K, Sumiyoshi K and Takano M 2007 *Nucl. Phys.* A **791** 232
[15] Mukherjee A 2009 *Phys. Rev.* C **79** 045811
[16] Illarionov A Yu, Fantoni S, Pederiva F. Gandolfi S and Schmidt K E 2012 *Phys. At. Nucl.* **75** 866
[17] Togashi H and Takano M 2013 *Nucl. Phys.* A **902** 53
[18] Schmidt K E and Pandharipande V R 1979 *Phys. Lett.* B **87** 11
Mukherjee A and Pandharipande V R 2007 *Phys. Rev.* C **75** 035802
[19] Takano M and Yamada M 1992 *Prog. Theor. Phys.* **88** 1131
[20] Akmal A and Pandharipande V R 1997 *Phys. Rev.* C **56** 2261
[21] Pudliner B S, Pandharipande V R, Carlson J, Pieper S C and Wiringa R B 1997 *Phys. Rev.* C **56** 1720
Wiringa R B and Pieper S C 2002 *Phys. Rev. Lett.* **89** 182501
[22] Carlson J, Pandharipande V R, Wiringa R B, 1983 *Nucl. Phys.* A **401** 59
[23] Takano M, Kaneko T and Yamada M 1997 *Prog. Theor. Phys.* **97** 569
[24] Wiringa R B, Stoks V G J and Schiavilla R 1995 *Phys. Rev.* C **51** 38

Application of extended random-phase approximation with ground-state correlations to collective excitations of ^{16}O

M Tohyama

Kyorin University School of Medicine, Mitaka, Tokyo 181-8611, Japan

E-mail: `tohyama@ks.kyorin-u.ac.jp`

Abstract. The ground-state correlation effects on the octupole excitation of ^{16}O are studied using the extended random phase approximation (ERPA) derived from the time-dependent density-matrix theory. ERPA includes the effects of ground-state correlations through the fractional occupation of single-particle states and a correlated two-body density matrix. The coupling to two particle - two hole configurations is also included in ERPA. It is found that the first 3^- state calculated in the random phase approximation (RPA) is shifted upward when the self-energy contributions due to ground-state correlations are included in particle - hole pairs and that the coupling to two particle - two hole states plays a role in shifting the first 3^- state down to the right position.

1. Introduction

The random phase approximation (RPA) based on the Hartree-Fock (HF) ground state has extensively been used to study nuclear collective excitations. It is generally considered that the HF + RPA approach is the most appropriate for doubly closed-shell nuclei such as ^{16}O, ^{40}Ca, ^{48}Ca and ^{208}Pb for which the HF theory would give a good description of the ground states. However, it has also been known that the ground state of ^{16}O actually is a highly correlated state (see Refs.[1, 2] for recent theoretical studies.) This indicates the necessity of using beyond RPA theories to study collective excitations of ^{16}O, though the HF + RPA approach has commonly been used for ^{16}O. In this report we investigate how the octupole excitation of ^{16}O, which has a collective low-lying state, is affected by ground-state correlations, using an extended RPA (ERPA) that has been derived from the time-dependent density-matrix theory (TDDM) [3, 4]. The advantage of ERPA is that it not only includes the effects of ground-state correlations but also describes the coupling to two particle - two hole configurations. We show that the first octupole state in ^{16}O is quite sensitive to the ground-state correlation effects and the coupling to the higher amplitudes.

2. Formulation

The TDDM consists of the coupled equations of motion for the one-body density matrix $n_{\alpha\alpha'}$ (the occupation matrix) and the correlated part of the two-body density matrix $C_{\alpha\beta\alpha'\beta'}$ (the correlation matrix). These matrices are defined as

$$n_{\alpha\alpha'}(t) = \langle\Phi(t)|a^\dagger_{\alpha'}a_\alpha|\Phi(t)\rangle, \tag{1}$$

$$\begin{aligned} C_{\alpha\beta\alpha'\beta'}(t) &= \langle\Phi(t)|a^\dagger_{\alpha'}a^\dagger_{\beta'}a_\beta a_\alpha|\Phi(t)\rangle \\ &- (n_{\alpha\alpha'}(t)n_{\beta\beta'}(t) - n_{\alpha\beta'}(t)n_{\alpha\beta'}(t)), \end{aligned} \tag{2}$$

where $|\Phi(t)\rangle$ is the time-dependent total wavefunction $|\Phi(t)\rangle = \exp[-iHt/\hbar]|\Phi(t=0)\rangle$. The equations of motion for reduced density matrices form a chain of coupled equations known as the Bogoliubov-Born-Green-Kirkwood-Yvon (BBGKY) hierarchy. In TDDM the BBGKY hierarchy is truncated by replacing a three-body density matrix with anti-symmetrized products of the one-body and two-body density matrices. Thus TDDM goes one step further beyond the time-dependent HF theory (TDHF). The TDDM equation for $C_{\alpha\beta\alpha'\beta'}$ contains all effects of two-body correlations; particle - particle, hole - hole and particle - hole correlations. The ground state in TDDM is given as a stationary solution of the TDDM equations. The stationary solution can be obtained using the gradient method [1]. This method is also used in the present work. The ERPA equations used here are derived by taking the small amplitude limit of TDDM as RPA corresponds to the small amplitude limit of TDHF and are written in matrix form for the one-body and two-body amplitudes $x^\mu_{\alpha\alpha'}$ and $X^\mu_{\alpha\beta\alpha'\beta'}$ [1]

$$\begin{pmatrix} A & C \\ B & D \end{pmatrix}\begin{pmatrix} x^\mu \\ X^\mu \end{pmatrix} = \omega_\mu \begin{pmatrix} S_1 & T_1 \\ T_2 & S_2 \end{pmatrix}\begin{pmatrix} x^\mu \\ X^\mu \end{pmatrix}, \tag{3}$$

where the matrix on the left-hand side is the Hamiltonian matrix and that on the right-hand side the norm matrix. The one-body sector of Eq. (3) $Ax^\mu = \omega_\mu S_1 x^\mu$ is formally the same as the equation in the self-consistent RPA (SCRPA) of Refs. [5, 6, 7], which also includes the effects of ground-state correlations through $n_{\alpha\alpha'}$ and $C_{\alpha\beta\alpha'\beta'}$. We refer to the approximation $Ax^\mu = \omega_\mu S_1 x^\mu$ as the modified RPA (mRPA). To explain the role of the correlation matrix in the mRPA equation, we explicitly show the matrices A and S_1:

$$\begin{aligned} A(\alpha\alpha':\lambda\lambda') &= (\epsilon_\alpha - \epsilon_{\alpha'})(n_{\alpha'\alpha'} - n_{\alpha\alpha})\delta_{\alpha\lambda}\delta_{\alpha'\lambda'} \\ &+ (n_{\alpha'\alpha'} - n_{\alpha\alpha})(n_{\lambda'\lambda'} - n_{\lambda\lambda})\langle\alpha\lambda'|v|\alpha'\lambda\rangle_A \\ &- \delta_{\alpha'\lambda'}\sum_{\gamma\gamma'\gamma''}\langle\alpha\gamma|v|\gamma'\gamma''\rangle C_{\gamma'\gamma''\lambda\gamma} \\ &- \delta_{\alpha\lambda}\sum_{\gamma\gamma'\gamma''}\langle\gamma\gamma'|v|\alpha'\gamma''\rangle C_{\lambda'\gamma''\gamma\gamma'} \\ &+ \sum_{\gamma\gamma'}(\langle\alpha\gamma|v|\lambda\gamma'\rangle_A C_{\lambda'\gamma'\alpha'\gamma} + \langle\lambda'\gamma|v|\alpha'\gamma'\rangle_A C_{\alpha\gamma'\lambda\gamma}) \\ &- \sum_{\gamma\gamma'}(\langle\alpha\lambda'|v|\gamma\gamma'\rangle C_{\gamma\gamma'\alpha'\lambda} + \langle\gamma\gamma'|v|\alpha'\lambda\rangle C_{\alpha\lambda'\gamma\gamma'}), \end{aligned} \tag{4}$$

$$S_1(\alpha\alpha':\lambda\lambda') = (n_{\alpha'\alpha'} - n_{\alpha\alpha})\delta_{\alpha\lambda}\delta_{\alpha'\lambda'}, \tag{5}$$

where the subscript A means that the corresponding matrix is antisymmetrized and $n_{\alpha\alpha'}$ is assumed to be diagonal. The first two terms on the right-hand side of Eq. (4) are the same as those in the RPA equation, the next two terms with $C_{\alpha\beta\alpha'\beta'}$ describe the self-energy of the particle - hole (p-h) state due to ground-state correlations [5], and the last four terms with $C_{\alpha\beta\alpha'\beta'}$ may be interpreted as the modification of the p-h interaction caused by ground-state correlations [5]. If the correlated ground state is replaced by the HF ground state, mRPA and ERPA become the same as RPA and the second RPA (SRPA) [8], respectively. In SRPA calculations the so-called diagonal approximation where the correlations in matrix D are neglected has often been used [8]. In the ERPA calculations shown below all correlations in matrix D are included. It has been pointed out [9] that if the effects of ground-state correlations are included perturbatively in extended RPA theories, the energy-weighted sum rule (EWSR) is not satisfied. In our ERPA and mRPA the ground state is calculated non-perturbatively in TDDM and the one-body amplitudes $x^{\mu}_{\alpha\alpha'}$ include the p-p and h-h components in addition to the p-h and h-p components. Therefore, ERPA and mRPA satisfy EWSR [10].

3. Results

3.1. Calculational details

The occupation probability $n_{\alpha\alpha}$ and the correlation matrix $C_{\alpha\beta\alpha'\beta'}$ are calculated within TDDM using the gradient method [1] and the single-particle states, the $1p_{3/2}$, $1p_{1/2}$, $1d_{5/2}$ and $2s_{1/2}$ states for both protons and neutrons. For the calculations of the single-particle states we use the Skyrme III force, which has often been used for nuclear structure calculations. To reduce the dimension size, we only consider the two particle - two hole (2p-2h) and 2h-2p elements of $C_{\alpha\beta\alpha'\beta'}$. A simplified interaction which contains only the t_0 and t_3 terms of the Skyrme III force is used as the residual interaction. The spin-orbit force and Coulomb interaction are also omitted from the residual interaction. To avoid a cumbersome treatment of the rearrangement effects of a density-dependent force in extended RPA theories [11], we use the three-body version of the Skyrme interaction, $v_3 = t_3\delta^3(\boldsymbol{r}_1 - \boldsymbol{r}_2)\delta^3(\boldsymbol{r}_1 - \boldsymbol{r}_3)$, which gives the following density-dependent two-body residual interaction: $t_3\rho_n\delta^3(\boldsymbol{r} - \boldsymbol{r}')$, $t_3\rho\delta^3(\boldsymbol{r} - \boldsymbol{r}')/2$ and $t_3\rho_p\delta^3(\boldsymbol{r} - \boldsymbol{r}')$ for the proton-proton, proton-neutron and neutron-neutron interactions, respectively, where ρ_p, ρ_n and ρ are the proton, neutron and total densities, respectively. In the RPA, mRPA and ERPA calculations the one-body amplitudes $x^{\mu}_{\alpha\alpha'}$ are defined using a large number of single-particle states including those in the continuum. We discretize the continuum states by confining the wavefunctions in a sphere with radius 15 fm and take all the single-particle states with $\epsilon_\alpha \leq 50$ MeV and $j_\alpha \leq 11/2\hbar$. As the residual interaction, we use the same simple force as that used in the ground-state calculation. Since the residual interaction is not consistent with the effective interaction used in the calculation of the single-particle states, it is necessary to adjust its strength so that the collective states calculated in RPA come at right positions. We use the reduction factor $f = 0.62$ so that the spurious mode corresponding to the center-of-mass motion comes at zero excitation energy in RPA. The p-h interaction in the second term on the right-hand side of Eq. (4) is multiplied with this f. This rather large reduction of the strength is a direct consequence of inconsistency of the residual interaction but the reduction procedure has been found [1] to give an excitation energy of the giant quadrupole resonance of ^{16}O that is comparable to the result of a self-consistent RPA calculation. To define the two-body amplitudes $X^{\mu}_{\alpha\beta\alpha'\beta'}$, we use the small single-particle space consisting of the $1s_{1/2}$, $1p_{3/2}$, $1p_{1/2}$, $1d_{5/2}$, $2s_{1/2}$, $1d_{3/2}$, $2p_{3/2}$, $2p_{1/2}$ and $1f_{7/2}$ for both protons and neutrons. To reduce the number of the two-body amplitudes, we consider only the 2p-2h and 2h-2p components of $X^{\mu}_{\alpha\beta\alpha'\beta'}$ with $|\epsilon_\alpha + \epsilon_\beta - \epsilon_{\alpha'} - \epsilon_{\beta'}| \leq 60$ MeV. For the matrix elements of the residual interaction which couple to $X^{\mu}_{\alpha\beta\alpha'\beta'}$, we use the same residual interaction (with $f = 1$) as that used for the ground-state calculation because the single-particle space for $X^{\mu}_{\alpha\beta\alpha'\beta'}$ is much smaller than that for $x^{\mu}_{\alpha\alpha'}$.

3.2. Ground state of ^{16}O

The occupation probabilities in the ground state of ^{16}O calculated in TDDM are shown in Table 1. The deviation from the HF values ($n_{\alpha\alpha}$=1 or 0) is more than 10%, which means that the ground state of ^{16}O is strongly correlated. A recent shell-model calculation by Utsuno and Chiba [2] also gives a similar result for the ground state of ^{16}O. The correlation energy E_c in the ground state, which is defined by $E_c = \sum_{\alpha\beta\alpha'\beta'}\langle\alpha\beta|v|\alpha'\beta'\rangle C_{\alpha'\beta'\alpha\beta}/2$, is -23.7 MeV. A large portion of the correlation energy is compensated by the increase in the mean-field energy due to the fractional occupation of the single-particle states. The resulting energy gain due to the

Table 1. Single-particle energies ϵ_α and occupation probabilities $n_{\alpha\alpha}$ in the ground state of ^{16}O calculated in TDDM.

	ϵ_α [MeV]		$n_{\alpha\alpha}$	
orbit	proton	neutron	proton	neutron
$1p_{3/2}$	-18.3	-21.9	0.894	0.893
$1p_{1/2}$	-12.3	-15.7	0.868	0.865
$1d_{5/2}$	-3.8	-7.1	0.108	0.109
$2s_{1/2}$	1.1	-1.6	0.019	0.021

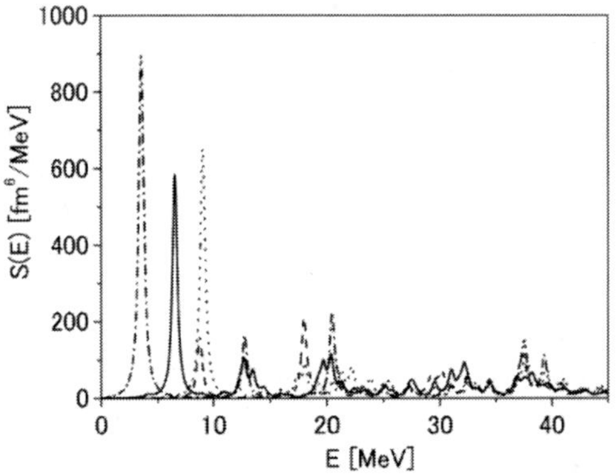

Figure 1. Strength functions calculated in RPA (chain line), SRPA (dashed line), mRPA (dotted line) and ERPA (full line) for the octupole excitation in ^{16}O. The distributions are smoothed with an artificial width $\Gamma = 0.5$ MeV.

ground-state correlations, which is given by the total energy difference between HF and TDDM, is 6.4 MeV. This value is much smaller than the value of the correlation energy itself $|E_c| = 23.7$ MeV.

3.3. Octupole excitation

The strength functions for the isoscalar octupole excitation calculated in RPA (chain line), SRPA (dashed line), mRPA (dotted line) and ERPA (full line) are shown in Figure 1. We used the excitation operator $r^3Y_{30}(\theta)$ to calculate the transition strength. In RPA the summed energy-weighted strength exhausts 97% of the EWSR value. The energy-weighted sum of the octupole strength in ERPA and mRPA is about 10% larger than that in RPA. However, $\langle r^{2L-2}\rangle$ (with $L = 3$) for the TDDM ground state is also larger than that for the HF ground state because the radius of the correlated ground state is slightly larger than that of the HF ground state. Consequently, 96% of the EWSR value is exhausted in ERPA and mRPA. The excitation energy of the first 3^- (3_1^-) state in RPA is 3.54 MeV, which is much smaller than the experimental data 6.13 MeV [12]. In SRPA the state corresponding to the 3_1^- state becomes an imaginary solution due to the coupling to the 2p-2h configurations. The results in RPA and SRPA suggest that the approaches based on the HF ground state cannot give a correct excitation energy of the 3_1^- state. The excitation energy of the 3_1^- state in mRPA is 9.1 MeV, which is much larger than that in RPA. This is explained by the self-energy contributions (see Eq. (4)) in the p-h pairs such as $(1p_{1/2})^{-1} \times 2d_{5/2}$ and $(1p_{3/2})^{-1} \times 2d_{5/2}$. The self-energy contributions increase the energy of the p-h pairs, reflecting the fact that the ground-state energy is lowered by the ground-state correlations. In ERPA the 3_1^- state is shifted downward to 6.56 MeV due to the coupling to the 2p-2h configurations. The results in mRPA and ERPA demonstrate that both the ground-state correlations and the coupling to the 2p-2h configurations play an important role in describing the properties of the 3_1^- state. The importance of the ground-state correlations and the coupling to 2p -2h configurations has also been found in other excitation modes [1, 13].

4. Summary

The effects of the correlations in the ground state of ^{16}O on the octupole excitation were studied using the extended RPA (ERPA). ERPA contains the ground-state correlation effects through the fractional occupation of single-particle states and the correlated part of the two-body density matrix. It also includes the coupling of the one particle - one hole amplitudes to two particle - two hole ones. It was found that the ground-state correlation effects on the first 3^- state are significant: the first 3^- state is shifted upward due to the self-energy contributions induced by the ground-state correlations. It was also found that the coupling to two particle - two hole states plays a role in producing the first 3^- at the right position. Our results demonstrate that the ground-state correlation effects should be properly taken into account in the study of nuclear collective excitations.

[1] Tohyama M 2007 *Phys. Rev.* C **75** 044310
[2] Utsuno Y and Chiba S 2011 *Phys. Rev.* C **83** 021301(R)
[3] Wang S J and Cassing W 1985 *Ann. Phys.* **159** 328
[4] Gong M and Tohyama M 1990 *Z. Phys.* A**335** 153
[5] Janssen D and Schuck P 1991 *Z. Phys.* A**339** 43
[6] Dukelsky J and Schuck P 1990 *Nucl. Phys.* A **512** 466
[7] Dukelsky J and Schuck P 1996 *Phys. Lett.* B **387** 233
[8] Drożdż S, Nishizaki S, Speth J and Wambach J 1990 *Phys. Rep.* **197** 1
[9] Adachi S and Lipparini E 1988 *Nucl. Phys.* A **489** 445
[10] Tohyama M and Schuck P 2007 *Eur. Phys. J.* A **32** 139
[11] Gambacurta D, Grasso M and Catara 2011 *J. Phys.* G **38** 035103
[12] Firestone R B 1996 *Table of Isotopes* (New York: John Wiley & Sons)
[13] Tohyama M 2013 *Phys. Rev.* C **87** 054330

Functional renormalization group approach to conventional theory of superfluidity and beyond

Y Tanizaki,[1,2] G Fejős,[2] and T Hatsuda[2,3]
[1]Department of Physics, The University of Tokyo, Tokyo 113-0033, Japan
[2]Theoretical Research Division, Nishina Center, RIKEN, Wako 351-0198, Japan
[3]Kavli IPMU (WPI), The University of Tokyo, Chiba 277-8583, Japan

E-mail: yuya.tanizaki@riken.jp

Abstract. Fermionic functional renormalization group (FRG) is applied to describe the superfluid phase transition of the two-component fermionic system with attractive contact interaction. Connection between the fermionic FRG approach and the Bardeen-Cooper-Schrieffer (BCS) theory with its Gorkov and Melik-Barkhudarov (GMB) correction is made clear, and the FRG flow of the fermion self-energy is also studied to go beyond the BCS+GMB theory. The superfluid transition temperature and the associated chemical potential are calculated in the region of the negative scattering length using fermionic FRG.

1. Introduction

Superfluidity in many-fermion systems is one of the central problems in condensed matter, atomic, nuclear and particle physics. Examples include liquid superfluid ^{3}He, electron superconductivity, cold atoms, nucleon superfluidity, color superconductivity, etc. [1]. We report our recent study on the application of fermionic functional renormalization group (f-FRG) method to the superfluid phase transition of two-component fermionic systems [2]. The purpose of this study is to develop a series of approximations without introducing the auxiliary bosonic field to make a firm connection between the non-perturbative FRG approach and the conventional Bardeen-Cooper-Schrieffer (BCS) theory with its Gorkov and Melik-Barkhudarov (GMB) correction. Furthermore, we wish to explore the role of the RG flow of the fermion self-energy to go beyond BCS+GMB theory. We note that such analyses can be best achieved by fermionic FRG without bosonization, because f-FRG can provide a systematic and unbiased study of interacting fermions [3, 4].

2. Fermionic FRG Formalism

We consider non-relativistic two-component fermions $\psi = (\psi_\uparrow, \psi_\downarrow)$ with a contact interaction:

$$S[\overline{\psi}, \psi] = \int_0^\beta \mathrm{d}\tau \int \mathrm{d}^3\boldsymbol{x} \left[\overline{\psi}\left(\partial_\tau - \frac{\nabla^2}{2m} - \mu\right)\psi + g\overline{\psi}_\uparrow\overline{\psi}_\downarrow\psi_\downarrow\psi_\uparrow\right], \tag{1}$$

where $\beta(= 1/T)$, μ, m and g are the inverse temperature, the chemical potential, the mass and the bare coupling constant, respectively. The classical action (1) is symmetric under global $U(1)$ phase rotation, $SU(2)$ spin rotation, and space-time translation.

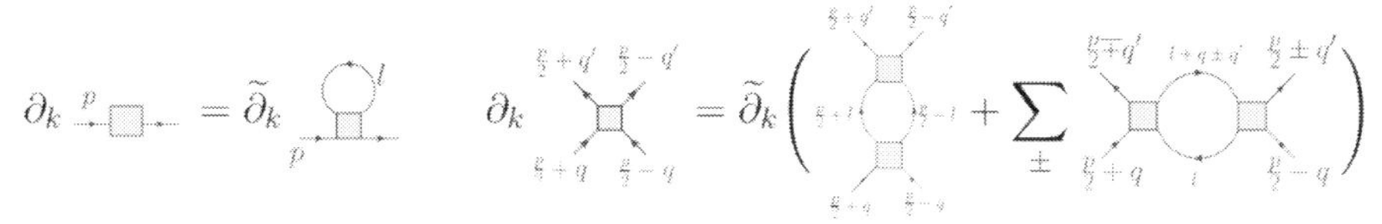

Figure 1. Flow equation of the self energy Σ_k and the four-point vertex $\Gamma_k^{(4)}$.

Following the idea of the FRG method [5], we define a scale-dependent average effective action $\Gamma_k[\overline{\psi}, \psi]$ as a one-particle-irreducible (1PI) effective action of $S[\overline{\psi}, \psi] + \overline{\psi} R_k \psi$, where R_k is an infrared (IR) regulating function. We consider the following Ansatz of Γ_k:

$$\begin{aligned}\Gamma_k[\overline{\psi}, \psi] &= \int_p^{(T)} \overline{\psi}(p)[G^{-1}(p) - \Sigma_k(p)]\psi(p) \\ &+ \int_{p,q,q'}^{(T)} \Gamma_k^{(4)}(p; q, q')\overline{\psi}_\uparrow(\frac{p}{2} + q)\overline{\psi}_\downarrow(\frac{p}{2} - q)\psi_\downarrow(\frac{p}{2} - q')\psi_\uparrow(\frac{p}{2} + q'),\end{aligned} \tag{2}$$

where $G^{-1}(p) = ip^0 + \frac{\boldsymbol{p}^2}{2m} - \mu$ is the inverse propagator with $p = (p^0, \boldsymbol{p})$, $\Sigma_k(p)$ is the self-energy, and $\Gamma_k^{(4)}(p; q, q')$ is the four-point vertex. Also, we adopt an abbreviated notation, $\int_p^{(T)} \equiv \int \frac{d^3\boldsymbol{p}}{(2\pi)^3} T \sum_{p^0}$. In this study, we neglect the momentum dependence of the self-energy and define its constant part by $\sigma_k \equiv \Re\Sigma_k(\pm\pi T, \boldsymbol{0})$. IR regulator $R_k(\boldsymbol{p})$ is chosen to suppress one-particle excitations around the Fermi surface with the excitation energy smaller than $k^2/2m$: $R_k(\boldsymbol{p}) = \left[\frac{k^2}{2m}\mathrm{sgn}(\xi(\boldsymbol{p})) - \xi(\boldsymbol{p})\right]\theta\left(\frac{k^2}{2m} - |\xi(\boldsymbol{p})|\right)$ with $\xi(\boldsymbol{p}) = \boldsymbol{p}^2/2m - \mu - \sigma_0$. Γ_k obeys the flow equation

$$\partial_k \Gamma_k = \frac{1}{2}\int_p^{(T)} \mathrm{Tr}\left[\frac{\partial_k R_k(p)}{\Gamma_k^{(2)}(p) + R_k(p)}\right]. \tag{3}$$

Our primary goal is to calculate the ratios T_c/ε_F and μ/ε_F as a function of the dimensionless constant $1/(k_F a_s)$: Here a_s is the scattering length between fermions, $k_F \equiv (3\pi^2 n)^{1/3}$ with n being the fermion number density and $\varepsilon_F \equiv k_F^2/2m$. For a second-order phase transition, the critical point $T = T_c$ can be determined from the Thouless criterion [6] by looking at the divergence of the fermion-fermion scattering matrix at the total momentum $p = 0$:

$$\left[\Gamma_{k=0}^{(4)}(p = 0)\right]^{-1} = 0 \quad \text{at } T = T_c. \tag{4}$$

Note that the number density n is related to T and μ through the number equation:

$$n = \langle\overline{\psi}\psi\rangle = 2\int_p^{(T)} \frac{-1}{G^{-1}(p) - \sigma_0}. \tag{5}$$

3. Flow equation of f-FRG for self-energy and four-point vertex

Taking the vertex expansion (2) of the scale dependent 1PI effective action $\Gamma_k[\overline{\psi}, \psi]$, a closed set of equations for the self-energy and four-point vertex can be derived from (3). They are given diagrammatically in Fig. 1. In order to simplify the flow equation of four-point vertex, we neglect its momentum dependence and regard it as an effective coupling constant. At the lowest momenta, our flow equations can be approximated as

$$\partial_k \sigma_k = \Re\tilde{\partial}_k \int_l^{(T)} \frac{\Gamma_k^{(4)}(\pm\pi T + l^0, \boldsymbol{l})}{G^{-1}(l) - \sigma_k + R_k(l)}, \tag{6}$$

Journal of Physics: Conference Series **529** (2014) 012027 doi:10.1088/1742-6596/529/1/012027

$$\begin{aligned}
\partial_k \Gamma_k^{(4)-1}(0) &= \tilde{\partial}_k \int_l^{(T)} \frac{1}{[G^{-1}(l) - \sigma_k + R_k(l)][G^{-1}(-l) - \sigma_k + R_k(-l)]} \\
&+ \tilde{\partial}_k \int_0^{8m\mu} \frac{\mathrm{d}|\boldsymbol{Q}|^2}{8m\mu} \int_l^{(T)} \frac{1}{[G^{-1} - \sigma_k + R_k](l)[G^{-1} - \sigma_k + R_k](Q+l)},
\end{aligned} \tag{7}$$

where transfer momenta $Q = (0, \boldsymbol{Q})$. Here, $\tilde{\partial}_k$ is the derivative of k acting only on the IR regulator R_k. On the r.h.s. of (7), the first and second terms refer to the particle-particle (PP) and particle-hole (PH) fluctuations, respectively. For the PH loop contribution of the flow of four-point coupling, the relative momenta is projected onto the Fermi surface. As already mentioned, in (6), the frequency of the fermion self-energy is restricted to the lowest values $\pm\pi T$. After performing the Matsubara sum, σ_k and σ_0 appear as a combination $\sigma_0 - \sigma_k$. Since the approximate particle-hole symmetry would make this combination small for $k < k_F$, where PH contribution will turn out to be already not significant, we take $\sigma_k = \sigma_0$ in the actual calculation of this contribution.

To take into account the momentum dependence in a minimal way, we make the following expansion of the four-point vertex in terms of center-of-mass momentum and keep first few terms:

$$\Gamma_k^{(4)-1}(p^0 + l^0, \boldsymbol{l}) \approx -Z_k^{-1} \left[i(l^0 + p^0) + S_k^{(1)} \cdot |\boldsymbol{l}| + S_k^{(2)} \cdot \boldsymbol{l}^2 + |\mu_k^B| \right], \tag{8}$$

where $|\mu_k^B| = -Z_k \Gamma_k^{(4)-1}(0)$. Here we adopt a hybrid approach in which $\Gamma_k^{-1}(0)$ is calculated from the flow equation, while its derivatives Z_k, $S_k^{(1,2)}$ with respect to the frequency and momentum are estimated by the solution of an approximated flow equation keeping only the PP-contribution (this actually coincides with the usual Random Phase Approximation). After applying the expansion (8), the flow equation of the self-energy becomes

$$\begin{aligned}
\partial_k \sigma_k &= \Re \tilde{\partial}_k \int_l^{(T)} \frac{1}{G^{-1}(l) - \sigma_k + R_k(l)} \frac{-Z_k}{i(l^0 \pm \pi T) + S_k^{(1)}|\boldsymbol{l}| + S_k^2 \boldsymbol{l}^2 + |\mu_k^B|}, \\
&= -\Re \tilde{\partial}_k \int_0^\infty \frac{dl l^2}{2\pi^2} \Big(n_B(\omega_k^B(\boldsymbol{l})) + n_F(\omega_k(\boldsymbol{l})) \Big) \frac{Z_k}{\omega_k(\boldsymbol{l}) - \omega_k^B(\boldsymbol{l}) \mp i\pi T},
\end{aligned} \tag{9}$$

where $\omega_k(\boldsymbol{l}) = \frac{\boldsymbol{l}^2}{2m} - \mu - \sigma_k + R_k(\boldsymbol{l})$, and $\omega_k^B(\boldsymbol{l}) = |\mu_k^B| + S_k^{(1)}|\boldsymbol{l}| + S_k^{(2)}\boldsymbol{l}^2$, with n_B being the Bose distribution function and n_F the Fermi distribution function.

4. Results

Let us briefly discuss the connection of f-FRG and the BCS+GMB theory. If we take into account only the PP correlation in the flow equation with neglecting the self energy (i.e. $\sigma_k \equiv 0$) and the PH loop in the four-point coupling flow, in the weakly coupled regime we reproduce the BCS transition temperature by applying the Thouless criterion: $T_c^{\mathrm{BCS}}/\varepsilon_F = 0.613 \exp(-\pi/(2k_F|a_s|))$. By taking into account the PH loop contribution (evaluating it at zero temperature) and still neglecting the self-energy, the transition temperature is reduced by a factor $(4e)^{1/3} \approx 2.2$, which is nothing but the GMB correction of the critical temperature: $T_c^{\mathrm{GMB}} = T_c^{\mathrm{BCS}}/2.2$.

On Fig.2, the flow of $-\partial_k(1/\Gamma_k^{(4)}(0))$ is plotted with and without particle-hole (PH) loop when the self-energy correction is neglected. The flows of the self-energy for $(k_F a_s)^{-1} = -2, -1, 0$ at T_c are shown in Fig.3. Fig.4 and Fig.5 show our numerical results of the critical temperature (T_c/ε_F) and the chemical potential (μ/ε_F) as a function of the inverse scattering length ($(k_F a_s)^{-1} < 0$),

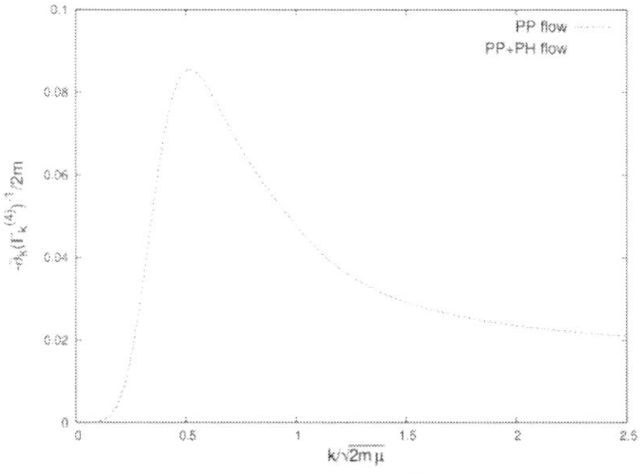

Figure 2. Derivative of $1/\Gamma_k^{(4)}(0)$ at $T/\mu = 0.06$ with and without the PH loop included.

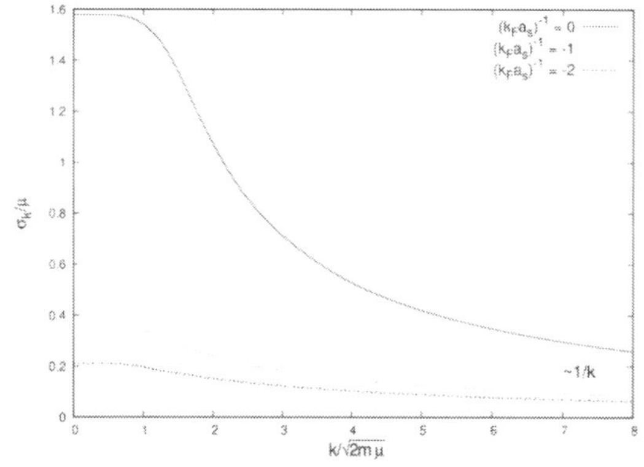

Figure 3. Self-energy σ_k as a function of k for $(k_F a_s)^{-1} = -2, -1, 0$ at $T = T_c$.

respectively, for three different levels of approximations: PP contains only the effect of particle-particle correlation, PP+PH just neglects the self-energy, and PP+PH+SE is our final result including self-energy (SE) as well.

We note that Fig.2 provides a physical interpretation of GMB correction of the BCS theory from a renormalization group point of view. When the self-energy is neglected, but the contribution of PH loop is included in the flow of four-point coupling, $\partial_k(1/\Gamma_k^{(4)}(0))$ is reduced only around the region $k \sim k_F$. Since the particle-particle loop contribution reproduces exactly the BCS result as the low-energy effective theory, this implies that the effective coupling of the BCS theory is weakened due to the existence of matter, and such screening occurs around intermediate energy scales $k^2/2m \sim \varepsilon_F$.

Considering the self-energy equation for large k ($\gtrsim k_F$), it can be shown that its flow can be well described by perturbation theory and that it behaves as $\sim 1/k$ (see Fig. 3). On the other hand, it almost stops for small k ($\lesssim k_F$): This is due to the fact that the approximate particle-hole symmetry valid for small k protects the shift of the Fermi level. Fig. 3 shows that the magnitude of the self-energy σ_0 becomes larger as the coupling becomes strong, while the saturation of σ_k at $k \sim k_F$ takes place for any coupling strength.

As can be seen from Fig. 4, PH correlation is the dominant source of the reduction of T_c/ε_F, and, as already discussed, its physical origin is the screening of the coupling strength. We should note that this does not imply the smallness of the self-energy correction. Since the flow of self-energy stops for $k \lesssim k_F$, it does not affect low-energy dynamics after shifting the Fermi level away from the chemical potential. The self-energy correction still leads to further small reduction of T_c/ε_F, since it makes ε_F (or the number density n) bigger, as the coupling strength becomes stronger towards the unitary regime, but the effect is rather limited because $\sigma_0 - \sigma_k$ around $k \sim k_F$ cannot be large due to the particle-hole symmetry, as discussed above.

Fig. 5 shows that, the effects of the PH correlation to μ/ε_F are opposite of what we find for T_c/ε_F. This is because the critical temperature T_c of PP approximation is higher than that of the PP+PH result for the same chemical potential μ. Therefore, the number of fermions in the PP case is larger than that of PP+PH. When the self-energy correction is taken into account, the number density n and hence the Fermi energy ε_F increase for given μ, therefore a considerable decrease of μ/ε_F arises.

Although it is beyond our scope to predict a quantitatively correct T_c in the unitary regime, it is still instructive to compare our result at unitarity, $(T_c/\varepsilon_F, \mu/\varepsilon_F) = (0.237, 0.367)$, with the previous FRG results using auxiliary bosonic field, $(T_c/\varepsilon_F, \mu/\varepsilon_F) \simeq (0.248, 0.51)$ [7]. For

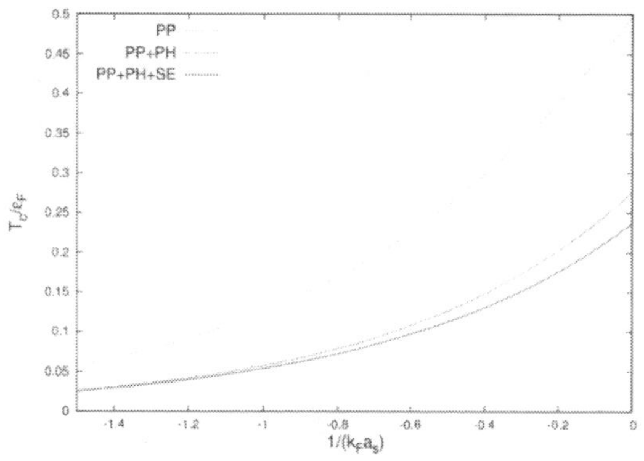

Figure 4. T_c/ε_F as a function of $(k_Fa_s)^{-1}$ in three different levels of approximations.

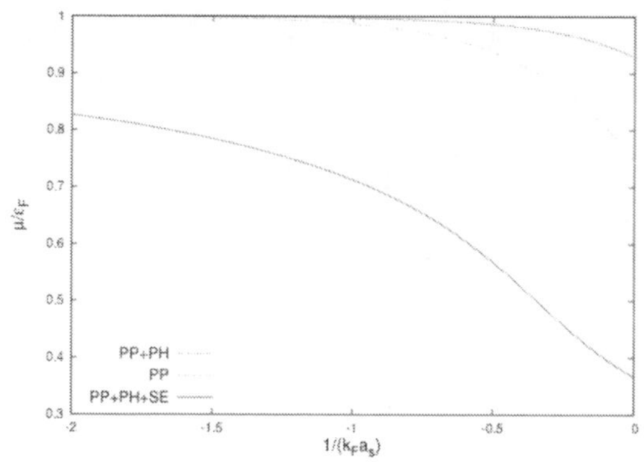

Figure 5. μ/ε_F as a function of $(k_Fa_s)^{-1}$. The meaning of each line is the same as Fig.4.

convenience of readers, we also quote some results of Monte Carlo simulations for $(T_c/\varepsilon_F, \mu/\varepsilon_F)$: $(0.152(7), 0.493(14))$ [8, 9], $(0.15(1), 0.43(1))$ [10], $(0.171(5), 0.429(9))$ [11].

5. Summary

We developed a fermionic FRG to describe the superfluid phase transition for two-component fermions with a contact interaction. By making vertex expansion of the 1PI effective action $\Gamma_k[\bar{\psi}, \psi]$ up to the four-point vertex and solving RG flow equations, we determined the critical temperature T_c in the regime of negative scattering lengths. In order to clarify the relation between the FRG approach and the BCS+GMB theory and to go beyond, we have taken into account the PP and PH correlations together with the self-energy correction, and the role of each is analyzed in details in their respective flow equation.

Resultant value of T_c/ε_F does not receive large correction from the self-energy except for the unitary regime $(1/(k_Fa_s) \to 0)$. On the other hand, μ/ε_F shows a large reduction by the self-energy correction due to the increase of the Fermi energy given μ. It shows that the self-energy correction is still comparable with the chemical potential even for $(k_Fa_s)^{-1} \lesssim -1$, however, its effect on T_c/ε_F is almost negligible. Extrapolation to unitarity gives $T_c/\varepsilon_F = 0.237$ and $\mu/\varepsilon_F = 0.367$.

Acknowledgements

Y. T. is supported by JSPS Research Fellowships for Young Scientists. G. F. is supported by the Foreign Postdoctoral Research program of RIKEN. T. H. is partially supported by RIKEN iTHES project. This work was partially supported by the Program for Leading Graduate Schools, MEXT, Japan.

References

[1] Cooper L N 2010 *BCS: 50 Years* 1st ed (World Scientific Publishing Company)
[2] Tanizaki Y, Fejős G and Hatsuda T 2014 *Prog. Theor. Exp. Phys.* **2014** 043I01
[3] Shankar R 1994 *Reviews of Modern Physics* **66** 129
[4] Metzner W, Salmhofer M, Honerkamp C, Meden V and Schönhammer K 2012 *Rev. Mod. Phys.* **84**(1) 299–352
[5] Wetterich C 1993 *Physics Letters B* **301** 90–94
[6] Thouless D 1960 *Annals of Physics* **10** 553–588

[7] Floerchinger S, Scherer M M and Wetterich C 2010 *Phys. Rev. A* **81**(6) 063619

[8] Burovski E, Prokof'ev N, Svistunov B and Troyer M 2006 *Phys. Rev. Lett.* **96**(16) 160402

[9] Burovski E, Kozik E, Prokof'ev N, Svistunov B and Troyer M 2008 *Phys. Rev. Lett.* **101**(9) 090402

[10] Bulgac A, Drut J E and Magierski P 2008 *Phys. Rev. A* **78**(2) 023625

[11] Goulko O and Wingate M 2010 *Phys. Rev. A* **82**(5) 053621

Journal of Physics: Conference Series **529** (2014) 012028 doi:10.1088/1742-6596/529/1/012028

Analytical Solution to the Dirac Equation for Few-Electron Ions of Transuranium Elements

Anton S Ulyanov and Alexander A Sadovoy

FSUE RFNC - All-Russian Research Institute of Experimental Physics, Mira Ave 37, Sarov 607190, Nizhny Novgorod region, Russia

E-mail: a.s.ulyanov@vniief.ru

Abstract.The paper presents relativistic generalization of the well-known method of multidimensional angular Coulomb functions. Two-component multidimensional angular functions have been constructed and they are used as a basis for expanding in series the many-electron wave function of the many-particle Dirac equation. A system of ordinary differential equations which factors are expressed in terms of matrix elements of various operators of the Dirac equation has been obtained for the expansion amplitudes. The binding energy of helium-, lithium-, beryllium- and carbon-like ions of transuranium elements has been calculated (Z = 92–101). Wave functions have been constructed for some states of transuranium ions. The developed method allows taking into account the many-particle effects during the calculation of relativistic many-electron systems which are described by the manyparticle Dirac equation with Coulomb interactions. The obtained results can be applied to specify the probability of transitions and particle paths in high energydensity processes.

Introduction

The paper is the next step forward in the studies described in [1-5]. The properties of few-electron systems are studied using a specific example of C-like ions of heavy elements by solving the Dirac equation for six electrons in the heavy nucleus field.

The developed relativistic option of the method of multidimensional angle Coulomb functions [6-8] allows finding an exact solution to the Dirac equation, which, in its turn, allows calculating probabilities of various physical processes using the available expressions for two-component wave functions, as well as Breit-Wigner corrections and corrections for radiation.

This paper presents the band 0^+ spectrum for C-like ions of elements with Z=92 to Z=101. The electron density profile has been obtained using the analytical wave function (WF) and the contribution from the lower WF component has been studied.

1. Hyperradial equations of the relativistic method of multidimensional angle Coulomb functions (MACF)

Moving electrons of a C-like ion in the central-symmetry electrical field of nucleus are described by the Dirac equation.

In the relativistic system of units, the Dirac equation for components of wave function $\Psi = \begin{pmatrix} \Phi \\ X \end{pmatrix}$ looks like

$$\left(E + \sum_{i=1}^{6} \frac{Z}{r_i} - \sum_{\substack{i>j \\ i=1}}^{6} \frac{1}{|\mathbf{r}_i - \mathbf{r}_j|} \right) \Phi = \sum_{i=1}^{6} c\boldsymbol{\sigma}_i \hat{\mathbf{p}}_i X, \tag{1.1}$$

$$\left(E + 12 + \sum_{i=1}^{6} \frac{Z}{r_i} - \sum_{\substack{i>j \\ i=1}}^{6} \frac{1}{|\mathbf{r}_i - \mathbf{r}_j|} \right) X = \sum_{i=1}^{6} c\boldsymbol{\sigma}_i \hat{\mathbf{p}}_i \Phi. \tag{1.2}$$

The system's binding energy in these equations is $E = \varepsilon - 6 < 0$, where ε is the system's total energy (mc^2).

To solve the equation above with MACF method, it is necessary to represent the WF components in terms of a series of multidimensional angle functions (MAF).

With minimum approximation, the wave function of state 0^+ of C-like ions in the spinor representation corresponding to electron configuration $\left(1s^2 2s^2 1p_{1/2}^2\right)$ has the form:

$$\Psi = \begin{pmatrix} \Phi \\ X \end{pmatrix} = \frac{1}{\rho^{17/2}} \begin{pmatrix} MU \\ NW \end{pmatrix}, \tag{1.3}$$

where $\rho = \sum_{i=1}^{A} |\vec{r}_i|$ is collective variable, M and N are amplitudes of the MAF factorization of the upper and lower WF components, respectively, and MAFs are Slater determinants:

$$U = \frac{N_K}{(-it\rho)^2} \begin{pmatrix} \psi_1(1) & \psi_1(2) & \dots & \psi_1(6) \\ \psi_2(1) & \psi_2(2) & \dots & \psi_2(6) \\ \vdots & \vdots & \vdots & \vdots \\ \psi_6(1) & \psi_6(2) & \dots & \psi_6(6) \end{pmatrix}, W = \frac{N_K}{(-it\rho)^2} \begin{pmatrix} \xi_1(1) & \xi_1(2) & \dots & \xi_1(6) \\ \xi_2(1) & \xi_2(2) & \dots & \xi_2(6) \\ \vdots & \vdots & \vdots & \vdots \\ \xi_6(1) & \xi_6(2) & \dots & \xi_6(6) \end{pmatrix}, \tag{1.4}$$

where N_K is normalization constant and orthonormal basis functions ψ and ξ, which are associated with an electron on a certain shell with a set of quantum numbers $\{n, j, l, m\}$ and, in general, have the form:

$$\psi_{jlm} = \sqrt{\frac{1}{\Gamma(3+n)}} \Omega_{jlm} L_n^2(-itr), \quad \xi_{jl'm} = (-1)^{l-j+\frac{1}{2}} \sqrt{\frac{1}{\Gamma(3+n)}} \Omega_{jl'm} L_n^2(-itr). \tag{1.5}$$

In the expressions above, the following symbols are used for the basis functions: L_n^2 are Laguerre polynomials; Ω_{jlm} are spinor spherical functions, and $l' = 2j - l$. For the system of interest, the set of basis functions has the form:

$$\psi_1=\frac{\Omega_{\frac{1}{2},0,-\frac{1}{2}}L_0^2(-itr)}{\sqrt{\Gamma(3)}},\ \psi_2=\frac{\Omega_{\frac{1}{2},0,\frac{1}{2}}L_0^2(-itr)}{\sqrt{\Gamma(3)}},\ \psi_3=\frac{\Omega_{\frac{1}{2},0,-\frac{1}{2}}L_1^2(-itr)}{\sqrt{\Gamma(4)}},\ \psi_4=\frac{\Omega_{\frac{1}{2},0,\frac{1}{2}}L_1^2(-itr)}{\sqrt{\Gamma(4)}},$$
$$\psi_5=\frac{\Omega_{\frac{1}{2},1,\frac{1}{2}}L_0^2(-itr)}{\sqrt{\Gamma(3)}},\ \psi_6=\frac{\Omega_{\frac{1}{2},1,-\frac{1}{2}}L_0^2(-itr)}{\sqrt{\Gamma(3)}}, \tag{1.6}$$

$$\xi_1=\frac{\Omega_{\frac{1}{2},1,-\frac{1}{2}}L_0^2(-itr)}{\sqrt{\Gamma(3)}},\ \xi_2=\frac{\Omega_{\frac{1}{2},1,\frac{1}{2}}L_0^2(-itr)}{\sqrt{\Gamma(3)}},\ \xi_3=\frac{\Omega_{\frac{1}{2},1,-\frac{1}{2}}L_1^2(-itr)}{\sqrt{\Gamma(4)}},\ \xi_4=\frac{\Omega_{\frac{1}{2},1,\frac{1}{2}}L_1^2(-itr)}{\sqrt{\Gamma(4)}},$$
$$\xi_5=\frac{-\Omega_{\frac{1}{2},0,\frac{1}{2}}L_0^2(-itr)}{\sqrt{\Gamma(3)}},\ \xi_6=\frac{-\Omega_{\frac{1}{2},0,-\frac{1}{2}}L_0^2(-itr)}{\sqrt{\Gamma(3)}}. \tag{1.7}$$

Normalization of WF is defined as

$$\int \Psi^{\dagger}\Psi d\tau_{18}=1, \tag{1.8}$$

where volume component $d\tau_{18}=\rho^{15}d\rho d\Omega_{18}=\frac{1}{2\pi}\int\limits_{-\infty}^{\infty}e^{-it\rho}dt\int e^{it\sum_{i=1}^{6}r_i}\prod_{i=1}^{6}d\vec{r}_i d\rho$, for the normalization constant we obtain

$$N_K^2=\frac{\Gamma(22)}{6!}. \tag{1.9}$$

If we substitute expansions of the WF components in terms of MAF to Eqs.(1.1) and (1.2), and then multiply the resultant equations by MAF on the left-hand side, we obtain the following system of equations for the WF expansion amplitudes

$$M'-\frac{a_1}{\rho}M-\frac{1}{6}\left(E+12+\frac{\alpha(c_1Z-d_1)}{\rho}\right)N=0, \tag{1.10}$$

$$N'+\frac{a_2}{\rho}N+\frac{1}{6}\left(E+\frac{\alpha(c_2Z-d_2)}{\rho}\right)M=0. \tag{1.11}$$

Here, coefficients a_1, a_2, c_1, c_2, d_1, and d_2 can be found by calculating matrix elements of potential energy and momentum operator for the upper and lower WF components; $\alpha=\frac{e^2}{\hbar c}$ is constant of thin structure.

The main difficulty in reducing the original Dirac equation in 18-dimensional space to the first-order system of ordinary differential equations for amplitudes of the wave function expansion in terms of two-component angle functions consists in the calculation of matrix elements from various operators of the Dirac equation using the multidimensional angle functions above. The techniques developed for this purpose have been described in [4].

2. Matrix elements of the operator of electron-nucleus interaction

For the upper WF component, matrix element (ME) of electron-nucleus interaction has the form

$$P_{e-n}=\langle U^{+}|\sum_{i=1}^{6}\frac{Z}{r_i}|U\rangle, \tag{2.1}$$

and ME for the lower WF component looks like

$$Q_{e-n} = \left\langle W^{+} \left| \sum_{i=1}^{6} \frac{Z}{r_i} \right| W \right\rangle . \tag{2.2}$$

Matrix elements P_{e-n} and Q_{e-n} are calculated in the same way and because of this we demonstrate here the calculation of matrix element P_{e-n} of the upper WF component.

In view of the fact that the potential energy operator $\sum_{i=1}^{6} \frac{Z}{r_i}$ is a single-particle operator, we can reduce equation (2.1) to the sum over subshells [9] and, hence, we obtain

$$P_{e-n}(\rho) = \sum_{\omega=1}^{6} \left\langle \psi_{\omega}^{*} \left| \frac{Z}{r} \right| \psi_{\omega} \right\rangle . \tag{2.3}$$

With regard to normalization and orthonormalized spin-angle functions the expression for matrix element in equation (2.3) can be represented in its general form (for arbitrary basis functions):

$$\left\langle \psi_{\omega}^{*} \left| \frac{1}{r} \right| \psi_{\omega} \right\rangle = \frac{\Gamma(3A+2K)}{A!} \frac{1}{2\pi\rho^{3A+2K-1}} \frac{A!}{\Gamma(3+n_{\omega})} \int_{-\infty}^{+\infty} \frac{e^{-it\rho}}{(-it)^{3A+2K-3}} dt \int e^{itr} \left[L_{n_{\omega}}^{2}(-itr) \right]^{2} r dr. \tag{2.4}$$

Hereinafter, ω is the basis function's number denoting the corresponding set of quantum numbers. Thus, the principal quantum number is $n_{\omega} = 0$ for basis functions ψ_1, ψ_2, ψ_5, and ψ_6 and $n_{\omega} = 1$ for basis functions ψ_3 and ψ_4 (see (1.6)).

Laguerre polynomial in equation (2.4) must be represented as a sum of Laguerre polynomials of a lower power [10]:

$$L_{n_{\omega}}^{2}(-itr) = \sum_{m_{\omega}=1}^{n_{\omega}} L_{m_{\omega}}^{1}(-itr). \tag{2.5}$$

Owing to the orthogonal nature of Laguerre polynomials the integral in ME of electron-nucleus interaction can be represented as

$$\begin{aligned} \int e^{itr} \left[L_{n_{\omega}}^{2}(-itr) \right]^{2} r dr &= \frac{1}{(-it)^{2}} \sum_{m_{\omega}=0}^{n_{\omega}} \frac{\Gamma(2+m_{\omega})}{\Gamma(1+m_{\omega})} = \frac{1}{(-it)^{2}} \sum_{m_{\omega}=0}^{n_{\omega}} (1+m_{\omega}) = \\ &= \frac{1}{(-it)^{2}} \frac{(n_{\omega}+2)(n_{\omega}+1)}{2}. \end{aligned} \tag{2.6}$$

Thus, for ME of electron-nucleus interaction we obtain

$$\left\langle \psi_{\omega}^{*} \left| \frac{1}{r} \right| \psi_{\omega} \right\rangle = \frac{\Gamma(3A+2K)}{\Gamma(3A+2K-1)} \frac{(n_{\omega}+1)(n_{\omega}+2)}{2\Gamma(3+n_{\omega})} \frac{1}{\rho} = \frac{(3A+2K-1)}{2\Gamma(n_{\omega}+1)} \frac{1}{\rho}. \tag{2.7}$$

As a result, for all MEs of electron-nucleus interaction of C-like ions in configuration $(1s^2 2s^2 1p_{1/2}^2)$ we obtain

$$\langle \psi_{\omega} | \frac{Ze^2}{r} | \psi_{\omega} \rangle = \langle \xi_{\omega} | \frac{Ze^2}{r} | \xi_{\omega} \rangle = \frac{21Ze^2}{2\rho}, \quad \omega = 1,...,6 . \tag{2.8}$$

Upon summation over shells one can write

$$P_{e-n} = Q_{e-n} = \frac{63Ze^2}{\rho}. \tag{2.9}$$

3. Matrix elements of the kinetic energy operator

To obtain the differential equation system for amplitudes of the two-component WF expansion, it is required to calculate the following MEs of the kinetic energy operator:

$$B = \langle U | \sum_{i=1}^{6} \boldsymbol{\sigma}_i \hat{\mathbf{p}}_i | \mathrm{X} \rangle, \tag{3.1}$$

$$H = \langle W | \sum_{i=1}^{6} \boldsymbol{\sigma}_i \hat{\mathbf{p}}_i | \Phi \rangle. \tag{3.2}$$

With regard to the form of multidimensional angle functions U and W it seems easier to calculate MEs of the form

$$B = \left\langle U \left| \sum_{i=1}^{6} \boldsymbol{\sigma}_i \hat{\mathbf{p}}_i \right| \frac{N(\rho)}{\rho^{21/2}} \left(\rho^2 W\right) \right\rangle, \tag{3.3}$$

$$H = \left\langle W \left| \sum_{i=1}^{6} \boldsymbol{\sigma}_i \hat{\mathbf{p}}_i \right| \frac{M(\rho)}{\rho^{21/2}} \left(\rho^2 U\right) \right\rangle. \tag{3.4}$$

Since operator $\vec{\sigma}\vec{p}$ is a single-particle operator, equation (3.3) can be written as a sum over shells [9]:

$$B = \sum_{\omega} \left\langle \psi_{\omega} \left| \boldsymbol{\sigma}\hat{\mathbf{p}} \right| \frac{N(\rho)}{\rho^{21/2}} \xi_{\omega} \right\rangle. \tag{3.5}$$

Similarly, equation (3.4) is written in the form:

$$H = \sum_{\omega} \left\langle \xi_{\omega} \left| \boldsymbol{\sigma}\hat{\mathbf{p}} \right| \frac{M(\rho)}{\rho^{21/2}} \psi_{\omega} \right\rangle. \tag{3.6}$$

Using the explicit form of momentum operator $\hat{\mathbf{p}} = -i\hbar \frac{\partial}{\partial \mathbf{r}}$ and the definition of collective variable ρ, we can write Eqs. (3.5) and (3.6) as

$$B = \sum_{\omega} \left[\frac{N(\rho)}{\rho^{\frac{3A+2K-1}{2}}} \langle \psi_{\omega} | \boldsymbol{\sigma}\hat{\mathbf{p}} | \xi_{\omega} \rangle - \left(\frac{i\hbar}{\rho^{\frac{3A+2K-1}{2}}} \left\{ \frac{d}{d\rho} - \frac{3A+2K-1}{2\rho} \right\} N(\rho) \right) \left\langle \psi_{\omega} \left| \boldsymbol{\sigma} \frac{\mathbf{r}}{r} \right| \xi_{\omega} \right\rangle \right], \tag{3.7}$$

$$H = \sum_{\omega} \left[\frac{M(\rho)}{\rho^{\frac{3A+2K-1}{2}}} \langle \xi_{\omega} | \boldsymbol{\sigma}\hat{\mathbf{p}} | \psi_{\omega} \rangle - \left(\frac{i\hbar}{\rho^{\frac{3A+2K-1}{2}}} \left\{ \frac{d}{d\rho} - \frac{3A+2K-1}{2\rho} \right\} M(\rho) \right) \left\langle \xi_{\omega} \left| \boldsymbol{\sigma} \frac{\mathbf{r}}{r} \right| \psi_{\omega} \right\rangle \right]. \tag{3.8}$$

To calculate single-particle MEs in the last expressions, it is possible to use the following relation between spinors [11]:

$$\Omega_{jl'm} = i^{l-l'}\left(\boldsymbol{\sigma}\frac{\mathbf{r}}{r}\right)\Omega_{jlm}, \tag{3.9}$$

$$\begin{gathered}\vec{\sigma}\vec{p}\xi_{jl'm} = i^{l-l'}\sqrt{\frac{1}{\Gamma(3+n)}}\boldsymbol{\sigma}\hat{\mathbf{p}}\left(\boldsymbol{\sigma}\frac{\mathbf{r}}{r}\right)(-1)^{\frac{1+l-l'}{2}}L_n^2(-itr)\Omega_{jlm} = \\ = i^{2l-2l'+1}\sqrt{\frac{1}{\Gamma(3+n)}}\left\{\left(\hat{\mathbf{p}}\frac{\mathbf{r}}{r}\right)+i\boldsymbol{\sigma}\left[\hat{\mathbf{p}}\mathbf{r}\right]\right\}\frac{L_n^2(-itr)}{r}\Omega_{jlm}.\end{gathered} \tag{3.10}$$

Using the explicit form of the momentum operator the last expression can be reduced to the form:

$$\begin{gathered}\vec{\sigma}\vec{p}\xi_{jl'm} = i^{2l-2l'+4}\hbar\sqrt{\frac{1}{\Gamma(3+n)}}\left(\nabla\mathbf{r}+\mathbf{r}\nabla+\boldsymbol{\sigma}\hat{\mathbf{l}}\right)\frac{L_n^2(-itr)}{r}\Omega_{jlm} = \\ = -\hbar\sqrt{\frac{1}{\Gamma(3+n)}}\left(3+r\frac{\partial}{\partial r}+\boldsymbol{\sigma}\hat{\mathbf{l}}\right)\frac{L_n^2(-itr)}{r}\Omega_{jlm},\end{gathered} \tag{3.11}$$

where $\hat{\mathbf{l}} = \left[\mathbf{r}\hat{\mathbf{p}}\right]$ is the orbital momentum operator. The eigenvalues of product $\left(\boldsymbol{\sigma}\hat{\mathbf{l}}\right) = 2\mathbf{s}\hat{\mathbf{l}}$ are

$$2\mathbf{s}\hat{\mathbf{l}} = \left(\mathbf{j}^2-\mathbf{l}^2-\mathbf{s}^2\right) = j(j+1)-l(l+1)-\frac{3}{4} = \begin{cases} j-\frac{1}{2}, & l = j-\frac{1}{2}, \\ -j-\frac{3}{2}, & l = j+\frac{1}{2}. \end{cases}$$

Equation (3.11) can be rewritten to obtain

$$\vec{\sigma}\vec{p}\xi_{jl'm} = -\hbar\sqrt{\frac{1}{\Gamma(3+n)}}\frac{1-\kappa_{jl}}{r}\psi_{jlm}, \tag{3.12}$$

where

$$\kappa_{jl} = \begin{cases} -(l+1), & j = l+\frac{1}{2}, \\ l, & j = l-\frac{1}{2}. \end{cases} \tag{3.13}$$

It should be noted that in equation (3.12) we omit the term obtained during the Laguerre polynomial differentiation, because this term is proportional to the Laguerre polynomial, which has one unit less power, and it gives zero during integration.
Similar approach allows obtaining

$$\vec{\sigma}\vec{p}\psi_{jlm} = \hbar\sqrt{\frac{1}{\Gamma(3+n)}}\frac{1+\kappa_{jl}}{r}\xi_{jl'm}, \tag{3.14}$$

matrix elements $\left\langle \xi_\omega \middle| \boldsymbol{\sigma}\frac{\mathbf{r}}{r} \middle| \psi_\omega \right\rangle$ and $\left\langle \psi_\omega \middle| \boldsymbol{\sigma}\frac{\mathbf{r}}{r} \middle| \xi_\omega \right\rangle$ are trivially calculated using Eq.(3.9). The basis functions and properties of Pauli matrices are defined as

$$i\left\langle \xi_\omega \middle| \boldsymbol{\sigma}\frac{\mathbf{r}}{r} \middle| \psi_\omega \right\rangle = -\left\langle \xi_\omega \middle| \xi_\omega \right\rangle = -1, \tag{3.15}$$

$$i\left\langle \psi_\omega \left| \boldsymbol{\sigma}\frac{\mathbf{r}}{r} \right| \xi_\omega \right\rangle = \left\langle \psi_\omega | \psi_\omega \right\rangle = 1. \tag{3.16}$$

Hence, equations(3.7) and (3.8) can be rewritten to obtain

$$B = -\frac{\hbar}{\rho^{\frac{3A+2K-1}{2}}} \sum_\omega \left[\frac{dN(\rho)}{d\rho} - \frac{3A+2K-1}{2\rho} N(\rho) + \left\langle \psi_\omega \left| \frac{1-\kappa_\omega}{r} \right| \psi_\omega \right\rangle N(\rho) \right], \tag{3.17}$$

$$H = \frac{\hbar}{\rho^{\frac{3A+2K-1}{2}}} \sum_\omega \left[\frac{dM(\rho)}{d\rho} - \frac{3A+2K-1}{2\rho} M(\rho) - \left\langle \xi_\omega \left| \frac{1-\kappa_\omega}{r} \right| \xi_\omega \right\rangle M(\rho) \right]. \tag{3.18}$$

With regard to the expression obtained for matrix element in the previous section,

$$\left\langle \psi_\omega^* \left| \frac{1}{r} \right| \psi_\omega \right\rangle = \left\langle \xi_\omega^* \left| \frac{1}{r} \right| \xi_\omega \right\rangle = \frac{(3A+2K-1)}{2\Gamma(n_\omega)} \frac{1}{\rho} \tag{3.19}$$

for equations (3.17) and (3.18) one can write

$$B = -\frac{\hbar}{\rho^{\frac{3A+2K-1}{2}}} \sum_\omega \left[\frac{dN(\rho)}{d\rho} - \frac{\kappa_\omega}{\rho} \left(\frac{3A+2K-1}{2\Gamma(n_\omega)} \right) N(\rho) \right], \tag{3.20}$$

$$H = \frac{\hbar}{\rho^{\frac{3A+2K-1}{2}}} \sum_\omega \left[\frac{dM(\rho)}{d\rho} - \frac{2+\kappa_\omega}{\rho} \left(\frac{3A+2K-1}{2\Gamma(n_\omega)} \right) M(\rho) \right]. \tag{3.21}$$

Table 1 gives values of coefficient κ for several states.

Table 1. The values of coefficient κ for various states.

ω	$1s_{1/2}$	$2s_{1/2}$	$1p_{1/2}$	$1p_{3/2}$
κ_ω	-1	-1	1	-2

For C-like ions with electron configuration $\left(1s^2 2s^2 1p_{1/2}^2\right)$ the summation over all shells gives us

$$B = -\frac{2\hbar}{\rho^{21/2}} \frac{21}{2\rho} N(\rho) - \frac{6\hbar}{\rho^{21/2}} \frac{dN(\rho)}{d\rho}, \tag{3.22}$$

$$H = -\frac{10\hbar}{\rho^{21/2}} \frac{21}{2\rho} M(\rho) + \frac{6\hbar}{\rho^{21/2}} \frac{dM(\rho)}{d\rho}. \tag{3.23}$$

4. Matrix elements of the operator of electron-electron interaction

Direct expansion of determinants shows that MEs of electron-electron interaction for the upper WF component of C-like ions have the form:

$$\left\langle U \left| \sum_{\substack{i=1\\i<j}}^{6} \frac{1}{|\mathbf{r}_i - \mathbf{r}_j|} \right| U \right\rangle = \sum_{\substack{i=1\\i<j}}^{6} \left\langle \psi_i^*(1)\psi_j^*(2) \left| \frac{1}{|\mathbf{r}_i - \mathbf{r}_j|} \right| \psi_i(1)\psi_j(2) - \psi_i(2)\psi_j(1) \right\rangle. \tag{4.1}$$

Thus, it is necessary to consider the summation over six shells.

This paper describes in details the calculation of only the first term in the sum (4.1). The rest of terms are calculated in the same way and here we give only final results of such calculations for these terms.

The orthogonal nature of the basis functions corresponding to electrons of shell $1s$ is due to the orthogonal nature of spin-functions and, hence, the exchange part of the matrix element of interest equals zero.

With regard to normalization, the direct-interaction part of the first term has the form:

$$\begin{aligned}&\langle 1s_-(1)1s_+(2)\rangle = \\ &= \frac{\Gamma(22)}{2\pi\rho^{21}} \int_{-\infty}^{+\infty} \frac{e^{-it\rho}}{(-it)^{4+12}} dt \int e^{it(r_1+r_2)} \psi_1^*(1)\psi_2^*(2) \frac{1}{|\vec{r}_1 - \vec{r}_2|} \psi_1(1)\psi_2(2) d\vec{r}_1 d\vec{r}_2.\end{aligned} \tag{4.2}$$

If we use multipole expansion

$$\frac{1}{|\mathbf{r}_i - \mathbf{r}_j|} = \sum_{qv} \frac{4\pi}{2q+1} Y_{qv}^*(\Omega_1) Y_{qv}(\Omega_2) \frac{r_<^q}{r_>^{q+1}}, \tag{4.3}$$

then for the angular part of integral (4.2) one can write

$$\int \Omega^*_{\frac{1}{2}0-}(1)\Omega^*_{\frac{1}{2}0-}(2) \sum_{qv} \frac{4\pi}{2q+1} Y_{qv}^*(1) Y_{qv}(2) \Omega_{\frac{1}{2}0-}(1)\Omega_{\frac{1}{2}0-}(2) d\Omega_1 d\Omega_2. \tag{4.4}$$

In view of the orthonormalized nature of spinors it becomes evident that only terms with $q = 0$ differ from zero and integral taken over the angular part of (4.2) equals 1.

So, the expression for matrix element of $\langle 1s_-(1)1s_+(2)\rangle$ can be transformed to obtain

$$\langle 1s_-(1)1s_+(2)\rangle = \frac{1}{2\pi\rho^{21}} \frac{\Gamma(22)}{(\Gamma(3))^2} \int \frac{e^{-it\rho}}{(-it)^{16}} dt \int r_1^2 dr_1 e^{it(r_1+r_2)} \left[\int_0^{r_1} \frac{r_2^2 dr_2}{r_1} + \int_{r_1}^{\infty} \frac{r_2^2 dr_2}{r_2} \right]. \tag{4.5}$$

The integral over coordinates r_1 and r_2 has the following value:

$$\int r_1^2 dr_1 e^{it(r_1+r_2)} \left[\int_0^{r_1} \frac{r_2^2 dr_2}{r_1} + \int_{r_1}^{\infty} \frac{r_2^2 dr_2}{r_2} \right] = \frac{1}{(-it)^5} \frac{5}{4}. \tag{4.6}$$

As a result, for (4.5) one can write

$$\langle 1s_-(1)1s_+(2)\rangle = \frac{1}{2\pi\rho^{21}}\frac{\Gamma(22)}{(\Gamma(3))^2}\int\frac{e^{-it\rho}}{(-it)^{21}}dt\frac{5}{4} = \frac{\Gamma(22)}{(\Gamma(3))^2\Gamma(21)}\frac{5}{4} = \frac{105}{16}\frac{1}{\rho}. \quad (4.7)$$

Table 2 gives the values of matrix elements for the operator of electron-electron interaction for the basis MAFs of the upper WF component of C-like ion, which have been calculated by the method shown in this section.

Table 2. Matrix elements of the operator of electron-electron interaction for the basis MAFs of the upper WF component.

Numbers of shells i, j	Direct-interaction part	Exchange interaction part	Numbers of shells i, j	Direct-interaction part	Exchange interaction part
1-2 $\langle 1s_-(1)1s_+(2)\rangle$	$\frac{105}{16}\frac{1}{\rho}$	0	2-6 $\langle 1s_+(1)1p_+(2)\rangle$	$\frac{105}{16}\frac{1}{\rho}$	$-\frac{7}{16}\frac{1}{\rho}$
1-3 $\langle 1s_-(1)2s_-(2)\rangle$	$\frac{21}{4}\frac{1}{\rho}$	$-\frac{21}{16}\frac{1}{\rho}$	3-4 $\langle 2s_-(1)2s_+(2)\rangle$	$\frac{147}{32}\frac{1}{\rho}$	0
1-4 $\langle 1s_-(1)2s_+(2)\rangle$	$\frac{21}{4}\frac{1}{\rho}$	0	3-5 $\langle 2s_-(1)1p_-(2)\rangle$	$\frac{21}{4}\frac{1}{\rho}$	$-\frac{7}{48}\frac{1}{\rho}$
1-5 $\langle 1s_-(1)1p_-(2)\rangle$	$\frac{105}{16}\frac{1}{\rho}$	$-\frac{7}{16}\frac{1}{\rho}$	3-6 $\langle 2s_-(1)1p_+(2)\rangle$	$\frac{21}{4}\frac{1}{\rho}$	$-\frac{7}{24}\frac{1}{\rho}$
1-6 $\langle 1s_-(1)1p_+(2)\rangle$	$\frac{105}{16}\frac{1}{\rho}$	$-\frac{7}{8}\frac{1}{\rho}$	4-5 $\langle 2s_+(1)1p_-(2)\rangle$	$\frac{21}{4}\frac{1}{\rho}$	$-\frac{7}{24}\frac{1}{\rho}$
2-3 $\langle 1s_+(1)2s_-(2)\rangle$	$\frac{21}{4}\frac{1}{\rho}$	0	4-6 $\langle 2s_+(1)1p_+(2)\rangle$	$\frac{21}{4}\frac{1}{\rho}$	$-\frac{7}{48}\frac{1}{\rho}$
2-4 $\langle 1s_+(1)2s_+(2)\rangle$	$\frac{21}{4}\frac{1}{\rho}$	$-\frac{21}{16}\frac{1}{\rho}$	5-6 $\langle 1p_+(1)1p_-(2)\rangle$	$\frac{105}{16}\frac{1}{\rho}$	0
2-5 $\langle 1s_+(1)1p_-(2)\rangle$	$\frac{105}{16}\frac{1}{\rho}$	$-\frac{7}{8}\frac{1}{\rho}$	Σ	$\frac{2555}{32}\frac{1}{\rho}$	

MEs of electron-electron interaction for MAF of the upper WF component have the form

$$\left\langle W\left|\sum_{\substack{i=1\\i<j}}^{6}\frac{1}{|\vec{r}_i-\vec{r}_j|}\right|W\right\rangle = \sum_{\substack{i=1\\i<j}}^{6}\left\langle \xi_i^*(1)\xi_j^*(2)\left|\frac{1}{|\vec{r}_1-\vec{r}_2|}\right|\xi_i(1)\xi_j(2)-\xi_i(2)\xi_j(1)\right\rangle. \quad (4.8)$$

The terms of the sum above for the lower component are integrated similarly to those for MAF of the upper WF component. Table 3 gives the obtained values of MEs.

Table 3. Matrix elements of the operator of electro-electron interaction for the basis MAFs of the lower WF component.

Numbers of shells i, j	Direct-interaction part	Exchange interaction part	Numbers of shells i, j	Direct-interaction part	Exchange interaction part
1-2 $\langle 1s_-(1)1s_+(2)\rangle$	$\frac{105}{16}\frac{1}{\rho}$	0	2-6 $\langle 1s_+(1)1p_+(2)\rangle$	$\frac{105}{16}\frac{1}{\rho}$	$-\frac{7}{16}\frac{1}{\rho}$
1-3 $\langle 1s_-(1)2s_-(2)\rangle$	$\frac{21}{4}\frac{1}{\rho}$	$-\frac{21}{16}\frac{1}{\rho}$	3-4 $\langle 2s_-(1)2s_+(2)\rangle$	$\frac{147}{32}\frac{1}{\rho}$	0
1-4 $\langle 1s_-(1)2s_+(2)\rangle$	$\frac{21}{4}\frac{1}{\rho}$	0	3-5 $\langle 2s_-(1)1p_-(2)\rangle$	$\frac{21}{4}\frac{1}{\rho}$	$-\frac{7}{48}\frac{1}{\rho}$
1-5 $\langle 1s_-(1)1p_-(2)\rangle$	$\frac{105}{16}\frac{1}{\rho}$	$-\frac{7}{16}\frac{1}{\rho}$	3-6 $\langle 2s_-(1)1p_+(2)\rangle$	$\frac{21}{4}\frac{1}{\rho}$	$-\frac{7}{24}\frac{1}{\rho}$
1-6 $\langle 1s_-(1)1p_+(2)\rangle$	$\frac{105}{16}\frac{1}{\rho}$	$-\frac{7}{8}\frac{1}{\rho}$	4-5 $\langle 2s_+(1)1p_-(2)\rangle$	$\frac{21}{4}\frac{1}{\rho}$	$-\frac{7}{24}\frac{1}{\rho}$
2-3 $\langle 1s_+(1)2s_-(2)\rangle$	$\frac{21}{4}\frac{1}{\rho}$	0	4-6 $\langle 2s_+(1)1p_+(2)\rangle$	$\frac{21}{4}\frac{1}{\rho}$	$-\frac{7}{48}\frac{1}{\rho}$
2-4 $\langle 1s_+(1)2s_+(2)\rangle$	$\frac{21}{4}\frac{1}{\rho}$	$-\frac{21}{16}\frac{1}{\rho}$	5-6 $\langle 1p_+(1)1p_-(2)\rangle$	$\frac{105}{16}\frac{1}{\rho}$	0
2-5 $\langle 1s_+(1)1p_-(2)\rangle$	$\frac{105}{16}\frac{1}{\rho}$	$-\frac{7}{8}\frac{1}{\rho}$	Σ	$\frac{2555}{32}\frac{1}{\rho}$	

5. The solution to the Dirac equation for C-like ions

With regard to the obtained ME values, the system of differential equations (1.10) and (1.11) describing C-like ions in state $\left(1s^2\, 2s^2 1p_{1/2}^2\right)$ has the form:

$$M' - \frac{35}{2\rho}M - \frac{1}{6}\left(E + 12 + \frac{\alpha\left(63Z - 2555/32\right)}{\rho}\right)N = 0, \tag{5.1}$$

$$N' + \frac{7}{2\rho}N + \frac{1}{6}\left(E + \frac{\alpha\left(63Z - 2555/32\right)}{\rho}\right)M = 0. \tag{5.2}$$

The solution to the equation system above is given in [1] and, in general case, it may be written as

$$M(r) = \sqrt{E + 2A}\cdot e^{-\lambda\rho/A}\left(\frac{2\lambda\rho}{A}\right)^{\gamma-1}\times$$
$$\times\left(A_1F\left(-n_r, 2\gamma-1-(a_1-a_2), \frac{2\lambda\rho}{A}\right) + B_1F\left(-n_r+1, 2\gamma-1-(a_1-a_2), \frac{2\lambda\rho}{A}\right)\right), \tag{5.3}$$

$$N(r) = -\sqrt{-E}e^{-\lambda\rho/A}\left(\frac{2\lambda\rho}{A}\right)^{\gamma-1} \times \\ \times\left(A_1 F\left(-n_r, 2\gamma-1-(a_1-a_2), \frac{2\lambda\rho}{A}\right) - B_1 F\left(-n_r+1, 2\gamma-1-(a_1-a_2), \frac{2\lambda\rho}{A}\right)\right). \tag{5.4}$$

Normalization constants A_1 and B_1 of degenerated hypergeometric functions can be found from the requirement of WF normalization to 1.

In the expressions above, A is the number of electrons; $\lambda = \sqrt{-E(E+2A)}$; $\gamma = 1 + \frac{a_1 - a_2}{2} + \left(\frac{(a_1+a_2)^2}{4} - \frac{(c_1 Z - d_1)(c_2 Z - d_2)\alpha^2}{A^2}\right)^{1/2}$; a_1, a_2, c_1, c_2, d_1, and d_2 are coefficients, which have been obtained in the calculation of MEs of the operators of kinetic energy, electron-nucleus and electron-electron interactions, respectively; n_r is radial quantum number.

Calculations of MEs of the electron-nucleus and electro-electron interactions demonstrate that

$$c_1 = c_2 = c,\ d_1 = d_2 = d \tag{5.5}$$

for C-like ions. Owing to the requirement of WF finiteness at infinity (this requirement is met, if n_r takes integral positive values) we obtain the following expression for binding energy:

$$E = A\left(\left(1 + \frac{\alpha^2(cZ-d)^2}{\left(An_r + \sqrt{A^2(a_1+a_2)^2/4 - \alpha^2(cZ-d)^2}\right)^2}\right)^{-1/2} - 1\right). \tag{5.6}$$

Table 4 gives the binding energy values for C-like ions of heavy elements with various values of quantum number n_r. The first four columns of this table ($n_r = 0,...,3$) show spurious levels occurred because of the Coulomb interaction taken into account. The binding energy of the ground state of C-like ions is given in the fifth column ($n_r = 4$). The binding energy of the excited state is in the sixth column ($n_r = 5$).

Table 4. The binding energy (KeV) of C-like ions in heavy elements.

n_r	0	1	2	3	4	5
Z						
92	-768.293	-652.991	-559.479	-483.163	-420.429	-368.462
93	-788.255	-670.301	-574.504	-496.241	-431.857	-378.495
94	-808.616	-687.975	-589.855	-509.609	-443.542	-388.754
95	-829.386	-706.024	-605.543	-523.276	-455.492	-399.247
96	-850.576	-724.458	-621.577	-537.252	-467.714	-409.981
97	-872.199	-743.290	-637.970	-551.548	-480.220	-420.964
98	-894.268	-762.532	-654.733	-566.173	-493.017	-432.205
99	-916.797	-782.197	-671.879	-581.139	-506.117	-443.713
100	-939.800	-802.301	-689.421	-596.459	-519.530	-455.497
101	-963.292	-822.857	-707.373	-612.146	-533.268	-467.570

If quantum number n_r is defined as a sum of quantum numbers of occupied shells, $n_r = 4$ is obtained for the ground state of C-like ions. Such approach is justified by the agreement achieved between the calculated binding energy values and data from other literary sources (Table 5).

Table 5 gives the binding energy values of uranium ions with ionization degree from 86 (C-like ion) to 90 (He-like ion) calculated using various methods.

Table 5. The binding energy (KeV) of various uranium ions.

Electron configuration	U^{90+} ($1s^2$)	U^{90+} ($1s^1 2s^1$)	U^{89+} ($1s^2 2s^1$)	U^{88+} ($1s^2 2s^2$)	U^{86+} ($1s^2 2s^2 2p_{1/2}^2$)
MACF	-262.45	-165.49	-281.24	-336.52	-420.42
[12]	-264.14	-166.15	-298.22	-332.29	-400.31
ATOM [13]	-256.25	-159.26	-288.34	-319.93	-380.44
GRASP [14, 15]	-262.67	-162.61	-295.58	-328,08	-

Note that there is a satisfactory agreement between the binding energy values obtained with MACF method and some other methods.

6. Electron density

The analytical solutions to the multiple-electron Dirac equation given above are used to calculate the electron density of a C-like uranium ion.

The electron density of A-electron ions is calculated from expression

$$n(\mathbf{r}) = \int \Psi^{\dagger} \sum_{i=1}^{A} \delta(\mathbf{r} - \mathbf{r}_i) \Psi d\mathbf{r}_1 d\mathbf{r}_2 . \tag{6.1}$$

With zero approximation by the relativistic MACF method the electron density can be represented as a sum density of electrons of the upper and lower components of the bi-spinor wave function:

$$n(\mathbf{r}) = \rho_1(\mathbf{r}) + \rho_2(\mathbf{r}), \tag{6.2}$$

where expressions for the terms corresponding to the upper and lower components of the bi-spinor wave function look like

$$\rho_1(r) = \frac{\Gamma(3A+2K)}{2\pi} \int \frac{M^2(\rho)}{\rho^{3A+2K-1}} \int_{-\infty}^{\infty} \frac{e^{-it(\rho-r)}}{(-it)^{3A+2K-3}} \sum_{\omega} \psi_{\omega}^2(-it\vec{r}) dt d\rho , \tag{6.3}$$

$$\rho_2(r) = \frac{\Gamma(3A+2K)}{2\pi} \int \frac{N^2(\rho)}{\rho^{3A+2K-1}} \int_{-\infty}^{\infty} \frac{e^{-it(\rho-r)}}{(-it)^{3A+2K-3}} \sum_{\omega} \xi_{\omega}^2(-it\vec{r}) dt d\rho . \tag{6.4}$$

For C-like ions (A=6, K=4), we obtain with regard to the basis functions that

$$n(\mathbf{r}) = \frac{\Gamma(22)}{4\pi\Gamma(19)} \int_{r}^{\infty} \left(M^2(\rho) + N^2(\rho)\right) \frac{(\rho - r)^{16}}{\rho^{21}} \left[5\rho^2 - 46r\rho + 143r^2\right] d\rho . \tag{6.5}$$

Figure 1 illustrates results of calculations using MACF method for the electron density of a C-like uranium ion in its ground state.

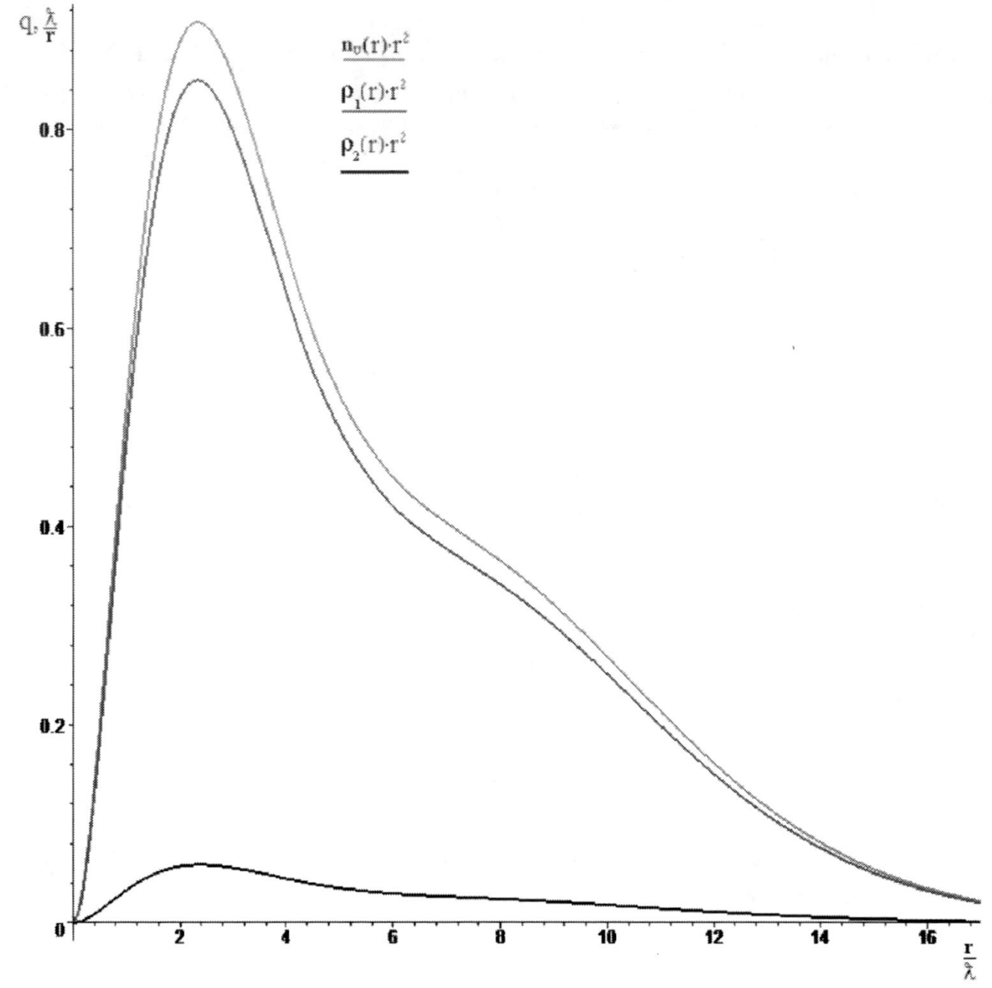

Figure 1. The electron density of C-like ions of uranium (ƛ is Compton constant, q is electron linear density)

To check the results obtained for validity, the total electron charge of a C-like ion has been calculated:

$$4\pi\int_0^\infty n(r)r^2dr = 6. \tag{6.6}$$

Conclusion

The closed-form analytic expressions for the wave function and binding energy of state 0^+ of C-like ions of elements with Z=92 to Z=101 have been found with minimum approximation by the relativistic MACF method.

The method allows calculating the spectrum and WF levels differing in parity and moments in the form of bands with the given quantum numbers. There are limitations in the calculation of transitions between different levels of photoionization, ionization by an electron collision, etc., because at present time we have the wave functions and binding energies calculated only for several states.

The studied properties of C-like and other type ions of transuranium elements allow specifying the parameters of high-temperature plasma (several KeVs) of transuranium elements. Ions with few electrons prevail in such plasmas.

The matrix elements calculated will be used to carry out calculations for other ions of heavy elements.

The paper illustrates the presence of spurious states for the Dirac operator with Coulomb interaction, which occur owing to its unboundedness.

The developed method opens prospects for studying the effect of mixed configurations on the calculation of transition probabilities and also allows considering the process of exciting a nucleus due to electron transitions.

References

[1] Sadovoy A A and Ulyanov A S 2007 A new method to calculate properties of He-like ions of transuranium elements *VANT Ser.: Theor. and Appl. Phys.* **2–3** 58

[2] Sadovoy A A and Ulyanov A S 2008 The electron density of highly ionized ions of transuranium elements *VI Sci. and Technol. Conf. «Youth in Science»* (Sarov, Russia, 2007).

[3] Neznamov V P, Sadovoy A A and Ulyanov A S Calculation of matrix elements in the Foldy-Wouthuysen representation 2009 *VANT Ser.: Theor. and Appl. Phys.* **3** 18

[4] Sadovoy A A and Ulyanov A S 2010 Studying the properties of He- and C-like ions of heavy elements using the analytical solution to the Dirac equation *15 N.Novgorod session of young scientists. Proc. of young scientists in natural-science disciplines* (N.Novgorod, Russia, 2010)

[5] Sadovoy A A and Ulyanov A S 2011 Some features of the few-electron Dirac equation and description of ions with high ionization degree. *16 N.Novgorod session of young scientists. Proc. of young scientists in natural-science disciplines* (N.Novgorod, Russia, 2011)

[6] Sadovoy A A 1994 *The Multidimensional Angular* Function *Methods in Theoretical and Applied Physics* (Arzamas-16: VNIIEF Publishers)

[7] Sadovoy A A 2000 The Multidimensional Angular Coulomb Function Method in Atomic and Molecular Physics *Few-Body Systems Suppl.* **12** 66

[8] Sadovoy A A and Ulyanov A S 2008 *New methods to solve problems of few bodies in atomic, molecular and nuclear physics: A book of scientific papers* ed A A Sadovoy (Sarov: FSUE «RFNC-VNIIEF»)

[9] Bete G 1965 *Quantum mechanics* (Moscow: Physmathgiz)

[10] Gradshtein I S and Ryshik I M 1963 *Tables of integrals, sums, series and products* (Moscow: Physmathgiz)

[11] Berestetsky V V, Lifshits E M and Pitayevsky A P 1968 *Relativistic quantum theory P.1* (Moscow: Science)

[12] Desclaux J P Relativistic 1973 Dirac-Fock Expectation Values for Atoms with Z=1 to Z=120 *Atomic Data and Nuclear Data Tables* **12 N. 4** 311

[13] Philippov A V, Povyshev V M, Sadovoy A A, Shevelko V P, Shirkov G D, Vasina E G and Vatulin V V 2002 Electron-impact ionization cross sections of Ti, Kr, Sn, Ta, U atoms and their ions in the electron energy range from the threshold up to 200 KeV. Part 2. *JINR* (*Preprint* E9-2002-5)

[14] Grant I P, McKentzie B J, Norrington P H, Mayers D F and Pyper N C 1980 An atomic multiconfiguration Dirac-Fock package *Comp. Phys. Comm.* **21** 207

[15] Dyall K G, Grant I P, Johnson C T,Parpia F A and Plummer E P 1989 GRASP: A general-purpose relativistic atomic structure program *Comp. Phys. Comm.* **55** 425

Journal of Physics: Conference Series **529** (2014) 012029 doi:10.1088/1742-6596/529/1/012029

Alpha Cluster Structures and Isoscalar Monopole Excitations in Light Nuclei

T Yamada

Lab. of Physics, Kanto Gakuin University, Yokohama 236-8501, Japan

E-mail: yamada@kanto-gakuin.ac.jp

Abstract. Isoscalar monopole excitations to cluster states in light nuclei are in general strong as to be comparable with the single particle strength. Their experimental strengths share about 20 % of the sum rule value in the case of ^{4}He, ^{11}B, ^{12}C, and ^{16}O etc. We discuss the isoscalar monopole strength function in ^{16}O, which is typical in light nuclei, is discussed up to $E_x \sim 40$ MeV. It is found that 1) two different types of monopole excitations exist in ^{16}O; one is the monopole excitation to cluster states which is dominant in the lower energy part, and the other is the monopole excitation of the mean-field type such as one-particle one-hole ($1p1h$) which is attributed mainly to the higher energy part, and 2) this character of the monopole excitations originates from the fact that the ground state of ^{16}O with the dominant doubly closed shell structure has a duality of the mean-field-type as well as alpha-clustering character.

1. Introduction

Isoscalar monopole excitation is related to a density fluctuation in nuclei. The typical example is the isoscalar giant monopole resonance (IS MR) observed as a single bump in medium and heavy nuclei, which exhausts almost 100 % of the energy weighted sum rule value. It is interesting to study what happens for the IS MR in lighter nuclei. When the nuclear masses decrease from medium nuclei to light nuclei, the isoscalar monopole strengths are in general fragmented. In this paper, we concentrate on the case of ^{16}O as a typical example. The histogram in Fig. 1 shows the experimental isoscalar monopole strength function in ^{16}O [1] (the date below $E_x \sim 10$ MeV is absent due to the experimental condition). The monopole strengths split over wide energy region, and one can see discrete peaks in low energy ($E_x \lesssim 16$ MeV) and gross three-bump structure in higher energy region ($16 \lesssim E_x \lesssim 40$ MeV). The experimental data is compared with the relativistic RPA (RRPA) calculation [2]. In order to match their calculation to the experimental centroid, the calculated strength function was shifted down in energy by 4.2 MeV and furthermore they normalized it by multiplying the RRPA curve by a factor of 0.25 [1], although their calculation failed to reproduce the 0^+ states found in the low energy region. In the nonrelativistic RPA and second-order RPA (SRPA) calculations for ^{16}O [3, 4] a significant discrepancy is also revealed as compared with the experimental data, in particular, in the low energy region, although the gross structures at the higher energy region in the RPA calculations are in rather good agreement with the data.

Recently the structure study of ^{16}O has made a great advance up to $E_x \simeq 16$ MeV around the 4α disintegration threshold. The six lowest 0^+ states of ^{16}O, up to $E_x \simeq 16$ MeV, including the ground state, have for first time been reproduced very well within the framework of the 4α

Journal of Physics: Conference Series **529** (2014) 012029 doi:10.1088/1742-6596/529/1/012029

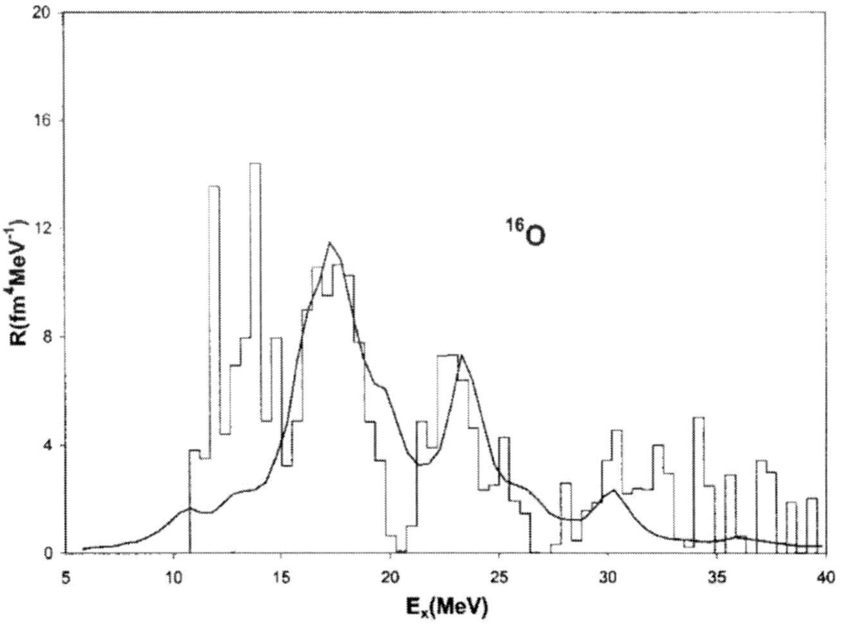

Figure 1. Experimental isoscalar monopole strength function of ^{16}O [1] is shown by the histogram. The real line is the calculated result by the relativistic RPA calculation [2] (see text). The date below $E_x \sim 10$ MeV is absent due to the experimental condition.

orthogonality condition model (OCM) [5] (see Fig. 2) combined with the aussian expansion method (EM) [6]. The OCM is a semi-microscopic cluster model, which is an approximation of R M (resonating group method) and is extensively described in Ref. [7]. Many successful applications of OCM to ordinary nuclei as well as hypernuclei are reported in Refs. [8, 9]. The six 0^+ states have the following characteristic structures [5] 1) the ground state (0_1^+) has dominantly a doubly-closed-shell structure, 2) the 0_2^+ state at $E_x = 6.05$ MeV and the 0_3^+ state at $E_x = 12.05$ MeV have mainly α+^{12}C structures where the α-particle orbits around the ^{12}C(0_1^+) core in an S-wave and around the ^{12}C(2_1^+) core in a D-wave, respectively, 3) the 0_4^+ ($E_x = 13.6$ MeV) and 0_5^+ ($E_x = 14.1$ MeV) states mainly have α+^{12}C(0_1^+) structure with higher nodal behavior and α+^{12}C(1^-) structure, respectively, and 4) the 0_6^+ state at 15.1 MeV is a strong candidate of the 4α condensate [11], $(0S)_\alpha^4$, with the probability of 61 %. The monopole strengths and decay widths are well reproduced, together with the r.m.s. radius of the ground state (see Table 1). We see that the monopole strengths in Table 1 are comparable with the single particle strength ($\sim$4 fm^2) [12].

The purpose of the present paper is to study whether the 4α OCM can reproduce the experimental isoscalar monopole strength function in the low energy region up to $E_x \simeq 16$ MeV in ^{16}O. It is noted that this energy region is difficult to be treated in the mean-field theory as mentioned above.

2. Isoscalar monopole strength function of ^{16}O with 4α OCM

Figure 3 shows the calculated isoscalar monopole strength function of ^{16}O with the 4α OCM, where we use the calculated monopole matrix elements and the calculated decay widths for the six 0^+ states up to $E_x \simeq 16$ MeV obtained by the 4α OCM calculation, also the

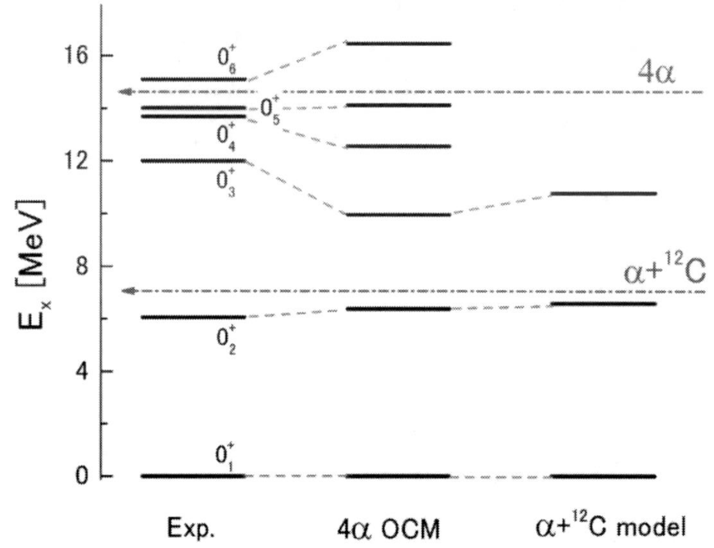

Figure 2. (Color online) Comparison of energy spectra among experiment, the 4α OCM calculation [5], and the α+^{12}C model calculation [10], where the α+^{12}C and 4α thresholds are shown. Experimental data are taken from Ref. [14] and from Ref. [15] for the 0_4^+ state.

experimental excitation energies for the six 0^+ states are employed [13]. We can see a rather good correspondence with the experimental data. The fine structures in the calculated strength function, i.e. one peak at $E_x = 12.1$ MeV (corresponding to the 0_3^+ state), one shoulder-like peak at $E_x = 13.8$ MeV (0_4^+), two peaks at $E_x = 14.1$ MeV (0_5^+) and 15.1 MeV (0_6^+), are well reproduced. It is noted that the date below $E_x \sim 10$ MeV in Figs. 1 and 3 is absent due to the experimental condition. Thus, the signal at 6.05 MeV is not seen by the experiment.

From the present results, one concludes that there exist two features of the isoscalar monopole excitations of ^{16}O, i.e. the monopole excitation to cluster states dominates the low energy region ($E_x \lesssim 16$ MeV), sharing about 20 % of the EWSR, while that to the $1p1h$-type states looks likely to be predominant at the higher energy region.

3. Why are alpha cluster states excited by the isoscalar monopole transitions?

It is instructive to discuss the mechanism of why the five α cluster states (0_2^+, 0_3^+, 0_4^+, 0_5^+, and 0_6^+) of ^{16}O are excited relatively strongly from the shell-model-like ground state by the isoscalar monopole transition. First we discuss a dual nature in the ground state wave function of ^{16}O , and then demonstrate an interesting role of the monopole operator.

The wave function of the ^{16}O ground state has the dominant $(0s)^4(0p)^{12}$ configuration, corresponding to the SU(3) $(\lambda, \mu) = (0, 0)$ wave function. According to the Bayman-Bohr theorem [17], this doubly closed shell model wave function is mathematically equivalent to a single α+^{12}C cluster model wave function as well as a single 4α cluster wave function with the

Table 1. Excitation energies (E_x), charge r.m.s. radii (R_c), E0 transition matrix elements [M(E0)], and particle decay widths (Γ) of the 0^+ states in ^{16}O obtained by the 4α OCM calculation [5, 13], together with the experimental data [14, 15]. They are given in the unit of MeV, fm, fm^2, and MeV, respectively. The experimental monopole matrix elements are obtained by the ^{16}O(e,e') reaction [14]. $P^{\rm e.w.}$ represents the percentage of the energy weight strength to the isoscalar-monopole energy weighted sum rule value. The finite size effects of α particle and ^{12}C are taken into account in estimating R_c with the 4α OCM (see Ref. [16] for details).

	4α OCM				Experiment				
	E_x	R_c	M(E0)	Γ	E_x	R_c	M(E0)	$P^{\rm e.w.}$	Γ
0_1^+	0.00	2.7			0.00	2.70			
0_2^+	6.37	3.0	3.9		6.05		3.55 ± 0.21	3.5%	
0_3^+	9.96	3.1	2.4		12.05		4.03 ± 0.09	8.9%	
0_4^+	12.56	4.0	2.4	0.60	13.60		no data		0.6
0_5^+	14.12	3.1	2.6	0.20	14.01		3.3 ± 0.7	6.9%	0.185
0_6^+	16.45	5.6	1.0	0.14	15.10		no data		0.166

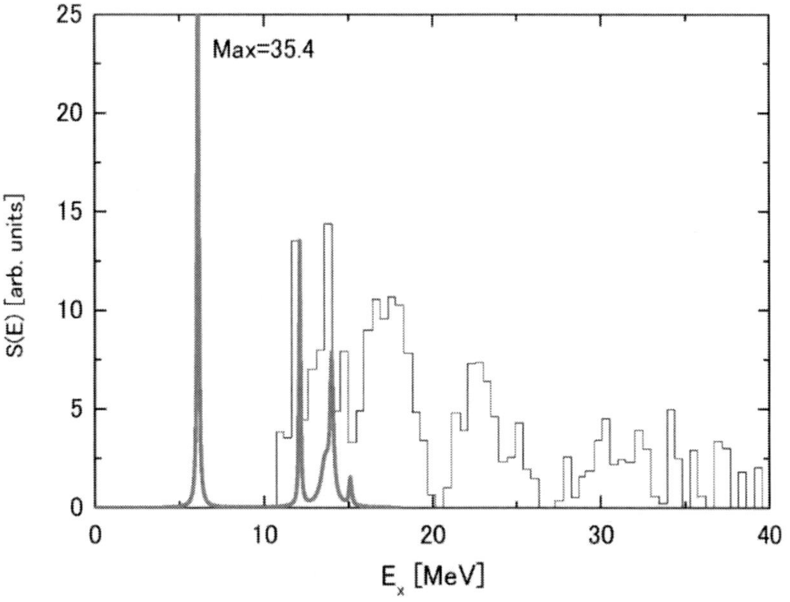

Figure 3. (Color online) Calculated isoscalar monopole strength functions of ^{16}O with the 4α OCM (bold line) [13] and experimental data (thin line see Fig. 1 [1]). The date below $E_x \sim 10$ MeV is absent due to the experimental condition. Thus, the signal at 6.05 MeV is not seen by the experiment.

total harmonic oscillator quanta $Q = 12$ [12, 13],

$$\det|(0s)^4(0p)^{12}| = N_0 \times \mathcal{A}\left\{\left[u_{40}(\boldsymbol{\xi}_3, 3\nu)\phi_{L=0}(^{12}\mathrm{C})\right]_{J=0}\phi(\alpha)\right\}\phi_{\rm cm}(\boldsymbol{R}_{\rm cm}), \tag{1}$$

$$= N_2 \times \mathcal{A}\left\{\left[u_{42}(\boldsymbol{\xi}_3, 3\nu)\phi_{L=2}(^{12}\mathrm{C})\right]_{J=0}\phi(\alpha)\right\}\phi_{\rm cm}(\boldsymbol{R}_{\rm cm}), \tag{2}$$

$$= N_0 \times \mathcal{A}\Big\{\left[u_{40}(\boldsymbol{\xi}_3, 3\nu)\left[u_{40}(\boldsymbol{\xi}_2, \frac{8}{3}\nu)u_{40}(\boldsymbol{\xi}_1, 2\nu)\right]_{L=0}\right]_{J=0}$$

$$\times\phi(\alpha_1)\phi(\alpha_2)\phi(\alpha_3)\phi(\alpha_4) \quad \phi_{\rm cm}(\boldsymbol{R}_{\rm cm}), \tag{3}$$

where $\phi_{\rm cm}$ denotes the wave function of the center-of-mass motion of ^{16}O, and $\boldsymbol{\xi}_k$ $(k = 1 \sim 3)$ stand for the Jacobi coordinates between the clusters. $\phi(\alpha)$ and $\phi_L(^{12}\text{C})$ represent, respectively, the internal wave function of the α cluster and that of ^{12}C with the angular momentum of L. $N_L(N_0)$ is the normalization factor. The relative wave function between the α and ^{12}C clusters in Eqs. (1) and (2) is described by the harmonic oscillator wave function $u_{QL}(\boldsymbol{\xi}, \beta)$ with $Q = 4$. Eqs. (1) and (2) mean that the doubly closed shell model wave function has an α+^{12}C cluster degree of freedom. In addition, Eq. (3) demonstrates that the doubly closed shell model wave function also possesses a 4α cluster degree of freedom. Thus the ground state of ^{16}O has the mean-field of degree of freedom as well as the cluster degree of freedom. We call this the dual nature.

On the other hand, the isoscalar monopole operator of ^{16}O can be decomposed into the internal part (for the constituent clusters) and relative part (with respect to the clusters) as follows [12, 13]

$$\mathcal{O}^{\rm IS}_{\rm E0}(^{16}\text{O}) = \sum_{i=1}^{16} (\boldsymbol{r}_i - \boldsymbol{R}_{\rm cm})^2 = \mathcal{O}^{\rm IS}_{\rm E0}(\alpha) + \mathcal{O}^{\rm IS}_{\rm E0}(^{12}\text{C}) + 3\boldsymbol{\xi}_3^2 \tag{4}$$

$$= \sum_{k=1}^{4} \mathcal{O}^{\rm IS}_{\rm E0}(\alpha_k) + \sum_{k=1}^{4} 4(\boldsymbol{R}_{\alpha_k} - \boldsymbol{R}_{\rm cm})^2, \tag{5}$$

where $\boldsymbol{R}_k$ denotes the c.o.m coordinate of the k-th α cluster in the 4α system. The reasons why the 0_2^+ and 0_3^+ states of ^{16}O with the $\alpha + {}^{12}$C cluster structure are excited by the monopole transition are given as follows 1) the ground state of ^{16}O is of the SU(3) $(\lambda, \mu) = (0, 0)$ nature, which has the α-clustering degree of freedom as discussed above, 2) the relative part in the monopole operator referring to the α+^{12}C relative motion, $3{\boldsymbol{\xi}_3}^2$, in Eq. (4) can activate the α-cluster degree of freedom in the ground state, and then, 3) the $\alpha + {}^{12}$C states are excited by the monopole operator. It should be reminded that the contribution from the other parts of the monopole operator becomes significantly smaller for the α+^{12}C cluster states, and the α-cluster-type ground state correlation significantly enhances the monopole strength compared with the case of the ^{16}O ground state being the pure SU(3) $(0, 0)$ wave function [12].

On the other hand, the reason why the 0_6^+ state with the 4α-gas-like character is excited by the monopole transition can be also understood from the property of the ground state of ^{16}O. The doubly closed shell-model wave function is mathematically equivalent to the single 4α cluster wave function with $Q = 12$ in Eq. (3). This equation means that the ground state of ^{16}O inherently has a 4α-cluster degree of freedom. The relative part (or second term) of the monopole operator in Eq. (5) can excite the relative motion among the 4α particles. In other words, the monopole operator has an ability to populate democratically 4α particles by $2\hbar\omega$ with respect to the center-of-mass coordinate of ^{16}O. The resultant state, thus, has some amount of the overlap with the 4α-gas-like state or $\alpha + {}^{12}\text{C}(0_2^+)$, i.e. 0_6^+, with the 4α-condensate-like structure [5]. It is noted that the mechanism of the 4α-gas-like state being populated by the monopole transition is similar to that of the Hoyle state (0_2^+) with the 3α-gas-like structure being excited by the monopole transition, although the ground state of ^{12}C has a shell-model-like compact structure with the main configuration of SU(3) $(\lambda, \mu) = (0, 4)$ [12].

As for the 0_5^+ state, its main configuration is α+$^{12}\text{C}(1_1^-)$. According to the Bayman-Bohr theorem [17], the SU(3) $(0, 0)$ state of ^{16}O has no component of the α+$^{12}\text{C}(1_1^-)$ channel. However, the monopole strength to the 0_5^+ state is as large as 3 fm^2 [13]. This is the reason that the 0_5^+ state has small but important components of the α+$^{12}\text{C}(0_1^+, 2_1^+, 0_2^+)$ configurations. Since these three configurations can be excited from the ground state of ^{16}O by the monopole operator as discussed above, their respective contributions are coherently added to provide the relatively large monopole strength to the 0_5^+ state. On the other hand, the situation of the

0_4^+ state is similar to the case of the 0_5^+ state. The 0_4^+ state has also small but non negligible components of the α+^{12}C(0_1^+, 2_1^+, 0_2^+) configurations, which contribute to the monopole strength for the 0_4^+ state.

4. Summary

We have investigated the monopole strength function in the low energy region up to $E_x \simeq 16$ MeV within the framework of 4α OCM. It was found that the fine structures at the low energy region up to $E_x \simeq 16$ MeV in the experimental monopole strength function obtained by the ^{16}O(α, α') experiment is rather satisfactorily reproduced within the 4α OCM framework. On the contrary, mean-field calculations have encountered difficulties to reproduce the fine structures of the monopole strength function at the low energy region as well as the monopole matrix elements for the 0_2^+ ($E_x = 6.05$ MeV), 0_3^+ ($E_x = 12.05$ MeV), and 0_5^+ ($E_x = 14.01$ MeV) states obtained in the ^{16}O(e, e') experiment. Our present results indicate that the isoscalar monopole excitation is useful to search for cluster states in light nuclei. Since the existence of the two different types of the monopole excitations and the duality of the ground state discussed in this paper are general in light nuclei [12, 13], it is highly hoped to study systematically the isoscalar monopole excitations in $4n$ and neutron-rich nuclei as well as hypernuclei.

The author thanks . Funaki, T. Myo, H. Horiuchi, K. Ikeda, . Röpke, P. Schuck, and A. Tohsaki, for useful discussion, comments and collaborations.

References

[1] Lui Y -W, Clark H L, and Youngblood D H, 2001 *Phys. Rev.* C **64** 064308
[2] Ma Z et al., 1997 *Phys. Rev.* C **55** 2385
[3] Gambacurta D et al., 2010 *Phys. Rev.* C **81** 054312
[4] Avez B and Simenel C, 2013 *Eur. Phys. J.* A **49** 76
[5] Funaki Y et al., 2008 *Phys. Rev. Lett.* **101** 082502
[6] Kamimura M, 1988 *Phys. Rev.* A **38** 621; Hiyama E et al., 2003 *Prog. Part. Nucl. Phys.* **51** 223
[7] Saito S, 1968 *Prog. Theor. Phys.* **40** 893; 1969 **41** 705; 1977 *Prog. Theor. Phys. Suppl.* No. 62 11
[8] Ikeda K et al., 1980 *Prog. Theor. Phys. Suppl.* **68**
[9] Hiyama E and Yamada T, 2009 *Prog. Par. Nucl. Phys.* **63** 339
[10] Suzuki Y, 1976 *Prog. Theor. Phys.* **55** 1751; 1976 **56** 111
[11] Yamada T et al., 2012 *Lecture Notes in Physics (Springer-Verlag, Berlin)* **848** 229
[12] Yamada T et al., 2008 *Prog. Thor. Phys.* **120** 1139
[13] Yamada T et al., 2012 *Phys. Rev.* C **85** 034315
[14] Ajzenberg-Selove F, 1986 *Nucl. Phys.* A **460** 1; Tilley D R et al., 1993 *Nucl. Phys.* A **564** 1; Angeli I, 2004 *At. Data Nucl. Data Tables* **87** 185
[15] Wakasa T et al., 2007 *Phys. Lett.* B **653** 173
[16] Yamada T and Schuck P, 2005 *Euro. Phys. J.* A **26**,185
[17] Bayman B F and Bohr A, 1958/59 *Nucl. Phys.* **9** 596

Exciton formation in strongly correlated electron-hole systems near the semimetal-semiconductor transition

B Zenker[1], **D Ihle**[2], **F X Bronold**[1] **and H Fehske**[1]

[1]Institut für Physik, Ernst-Moritz-Arndt-Universität Greifswald, D-17489 Greifswald, Germany

[2]Institut für Theoretische Physik, Universität Leipzig, D-04109 Leipzig, Germany

E-mail: zenker@physik.uni-greifswald.de, fehske@physik.uni-greifswald.de

Abstract. The region surrounding the excitonic insulator phase is a three-component plasma composed of electrons, holes, and excitons. Due to the extended nature of the excitons, their presence influences the surrounding electrons and holes. We analyze this correlation. To this end, we calculate the density of bound electrons, the density of electrons in the correlated state, the momentum-resolved exciton density, and the momentum-resolved density of electron-hole pairs that are correlated but unbound. We find qualitative differences in the electron-hole correlations between the weak-coupling and the strong-coupling regime.

1. Introduction

The semimetal-semiconductor (SM-SC) transition is of particular interest, since the excitonic insulator (EI) might be realized in its vicinity at low enough temperatures [1, 2, 3, 4, 5]. Promising materials for the experimental verification are $TmSe_{0.45}Te_{0.55}$ [6], $1T$-$TiSe_2$ [7], Ta_2NiSe_5 [8], or a double bilayer graphene system [9]. The EI constitutes an exciton condensate as an equilibrium phenomenon in contrast to optically created exciton condensates [10]. Approaching, respectively, the EI from the SM or the SC side, a BCS-type transition of electron-hole pairs or a Bose-Einstein condensation (BEC) of preformed excitons occurs. Hence, the EI is discussed in view of a BCS-BEC crossover [11, 12, 13]. A different nature of the electron-hole pairs below the critical temperature $T_{\rm EI}$ might also show up in terms of precursor effects above $T_{\rm EI}$ [13]. Close to $T_{\rm EI}$ free excitons appear possibly in a large number, and the region surrounding the EI, in Ref. [11] termed "halo", constitutes a three-component plasma consisting of electrons, holes, and excitons. Whereas the EI transition scenario itself and the presence of the halo phase above $T_{\rm EI}$ was analyzed in previous works [11, 12, 13, 14, 15], the correlation effects caused by the excitons have not been studied so far. We address this issue in this work.

The Falicov-Kimball model extended by a finite f-bandwidth, in the following shortly denoted as extended Falicov-Kimball model (EFKM), is the minimal model describing the SM-SC transition and the EI formation [12, 16, 17, 18]. Expressing the orbital degree of freedom by a pseudospin variable σ, the EFKM can be written as

$$H = \sum_{\mathbf{k},\sigma} \varepsilon_{\mathbf{k}\sigma} c^{\dagger}_{\mathbf{k}\sigma} c_{\mathbf{k}\sigma} + U \sum_{i} n_{i\uparrow} n_{i\downarrow}, \qquad (1)$$

where $c^{(\dagger)}_{\mathbf{k}\sigma}$ annihilates (creates) an electron with momentum $\mathbf{k}$ in the band labeled by $\sigma = \uparrow, \downarrow$, and $n_{i\sigma} = c^{\dagger}_{i\sigma} c_{i\sigma}$ characterizes the σ-electron occupation of the Wannier site i. Hereafter, $\sigma = \uparrow$ denotes the valence band and $\sigma = \downarrow$ denotes the conduction band. The bare band dispersions are given by $\varepsilon_{\mathbf{k}\sigma}$, the Coulomb interaction strength is denoted by U, and N is the total number of lattice sites. In order to investigate the SM-SC transition we consider the EFKM at half filling,

$$n_{\uparrow} + n_{\downarrow} = 1, \tag{2}$$

where $n_{\sigma} = \frac{1}{N} \sum_{\mathbf{k}} \langle c^{\dagger}_{\mathbf{k}\sigma} c_{\mathbf{k}\sigma} \rangle$.

The scenario obtained within mean-field approximation agrees qualitatively with the result from more accurate approaches [14, 15, 19, 20, 21]. The mean-field ground-state phase diagram is even in quantitative accordance with the constraint path Monte Carlo data [22]. However, correlation effects are widely ignored by mean-field approaches, and for the detailed analysis of the bound states in the normal phase a more sophisticated approach is necessary. In this work we will apply the projection technique for Green functions developed by Plakida [23] to describe the correlations in the "electron-hole-exciton system" in terms of a self-energy.

The paper is organized as follows. In Sec. 2 we outline the method. In Sec. 3 we analyze the electron-hole correlations and determine the excitonic susceptibility, the density of bound electrons, and the density of electrons that are correlated but not bound. Our numerical results are presented in Sec. 4. Section 5 summarizes this work.

2. Projection technique for Green functions

The projection approach starts with the decomposition of the time derivatives of the electron operators into a part proportional to themselves and an irreducible part (throughout the paper we set $\hbar = 1$),

$$i \frac{d}{dt} c_{\mathbf{k}\sigma} \equiv i\dot{c}_{\mathbf{k}\sigma} = \bar{\varepsilon}_{\mathbf{k}\sigma} c_{\mathbf{k}\sigma} + i\dot{c}^{(\mathrm{ir})}_{\mathbf{k}\sigma} \quad \text{with} \quad \langle [i\dot{c}^{(\mathrm{ir})}_{\mathbf{k}\sigma}, c^{\dagger}_{\mathbf{k}\sigma}]_{+} \rangle = 0. \tag{3}$$

Introducing the spectral moments $M_{\mathbf{k}\sigma} = \langle [c_{\mathbf{k}\sigma}, c^{\dagger}_{\mathbf{k}\sigma}]_{+} \rangle$ and $M'_{\mathbf{k}\sigma} = \langle [i\dot{c}_{\mathbf{k}\sigma}, c^{\dagger}_{\mathbf{k}\sigma}]_{+} \rangle$, we have $M_{\mathbf{k}\sigma} = 1$ and $M'_{\mathbf{k}\sigma} = \bar{\varepsilon}_{\mathbf{k}\sigma} = \varepsilon_{\mathbf{k}\sigma} + U n_{-\sigma}$.

To derive an exact expression for the self-energy, we consider the equation of motion for the Green function $G_{\mathbf{k}\sigma}(t - t') = \langle\langle c_{\mathbf{k}\sigma}(t); c^{\dagger}_{\mathbf{k}\sigma}(t') \rangle\rangle$. Differentiating $G_{\mathbf{k}\sigma}(t - t')$ successively with respect to time t and t', we obtain a system of equations which in the Fourier representation reads

$$(\omega - \bar{\varepsilon}_{\mathbf{k}\sigma}) \langle\langle c_{\mathbf{k}\sigma}; c^{\dagger}_{\mathbf{k}\sigma} \rangle\rangle_{\omega} = 1 + \langle\langle i\dot{c}^{(\mathrm{ir})}_{\mathbf{k}\sigma}; c^{\dagger}_{\mathbf{k}\sigma} \rangle\rangle_{\omega}, \tag{4}$$

$$(\omega - \bar{\varepsilon}_{\mathbf{k}\sigma}) \langle\langle i\dot{c}^{(\mathrm{ir})}_{\mathbf{k}\sigma}; c^{\dagger}_{\mathbf{k}\sigma} \rangle\rangle_{\omega} = T_{\mathbf{k}\sigma}(\omega), \tag{5}$$

with the scattering matrix

$$T_{\mathbf{k}\sigma} = \langle\langle i\dot{c}^{(\mathrm{ir})}_{\mathbf{k}\sigma}; -i\dot{c}^{\dagger\,(\mathrm{ir})}_{\mathbf{k}\sigma} \rangle\rangle_{\omega}, \tag{6}$$

where $i\dot{c}^{(\mathrm{ir})}_{\mathbf{k}\sigma} = \frac{U}{N} \sum_{\mathbf{q}} c_{\mathbf{k}+\mathbf{q}\sigma} \rho_{-\mathbf{q}-\sigma} - U n_{-\sigma} c_{\mathbf{k}\sigma}$ and $\rho_{\mathbf{q}\sigma} = \sum_{\mathbf{k}} c^{\dagger}_{\mathbf{k}\sigma} c_{\mathbf{k}+\mathbf{q}\sigma}$. Introducing the zeroth-order Green function $G^{(0)}_{\mathbf{k}\sigma}(\omega) = (\omega - \bar{\varepsilon}_{\mathbf{k}\sigma})^{-1}$, we can rewrite Eqs. (4) and (5) as

$$G_{\mathbf{k}\sigma}(\omega) = G^{(0)}_{\mathbf{k}\sigma}(\omega) + G^{(0)}_{\mathbf{k}\sigma}(\omega) T_{\mathbf{k}\sigma}(\omega) G^{(0)}_{\mathbf{k}\sigma}(\omega). \tag{7}$$

The self-energy is defined by the Dyson equation

$$G_{\mathbf{k}\sigma}(\omega) = G^{(0)}_{\mathbf{k}\sigma}(\omega) + G^{(0)}_{\mathbf{k}\sigma}(\omega) \Sigma_{\mathbf{k}\sigma}(\omega) G_{\mathbf{k}\sigma}(\omega). \tag{8}$$

Journal of Physics: Conference Series **529** (2014) 012030 doi:10.1088/1742-6596/529/1/012030

From Eqs. (7) and (8) we get the relation between the self-energy and the scattering matrix,

$$T_{\mathbf{k}\sigma}(\omega) = \Sigma_{\mathbf{k}\sigma}(\omega) + \Sigma_{\mathbf{k}\sigma}(\omega)G^{(0)}_{\mathbf{k}\sigma}(\omega)T_{\mathbf{k}\sigma}(\omega). \tag{9}$$

This equation shows that the self-energy is the "proper part" of the scattering matrix (6), which has no parts connected by a single zeroth-order Green function, i.e.,

$$\Sigma_{\mathbf{k}\sigma}(\omega) = \langle\langle i\dot{c}^{(\mathrm{ir})}_{\mathbf{k}\sigma}; -i\dot{c}^{\dagger\,(\mathrm{ir})}_{\mathbf{k}\sigma}\rangle\rangle^{(p)}_{\omega}. \tag{10}$$

The spectral representation of the self-energy is given by

$$\Sigma_{\mathbf{k}\sigma}(\omega) = \int_{-\infty}^{\infty} \frac{d\bar{\omega}}{2\pi}\frac{\Gamma_{\mathbf{k}\sigma}(\bar{\omega})}{\omega-\bar{\omega}} \tag{11}$$

with

$$\Gamma_{\mathbf{k}\sigma}(\bar{\omega}) = (e^{\beta\bar{\omega}}+1)\int_{-\infty}^{\infty} dt\, e^{i\bar{\omega}t}\,\frac{U^2}{N}K_{\mathbf{k}\sigma}(t), \tag{12}$$

$$K_{\mathbf{k}\sigma}(t) = \frac{1}{N}\sum_{\mathbf{q},\mathbf{q}'}\langle \rho_{\mathbf{q}'-\sigma}(0)c^{\dagger}_{\mathbf{k}+\mathbf{q}'\sigma}(0)c_{\mathbf{k}+\mathbf{q}\sigma}(t)\rho_{-\mathbf{q}-\sigma}(t)\rangle^{(p)}. \tag{13}$$

Then the spectral function of the electrons reads

$$A_{\mathbf{k}\sigma}(\omega) = -2\mathrm{Im}G_{\mathbf{k}\sigma}(\omega) = -2\frac{\Sigma''_{\mathbf{k}\sigma}(\omega)}{\left[\omega-\bar{\varepsilon}_{\mathbf{k}\sigma}-\Sigma'_{\mathbf{k}\sigma}(\omega)\right]^2+\left[\Sigma''_{\mathbf{k}\sigma}(\omega)\right]^2}, \tag{14}$$

where we have introduced the shorthand notation $\Sigma'_{\mathbf{k}\sigma}(\omega) = \mathrm{Re}\Sigma_{\mathbf{k}\sigma}(\omega)$ and $\Sigma''_{\mathbf{k}\sigma}(\omega) = \mathrm{Im}\Sigma_{\mathbf{k}\sigma}(\omega)$. The renormalized dispersion is given by

$$\tilde{\varepsilon}_{\mathbf{k}\sigma} = \bar{\varepsilon}_{\mathbf{k}\sigma} + \Sigma'_{k\sigma}(\omega)\big|_{\omega=\tilde{\varepsilon}_{\mathbf{k}\sigma}}. \tag{15}$$

Let us emphasize that the equations up to Eq. (15) are exact. However, approximations become necessary if $\Sigma_{\mathbf{k}\sigma}$ is calculated.

3. Electron-hole correlations

In Eq. (13), a three-particle correlation function enters the self-energy. To proceed, we perform a two-time decoupling, because the "proper part" of the correlation function (13) must be considered. Following the Martin-Schwinger decoupling scheme [24], we identify four contributions,

$$K_{\mathbf{k}\sigma}(t) = K^{(1)}_{\mathbf{k}\sigma}(t) + K^{(2)}_{\mathbf{k}\sigma}(t) + K^{(3)}_{\mathbf{k}\sigma}(t) + K^{(4)}_{\mathbf{k}\sigma}(t), \tag{16}$$

with

$$K^{(1)}_{\mathbf{k}\sigma}(t) = -\frac{2}{N}\sum_{\mathbf{k}',\mathbf{q}}\langle c_{\mathbf{k}'-\sigma}(0)c^{\dagger}_{\mathbf{k}'-\sigma}(t)\rangle\langle c^{\dagger}_{\mathbf{k}'-\mathbf{q}-\sigma}(0)c_{\mathbf{k}'-\mathbf{q}-\sigma}(t)\rangle\langle c^{\dagger}_{\mathbf{k}+\mathbf{q}\sigma}(0)c_{\mathbf{k}+\mathbf{q}\sigma}(t)\rangle, \tag{17}$$

$$K^{(2)}_{\mathbf{k}\sigma}(t) = \frac{1}{N}\sum_{\mathbf{k}',\mathbf{q},\mathbf{q}'}\langle c_{\mathbf{k}'+\mathbf{q}'-\sigma}(0)c^{\dagger}_{\mathbf{k}+\mathbf{q}'\sigma}(0)c_{\mathbf{k}+\mathbf{q}\sigma}(t)c^{\dagger}_{\mathbf{k}'+\mathbf{q}-\sigma}(t)\rangle\langle c^{\dagger}_{\mathbf{k}'-\sigma}(0)c_{\mathbf{k}'-\sigma}(t)\rangle, \tag{18}$$

$$K^{(3)}_{\mathbf{k}\sigma}(t) = \frac{1}{N}\sum_{\mathbf{k}',\mathbf{q},\mathbf{q}'}\langle c^{\dagger}_{\mathbf{q}'-\sigma}(0)c_{\mathbf{q}+\mathbf{q}'-\sigma}(0)c^{\dagger}_{\mathbf{k}'-\sigma}(t)c_{\mathbf{k}'-\mathbf{q}-\sigma}(t)\rangle\langle c^{\dagger}_{\mathbf{k}+\mathbf{q}\sigma}(0)c_{\mathbf{k}+\mathbf{q}\sigma}(t)\rangle, \tag{19}$$

$$K^{(4)}_{\mathbf{k}\sigma}(t) = \frac{1}{N}\sum_{\mathbf{k}',\mathbf{q},\mathbf{q}'}\langle c^{\dagger}_{\mathbf{k}'-\mathbf{q}'-\sigma}(0)c^{\dagger}_{\mathbf{k}+\mathbf{q}'\sigma}(0)c_{\mathbf{k}+\mathbf{q}\sigma}(t)c_{\mathbf{k}'-\mathbf{q}-\sigma}(t)\rangle\langle c_{\mathbf{k}'-\sigma}(0)c^{\dagger}_{\mathbf{k}'-\sigma}(t)\rangle. \tag{20}$$

Since we are concentrating on excitonic electron-hole fluctuations, described by the term $K^{(2)}_{\mathbf{k}\sigma}(t)$, we further decouple $K^{(3)}_{\mathbf{k}\sigma}(t)$ and $K^{(4)}_{\mathbf{k}\sigma}(t)$ into products of single-particle correlation functions. This leads to $K^{(3)}_{\mathbf{k}\sigma}(t) = K^{(4)}_{\mathbf{k}\sigma}(t) = -\frac{1}{2}K^{(1)}_{\mathbf{k}\sigma}(t)$. Then, $K_{\mathbf{k}\sigma}(t) = K^{(2)}_{\mathbf{k}\sigma}(t)$, and we end up with

$$\Gamma_{\mathbf{k}\sigma}(\bar{\omega}) = -\frac{2U^2}{N\pi}\sum_{\mathbf{k}'}\int_{-\infty}^{\infty} d\omega'\big[f(\omega') + p(\omega' - \bar{\omega})\big]\,\mathrm{Im}\chi^{-\sigma,\sigma}_{\mathbf{k}-\mathbf{k}'}(\bar{\omega} - \omega')\,\mathrm{Im}G_{\mathbf{k}'-\sigma}(\omega'), \tag{21}$$

where $f(\omega)$ is the Fermi function, $p(\omega)$ is the Bose function, and

$$\chi^{-\sigma,\sigma}_{\mathbf{q}}(\omega) = \frac{1}{N}\sum_{\mathbf{k},\mathbf{k}'}\langle\langle c^{\dagger}_{\mathbf{k}-\sigma}c_{\mathbf{k}+\mathbf{q}\sigma};\, c^{\dagger}_{\mathbf{k}'+\mathbf{q}\sigma}c_{\mathbf{k}'-\sigma}\rangle\rangle_{\omega} \tag{22}$$

is the excitonic susceptibility. We will concentrate on the presence of excitons and their effect on the unbound electrons by including only the according contribution in Eq. (22).

3.1. Excitonic susceptibility

The ladder approximation [25] of the excitonic susceptibility, illustrated in Fig. 1, is suitable to describe electron-hole bound states. Taking into account only the nearly-free part of the electron Green function, $G^{\mathrm{nf}}_{\mathbf{k}\sigma}(\omega) = (\omega - \tilde{\varepsilon}_{\mathbf{k}\sigma})^{-1}$, we find

$$\chi^{-\sigma,\sigma}_{\mathbf{q}}(\omega) = \frac{\tilde{\chi}^{-\sigma,\sigma}_{\mathbf{q}}(\omega)}{U\tilde{\chi}^{-\sigma,\sigma}_{\mathbf{q}}(\omega) + 1}, \tag{23}$$

where $\tilde{\chi}^{-\sigma,\sigma}_{\mathbf{q}}(\omega) = N^{-1}\sum_{\mathbf{k}}[f(\tilde{\varepsilon}_{\mathbf{k}-\sigma}) - f(\tilde{\varepsilon}_{\mathbf{k}+\mathbf{q}\sigma})]/(\omega + \tilde{\varepsilon}_{\mathbf{k}-\sigma} - \tilde{\varepsilon}_{\mathbf{k}+\mathbf{q}\sigma})$.

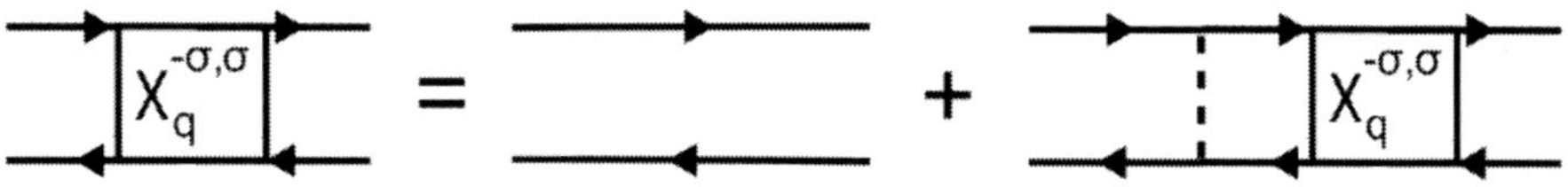

Figure 1. Ladder approximation for the excitonic susceptibility $\chi^{-\sigma,\sigma}_{\mathbf{q}}$. The solid lines represent the electron Green functions, and the dashed line represents the Coulomb attraction.

The pole of the excitonic susceptibility located outside the electron-hole continuum, $\omega^{-\sigma,\sigma}_{X}(\mathbf{q})$ with $|\omega^{-\sigma,\sigma}_{X}(\mathbf{q})| < |\omega^{-\sigma,\sigma}_{C}(\mathbf{q})| = \min_{\mathbf{k}}|\tilde{\varepsilon}_{\mathbf{k}+\mathbf{q}\sigma} - \tilde{\varepsilon}_{\mathbf{k}-\sigma}|$, describes the bound state of an electron from the σ-band with a hole from the band with index $-\sigma$. Note that $\omega^{-\sigma,\sigma}_{X}(\mathbf{q})$ increases with increasing temperature. However, for very high temperatures $\chi^{-\sigma,\sigma}_{\mathbf{q}}$ does not exhibit a pole outside the continuum: consequently excitons do not exist.

3.2. Density of bound electrons

Since an exciton contains one electron, the number of bound electrons equals the number of excitons. Chosen $\sigma =\downarrow$ to be the index for the conduction band, the exciton dispersion is

$$\omega_X(\mathbf{q}) = \omega^{\uparrow,\downarrow}_{X}(\mathbf{q}) = -\omega^{\downarrow,\uparrow}_{X}(\mathbf{q}). \tag{24}$$

The binding energy of an exciton can be obtained from $E^X_{\mathrm{B}}(\mathbf{q}) = \omega_C(\mathbf{q}) - \omega_X(\mathbf{q})$, where $\omega_C(\mathbf{q}) = \omega^{\uparrow,\downarrow}_{C}(\mathbf{q})$.

Provided that $\omega_X(\mathbf{q})$ exists, we follow Ref. [13] and obtain the momentum-resolved exciton density as

$$N_X(\mathbf{q}) = Z(\omega_X, \mathbf{q})p(\omega_X), \tag{25}$$

where the spectral weight of the excitons becomes

$$Z(\omega_X, \mathbf{q}) = \left[\frac{U^2}{N}\sum_{\mathbf{k}} \frac{f(\tilde{\varepsilon}_{\mathbf{k}\uparrow}) - f(\tilde{\varepsilon}_{\mathbf{k}+\mathbf{q}\downarrow})}{(\omega_X + \tilde{\varepsilon}_{\mathbf{k}\uparrow} - \tilde{\varepsilon}_{\mathbf{k}+\mathbf{q}\downarrow})^2}\right]^{-1} = Z^{\uparrow,\downarrow}(\omega_X^{\uparrow,\downarrow}, \mathbf{q}) = Z^{\downarrow,\uparrow}(\omega_X^{\downarrow,\uparrow}, \mathbf{q}), \tag{26}$$

and the exciton density (density of bound electrons) is given by

$$n_X = \frac{1}{N}\sum_{\mathbf{q}} N_X(\mathbf{q}). \tag{27}$$

3.3. Density of correlated electrons

Expanding the spectral function, Eq. (14), for small damping [26], we can decompose the density into a nearly-free part (with renormalized band dispersion) and a part, where the exciton formation enters,

$$n_\sigma = \frac{1}{N}\sum_{\mathbf{k}} \int_{-\infty}^{\infty} \frac{d\omega}{2\pi} A_{\mathbf{k}\sigma}(\omega) f(\omega) = n_\sigma^{\mathrm{nf}} + n_\sigma^{\mathrm{corr}}, \tag{28}$$

where $n_\sigma^{\mathrm{nf}} = \frac{1}{N}\sum_{\mathbf{k}} f(\tilde{\varepsilon}_{\mathbf{k}\sigma})$, and the density of the correlated electrons reads

$$n_\sigma^{\mathrm{corr}} = \mathrm{sgn}(\omega_X^{-\sigma,\sigma}) \frac{U^2}{N^2} \sum_{\mathbf{k},\mathbf{k}'} E^{-\sigma,\sigma}(\mathbf{k},\mathbf{k}') F^{-\sigma,\sigma}(\mathbf{k},\mathbf{k}'). \tag{29}$$

Straightforward calculation shows that

$$E^{\uparrow,\downarrow}(\mathbf{k},\mathbf{k}') = \frac{Z(\omega_X, \mathbf{k}-\mathbf{k}')}{(\omega_X + \tilde{\varepsilon}_{\mathbf{k}'\uparrow} - \tilde{\varepsilon}_{\mathbf{k}\downarrow})^2} = E^{\downarrow,\uparrow}(\mathbf{k}',\mathbf{k}), \tag{30}$$

$$F^{\uparrow,\downarrow}(\mathbf{k},\mathbf{k}') = f(\tilde{\varepsilon}_{\mathbf{k}'\uparrow}) f(\tilde{\varepsilon}_{\mathbf{k}\downarrow}) - f(\tilde{\varepsilon}_{\mathbf{k}\downarrow}) - f(\tilde{\varepsilon}_{\mathbf{k}\downarrow}) p(\omega_X^{\uparrow,\downarrow}) + f(\tilde{\varepsilon}_{\mathbf{k}'\uparrow}) p(\omega_X^{\uparrow,\downarrow}) = F^{\downarrow,\uparrow}(\mathbf{k}',\mathbf{k}). \tag{31}$$

Now, it is easy to see that $n_\uparrow^{\mathrm{corr}} = -n_\downarrow^{\mathrm{corr}}$, and the chemical potential is exclusively determined by the nearly-free part of the densities (see also [11]):

$$n_\uparrow + n_\downarrow = n_\uparrow^{\mathrm{nf}} + n_\downarrow^{\mathrm{nf}} = 1. \tag{32}$$

Note that $n_\downarrow^{\mathrm{corr}} \neq n_X$, i.e., the considered correlation effects go beyond the simple binding of electrons (see below). Moreover, the presence of excitons leads to a decrease of the density of valence-band electrons, because $n_\uparrow^{\mathrm{corr}} < 0$.

4. Numerical results

Let us first recapitulate the basic properties of the EFKM. Thereby we restrict ourselves to a square lattice and a direct band-gap situation, i.e., valence-band maximum and conduction-band minimum lie at the Brillouin-zone center. We choose $t_\downarrow$ to be the unit of energy, and the other model parameters are $E_\downarrow = 0$, $E_\uparrow = -2.4$, and $t_\uparrow = -0.8$. In a first approximation we neglect the exciton-induced band renormalization in the numerics, i.e., $\tilde{\varepsilon}_{\mathbf{k}\sigma} = \bar{\varepsilon}_{\mathbf{k}\sigma}$, where the mean-field Hartree-shift is contained in $\bar{\varepsilon}_{\mathbf{k}\sigma}$.

The middle panel of Fig. 2 shows the phase diagram of the EFKM in the U-T plane. The EI separates the SM phase and the SC phase at low temperatures. Valence band and conduction band overlap in a SM and, as a result, a well-defined Fermi surface exists. In this case, only finite momentum excitons can develop, see Fig. 2 (left panel). On the other hand, in a SC the valence band and the conduction band are separated by an energy gap, and excitons with an

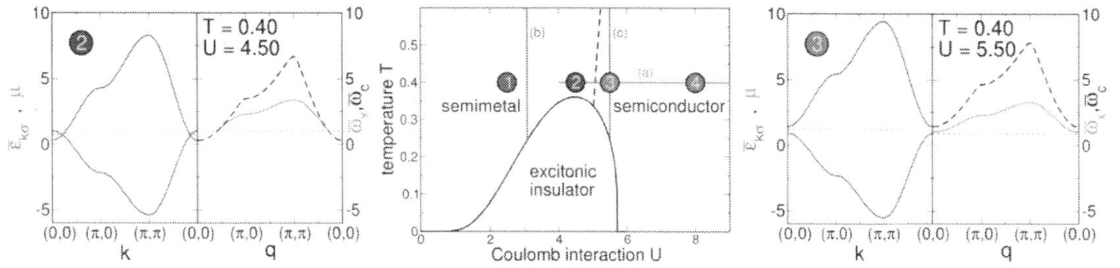

Figure 2. (Middle panel) Phase diagram of the EI phase in the half-filled EFKM for a square lattice. In Fig. 3 we analyze the density of bound electrons and the density of correlated electrons along the red and blue lines (a), (b), and (c). The points 1, 2, 3, and 4 mark the points where we analyze the momentum-resolved exciton density and the momentum-resolved density of correlated but unbound electron-hole pairs in Fig. 4. (Left panel) Band structure and exciton dispersion for a semimetal. (Right panel) Band structure and exciton dispersion for a semiconductor. In the outer panels the renormalized band dispersions $\bar{\varepsilon}_{\mathbf{k}\sigma}$, the chemical potential μ, the exciton energies $\bar{\omega}_X = \omega_X + \max_{\mathbf{k}}(\bar{\varepsilon}_{\mathbf{k}\uparrow})$, and the boundary of the electron-hole continuum $\bar{\omega}_C = \omega_C + \max_{\mathbf{k}}(\bar{\varepsilon}_{\mathbf{k}\uparrow})$ are shown. The blue dotted line shows $\max_{\mathbf{k}}(\bar{\varepsilon}_{\mathbf{k}\uparrow})$. The binding energy of an exciton is given by $E_B^X(\mathbf{q}) = \bar{\omega}_C(\mathbf{q}) - \bar{\omega}_X(\mathbf{q})$.

arbitrary momentum can form, in particular zero-momentum excitons, which undergo a BEC at the SC-EI transition, see Fig. 2 (right panel).

Electrons and holes may bind to excitons, and subsequently the remaining unbound electrons and holes may scatter as well on these quasiparticles. Both effects, the exciton binding and the influence on the unbound electrons, are considered in n_σ^{corr}. How strongly the electrons are affected is mainly determined by the spatial extension of the excitons, and the underlying band structure.

Let us start by keeping the temperature constant and raising U, crossing thereby the SM-SC transition, see Fig. 3(a). On the SM side, excitons are loosely bound objects with a relatively large extension. As a result, many electrons are influenced by a single exciton. An increasing Coulomb attraction U causes a tighter binding of electrons and holes. Consequently the exciton radius shrinks, and $n_\downarrow^{\mathrm{corr}}/n_X$ decreases.

At the SM-SC transition both the density of bound electrons and the density of correlated electrons reach their maximum. The kink of n_X and $n_\downarrow^{\mathrm{corr}}$ will be less pronounced for a three-dimensional system, c.f. Ref. [13].

In the strong coupling regime, $n_\downarrow^{\mathrm{corr}}$ tends to n_X. Here, the large Coulomb attraction gives rise to a very small distance between the bound electrons and holes: the Frenkel-limit of excitons is realized. Since there is only a very limited number of unbound electrons and holes, the correlation effects—apart from those leading to the formation of excitons by itself—are rather weak.

Let us now analyze temperature effects. On the SM side, excitons are less important for the EI formation [13] and only a few electrons are bound or correlated, see Fig. 3(b). With increasing temperature the broadening of the Fermi and Bose function will provide more electrons and holes to form excitons, which overcompensates the exciton dissociation. However, for very high temperatures the situation reverses, and the exciton dissociation dominates: n_X and $n_\downarrow^{\mathrm{corr}}$ decrease with growing temperature until excitons disappear completely at $T_X = 6.7$ for $U = 3.08$ [not shown in Fig. 3(b)]. The higher kinetic energy at high temperatures enables the electrons to avoid correlations, and the ratio $n_\downarrow^{\mathrm{corr}}/n_X$ decreases with increasing temperature.

On the SC side, other effects prevail, see Fig. 3(c). The SC-EI transition at T_{EI} is driven

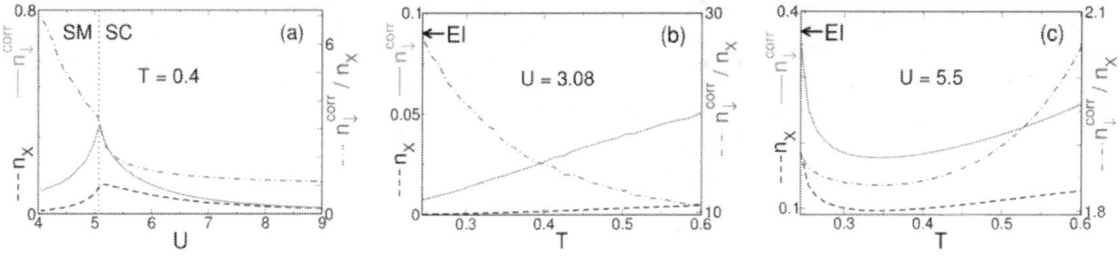

Figure 3. Density of bound electrons n_X (black, dashed line), density of correlated electrons $n_\downarrow^{\rm corr}$ (red, solid line), and the ratio between them $n_X/n_\downarrow^{\rm corr}$ (blue, dot-dashed line) (a) as a function of the Coulomb interaction U ($T = 0.4$), crossing thereby the semimetal-semiconductor transition , (b) as a function of temperature for a semimetal ($U = 3.08$), and (c) as a function of temperature for a semiconductor ($U = 5.5$). The panels (a), (b), (c) correspond to the lines indicated in the middle panel of Fig. 2.

by the BEC of zero-momentum excitons. Close to $T_{\rm EI}$ the proliferation of $q = 0$ excitons leads to a large density of electrons that are bound or correlated. In contrast to the SM side, the ratio $n_\downarrow^{\rm corr}/n_X$ increases with increasing temperature. Here, the shrinking of the band gap makes conduction-band electrons available which can participate in the correlations. Note that $n_\downarrow^{\rm corr}/n_X$ is now significantly smaller than on the SM side, suggesting that the exciton formation resembles more the creation of local electron-hole pairs. For $U = 5.5$ [Coulomb strength in Fig. 3(c)] the SC-SM transition is crossed at $T_{\rm SC-SM} = 0.74$, and the total density of excitons vanishes at $T_X = 12.5$.

Although the constraint for the chemical potential reduces to the mean-field condition, the electron densities differ significantly from their mean-field values, see Eq. (28).

Having examined the composition of the halo phase, we turn to the momentum distribution of bound and unbound electron-hole pairs. By the replacement $\mathbf{k}' \to \mathbf{k}+\mathbf{q}$ in Eq. (29) we can write $n_\downarrow^{\rm corr} = N^{-1}\sum_{\mathbf{q}} N_\downarrow^{\rm corr}(\mathbf{q})$, with

$$N_\downarrow^{\rm corr}(\mathbf{q}) = \frac{U^2}{N}\sum_{\mathbf{k}} Z(\omega_X,\mathbf{q})\frac{f(\tilde\varepsilon_{\mathbf{k}+\mathbf{q}\downarrow})[f(\tilde\varepsilon_{\mathbf{k}\uparrow})-1]}{(\omega_X+\tilde\varepsilon_{\mathbf{k}\uparrow}-\tilde\varepsilon_{\mathbf{k}+\mathbf{q}\downarrow})^2} + p(\omega_X). \tag{33}$$

We separate $N_\downarrow^{\rm corr}(\mathbf{q})$ into bound pairs (excitons) and pairs which contain conduction-band electrons and valence-band holes that are correlated but unbound. We will shortly denote these objects as correlated pairs. The momentum-resolved exciton density $N_X(\mathbf{q})$ is given by Eq. (25), and the momentum-resolved density of correlated pairs is given by

$$D_X(\mathbf{q}) = N_\downarrow^{\rm corr}(\mathbf{q}) - N_X(\mathbf{q}). \tag{34}$$

Excitons can produce correlated pairs with the same momentum only, see Fig. 4. The ratio $D_X(\mathbf{q})/N_X(\mathbf{q})$ modulates with $\mathbf{q}$, suggesting that the size of the excitons depends on their momentum.

To analyze which electrons become correlated in the course of exciton formation, we resolve $n_\downarrow^{\rm corr}$ according to the momenta of the electrons, $n_\downarrow^{\rm corr} = N^{-1}\sum_{\mathbf{k}} n_{\mathbf{k}\downarrow}^{\rm corr}$, with

$$n_{\mathbf{k}\downarrow}^{\rm corr} = \frac{U^2}{N}\sum_{\mathbf{q}} Z(\omega_X,\mathbf{q})\frac{p(\omega_X)\big[f(\tilde\varepsilon_{\mathbf{k}-\mathbf{q}\uparrow})-f(\tilde\varepsilon_{\mathbf{k}\downarrow})\big]+f(\tilde\varepsilon_{\mathbf{k}\downarrow})\big[f(\tilde\varepsilon_{\mathbf{k}-\mathbf{q}\uparrow})-1\big]}{(\omega_X+\tilde\varepsilon_{\mathbf{k}-\mathbf{q}\uparrow}-\tilde\varepsilon_{\mathbf{k}\downarrow})^2}. \tag{35}$$

The absolute values of $n_{\mathbf{k}\downarrow}^{\rm corr}$ can differ substantially. To compare the situations on the SM side with that on the SC side, we scale $n_{\mathbf{k}\downarrow}^{\rm corr}$ with its largest contribution, $\bar n_{\mathbf{k}\downarrow}^{\rm corr} = n_{\mathbf{k}\downarrow}^{\rm corr}/\max_{\mathbf{k}}(n_{\mathbf{k}\downarrow}^{\rm corr})$.

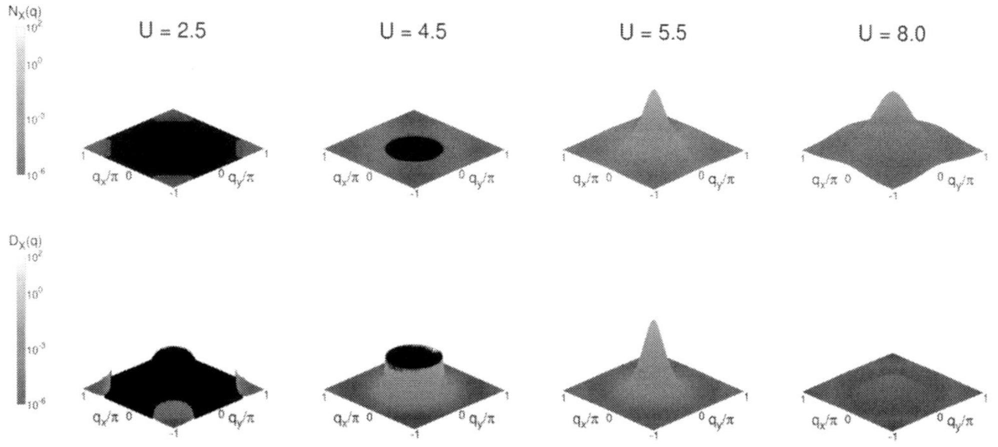

Figure 4. (Upper panels) Momentum-resolved exciton densities $N_X(\mathbf{q})$ and (lower panels) the corresponding momentum-resolved densities of correlated pairs $D_X(\mathbf{q})$ for $T = 0.4$ (cf. with marks in Fig. 2). The z-axis is equal within a vertical subfigure, but differs between the subfigures.

On the SM side, $\bar{n}^{\rm corr}_{\mathbf{k}\downarrow}$ shows a complicated structure. Let us recall that in a SM only finite-momentum excitons and correlated pairs exist, see Fig. 4(a). Here, the nesting property for finite momenta is crucial. Electrons become correlated only if holes with an adequate momentum are available. This is only the case at certain momenta close to the Fermi surface, see Fig. 5 ($U = 2.5$). Moreover, $\bar{n}^{\rm corr}_{\mathbf{k}\downarrow}$ is sharply peaked, that is, the mean distance between the correlated electrons is large.

On the SC side, the zero-momentum correlations are most relevant (see $U = 8.0$ in Fig. 4) and electrons close to the conduction-band minimum are predominantly correlated, as can be seen in Fig. 5 ($U = 8.0$). The structure of $\bar{n}^{\rm corr}_{\mathbf{k}\downarrow}$ is broader than on the SM side, suggesting a more local nature of the correlations. This corresponds to the situation within the EI phase, see the coherence length in Ref. [13].

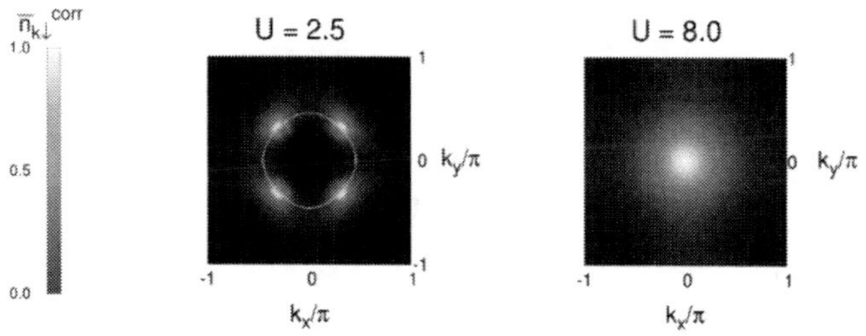

Figure 5. Momentum-resolved density of correlated electrons in case of (left panel) a semimetal and (right panel) a semiconductor for $T = 0.4$. The yellow, dotted line in the left panel shows the Fermi surface.

5. Conclusions

In this work we studied the normal phase close to the excitonic insulator (EI) phase, the so-called "halo" phase. We focused on the correlations that are induced by the presence of excitons and calculated the self-energy by the projection technique for Green functions. This enabled us to separate the electron densities into a nearly-free part, which is in principle uncorrelated (only the band dispersion is renormalized), and a correlation part, where the influence of the excitons enters.

On the semimetal (SM) side excitons are loosely bound objects, whose radius is relatively large. Here, bound electron-hole pairs (excitons) only exist with finite momentum, and correlated electrons and holes form due to the presence of these excitons. The SM-EI transition is driven by critical electron-hole fluctuations. Due to the nesting property of the Fermi surface for finite momenta, correlated electrons are concentrated at four distinguished points close to the Fermi momenta.

With increasing Coulomb interaction the transition from the SM to the semiconductor (SC) is crossed. Thereby, the band structure and consequently the electron-hole pair spectrum changes dramatically.

In a SC no Fermi surface exists, and the electrons close to the conduction-band minimum are involved in the electron-hole correlations. The excitons on the SC side are tighter bound, which suggests that their radius is smaller than on the SM side. Besides those with finite momentum, zero-momentum electron-hole pairs exist, which represent the largest contribution to the electron-hole pair spectrum. The BEC of these excitons drive the SC-EI transition.

The qualitative difference in the electron-hole correlations on the SM side and on the SC side anticipates the BCS-BEC crossover within the EI phase.

Let us finally stress that the present work exclusively addresses the formation of excitons and the question how they affect unbound charge carriers. The exciton-exciton interaction and particularly the formation of biexcitons is another interesting subject, which deserves further study in the framework of our approach. The formation of excitonic molecules could be relevant near the SM-SC transition, where a collective biexcitonic phase might possibly compete with the EI [27].

Acknowledgments

This work was supported by the Deutsche Forschungsgemeinschaft through SFB 652, project B5.

References

[1] Mott N F 1961 *Philos. Mag.* **6** 287
[2] Knox R 1963 *Solid State Physics* ed Seitz F and Turnbull D (New York: Academic Press) p Suppl. 5 p. 100
[3] Keldysh L V and Kopaev H Y V 1965 *Sov. Phys. Sol. State* **6** 2219
[4] Jérome D, Rice T M and Kohn W 1967 *Physical Review* **158** 462
[5] Halperin B I and Rice T M 1968 *Rev. Mod. Phys.* **40** 755
[6] Bucher B, Steiner P and Wachter P 1991 *Phys. Rev. Lett.* **67** 2717
[7] Cercellier H, Monney C, Clerc F, Battaglia C, Despont L, Garnier M G, Beck H, Aebi P, Patthey L, Berger H and Forró L 2007 *Phys. Rev. Lett.* **99** 146403
[8] Wakisaka Y, Sudayama T, Takubo K, Mizokawa T, Arita M, Namatame H, Taniguchi M, Katayama N, Nohara M and Takagi H 2009 *Phys. Rev. Lett.* **103** 026402
[9] Perali A, Neilson D and Hamilton A R 2013 *Phys. Rev. Lett.* **110** 146803
[10] Littlewood P B, Eastham P R, Keeling J M J, Marchetti F M, Simons B D and Szymanska M H 2004 *J. Phys. Condens. Matter* **16** S3597
[11] Bronold F X and Fehske H 2006 *Phys. Rev.* B **74** 165107
[12] Ihle D, Pfafferott M, Burovski E, Bronold F X and Fehske H 2008 *Phys. Rev.* B **78** 193103
[13] Zenker B, Ihle D, Bronold F X and Fehske H 2012 *Phys. Rev.* B **85** 121102(R)
[14] Zenker B, Ihle D, Bronold F X and Fehske H 2011 *Phys. Rev.* B **83** 235123
[15] Phan N V, Fehske H and Becker K W 2011 *Europhys. Lett.* **95** 17006

[16] Batista C D 2002 *Phys. Rev. Lett.* **89** 166403
[17] Batista C D, Gubernatis J E, Bonča J and Lin H Q 2004 *Phys. Rev. Lett.* **92** 187601
[18] Schneider C and Czycholl G 2008 *Eur. Phys. J.* B **64** 43
[19] Zenker B, Ihle D, Bronold F X and Fehske H 2010 *Phys. Rev.* B **81** 115122
[20] Phan N V, Becker K W and Fehske H 2010 *Phys. Rev.* B **81** 205117
[21] Seki K, Eder R and Ohta Y 2011 *Phys. Rev.* B **84** 245106
[22] Farkašovský P 2008 *Phys. Rev.* B **77** 155130
[23] Plakida N M 2011 *Springer Series in Solid-State Sciences* **171** 173
[24] Martin P C and Schwinger J 1959 *Phys. Rev.* **115** 1342
[25] Kadanoff L P and Baym G 1962 *Quantum Statistical Mechanics* (Reading, Massachusetts: Benjamin/Cumming Publishing Company)
[26] Kraeft W D, Kremp D, Ebeling W and Röpke G 1986 *Quantum Statistics of Charged Particle Systems* (Akademie-Verlag Berlin)
[27] Maezono R, Ríos P L, Ogawa T and Needs R J 2013 *Phys. Rev. Lett.* **110** 216407

AUTHOR INDEX

Institute of Physics
Dirac House, Temple Back
Bristol BS1 6BE UK

CURRAN ASSOCIATES INC
proceedings.com